DATE DUE			
Dec 15 '76			
Mar 14 7 7			
Dec 13 '77			
Apr 26 7 9			
Mar 13 '80			

Gene Expression

Volume 2

Gene Expression

Volume 2
Eucaryotic Chromosomes

Benjamin Lewin

Editor, *Cell*

A Wiley–Interscience Publication

JOHN WILEY & SONS

LONDON   NEW YORK   SYDNEY   TORONTO

575.21
L 58g
97872
Sept. 1976

Library of Congress Cataloging in Publication Data:

Lewin, Benjamin M.
Gene expression.

"A Wiley-Interscience publication."
Includes bibliographical references.

CONTENTS: v. 1. Bacterial genomes.—v. 2. Eucaryotic
chromosomes.
1. Gene expression. 2. Molecular genetics.
I. Title.

QH450.L48   575.2′1   73–14382
ISBN 0 471 53164 2 (v. 2)
ISBN 0 471 53166 9 (pbk., v. 2)

Printed in Great Britain by William Clowes & Sons Ltd.,
London, Colchester and Beccles.

For both my Anns

PREFACE

My purpose in writing this book has been to attempt to draw together some of the many disparate threads which contribute to our views of how genes are expressed in eucaryotic cells. Although our knowledge of eucaryotic gene expression is at present only tentative, sufficient advances have been made in recent years to suggest a preliminary conceptual framework which may be useful in thinking about the systems which are used to express and control genes in eucaryotic cells. What might be termed the approach of molecular biology to eucaryotic cells has perhaps reached the stage where we can begin to formulate relevant questions even if the nature of the answers, and sometimes of the means even to investigate them, is not yet apparent.

This book therefore presents a view of gene expression in higher cells; and attempts to pose some of the questions which we should at present ask in cell biology and to suggest some of the lines along which solutions may come. I have been as concerned to set out possible concepts of how gene expression may be controlled as to relate the facts which we have so far gathered. Where paradoxes exist which cannot yet be reconciled, I have tried to give a balanced account of ways to view them. Because of the fluid nature of cell biology, this book thus constitutes more a personal view than the volume of Gene Expression concerning bacterial systems.

An issue crucial to the definition of eucaryotic gene expression is clearly the structure of the genetic material itself. The first part of this book is therefore devoted to a discussion of the nature of the components of eucaryotic chromosomes and the interactions which may establish their compact structure. An obvious discrepancy lies between our ability to define the overall ultrastructure of the chromosome in the light and electron microscope and our knowledge of the interactions between DNA and proteins which constitute the molecular structure. The gap between these two levels of study has yet to be bridged by a model to account for the tight coiling of DNA within the chromosome.

The first chapter considers the ultrastructure of the chromosome in terms of the basic fibres of which it seems to be composed. This establishes that each chromosome possesses one very long thread of DNA, associated with proteins, and subject to more than one degree of supercoiling. Because the eucaryotic cell passes through several stages in its life cycle which may see changes in the state of its genetic material, the second chapter concerns molecular

vii

changes in chromosome function, structure and synthesis during interphase and discusses the behaviour of chromosomes mediated by their centromeres at mitosis and meiosis.

Understanding gene expression at the molecular level demands definition of the proteins present in the chromosome and of the organization of DNA sequences. In the next chapter, devoted to the protein components, it is possible to give a fairly detailed account of histone structures; the restricted number of histone classes and the evolutionary conservation of histone sequences suggest strongly that histones comprise a structural and not a regulatory component of the chromosome. We can be much less definite about the non-histone proteins; although at present it seems likely that their numbers and variation in tissues and species may be greater, it is not yet clear whether they provide the source of specificity for gene transcription.

Organization of the DNA component of the chromosome, the topic of the next chapter, cannot be well defined. We can classify sequences of DNA by their degree of repetition in the haploid genome. The unique sequences present in only one copy must presumably include the genes; and the very highly repetitive fraction which often forms a satellite appears to have a structural role since it is not transcribed and is located in centromeric heterochromatin. But we have little idea about the functions of the fraction of intermediate repetition, some at least of which is transcribed and is found also in the messengers of polysomes. Only rather crude approaches at present seem possible towards defining the organization and functions of the unique and intermediate sequence components; and it remains difficult to account for the apparent excess of DNA in the genome compared with the quantity which appears to be necessary to code for proteins. Although the general structures of euchromatin and heterochromatin can be delineated and correlated with some segregation of both DNA sequences and proteins, the structure of euchromatin itself, in particular any definition of the specificity with which sequences of DNA interact with proteins, remains elusive.

The second part of the book is concerned directly with the problems of how genes in eucaryotic chromosomes are expressed. The fate of the RNA transcription product, discussed in the fifth chapter, is becoming clear. Both ribosomal and messenger RNAs are transcribed in the form of large precursors, much of which is degraded when the mature molecules are cleaved and transported from nucleus to cytoplasm. One problem of some urgency is to extend our studies of the mechanism of processing giant heterogeneous nuclear RNA, which seems to involve the addition of sequences of poly-adenylic acid, in order to define the specificity of this reaction. That some messengers coding for particular proteins can now be identified marks an advance from studies of RNA populations which may lead to an understanding of whatever control mechanisms lie between the stages of transcription and translation.

That eucaryotic gene expression is controlled primarily at transcription is a common working assumption; but how precise this control may be and how it relates to the degradation of the immediate products of transcription remain to be defined. The critical problem examined in chapter six is the nature of the unit of transcription: is it a gene, more than one gene, or a sequence in part coding protein and in part serving other functions? Some of each unit of transcription is clearly degraded in the cytoplasm, but it is far from sure that all the sequences which enter the cytoplasm then serve as templates for protein synthesis. Studies of the organization of the transcription apparatus, of the puffs extruded by polytene chromosomes of insect salivary glands, and of the transcription of chromatin in vitro represent three different approaches to this problem. This chapter concludes with an account of formal models for how gene expression may be controlled. Because the genetic analysis of regulator mutations and biochemical isolation of specific control components is not possible with eucaryotic chromosomes, these models are speculative, in contrast with the detailed analysis of bacterial control systems presented in the first volume of Gene Expression.

The general nature of the interactions between nucleus and cytoplasm, a common starting point for the consideration of genetic development, is the topic of the last chapter. Much of this discussion is concerned with the striking experiments made possible by the technique of somatic cell hybridization, which is proving useful not only for studies of cellular functions but also for providing a means for chromosome if not fine genetic mapping.

Any book written in cell biology today must of necessity be more selective than its author would wish; so many cell systems are subjects of study that it is sometimes difficult to know which results are of general relevance and which features may be characteristic only of certain organisms. To maintain coherence in this book, discussion has for the most part been restricted to mammalian cells. I have tried to develop a framework to explain the general nature of the processes by which genes are expressed and controlled rather than to discuss some of the very intriguing systems which may prevail in certain organisms.

As the second volume of Gene Expression, this book is concerned with those problems particular to the cells of higher organisms. Since many of the mechanisms of protein and nucleic acid synthesis appear to be similar in all organisms, the first volume discusses the enzyme systems which undertake protein, RNA and DNA synthesis in both bacteria and eucaryotic cells. This book is in general concerned with events which influence gene expression within the mammalian cell; because of the diversity of the systems used to investigate developmental programmes, it would be impossible to include an adequate account of interactions between cells. Perhaps a future volume may be able to consider the pathways used to control sequential gene action in embryonic development. It is my belief that extension of the concepts first suggested in the molecular biology of bacteria will prove of increasing importance in the biology of eucaryotic cells; and I think we can see already the possible evolution of a single discipline of gene biology concerned with the expression of genes in all classes of cell. I hope that the two volumes of Gene Expression may contribute to this development.

The recent nature of the research discussed in this book is illustrated by the references; almost all derive from this decade or the late sixties. I have followed the same general policy in citing references as in the first volume; because of the several different fields upon which this volume draws, it is perhaps more important, although also more difficult, to indicate those references which may prove seminal for cell biology.

Although I have acknowledged in the first volume the invaluable help which I have received from my father, Dr. Sherry Lewin, both in the writing of Gene Expression and in earlier years, it is once again a pleasure to thank him for initiating me into the world of chromosomes and in particular for suggesting many improvements in this volume. I should like also once again to thank my wife, Ann, for encouraging and assisting me in the writing of both books.

BENJAMIN LEWIN

CONTENTS

ORGANIZATION OF THE GENETIC APPARATUS

3   Protein Components of the Chromosome

4   Sequences of Eucaryotic DNA

EXPRESSION OF GENETIC INFORMATION

7 Interactions between Nucleus and Cytoplasm

Organization of the Genetic Apparatus

CHAPTER 1

Structure of the Chromosome

Distribution of DNA between Chromosomes

Linkage Groups in the Genome

That the chromosomes observed in the nuclei of cells of higher organisms carry the genetic information was realised in the last century. The parallel between the visible behaviour of chromosomes and that demanded of genes by Mendel's laws is striking. Somatic cells of each species have a characteristic content of genetic material which consists of two copies—homologues—of each chromosome (with the exception of the sex chromosomes for which one sex may carry two X chromosomes and the other in turn possesses one X and one Y chromosome). The same diploid number of chromosomes is found in all the somatic tissues of any given eucaryote.

Germ cells are produced by a meiotic division in which the number of chromosomes is halved, so that each sperm or egg receives only the haploid complement consisting of one copy of each chromosome. The different members of the two haploid sets of the parent enter the germ cells independently so that although each germ cell must receive one of the two parental copies of each homologue, there is an equal chance that either will be inherited. Union of sperm and egg to form the next generation reestablishes the diploid complement by bringing together the haploid parental and maternal chromosome sets. Each zygote therefore obtains one copy of each chromosome from each parent. This diploid complement is maintained in the subsequent development of the organism through a series of mitotic divisions in which the chromosomes are replicated and the cell then divides to give two daughter cells, each containing the normal diploid complement. When this organism in turn generates germ cells, each sperm or egg obtains a haploid complement comprising some chromosomes derived from the maternal and some from the paternal set.

That each chromosome is divided into a large number of genes—all roughly of the same order of size—is suggested by the correlation between the number and sizes of linkage groups in the genome and the appearance of the chromosomes. Each homologue contains the same gene loci, although different versions—alleles—of a gene may be present at the corresponding loci on

1

the two homologues of a diploid chromosome set. Each diploid organism thus contains one paternal and one maternal copy of each gene. Meiosis therefore ensures that paternal and maternal alleles must segregate into different germ cells and that genes located on different chromosomes are inherited independently in the manner observed by Mendel.

Those genes located on the same chromosome, however, can be separated only by a recombination event demanding the physical exchange of material between homologues. Two genes located on the same chromosome therefore tend to remain linked together in inheritance; the recombination frequency with which they separate depends upon their distance apart. Exchange between two genes located an appreciable distance from each other on a chromosome may be sufficiently frequent for their inheritance to become independent, as though they were located on different chromosomes. However, each gene must be linked to other gene loci located between them and these loci may in turn be linked to each other. In this way, a genetic map may be extended to form a linkage group which includes genes too far apart from each other to display linkage directly. Each linkage group represents the genetic material of one chromosome; the length of eucaryotic chromosomes means that a large number of mutants is required to construct such a map. However, in those organisms where enough genes have been identified and placed into linkage groups, the genetic map corresponds well with the physical picture of chromosomes; the number and the relative sizes of the linkage groups are the same as those of the chromosomes.

Cytological studies of chromosomes have, of course, been confined largely to observations of their behaviour during cell division. Although interphase, to use the cytological terminology for the period between mitoses, is a misleading term because most of the synthetic activities of the cell take place during this period, an interphase cell appears under the microscope to be quiescent compared with the spectacular changes in cell morphology which take place at mitosis. An interphase nucleus is bounded by a nuclear membrane and the twisted filaments of interphase chromatin are distributed through the nucleoplasm. Individual chromosomes cannot be distinguished, although chromocentres can be seen where the chromatin appears to be more condensed. A nucleolus—or more than one—is often visible and some of the chromocentres may adhere to it.

At early prophase, individual chromosomes can be distinguished as they appear as extended threads. Each chromosome appears to consist of two coiled threads, termed chromatids; these are simply the duplicated copies of the chromosome which have not yet separated. Later in prophase, the chromosomes become short, compact rods. A spindle forms between the two centrioles located at the opposite poles of the cell and as the cell enters metaphase the chromosomes line up along the equatorial plane of the spindle. The form of the spindle may vary with the cell type but constitutes a change in the structure of the cell in which fibres run between the two centrioles, some of them attaching to the chromosomes. At anaphase, the sister chromatids (duplicate chromosomes) separate and move to opposite poles of the cell along the line of the spindle fibres. At telophase, two new nuclei form around the separated sets of daughter chromosomes. The distinction between the phases of mitosis, such as prophase and metaphase, is arbitrary in the sense that mitosis is a continuous process and there are no discontinuous steps during division of the cell into two. These terms therefore serve simply to identify the approximate stages of mitosis which chromosomes have reached.

The status of the cell itself suffers drastic changes during mitosis. As prophase proceeds, the chromosomes approach the nuclear envelope; at the beginning of metaphase the membrane disintegrates so that there is no compartmentalization of chromosomes and the contents of the nucleus and cytoplasm become intermingled. This state of affairs is maintained until telophase when nuclear membranes form around the two daughter sets of chromosomes. Nuclear reconstruction appears to be prophase in reverse. Of course, cell division demands not only that the chromosomes must be reproduced and then segregated, but also that the other components of the cell—such as mitochondria and ribosomes—are gained in roughly equal proportions by the two daughter cells.

Chromosomes appear in their most distinctive forms during metaphase and anaphase when they are highly condensed and stain intensely with basic dyes such as feulgen. It is only at this point in the cell cycle that the features of the different members of the chromosome set can be distinguished. Chromosomes are therefore commonly classified by their appearance at metaphase, when their most prominent feature is the position of the primary constriction at the point where the "arms" of the duplicate chromosomes meet. This identifies the centromeric region; the centromere of each chromosome attaches to the spindle fibres and is responsible for the movement of the chromosomes towards the poles at anaphase. In mitosis, the two chromosomes of each pair separate when their centromere appears to divide, the two daughter centromeres moving towards opposite poles. In meiosis, the centromeres display a related, but not identical, behaviour which ensures the even distribution of chromosomes to germ cells.

Because the centromere appears to move to the pole with the rest of the chromosome following, the appearance of a chromosome at metaphase depends upon the location within it of the centromere. Chromosomes in which the centromere is located at one end take a rod like structure—they are telocentric. Acrocentric chromosomes appear similar but possess a small arm beyond the centromere. (There has been some controversy as to whether chromosomes may be truly telocentric or must always have a short arm, even if this is too small to detect cytologically.) Metacentric chromosomes have arms of about equal sizes because their centromeres are centrally located, and therefore take up a "V" shape at metaphase; when the two arms are of unequal length, the product is a submetacentric chromosome which is "L" shaped. The centromere may therefore be located at any point along the length of the chromosome (with the possible exception of the very end itself); the classification of chromosomes by appearance, although useful for helping to distinguish the different chromosomes of a set, probably has little functional significance.

Chromosomes may also have secondary constrictions at characteristic loci. A secondary constriction is usually associated with the formation of a nucleolus; these constrictions are often described as the nucleolar organisers. There are commonly two such sites, one on each of the homologues carrying the genes for ribosomal RNA, in each somatic nucleus. The function of other secondary constrictions is not always apparent.

The termini of chromosomes—the telomeres—also appear to possess characteristic structural features. When chromosomes are broken by X-rays, the resulting segments may fuse together again; but they cannot do so with the telomeres of unbroken chromosomes. This suggests that the telomere may comprise some particular structure which precludes its attachment to other chromosomes or parts of chromosomes.

Most chromosomes fail to exhibit any underlying structure during the compact stages of metaphase and anaphase, but in the less compact stages of mitosis a coiled fibre can be seen within the chromosome—this has been termed the chromonema. Agents which separate the coils—such as hot water, alkali, KCN—make this structure more visible. In prophase of both mitosis and meiosis, the chromonema may show alternating wide and narrow regions—this gives the chromosome the appearance of beads on a string. The bead like structures have been termed chromomeres and the position of each chromomere is relatively constant for each chromosome. Chromomere patterns can be seen most clearly in meiotic chromosomes. However, it is doubtful whether any functional significance should be attributed to these structures—they probably represent the local manner of coiling of the chromosome.

Some chromosome regions remain condensed during interphase; these have been termed heterochromatin and compare with the much greater part of the length of chromatin which uncoils and swells during the same period—the euchromatin. Heterochromatin often appears to be in close contact with the nucleolus. So far as can be seen from the light microscope, heterochromatin and euchromatin appear to represent different extremes of folding of the chromosome structure, for in chromosomes which have both heterochromatic and euchromatic regions the chromonema appears to run continuously between the two classes of chromatin.

Chromosomes behave during meiosis in a manner which generally resembles their behaviour during mitosis. Meiosis involves one replication but two subsequent divisions of the chromosomes to generate haploid germ cells. An important feature of this process is that homologous chromosomes pair longitudinally to form bivalents; since each chromosome has already been replicated to form two chromatids, each bivalent consists of four chromatids— this is the derivation of the term tetrad. Each chromatid has a single pairing partner at any particular point, although partners may be exchanged along the length of the bivalent so that one chromatid pairs with two or three of the others overall. The sites where partners are exchanged are the chiasmata and it is this exchange which provides the physical basis for genetic recombination between paternal and maternal sets of genes. Chiasma formation takes place during the extended prophase of the first meiotic division. Completion of this division, which subsequently passes through the same stages as a mitosis, separates the homologous chromosomes to give two diploid sets. At the second division, the chromatids of each of these chromosomes in turn separate to give four haploid cells. The sequence of chromosome division is dictated by the behaviour of the centromeres, which must cause the chromatids first to separate into pairs and then into individuals in the two divisions.

The visible morphological features of chromosomes, and the activities which they control, must depend upon the structure of the chromosome as a nucleoprotein particle. The general problem of chromosome structure is to explain how its DNA is folded into a compact structure by its association with proteins; we must account for the forms which the chromosome takes during the cell cycle and explain how its degree of condensation is controlled. Although chromosome structure may appear to be diffuse during interphase, it is no doubt specific at the molecular level and any model for the chromosome must include a structure for interphase chromatin as well as for the condensed metaphase chromosomes. In addition to these general problems, we must also account in molecular terms for the characteristic features, such as constrictions, which chromosomes may display in metaphase. In particular, it is important to define the structure and function of the centromere at both mitosis and meiosis in terms of the specific sequences of DNA and proteins which may be present. Another important question concerns the features which allow homologous chromosomes to recognise each other at meiosis; this ability implies that each homologue may possess a distinct surface structure.

Genetic Material of Eucaryotes

That chromosomes consist of both nucleic acid and protein has been known for some considerable length of time. But it is only more recently that their genetic component has been identified with DNA. And it is yet more recently still that we have come to realise the complexity, both biochemical and genetic, of eucaryotic DNA. One of the most remarkable features of eucaryotes is the enormous amount of DNA which is contained in the nucleus of each cell. The diploid nucleus of a human cell, for example, contains some 5·6 picograms

$(3.5 \times 10^{12}$ daltons) of DNA. Organized in a linear duplex, this DNA would stretch for some 174 cms. Within the cell, it must therefore be compressed many times in length. Another way to view the problem of packaging DNA is to calculate the total length which the DNA of an average human would occupy if stretched end to end in duplex form; the 100 grams of DNA would stretch for some 2.5×10^{10} kilometres—more than one hundred times the distance from the earth to the sun!

The DNA of a eucaryotic cell is not, of course, a single duplex molecule, for it is organised into many individual chromosomes. Although eucaryotic chromosomes vary greatly in size, the quantity of DNA in even a small chromosome may be considerable. The smallest of the 23 human chromosomes contains some 0.046 picograms $(3 \times 10^{10}$ daltons), equivalent to a linear duplex length of 1.4 cms; and the largest contains some 0.235 picograms $(1.5 \times 10^{11}$ daltons), equivalent to a linear duplex of 7.3 cms. But the lengths of these chromosomes, observed at metaphase, are of the order of $2-10 \times 10^{-4}$ cms. Compression in the length of genetic material is common to all organisms and viruses, although a greater amount of DNA is involved in eucaryotes—even the smallest human chromosome contains ten times the amount of DNA in E.coli, whose chromosome comprises 2.8×10^9 daltons $(4.6 \times 10^6$ base pairs) in a cell of diameter about 10^{-4} cms. Table 1.1 compares the relative amounts of DNA in various species.

This enormous compression of DNA raises several topological problems. First of all, how is DNA physically arranged within a chromosome and how much precision is there in its organization? Is there only a general structure for most chromosome regions, or do particular sequences of DNA always interact with specific proteins? How does this structure change during the cell cycle? A second set of questions concerns the expression of nuclear DNA. Assuming that its replication and transcription follow the general rules established in bacteria, how does the double helix unwind within the restrictions of the nuclear volume so that it may be replicated; and how do the daughter duplex strands separate to form the replica chromosomes which are distributed to daughter cells at mitosis? How do specific sequences of DNA become accessible to RNA polymerase for transcription?

Another important problem raised by the amount of DNA in the eucaryotic genome is whether all its sequences code for protein. Taking 10^6 daltons as the approximate size of a gene—this corresponds to a length of some 1500 base pairs which could code for a protein of about 50,000 daltons—the haploid human genome could specify 1.75×10^6 genes, almost two million. Although it is impossible to predict with much precision the number of proteins for which the genome codes, this number is greater by one or two orders of magnitude than estimates for the number of genes. Allowing for the proteins which may be involved in embryonic as well as in adult stages of development, arguments about the number of genes compatible with the known rates of mutation and evolution suggest that there may be of the order of 50,000 genes in the human genetic material (see Chapter 4).

This discrepancy is so great that we are compelled to conclude that, if the genetic arguments are valid, much of the genetic material cannot be equated with genes coding for proteins which can be identified by mutation. Two general types of solution have been proposed for this problem. One is to suppose that there is only one copy in the haploid genome of the gene coding for any parti-

Table 1.1: content of DNA in the haploid genomes of viruses and species of cell. Viruses usually possess 2–100 $\times$ 10^6 daltons of DNA and bacteria of the order of 2×10^9 daltons. Mammals usually possess about 2×10^{12} daltons, about 1000 times the DNA content of bacteria. Other eucaryotes range widely: unicellular organisms usually display about 10 times the bacterial DNA content and multicellular organisms may have from 100 times the bacterial DNA content to values of more than an order of magnitude in excess of the mammalian content, that is in excess of 10^4 times the bacterial content. Data of Cairns (1966), DuPraw (1970), McCarthy (1969), Rasch et al. (1971), Stebbins (1966) and Thomas and McHattie (1967).

		haploid genome size in		
	species	picograms	daltons	base pairs
phages:	ϕX 174		$1 \cdot 6 \times 10^6$	5,500
	lambda		30×10^6	45,000
	T4		130×10^6	200,000
animal virus:	adenovirus 12		14×10^6	21,000
procaryotes:	Mycoplasma gallisepticum		$0 \cdot 2 \times 10^9$	$0 \cdot 3 \times 10^6$
	B. subtilis		$1 \cdot 3 \times 10^9$	$2 \cdot 1 \times 10^6$
	E. coli		$2 \cdot 8 \times 10^9$	$4 \cdot 6 \times 10^6$
unicellular eucaryote:	S. cerevisiae		$6 \cdot 5 \times 10^9$	10×10^6
higher organisms:	D. melanogaster	0·18	$0 \cdot 11 \times 10^{12}$	$0 \cdot 18 \times 10^9$
	Zea mays	15	10×10^{12}	15×10^9
	X. laevis	22	14×10^{12}	22×10^9
	mouse	2·3	$1 \cdot 5 \times 10^{12}$	$2 \cdot 3 \times 10^9$
	man	2·8	$1 \cdot 8 \times 10^{12}$	$2 \cdot 8 \times 10^9$
	(smallest chromosome)		$0 \cdot 03 \times 10^{12}$	$4 \cdot 5 \times 10^7$
	(largest chromosome)		$0 \cdot 15 \times 10^{12}$	$2 \cdot 3 \times 10^8$

cular protein. This implies that even if the number of coding genes is greater than we estimate, there must be a considerable excess of DNA which has some purpose—perhaps concerned with the control of gene activity—other than coding for proteins. An alternative theory is to suppose that much of the potential protein-coding DNA does in fact code for proteins, the reason there is so much of it being that genes are repeated in the genome. That is to say that there may be more than one copy of each gene, the exact number varying with the particular gene, but the degree of repetition being as large as one hundred fold.

A genetic reason for supposing that eucaryotes contain only one copy of each gene in the haploid genome is the relationship of the paternal and maternal gene sets in diploid organisms. Each parent appears to contribute only one copy of each gene; a mutation in the allele derived from either parent is therefore sufficient to control its genetic contribution. In those instances where gene products can be examined directly, a mutation in the allele of either parental set alters half of the corresponding protein molecules. If each haploid genome carried many copies of the gene, a single mutation should change only a small proportion of the protein products. The biochemical verification of mendelian genetics therefore confirms that each haploid genome contains only one copy of each particular gene.

To overcome this objection to accounting for the excess DNA in terms of multiple copies of coding genes, Callan (1967) and Whitehouse (1967) have proposed models for the chromosome in which one of the many copies of each gene is a "master" copy. According to these models, genetic studies follow the inheritance of the single copy of each master gene, for mutations are assumed to be effective only if they occur in this sequence. Mutation in the other "slave" copies of the gene should not be detected in genetic studies because their sequences must depend upon that of the master gene. Any mutations occurring in the slave genes are therefore corrected by comparison with the sequence of the master copy; but if a mutation occurs in the master gene, all the slave copies are altered to conform with its new sequence.

Enzymes with activities analogous to those engaged in repair of radiation damage to DNA would undertake the correction of slave copies to the master sequence. Such activities might be adequate to explain the inheritance of point mutations, but must encounter difficulties in explaining the correction of mutational events such as deletions, insertions, transpositions which influence longer sequences of DNA. And eucaryotic DNA does not possess identical sequences repeated the 2–100 times required of the slave genes. In those instances where individual genes can be assayed because their messenger RNAs have been isolated, the number of copies in the genome is usually small, probably corresponding to only one. It seems very likely, therefore, that most eucaryotic genes are represented only once in the haploid genome.

Nucleic acid hybridization studies show that although much of the eucaryotic genome consists of unique sequences of DNA, a considerable proportion represents varying degrees of repetition of sequences related to, but not identical with, each other. Since these sequences are not identical, they cannot represent repeated copies of genes (and their frequency of repetition is too great, usually in excess of 1000 times). Although some of the repetitive sequences are transcribed into RNA, it is not clear to what extent they are represented in the messenger population and it is doubtful whether they provide any appreciable proportion of the sequences coding for protein. It is often supposed that repetitive sequences may include control elements.

The remaining sequences of DNA, comprising the nonrepeated component, presumably include single copies of individual genes; these must represent most of the coding functions of the genome. But since this component itself contains sufficient DNA to code for several hundred thousand proteins, the discrepancy between genetic estimates of gene number and biochemical potential remains to be resolved. If most or all of the nonrepetitive DNA codes for proteins, the genetic estimates must be incorrect by about an order of magnitude; but if the gene number is low, much of the nonrepetitive as well as repetitive DNA must have some function other than coding for protein (see Chapter 4).

Of course, not all the genetic information of a eucaryotic cell is contained in the DNA of the chromosomes in the nucleus, for some of the cytoplasmic organelles also contain DNA. Examples of non-mendelian inheritance are generated when cells are heterozygous for genes carried on mitochondrial or chloroplast DNA. In this case, reproduction does not take place by the precise segregation of mitosis and meiosis but by a more approximate distribution of the organelles in the cell. And the egg, of course, provides the sole supply of mitochondria for a developing embryo; the sperm contributes only a haploid set of nuclear genetic information.

The amount of DNA in cytoplasmic organelles is quite small. Mitochondrial DNAs of animal cells are usually found as circular molecules of duplex DNA of about 1×10^7 daltons; this might code for some 30 proteins, many of which are likely to be part of the protein synthetic apparatus of the organelle. The mitochondria of other eucaryotes, such for example as yeast, may be much larger; and chloroplast DNA may also carry an appreciable amount of genetic information, in some instances as much as a bacterial genome. Systems for the transcription of RNA from the DNA of mitochondria and chloroplasts and for its translation into protein generally appear to be more similar to the synthetic processes of bacteria than to the surrounding cytoplasm. Centrioles also appear to contain DNA, although we do not know whether it is transcribed and translated. The total contribution of these cytoplasmic DNAs to the genetic information of the cell is therefore small; in the main, it appears to be concerned with coding for specialised functions of the organelle which contains it (for review see Sager, 1972).

Continuity of DNA Threads in Chromosomes

Semi-Conservative Replication of the Chromosome

One of the strongest indications that the eucaryotic chromosome consists of a single linear duplex of DNA is its semi-conservative replication. Eucaryotic DNA is not amenable to the biochemical analysis which Meselson and Stahl

(1958) used to show that bacterial DNA is replicated semi-conservatively. However, tritiated thymidine can be incorporated into chromosomes during their replication and the segregation of the labelled daughter chromosomes can be followed through the subsequent cell divisions.

The diploid chromosome complement of the plant Vicia faba comprises some twelve large chromosomes which are particularly suitable for cytological analysis. Taylor, Woods and Hughes (1957) therefore used Vicia faba seedlings in a protocol in which the seeds were incubated in H^3-thymidine and then transferred to an unlabelled medium containing colchicine. This inhibitor of mitosis allows chromosomes to reproduce but prevents their movement apart and the subsequent formation of separate daughter cells. As colchicine treated cells accumulate in mitosis, therefore, chromosomes are found in the metaphase condition with sister chromatids (daughter chromosomes) spread apart instead of lying parallel to each other.

Because division of Vicia faba cells does not take place until some 8 hours after DNA synthesis has been completed, few if any of the nuclei which have incorporated their label into DNA should have time to pass through a division before transfer to colchicine. This means that virtually all the labelled nuclei are prevented from completing their cell division. After their transfer to medium containing colchicine, cells may be allowed to grow for various periods of time before fixation and autoradiography to determine the fate of the H^3 label. The type of metaphase plate found when the cells are removed from the medium depends upon the number of times the chromosome set has been replicated.

When the cells remain in colchicine for only 10 hours before they are fixed, all the metaphase plates contain 12 chromatid pairs; the cells have had time to replicate their chromosomes only once, in the presence of the thymidine label and before transfer to colchicine medium. Each of the 12 chromatid pairs consists of the sister chromatids (daughter chromosomes) produced by the replication, which are spread apart but remain attached at the centromere. All the chromosomes in these cells are labelled and the two sister chromatids of each pair are equally and uniformly labelled as illustrated in figure 1.1.

When the cells are incubated in colchicine for 34 hours, metaphase plates can be found containing 12, 24 or 48 pairs of sister chromatids. Those with 12 pairs are usually unlabelled; these therefore represent cells which have replicated their DNA only after the end of the period of labelling with H^3 thymidine. In the cells with 24 chromatid pairs, all the pairs are labelled; but only one of the two replica chromosomes of each pair carries the radioactive label. These chromosomes have duplicated once in the radioactive thymidine and once in its absence. As figure 1.1 shows, the second replication segregates the radioactive label incorporated in the first replication as would be predicted from semi-conservative synthesis of DNA. One strand of each daughter chromosome must have been labelled during the first replication; these strands separate during the second replication, in unlabelled medium, so that one of the two products receives a labelled strand and the other receives the unlabelled strand.

Only a few cells have been found with 48 chromatid pairs but in these cells one half of the chromatid pairs are usually labelled and the other half unlabelled. (The appearance of these cells shows some shortening in the usual

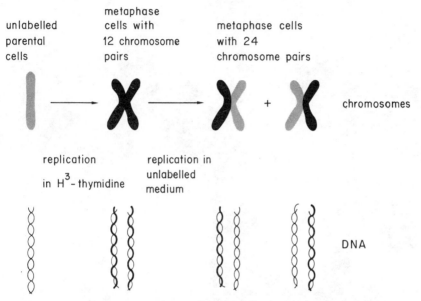

Figure 1.1: semi-conservative replication of eucaryotic chromosomes. Upper: incorporation of grains observed by autoradiography; lower: interpretation of structure of DNA contained in chromosomes. In the first metaphase after replication in labelled medium, all chromosomes show an equal label (dark regions); all their DNA molecules must be hybrid, consisting of one parental (unlabelled) and one newly synthesized (labelled) strand. In the second metaphase after replication, the labelled DNA strands segregate semi-conservatively in the unlabelled medium, so that each chromosome pair has one chromatid consisting of hybrid DNA and one chromatid consisting of unlabelled DNA.

24 hour cycle of Vicia faba seedlings.) These results are consistent with a further semi-conservative segregation of the labelled material.

In some of the cells with 24 or 48 pairs, a few of the replica chromosomes are labelled along only part of their length. But in these instances, the sister chromosome is labelled in the remaining portion. This is an apparent breach of semi-conservative replication, but is caused by an exchange of regions of sister chromosomes after they have been replicated (see Taylor, 1958).

Similar experiments have since been performed with many other cell types; the same results have been obtained in all instances. For example, Prescott

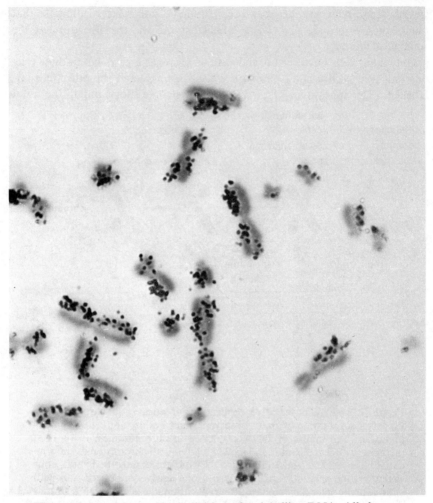

Figure 1.2: second metaphase division after labelling DNA. All chromosomes possess one labelled chromatid. Sister chromatid exchanges are
responsible for a switch of label between the two chromatids (see text).
Data of Marin and Prescott (1964): photograph kindly provided by
Professor David Prescott.

and Bender (1963a, b) found that an H^3-thymidine label given to human
leucocytes or Chinese hamster cells follows the predicted semi-conservative
segregation for the next four cell divisions. Incorporation of the label is
permanent and none is lost from the chromosomes during this time. Figure 1.2
shows a second metaphase division observed by Marin and Prescott (1964),
which corresponds to the stage at the far right of figure 1.1.

When RNA is labelled during interphase the label is lost at prophase when all the nuclear RNA is released into the cytoplasm of the cell. Nor is the segregation of labelled proteins semi-conservative; although labelled proteins are equally distributed to the two sister chromatids found at the first metaphase after addition of the label, all the radioactive protein is lost from the chromosomes during subsequent generations. This shows that the association of RNA with the chromosomes is transient and that the protein components must turn over, reinforcing the conclusion that the sole genetic component of the chromosome is DNA.

The simplest explanation for the observed inheritance of radioactive DNA is that each chromosome consists of one very long molecule of double stranded DNA (although this is folded into the compact structure of the chromosome). Replication by the same semi-conservative mechanism used in bacteria, in which the two daughter duplexes of DNA constitute the daughter chromosomes, generates the observed pattern of segregation of the radioactive label. Of course, it is also possible to explain these results in terms of chromosomes which contain many independent molecules of DNA; but such explanations demand the postulation of special mechanisms to ensure that the segregation of the several replicated strands follows the observed semi-conservative distribution. In general, this requires the presence in the chromosome of some component other than DNA—this has usually been assumed to be protein—which also segregates in a semi-conservative manner. However, it is clear that the chromosomal proteins and RNA are not associated with the genetic material in the manner this theory would require. These experiments therefore suggest that the DNA of each chromosome comprises one double stranded molecule of covalent integrity whose single strands are the sole units of inheritance.

Lampbrush Chromosomes

One of the most convincing lines of evidence which suggests that the eucaryotic chromosome comprises a continuous duplex of DNA has been derived from studies of a specialized chromosome system, the lampbrush chromosomes of amphibians. Oogenesis in amphibians may take a considerable length of time and during this lengthy meiosis the chromosomes become stretched out so that they may be observed in the light microscope. This greatly elongated state is maintained during the diplotene stage of meiosis and the chromosomes later revert to their usual size. We may therefore take the lampbrush chromosomes to represent an unfolded version of the normal state of the chromosome.

Lampbrush chromosomes may be between 500 μ and 800 μ long at diplotene although their length contracts to some 15–20 μ later in meiosis. They are found in the form of meiotic bivalents in which the two homologues of each chromosome pair are longitudinally associated with each other. As Callan

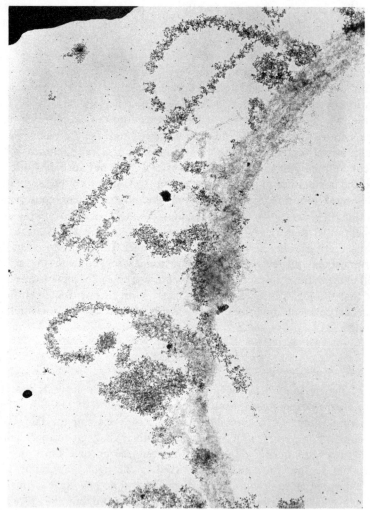

Figure 1.3 (a)

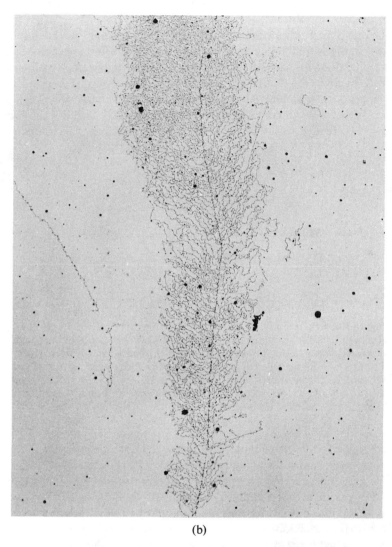

(b)

Figure 1.3: lampbrush chromosomes of oocytes of Triturus viridescens. (a) Portion of main axis with several lateral loops, × 5,550. (b) portion of loop from near intersection with axis, showing synthesis of RNA increasing in length with distance from axis. × 14,625. Photographs kindly provided by O. L. Miller, jr. and B. R. Beatty, Oak Ridge National Laboratory.

and Lloyd (1960) have shown, the axis of each chromosome of the paired structure consists of a series of closely spaced chromomeres connected to each other by very thin threads.

The lampbrush chromosomes take their name from the lateral loops present in the two homologues in pairs of similar size and appearance. Each loop consists of an axis surrounded by a matrix of ribonucleoprotein. As figure 1.3 shows, the matrix increases in length proceeding around the loop and appears to represent RNA molecules under synthesis which associate with proteins to form compact ribonucleoproteins. The size of each loop is inversely correlated with the size of the chromomere from which it projects,

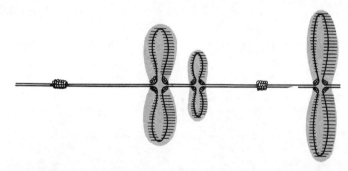

Figure 1.4: structure of lampbrush chromosomes. The axis of the chromosome consists of two duplex threads of DNA lying close together; although drawn separately here, they cannot be distinguished under the microscope. The DNA is tightly coiled at chromomeres. The loops which extrude from some of the chromomeres comprise an axis of duplex DNA surrounded by a matrix of ribonucleoprotein. The axis of the loop is continuous with the central axis of the chromosome. The two homologues are usually symmetrical.

which suggests that the loops represent DNA extended for transcription whereas the chromomeres contain more tightly coiled inactive lengths of DNA.

The lampbrush chromosome therefore appears to comprise a long extended thread of genetic material. Callan and McGregor (1958) showed that DNAase destroys the integrity of the chromosomes and by this means demonstrated that the axis of each loop is continuous with the central axis of each chromosome. Proteolytic enzymes or ribonuclease fail to disrupt the continuity of the chromosome. Figure 1.4 illustrates the structure of the lampbrush chromosome; it consists of long threads of genetic material coiled into chromomeres at intervals, loops projecting from some of these chromomeres.

By measuring the rate at which DNAase I introduces breaks into lampbrush chromosomes, Gall (1963) was able to show that each chromosome thread consists of a duplex of DNA. This enzyme independently attacks the two polynucleotide strands of a DNA duplex molecule at random sites. The rate

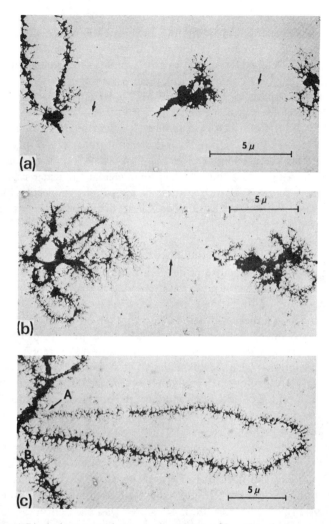

Figure 1.5: whole mount electron microscopy of lampbrush chromosomes. (a) main chromosome axis showing loops protruding from chromomeres, × 4,200. (b) main chromosome axis connecting loops, × 3,600. (c) lateral loop with typical ribonucleoprotein matrix showing transition from thin end of loop (A) to thick end (B). × 3,000. Data of Miller (1965); photograph kindly provided by Dr. O. L. Miller.

of breakage of the chromosome increases with the time for which it is exposed to DNAase; by assuming that a visible break in the chromosome is produced only when all its subunits are broken at or close to the same site, it is possible to show that two hits are needed to cause a break in a loop but four hits are needed in the axis where the two homologous chromosomes are paired together. This suggests that each chromosome consists of a single duplex of DNA, which is exposed in the loop but which is tightly paired with its partner duplex in the synapsed regions of the central axis.

Electron microscopy of preparations of lampbrush chromosomes spread on a water surface has confirmed this picture; see figure 1.5. Miller (1965) found that when the ribonucleoprotein matrix is removed from the loops, their axial material has a diameter which varies from 75–100Å at its thickest to 30–50Å at its thinnest. Trypsin reduces its diameter, which supports the idea that it constitutes a DNA double helix surrounded by protein. The fibres connecting the chromomeres are slightly wider, ranging from 50–75Å in the thinner regions to about 200Å in diameter at the wider regions; this accords with the idea that they consist of two duplex molecules of DNA, running parallel (although it is not possible to distinguish the two threads).

Classical Models for the Chromosome

Early models for chromosome structure generally supposed that each chromosome must consist of more than one molecule of DNA. One common class of model relied upon the idea that the integrity of the DNA duplex might be broken at linkers consisting of some other chromosome component, presumably protein. One reason for postulating the presence of linkers is to explain how very long sequences of DNA might unwind for replication. Some form of interruption—whether linkers or single strand breaks—must be made in eucaryotic DNA to allow each replicon to synthesize DNA; linkers providing frequent breaks in the covalent continuity of the double helix would therefore provide an opportunity for each molecule of DNA to untwist about its axis independently.

According to the model of Freese (1958) each chromosome might consist of many duplex molecules of DNA, linked end to end by a series of protein molecules. Coiling and uncoiling of the chromosome might be explained by the interactions of these protein molecules with each other. An earlier model proposed by Mirsky and Ris (1951) suggested that each chromosome might consist of a longitudinal backbone of protein to which individual DNA molecules are attached laterally. Taylor (1957) extended this model to allow semi-conservative replication by proposing that each chromosome might have two parallel backbones which segregate semi-conservatively, the lateral duplexes of DNA attaching alternately to each backbone. More complex variations of this model to include subsequent observations of chromosome structure have been considered by Taylor (1963).

More recent studies of chromosomes argue against models which invoke frequent interruptions in the integrity of the DNA double helix. An important objection to all models which insert protein linkers in the chromosome is that the integrity of the DNA has been found to be sensitive only to DNAase and not to ribonuclease or proteolytic enzymes. This is particularly evident in the lampbrush chromosomes where it is clear that the continuing element of the axis is DNA alone. Another objection is that only the DNA component of chromosomes is inherited semi-conservatively; models in which a backbone segregates semi-conservatively would demand that this pattern should be followed by at least some of the chromosomal protein.

Direct evidence that each chromosome contains DNA of a considerable length has been provided by autoradiographic studies of replication. Cairns (1966) observed labelled DNA strands of a continuous length of upto 500 μ in Hela cells; and Huberman and Riggs (1968) have observed even greater lengths of replicating DNA of upto 1800 μ in Chinese hamster cells. It is usually difficult to obtain unbroken lengths of DNA of greater than 1 mm (1000 μ), so it is not possible to confirm directly that the DNA of each chromosome is a continuous covalently linked double helix. However, DNA strands of upto 2 cms have been observed by Sasaki and Norman (1966) in autoradiography of human DNA; this length is sufficient to account for the entire content of DNA of some of the smaller human chromosomes. And by measuring the viscoelastic properties of DNA released from Drosophila chromosomes, Kavenoff and Ziman (1973) showed that each chromosome probably contains one very long DNA molecule.

One of the controversies about chromosome structure which continued for some time was the question of whether chromosomes are single stranded (unineme) or multistranded (polyneme) (reviewed by Prescott, 1970). By single stranded we mean that each chromosome consists of a single duplex of DNA, although this molecule must of course be coiled upon itself into the compact structure of the chromosome. A multistranded chromosome, by contrast, would consist of fibres containing many duplex molecules of DNA associated in parallel with each other and in principle running longitudinally along the chromosome, although once again these fibres should be coiled into a more compact structure. (The term polynemy is used to describe the situation when the basic fibre of the chromosome might contain many strands of duplex DNA; the term polyteny describes the situation achieved in certain tissues when chromosomes replicate many times but remain longitudinally associated instead of dividing.)

Earlier studies of chromosome structure have been reviewed by Taylor (1963). At anaphase, chromosomes appear under the light microscope to be composed of a single chromatid, forming a flexible rod which consists of a helically coiled chromonema. (Ohnucki (1968) has visualized the coiling of the chromonema by using a hypotonic salt solution to release some of the chromosome components.) When the chromosomes expand to form the interphase chromatin, they cannot be ascribed any structure by either light or conventional electron microscopy. When chromosomes can be individually distinguished again at the next prophase, each appears to consist of two chromatids, which continue to shorten during mitosis; each chromatid usually appears to consist of a single coiled chromonema.

The critical argument about chromosome structure has been whether the chromonema contains a single duplex of DNA or many duplex molecules associated in parallel. Support for multistranded models for the chromosome was lent by early studies with the electron microscope, which appeared to show a substructure for the chromosome thread consisting of many parallel fibres. Chromosomes are too thick for electron microscopy of sections in their mitotic state and interphase chromatin is difficult to use for this technique. Ris (1956) therefore used both the lampbrush chromosomes of amphibian oocytes and the pachytene chromosomes of some insects and plants. These prophase chromosomes consist of bundles of fibres which are helically coiled or twisted less regularly.

One early model viewed the fibrils as 500–600Å thick, consisting of two finer fibrils of about 200Å in diameter. The later observations of Ris and Chandler (1963) suggested that the principal constituent of the chromosome is a fibre about 250Å in diameter. Fibres of this order of size, with irregular diameters ranging from 200–300Å have been observed in a variety of preparations of both mitotic chromosomes and interphase chromatin by the technique of spreading chromosomes on a water surface before microscopy. The argument about chromosome structure has therefore resolved into a debate about the substructure of these fibres, which are now thought to constitute the principal thread in the chromatin (for review see DuPraw, 1970).

A typical multistranded model for the chromosome is that proposed by Ris and Chandler which suggests that each 250Å fibre consists of two fibres of 100Å in diameter wound around each other. Each 100Å fibre in turn consists of two parallel fibres of 40Å diameter, each of which contains a duplex of DNA surrounded by protein. The early evidence taken to support such models has been reviewed by Ris (1969). However, DuPraw (1970, 1972), following the results of Barnicott (1967), has pointed out that these apparent substructures may be produced as artefacts of aggregation and separation during the procedures followed to prepare the chromosomal material for microscopy. It is also possible that some reports of parallel double structures may represent observations of an abnormal situation in which the chromosomes replicate but do not divide. Phase contrast and interference microscopy of living cells do not reveal parallel substructures. Further support for the idea that each 200–300Å fibre contains only a single duplex molecule tightly packed within it is provided by the observation of DuPraw (1965a) that treatment with trypsin reduces the fibres to a single filament with the dimensions of a DNA duplex; treatment with DNAase degrades the filaments.

One problem in accepting the view that each chromosome consists of a single duplex of DNA is the apparent subchromatid exchange promoted by irradiation with X-rays. Treatment with X-rays during prophase or early metaphase causes aberrations which appear to result from damage to a subunit of the chromatid; for exchanges appear to take place between subunits of chromatids (that is replicated daughter chromosomes) as though each chromatid consists of two substructures. Figure 1.6 shows that if these apparent subchromatid exchanges are real, they should appear as exchanges of the full chromatid type at the second mitosis (and should influence both chromatids of a mitotic pair only at the third meiosis).

This prediction has been tested by Kihlman and Hartley (1967), who analysed breaks induced in chromosomes by the reagent 8-ethoxy-caffeine. After treatment, cells were incubated with colchicine so that although the

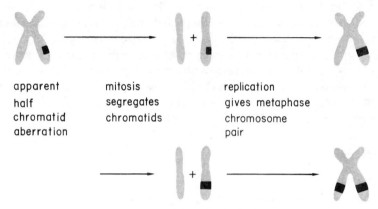

apparent	mitosis	replication
half	segregates	gives metaphase
chromatid	chromatids	chromosome
aberration		pair

Figure 1.6: segregation of apparent half chromatid aberrations. If the aberration represents damage to a subunit comprising half a chromatid, at the next metaphase it should appear as an aberration of one chromatid of the metaphase pair (upper). If the aberration really represents damage to the whole chromatid, both of the chromatids of the pair carrying the damage will display the aberration at the next metaphase (lower). The latter expectation has been confirmed by experiment (see text).

aberrant chromosomes could replicate, they were unable to separate. Exchanges which had appeared at the first mitosis to constitute subchromatid aberrations were displayed at the second mitosis as full chromosome exchanges influencing both sister chromatids. This implies that the appearance at the first mitosis of subchromatid exchanges is deceptive; these events appear to constitute an aberrant form of a full chromatid exchange which masks their true nature. Observations of damage to the chromosome at the subchromatid level had previously been thought to provide strong evidence for a subunit structure for the chromosome, but it now seems that they involve the basic chromatid thread and may best be interpreted in terms of the folded structure for a single duplex of DNA discussed in the next section (for review see DuPraw, 1970; Kihlman, 1971).

Several lines of evidence thus suggest that chromosomes are single stranded.

This is the most straightforward way to explain the semi-conservative segregation of DNA when chromosomes replicate, for this observation argues both against polynemy and against interruptions in the covalent integrity of the single duplex molecule. In the case of lampbrush chromosomes, although these are admittedly extended beyond their usual state, the axis of each chromosome constitutes a single duplex of DNA. This model is lent support by the isolation of considerable covalent lengths of eucaryotic DNA and by the demonstration that trypsin reduces chromosome fibres to the size of a DNA duplex and that this structure is susceptible to destruction only by DNAase.

Cytological observations of multistranded substructures for the chromatid may result from artefacts of preparation and do not represent genuine features of the chromosome. And if the chromosome were to consist of parallel subunits, then it should be possible to induce exchanges between them; however, it seems likely that all such apparent events can be explained by exchanges of the single thread constituting the chromatid (see below). These results implicate a 200–300Å fibre as the basic component of the chromosome, containing only a single coiled duplex of DNA.

Organization of Chromatin Fibres

Folded Fibre Model for the Chromosome

Some of the problems in obtaining preparations of chromosomes suitable for electron microscopy have been overcome by the development of the technique of whole mount microscopy in which the chromosomes are spread out for visualization. To obtain metaphase chromosomes, cells are incubated in colchicine, lysed in distilled water, and the chromosome preparation is floated on the surface of the water. It can then be picked up on a grid, washed or treated with reagents, and dried from liquid CO_2. Interphase chromatin can be treated in a similar manner.

Using preparations of human cells in metaphase, DuPraw (1966) found that the chromosome seems to consist of fibres with an average diameter of about 300Å, rather tightly and irregularly folded into the characteristic structure of the chromosome. Preparations of interphase chromatin seem to be composed of similar fibres, although the average diameter is less, usually about 230Å. DuPraw (1965a) obtained similar results with metaphase chromosomes of embryonic cells of the honeybee, which consist of fibres of about 250Å. Such fibres have since been observed in many preparations of metaphase chromosomes and interphase chromatin. A characteristic feature is their "bumpy" appearance, with the diameter commonly varying from as little as 200Å to as much as 500Å; this pattern may reflect the irregular twisting of the fibre.

The overall structure of the larger human chromosomes is preserved during

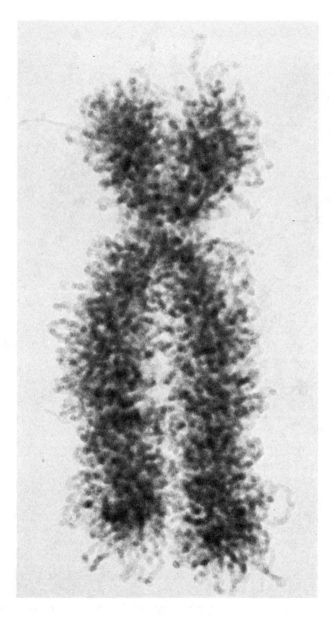

Figure 1.7: whole mount electron micrograph of human chromosome 2, × 44,460. Kindly provided by Professor E. J. DuPraw (1970); from *DNA and Chromosomes* (Holt, Rinehart and Winston).

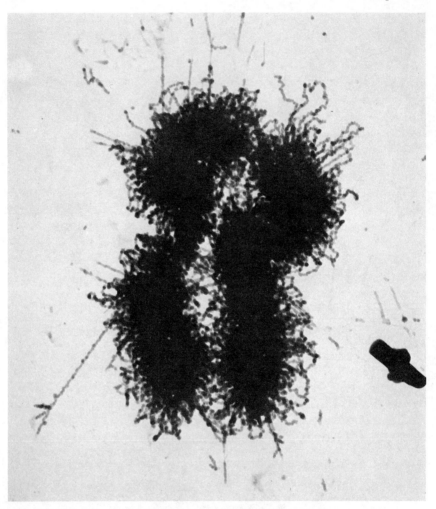

Figure 1.8: human chromosome 1 in a tightly packed conformation,
× 25,600. Kindly provided by Professor E. J. DuPraw (1970); from *DNA
and Chromosomes* (Holt, Rinehart and Winston).

whole mount microscopy, for they display their characteristic morphologies
on the microscope grid, in particular the primary constriction of the centromeric region. Figure 1.7 shows a micrograph of human chromosome 2 in a
"standard" conformation; figure 1.8 shows a more tightly packed preparation
of chromosome 1. Since no free ends are seen at the telomeres, each fibre
must double back upon itself at the end of the chromosome structure. Each
chromosome may consist of one long fibre irregularly folded along its length
in such a way that the coiled quaternary structure of the chromosome is

maintained. The entire mass of each chromosome can be accounted for by its folded fibre without addition of any other components.

The extent of folding of the fibre can be determined by comparing its total length with that of the chromosome itself. DuPraw and Bahr (1969) have determined the dry mass of complete chromosomes and of limited regions of their constituent fibres by quantitative electron microscopy. This technique calibrates the electron scattering of the fibres or chromosomes against a control of polystyrene spheres of known volume and weight. The weight of a fibre depends upon its diameter; the 200Å regions of human chromosome fibres have a weight of about $4 \cdot 5 \times 10^{-16}$ grams/micron; regions with a diameter of 500Å weigh some $29 \cdot 9 \times 10^{-16}$ grams/micron. An average fibre of interphase chromatin of diameter 230Å weighs about $6 \cdot 1 \times 10^{-16}$ g/μ; and the more dense average metaphase fibre regions of diameter 300Å weigh almost 8×10^{-16} g/μ.

These figures can be taken only as average values, for the diameters and dry mass values of the fibres are not uniform along their length but vary considerably. There are also appreciable variations from mitosis to mitosis of the dry mass values of different samples of the same chromosome; the range of values found for any particular chromosome—identified by its relationship to the other members of its set—seems to fall within extremes about $2 \cdot 5$ times apart. Figure 1.9 shows two different preparations of human chromosome 2. The extent of this variation is emphasized by the report of Bahr and Golomb (1971) that a study of 925 human chromosomes did not produce clusters corresponding to the different size groups but instead generated a continuous spectrum of chromosome sizes. Measurements on the fibres of chromosomes therefore identify only the range of dry mass values and do not yield an absolute value for any chromosome.

Since the amount of DNA in any particular homologue must be constant, this variation implies that different amounts of protein must be present in the chromosomes at different metaphases. But in any one metaphase plate the mass of each chromosome is proportional to its DNA content; the same general variation is therefore displayed by all the chromosomes of any one mitotic cell. It is possible that much of this variation may depend upon the different extents to which the fibres of interphase chromatin associate with additional proteins at each mitosis.

The mass of an individual chromosome can be compared directly with that of its constituent fibre by determining the electron scattering values of the entire chromatid and of a representative length of fibre. The ratio of total scattering to scattering per micron gives the total length of the fibre, which is usually of the order of 50 times the length of the chromosome itself at metaphase. In one set of measurements obtained by DuPraw and Bahr, the smallest and largest human chromosomes proved to have fibre lengths of 135 μ and 723 μ respectively, compared with their overall metaphase lengths of 2 μ and 11 μ.

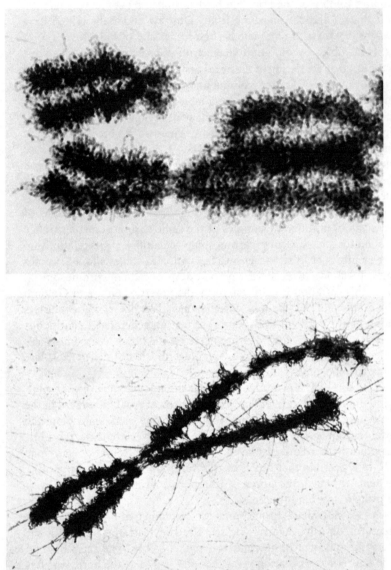

Figure 1.9: two different preparations of human chromosome 2. Left: a "heavy" conformation in which the chromosome has a high mass, × 8,225. Right: a "standard" conformation representing the most common appearance of the chromosome, × 18,200. The chromosome on the right of chromosome 2 is a member of the group 13–15. Kindly provided by Professor E. J. DuPraw (1970); from *DNA and Chromosomes* (Holt, Rinehart and Winston).

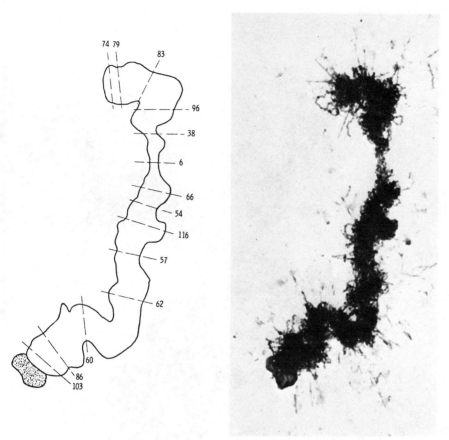

Figure 1.10: estimates of numbers of fibres from measurements of the cross sectional mass of human chromosome 4–5, × 16,200. Left: numbers of fibres at each point in chromosome. Right: whole mount electron micrograph. The number of fibres is about 60 in the body of the chromosome but rises at the telomeres where the fibres must double back and decreases at the primary constriction of the centromere. Kindly provided by Professor E. J. DuPraw (1970); from *DNA and Chromosomes* (Holt, Rinehart and Winston).

Another calculation which can be made is to measure the cross-sectional mass of the chromatids at various points along their length; this estimates the number of fibres lying side by side at each site. Each chromatid of the 4–5 human group contains about 405 μ of fibre and is about 5·8 μ long at metaphase. On average, there should therefore be some 70 equivalents of fibre lying side by side at any point. Measurements of cross-sectional mass such as those illustrated in figure 1.10 suggest that 60–70 fibres are found at most points, with the exceptions of the telomeres which have an increased number of fibres—

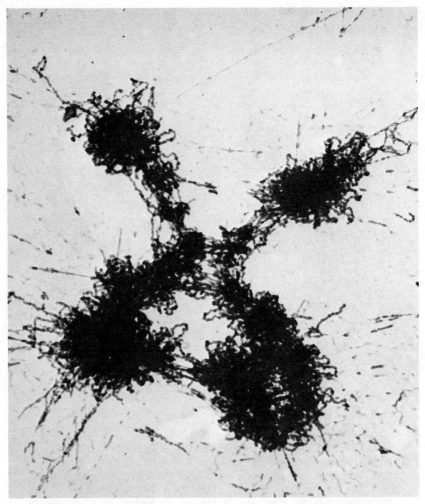

Figure 1.11: human chromosome of the 17–18 group, × 24,720. Fibres hold
the two chromatids together both at the centromere and between the arms.
Kindly provided by Professor E. J. DuPraw (1970); from *DNA and
Chromosomes* (Holt, Rinehart and Winston).

presumably because the fibre doubles back on itself at this point—and at the
centromere where the number is much reduced, corresponding to only about
six fibre equivalents. This reduction explains the presence of the primary
constriction. In general, cross sections of human chromosomes at metaphase
appear to have 50–100 fibre threads at most sites, with an increase at the telomere to the order of 100 threads, and a decrease at the centromere to about 40.
    Visualization of the centromere region suggests that its two functions reside
in different structural properties. Attachment of the chromosome to the

spindle is mediated by the kinetochore, a dense granule which is lost when the chromosomes are prepared for whole mount microscopy. The second function, maintaining the association of the two sister chromatids, seems to be a property of the chromosome fibre.

Some fibres appear to cross from one sister chromatid to the other at the centromere; others may reverse direction or continue along the chromosome

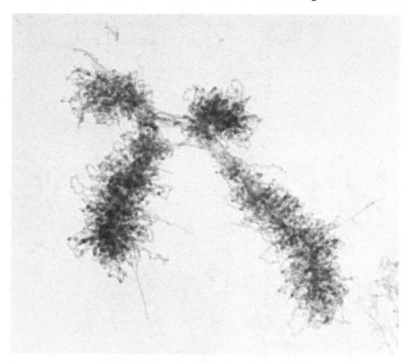

Figure 1.12: human chromosome of the 13–15 group, × 22,560. The two chromatids are held together by fibres which pass between the centromeres. Kindly provided by Professor E. J. DuPraw (1970); from *DNA and Chromosomes* (Holt, Rinehart and Winston).

arms, although only a reduced number passes through the constriction. A similar reduction in the number of fibres at the centromere has been observed in whole mount preparations by Abuelo and Moore (1969); and comparable observations by Comings and Okada (1970a) have been interpreted without the postulate that fibres cross from one chromatid to the other.

The fibres within the centromere region appear to be continuous with those located in the chromosome arms. DuPraw (1970) has suggested that the fibres running between the two sister chromatids may hold them together. Figure 1.11 shows a metaphase chromatid pair which appears to be held together both at the centromere and by fibres between the arms; figure 1.12 shows two sister chromatids apparently joined only at the centromere region. DuPraw proposed a model in which these interchromatid fibres contain sequences of DNA which have not yet been replicated, unlike the fibres in the arms of the two sister chromatids which contain the daughter duplex molecules of DNA produced by the last replication. As figure 1.13 shows, this implies that the

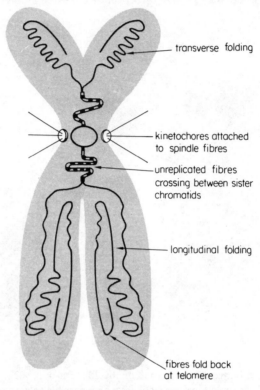

Figure 1.13: folded fibre model for the chromosome. The chromosome fibre may be both transversely and longitudinally folded in the chromosome arms. The broken lines indicate fibre segments which may not yet have been replicated and may therefore hold the sister chromatids together. After DuPraw (1970).

two sister chromatids can separate only after a burst of DNA synthesis to separate these regions.

According to this model, the centromere region contains both (a reduced number of) replicated fibres which pass from one arm of a chromatid to the other, and also unreplicated fibres which pass between the two sister chromatids. In addition, a kinetochore is present on each sister chromatid. This comprises a dense granule within the chromosome structure, surrounded by a less dense zone. In mammals, the kinetochore appears as a flat circular shape;

in plants it appears spherical. The kinetochores of any two sister chromatids face in opposite directions at metaphase and therefore pull the replica chromosomes to opposite poles (see page 86).

When the segregation of labelled chromosomes is followed through successive metaphases, two patterns are sometimes seen in individual chromosomes which do not conform to the predictions of semi-conservative replication (see

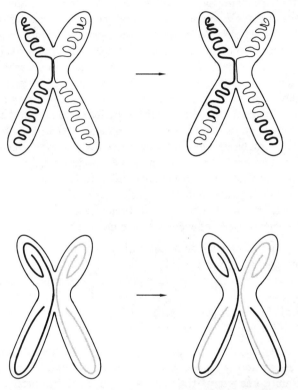

Figure 1.14: generation of chromosome aberrations according to folded fibre model. Upper: exchange of transversely folded fibres causes sister chromatid exchange. Lower: exchange at telomeres of longitudinally folded fibres causes equal isolabelling of both sister chromatids.

figure 1.2). In sister chromatid exchanges, although the entire longitudinal axis of the chromatid pair is labelled, the label switches from one chromatid to the other at some point along their length. Figure 1.14 shows that this can be achieved by a breakage and reunion between transversely folded chromosome fibres.

Isolabelling is found when both sister chromatids are labelled along the same part of their length. As DuPraw (1968) has observed, this can be achieved by a breakage and reunion of longitudinally folded chromosome fibres;

figure 1.14 illustrates the consequence of a breakage and exchange at the telomere where the fibre doubles back upon itself; this produces equal iso-labelling of both sister chromatids at the subsequent metaphase. (Previous explanations for isolabelling have been postulated in terms of multistranded structures for the chromosome, but the folded fibre model provides an explanation for a single fibre of DNA.)

The two folded chromosome structures shown in the figure are extremes and combinations of transverse and longitudinal packing can explain the types of chromatid exchange observed when a label segregates. Since most cell types show both sister chromatid exchanges and isolabelling, it is likely that both types of fibre folding are found in most chromosomes. In order to achieve the regular patterns of exchange observed between homologues, however, the mode of folding must be the same in each of the sister chromatids. This implies that although folding may be imprecise in the sense that the parameters of the fibre may vary and it appears to be irregularly packed, each homologue must follow characteristic general rules; and in any given metaphase, the packing of the two sister chromatids must be closely related.

By postulating that the chromosome fibre is folded in an appropriate manner, this model can account for the various aberrations which are produced by irradiation (for review of chromosome breaking and rejoining see Kihlman, 1971). Three general types of event are commonly observed:

chromosome aberrations which affect both sister chromatids of a metaphase pair in exactly the same way;
chromatid aberrations which influence only one chromatid of a pair at meta-phase;
subchromatid breaks or exchanges in which the aberration appears to extend across only half the width of one sister chromatid at metaphase (see figure 1.6).

The time at which cells are irradiated determines whether a chromosome or chromatid aberration is produced. Irradiation early in interphase, before the diploid chromosome set of the cell has replicated, produces chromosome aberrations because the damage suffered by any particular chromosome is faithfully reproduced when it is replicated; both daughter chromosomes—that is the sister chromatids—therefore show the same aberration at the subsequent mitosis. Irradiation later in interphase, when replication has already taken place so that the cell contains two diploid sets of chromosomes, influences each chromosome individually, so that only one of the two daughter chromosomes is likely to suffer any given damage. The observed chromosome and chromatid aberrations can be explained by breakage and reunion events between a chromosome fibre both transversely and longitudinally folded.

Chromatids sometimes appear to have a double structure in the light microscope, but Comings and Okada (1970b) have found that these structures are not seen in electron micrographs of whole mount preparations. This supports the idea that apparent substructures may be produced by the coiled structure of the chromatid and do not reflect any functional division. Apparent subchromatid breaks have often been taken as evidence for models which propose that the chromosome consists of more than one duplex of DNA. As figure 1.6 shows, however, such aberrations appear in fact to constitute a masked form of a full chromatid exchange involving the basic unit of the chromatid. DuPraw (1970) has pointed out that these events can be explained by breakage and reunion between adjacent segments of a single folded fibre; this explains why at subsequent mitoses they prove to be events influencing the chromosome itself.

Structure of Interphase and Metaphase Fibres

The exact conformation in three dimensions of the chromosome fibre in any given chromatid is not constant and must vary continuously from metaphase to interphase and, indeed, in successive division cycles. In this sense, each chromosome is a macromolecular structure which may assume any one of a number of conformations depending upon the conditions prevailing. The fibre comprising each chromosome seems to contain a single duplex of DNA, very tightly coiled upon itself and associated with protein. The compression of DNA within the length of the fibre is increased when the fibre contracts its length and increases its diameter in the transition from interphase to metaphase.

The basic unit of interphase chromatin appears to be a fibre of diameter about 230Å which, as figure 1.15 shows, has a bumpy appearance analogous to that of the fibres of metaphase chromosomes. Observing cells of the honeybee, DuPraw (1965b) noted that these fibres are often attached to the nuclear membrane as is apparent in the centre of the figure. Because this attachment protects the fibre during preparation for whole mount microscopy, it is possible to degrade individual fibre segments with trypsin. Such treatment gives the fibre the appearance of beads on a string, revealing a resistant core of the fibre with a diameter generally about 35–50Å but on occasion as low as 23Å. This supports the conclusion that each fibre contains only one coiled duplex of DNA.

Similar results have been obtained with metaphase chromosomes. Abuelo and Moore (1969) found that the chromosomes of human leucocytes in metaphase consist of fibres usually some 240 ± 50Å in diameter, but with the characteristic bumpy variations from diameters as small as 140–150Å to as large as 350Å. Trypsin loosens the packing of the fibres and produces thin patches of 25–50Å (a DNA duplex has a diameter of about 25Å); pronase has a similar effect. Treatment with DNAase breaks the fibres so that only short fragments are seen.

Although the length of fibre in each chromosome varies in individual specimens, it is usually of the order of 1/100 of the length of its DNA content.

The smallest and largest human chromosomes contain about 1·4 and 7·3 cms (14,000 and 73,000 μ) of DNA each; this is about 100 times greater than the fibre lengths of 135 μ and 723 μ reported by DuPraw and Bahr (1969). A greater packing ratio of DNA/fibre was found in a sample of chromosomes 13–15, which contained 3·1 cms of DNA organised into a fibre 229 μ long of average weight 21·6 × 10^{-16} gm/μ; this implies a packing ratio of about 140.

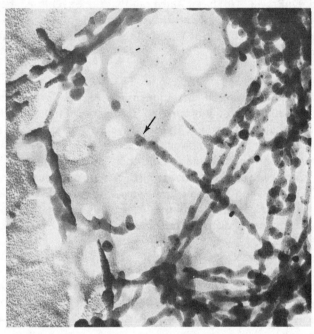

Figure 1.15: interphase chromatin of the honeybee, × 89,200. The arrow indicates where a fibre is attached to the nuclear membrane at the edge of an annulus. Kindly provided by Professor E. J. DuPraw (1970); from *DNA and Chromosomes* (Holt, Rinehart and Winston).

Fibres of metaphase chromosomes in general display greater diameters than those of interphase chromatin. DuPraw (1970) noted that diameters usually in excess of 300Å are characteristic for human metaphase chromosome fibres, compared with the 230Å diameter of fibres of interphase chromatin. Metaphase fibres therefore have a greater dry mass per unit length than interphase fibres; the packing ratio of DNA within the fibre is increased and their content of protein appears to double during the transition from interphase to metaphase.

Comparison of the mass of the fibre with the mass of DNA suggests that each fibre must contain a single duplex of DNA very tightly coiled upon itself; these measurements exclude models which suppose that the basic 230Å interphase fibre might consist of one or even many parallel duplex molecules of DNA, extended along its length and surrounded by protein. An extended duplex of DNA has a dry mass of some $3 \cdot 26 \times 10^{-18}$ g/μ; this is only $0 \cdot 5\%$ of the dry mass of interphase chromatin, $6 \cdot 1 \times 10^{-16}$ g/μ. Even four or eight duplex molecules of DNA extended in parallel would amount to only 2–4% of the dry mass of the fibre. However, DNA comprises about 30% of the mass of interphase chromatin, so that each fibre must contain about $1 \cdot 8 \times 10^{-16}$ grams of DNA per micron. This implies that each micron of interphase fibre must contain about 56 microns of DNA. The packing ratio of DNA within the fibre must therefore double at mitosis when the length of fibre contracts, increasing its diameter and mass per unit length.

DNA packing ratios in fibres of 230–300Å diameter cannot achieve values of 60–100 by a single order of supercoiling the duplex. Instead, DNA must be packed into a supercoiled supercoil. In other words, the DNA duplex is twisted into a supercoil; and this supercoil in turn must be supercoiled to give the basic fibre of 230–300Å diameter. The organization of this fibre then determines the overall structure of the chromosome and the chromonema seen in the light microscope represents multiple segments of folded fibre. Local changes in the fibre equivalent content along the length of the chromosome are reflected in the macrocoiling of the chromonema seen in the light microscope.

Biophysical studies of chromatin fibres have not in general proved very productive, although they have excluded models in which many duplex molecules of DNA are extended in parallel. However, Pardon et al. (1967) and Pardon and Wilkins (1972) obtained X-ray diffraction patterns of interphase chromatin preparations consistent with the presence of a superhelical structure of diameter 100Å and pitch 120Å; Pardon et al. (1973) reported that Chinese hamster metaphase chromosomes display similar diffraction patterns. The same type of supercoiling may therefore establish the structures of chromosomes in both interphase and mitosis. The electron microscopic studies of Lampert (1971) have suggested that the 100Å diameter fibre is supercoiled into the fibre of 250Å which provides the basic unit folded into the chromosome structure. Bram and Beerman (1971) and Bram (1971) have obtained data from X-ray and light scattering of calf thymus nucleohistone in solution which suggest that the chromosome fibres contain duplex DNA organised in a superhelical form, with supercoils of pitch 45Å and diameter 60Å.

These conclusions agree with the direct observation of two types of fibre in human chromosomes by DuPraw and Bahr (1969). In addition to the predominant 230–300Å fibres, regions containing narrower fibres can be visualized. These fibres appear to comprise strands of diameter 50–110Å and probably represent the first order of supercoiling of the DNA double helix. The two types of segment may alternate along one fibre and in some locations the wider 230–300Å fibre appears to unwind into the narrow 50–110Å fibre and then to wind up again into the broader structure. Gentle treatment with trypsin may allow the narrow fibre segments to be released from the wide fibre into which they are supercoiled.

Measurements of the cross sectional mass of the two types of fibre show that the narrow (type A) fibre segments have a dry mass per micron which is about 10–20% of that of the wide (type B) fibre segments. This implies that the overall packing ratio of DNA in metaphase chromosomes of 100 is achieved first by a 10 fold packing of DNA into the type A fibre and then by a further 10 fold packing of the type A fibre into the type B fibre.

The narrow fibres have a greater mass relative to their cross section than that of the wider fibres. The higher density of the type A fibres suggests that they contain a greater proportion of DNA than the type B fibres; in other words, some additional protein components are added when the type A nucleoprotein fibre is supercoiled into the type B fibre. These additional proteins may also play some role in maintaining the higher orders of coiling of the chromosome.

Because it is impossible to isolate individual chromosomes during interphase, the changes in the folding of the 230–300Å fibre which accompany the transition from interphase to metaphase cannot be measured quantitatively. However, the change in DNA/fibre packing ratios implies that the fibre must contract its length to about half of its previous value, at the same time increasing its diameter from 230–250Å to more than 300Å. Both interphase and metaphase chromosomes contain both type A and type B fibres, which suggests that the underlying hierarchical organization of the chromosome remains unchanged during condensation for cell division, although the packing of the type A fibres within the type B fibres, and perhaps of DNA within the type A fibre, increases appreciably.

In addition to the contraction of the fibres themselves, the amount of material included in the chromosome increases at mitosis. Huberman and Attardi (1966) found that DNA comprises about 30% of the weight of interphase chromatin but only about 15% of the weight of metaphase chromosomes. Since the amount of DNA in each chromosome must remain constant, this implies that the protein content doubles at metaphase. Most and perhaps all of the additional proteins fall into the class of non histones (see page 145).

The considerable range of values reported for the dry mass of corresponding chromosomes in different metaphase plates supports the idea that much of the increase in mass at mitosis may be due to variations in the amounts of the proteins added to the interphase fibres. An appreciable proportion of this material may comprise non-specific addition of cellular components made possible by the dissolution of the nuclear membrane. However, we may speculate that some of the additional proteins may be concerned with the packing of the type A into the type B fibres and may thus help to achieve a more compact structure for the chromosome.

The very large extent of variation between different preparations places an important limitation on the quantitative interpretation of the packing of DNA within fibres and of the fibre packing within the chromosome. Some of the variations observed in the structures and dimensions of chromosome fibres may be caused by different methods of preparation, as has been emphasized by Solari (1971). Together with the possibility that chromosomes prepared for microscopy may not represent exactly the structure which is maintained within the nucleus, these variations therefore argue that the folded fibre model—although the most attractive structure proposed for the eucaryotic chromosome—should for the present be viewed in qualitative rather than quantitative terms (for review of chromosome fibre structures see Huberman, 1973). The degrees of packing of DNA into fibres in vivo may thus take different values from those inferred from measurements in vitro, although probably of similar magnitude.

Euchromatin and Heterochromatin

The usual state of the genetic material in the nucleus of a eucaryotic cell at interphase is a dispersed network of fibres in which individual chromosomes cannot be recognised. The threads of each chromosome can first be distinguished at the beginning of mitosis and the chromosomes take up their characteristic condensed morphologies by metaphase. Following telophase, they return to the dispersed condition of interphase chromatin. This chromosomal material, the *euchromatin*, therefore passes through a cycle of condensation and dispersion in successive divisions of the cell.

Some of the genetic material, however, does not become dispersed at the end of mitosis but instead remains in its condensed state throughout interphase as well as division (first noted by Heitz, 1928). These sequences of *heterochromatin* stain densely with some dyes and appear to represent tightly coiled regions of the chromosome. Heterochromatin may take the form of *chromocentres* at interphase when different heterochromatic regions aggregate into dense clumps of genetic material. Chromocentres often tend to associate with the nucleolus.

In a study of interphase nuclei of salamander larvae with the electron microscope, Hay and Revel (1963) noted that the nucleoplasm consists of a fine network of fibrils containing opaque aggregates. The dispersed network comprises euchromatin; the dense aggregates are provided by the chromocentres of heterochromatin. The dense structures have the same filamentous structure as the dispersed network but their fibres are packed more tightly together. This supports the idea that heterochromatin is composed of the same nucleoprotein fibres as euchromatin, but represents a more compact state. The increased reaction of heterochromatin with feulgen stain, for example, is probably due simply to an increase in the concentration of DNA in the dense aggregates; spectrophotometric measurements show that the amount of DNA per unit area is greater in heterochromatic regions of the interphase nucleus than in euchromatin.

Interphase chromatin can be crudely separated by centrifugation into dense and light fractions and Frenster (1969) reported that each fraction has the appearance characteristic of the condensed and dispersed regions of interphase chromatin, respectively. The dense fraction of calf thymus lymphocytes shows very tightly packed clumps of chromatin in which few features can be distinguished under the electron microscope; the light fraction shows the characteristic network of euchromatin fibres. If the lymphocytes are incubated with radioactive precursors of RNA before extraction, most of the label is found in the dispersed fraction.

This supports the conclusion derived from autoradiography in vivo that heterochromatin is comparatively inactive in RNA synthesis; little incorporation of H^3-uridine is achieved by heterochromatic regions of the chromosomes. This idea correlates with the conclusion that heterochromatin is genetically inert and is not used to direct protein synthesis. Another distinctive property of heterochromatin is that it tends to be replicated later than euchromatin; autoradiography shows that DNA synthesis in heterochromatin begins later than that of euchromatin; and by the end of the DNA synthetic phase of the cell cycle heterochromatin alone may be completing replication (for review see Lima-de-Faria, 1969).

Eucaryotic cells contain two different classes of heterochromatin, which appear to share only the same characteristic condensed morphology and synthetic inactivity (for review see Brown, 1966). Facultative heterochromatin is produced by inactivation of one of the two X chromosomes in the cells of females. The formation of sex heterochromatin takes place at an early stage of embryogenesis, after which all the somatic daughter cells derived from the cell with the inactive sex chromosome contain one X chromosome which appears to be euchromatic and one which is heterochromatic. The heterochromatic X chromosome remains tightly coiled, is replicated late during the division cycle, and does not seem to be transcribed into RNA. In agreement with these biochemical properties, it behaves in genetic studies as though inert, for these cells exhibit gene functions coded only by the euchromatic X chromosome (see page 215).

Facultative heterochromatin represents a particular state of the chromosome for it does not occur at all stages of development; in Drosophila, for example, it is absent from early embryonic stages and appears only later. Lima-de-Faria (1969) has pointed out that late replication is associated with the state of facultative heterochromatin and not with the particular DNA sequences which it contains; in the grasshopper Melanoplus, the X chromosome of males is heterochromatic at meiotic prophase of spermatocytes and replicates later than the other chromosomes. But in spermatogonia, this chromosome is euchromatic and is replicated at the same time as the autosomes. A similar correlation has been observed in other systems.

Constitutive heterochromatin represents comparatively short regions of densely staining material which are found at the same position in *both* members of a homologous pair (for review see Yunis and Yasmineh, 1971, 1972). The locations of constitutive heterochromatin within a chromosome set are characteristic and any chromosome may have both heterochromatic and euchromatic regions. Heterochromatin is often observed at the centromeres and telomeres of mitotic chromosomes and at interphase these regions may aggregate to form chromocentres. Hsu et al. (1971) have observed that in the mouse—where heterochromatin is found at the centromeres of all the telocentric chromosomes—the interactions of regions of heterochromatin to form chromocentres vary with the cell type and its stage of development. Hsu and Arrhigi (1971) have noted that all mammals appear to have constitutive heterochromatin at the centromeres of their chromosomes, usually on both sides of the kinetochore but sometimes on one only. The next most common location for the heterochromatin is at the telomeres. Less frequently, bands of constitutive heterochromatin are found within the length of the chromosome.

Genetic analysis shows that constitutive heterochromatin has fewer genes per unit length than euchromatin and a reduced frequency of recombination. Little crossing over occurs within centromeric heterochromatin; for example, in Drosophila the genes on either side of the centromere tend to remain linked together. The genetic material contained within regions of centromeric heterochromatin is rich in DNA which consists of short sequences of nucleotides repeated many times (see page 186); these sequences do not seem to code for protein. The protein components of centromeric heterochromatin may differ from those of euchromatin (see page 210). The role of this region in chromosome function may therefore be structural, since it is clearly inactive in gene functions.

Although constitutive heterochromatin in most mammals is confined to short regions, the chromosomes of the field vole Microtus agrestis contain extensive lengths. Somatic cells of M. agrestis contain 50 chromosomes, but the two sex chromosomes are very much larger than the autosomes and can therefore be identified readily. Constitutive heterochromatin occupies the long arm of both the X and Y chromosomes and also one third of the short arm of the X chromosome. Lee and Yunis (1971a, b) found that these regions remain heterochromatic throughout development in all cell types and vary only in the extent of condensation shown in different cell types. This heterochromatin appears to consist of many fibres which are packed and folded into any one of several patterns; the degree of condensation and the pattern of folding depends on the cell type. The same fibres appear to run continuously between euchromatin and heterochromatin, reinforcing the conclusion that both regions consist of the same nucleoprotein components organised into structures of different density.

The condensed structure of heterochromatin may help to account for its generally inert nature, including its inactivity in RNA synthesis and its late replication. DuPraw (1970) has pointed out that the fibres of euchromatin have smaller diameters than those of heterochromatin. According to the folded fibre model, this implies that the packing of DNA within the fibres is increased in heterochromatic regions; and the fibres themselves are then folded together more tightly than in euchromatic regions. This means that it may be more difficult for enzymes such as RNA polymerase and DNA polymerase to obtain access to the DNA; these topological considerations may therefore play some role in restricting the synthetic activities of heterochromatin. Since there is no reason to suppose that the sequence of DNA directly influences its packing in the nucleoprotein fibre, and since the same sequences are in any case present in facultative heterochromatin and its homologous euchromatic counterpart, we must attribute the condensed nature of heterochromatin to the proteins which interact with DNA (although all attempts to identify different proteins in crude preparations of heterochromatin and euchromatin have so far failed).

Polytene Chromosomes of Dipteran Salivary Glands

The salivary glands—and also certain other tissues—of the larvae of dipteran flies such as Drosophila contain interphase nuclei whose chromosomes are displayed in a giant state in which they possess both larger diameter and greater length than usual. In Drosophila melanogaster, the total length of the giant chromosome set may be of the order of 2000 μ. When viewed under the electron microscope these chromosomes exhibit a series of *bands* which stain intensely, alternating with the more lightly staining *interband* regions. The largest bands extend for about 0·5 μ along the axis of the chromosome; the smallest have a length of about 0·05 μ and can be seen only under the electron microscope. The centromeres of all four chromosomes of Drosophila characteristically aggregate to form a chromocentre of heterochromatin, which in the male includes the entire Y chromosome. The chromosome set of Chironomus tentans salivary glands is shown in figures 6.4 and 6.5 (pages 342 and 343).

The pattern of bands remains the same in each particular strain of Drosophila and the constancy in their number and linear arrangement was first noted by Heitz and Bauer (1933) and Painter (1934). Rearrangements of genetic material—such as deletions, inversions, duplications—result in changes of band order and such observations led to the idea that the bands represent a cytological map of the chromosome. Cytological changes can be correlated with changes in the positions of genes on the genetic map and Bridges (1935) first equated the linear order of bands with the linear array of genes and showed that changes at one particular genetic locus always generate cytological changes only in one particular band.

That the bands contain a high concentration of DNA is shown by their intense staining with dyes such as feulgen reagent. An early controversy over whether the interbands also contain DNA was resolved by the development of more sensitive assays, which have shown that DNA is present between the bands, although at a lower concentration than within them (reviewed by Swift, 1962). Lezzi (1965) has shown that the interbands are susceptible to degradation by DNAase.

That the giant chromosomes represent an unusual state of the chromosomes found in diploid somatic cells (and not some novel reorganization of genetic material) has been confirmed by the results of Beerman and Pelling (1965). The chromosomes of Chironomus were labelled with H^3-thymidine at an early stage of embryogenesis and larvae allowed to grow on unlabelled medium and to develop salivary glands. The DNA originally labelled in the normal chromosomes appears in the giant chromosomes, exhibiting a label along the entire length of the chromosome. This implies that the giant chromosomes are derived by the usual replication of genetic material, but represent a more extended state of organization.

The giant chromosomes are produced by many successive replications of the normal diploid pair representing each chromosome; the replicas do not separate but remain attached to each other in their unusually extended state. By using a quantitative feulgen assay to follow the increase in DNA content of salivary gland nuclei during the development of larvae of D. melanogaster, Swift (1962) found that the nuclei start with the usual diploid content, 2C. The amount of DNA then increases in geometric steps to give cells containing 4C, 8C, 16C, 32C etc upto a maximum of 1024C. This suggests the concept that each giant chromosome consists of hundreds of identical copies of its genetic material, all of which are replicated in each successive cycle. The production of cells containing 1024C corresponds to 9 successive doublings of the diploid DNA content.

Each giant chromosome consists of many parallel fibres running longitudinally along its axis; an early estimate of the number of fibres of 1000–2000 was made by Beerman and Bahr (1954), using thin section techniques. Each of these fibres represents one copy of the duplex DNA–protein fibre which constitutes a normal chromosome (reviewed by Swift, 1962). The state of these chromosomes is therefore described as *polytene* (as compared with *polyneme*, which would describe a chromosome whose basic fibre consisted of many duplex molecules of DNA).

The number of chromosome doublings which are required to generate the polytene chromosomes can be calculated by comparing their DNA content with that of diploid cells. Daneholt and Edstrom (1967) extracted the DNA of individual polytene chromosomes of Chironomus tentans, hydrolysed it to nucleotides, and assayed the amount of product. DNA is distributed between the four chromosomes as expected from their relative lengths. Since the number of bands in each chromosome is also proportional to its length, this argues that the DNA content of the bands remains of the same order of size in all parts of the genome.

The largest set of polytene chromosomes isolated contained a total of 3360 picograms of DNA; this would be produced by 13 duplications of a diploid set of 0·41 picograms. The DNA content of the second largest chromosome set isolated was close to half of that of the largest set at 1776 picograms; this would correspond to 12 doublings of a diploid genome of 0·43 picograms. The geometric relationship between these two chromosome sets suggests that most or all of the DNA of the genome participates in each doubling. The estimated value of the DNA content of diploid nuclei was 0·5 picograms of DNA, which may therefore be too great.

There are about 2000 bands in Chironomus tentans. If most of the DNA is in the bands and little is in the interbands, this would mean that each haploid chromatid has about 10^{-16} grams of DNA in each band; the variation in band sizes therefore ranges from about 10^4 to 10^6 base pairs in the length of DNA. The average value of 10^5 base pairs could code for 30,000 amino acids (three million daltons of protein) which is very much larger than a single gene. Each band must therefore either contain many genes, or much of its DNA must have some function other than coding for protein.

In Drosophila, the amount of DNA in the polytene chromosomes is not a geometric multiple of the DNA content of the diploid chromosome complement. By comparing the response to feulgen of the DNA of polytene nuclei of salivary glands and ganglion cells with the content of diploid cells, Rudkin (1969) found that a progressive increase in DNA values follows the sequence: 1·63, 3·50, 6·58, 12·6. The best fit of this series to the replication of Drosophila DNA is given by assuming that some of the genome fails to replicate whilst the rest undergoes a series of successive doublings. The heterochromatin contained in the chromocentre makes up about 26% of the genome; if this DNA replicates only a small number of times compared with the duplications of the euchromatin, the increase in DNA content should be close to that observed experimentally. These data, and those of Rodman (1967), suggest that there may be 8, perhaps 9, doublings of the initial diploid set of chromosomes present. (These estimates of the degree of polyteny rely upon the ratios of the feulgen values of diploid and polytene cells and do not reveal the absolute amount of DNA in the genome.)

The amount of heterochromatin in the polytene nuclei can be assessed by measuring the size of the chromocentre (which includes all the centromeric heterochromatin and in males also encompasses the Y chromosome). When Berendes and Keyl (1967) compared the proportions of heterochromatin present in polytene nuclei and diploid cells of the ganglion of D. hydei, they observed that the polytene cells possess relatively less heterochromatin. The overall proportions of heterochromatin and euchromatin can be explained by supposing that replication of DNA in the chromocentre is turned off after two or three doublings, whereas that of the euchromatin passes through many more duplications. This idea is supported by measurements of the frequency in the polytene genome of sequences of DNA which are located in the heterochromatin (see page 189). In measurements of heterochromatic sequences in the X chromosome, Hennig (1972a, b) noted that the euchromatic part of the X chromosome suffers duplication to become polytene whereas heterochromatic sequences do not. This implies that the duplicated polytene strands must have suffered an interruption in their usual covalent integrity.

In addition to the very limited duplication of heterochromatin—in D. melanogaster much heterochromatin does not replicate at all—the cistrons coding for ribosomal RNA achieve only a limited degree of polyteny. In hybridization measurements of the frequency of these sequences, Hennig and Meer (1971) observed that genes for rRNA are appreciably under-represented in the polytene nuclei of D. hydei. By determining saturation values of hybridization with rRNA in polytene and diploid nuclei of two strains of D. melanogaster, Spear and Gall (1973) observed that 0·47% or 0·37% of the DNA of diploid cells of XX females represents ribosomal RNA, compared with 0·26% or 0·24% in the cells of XO flies. This is consistent with the location of the rRNA genes on the X chromosome. The salivary gland cells of either XX or XO flies, however, devote about 0·08% of their DNA to ribosomal RNA genes. Compared with the nine doublings of euchromatin, the rRNA genes suffer only 6–7 doublings to reach a final level which is the same whether only one (XO cells) or two (XX cells) nucleolar organisers were originally present. Under separate control, the heterochromatic regions of DNA undertake even fewer or no doublings. Formation of polytene strands therefore falls under at least three different controls.

Thin section techniques suggest that the polytene chromosomes of Drosophila salivary glands comprise a large number of fibres of diameter 125–500Å (according to the different estimates), with the fibres more tightly packed in the bands than in the interbands. Rae (1966) used whole mount electron microscopy to show that the fibres of salivary gland chromosomes of D. virilis are the typical type B chromatin fibres observed in diploid somatic cells. They have the usual bumpy appearance, varying in diameter from 180–290Å, with an average of about 230Å. The fibres appear to be extended in the interbands but folded tightly in the bands.

At any given point a polytene chromosome may contain from 1000–4000 of these fibres, organised in parallel, and DuPraw and Rae (1966) suggested that each fibre represents a "unit chromatid", that is one copy of the genetic material of that chromosome usually found in a chromatid at metaphase.

Figure 1.16 shows their folded fibre model for the polytene chromosome, with DNA concentrated in the bands by compact folding and extended longitudinally between them. Each fibre therefore runs continuously from one end of the chromosome to the other and contains a single duplex molecule of DNA. Since the total DNA content of the haploid genome of Drosophila (0·18 picograms) corresponds to about 5–6 cms in length, its division into the 2000 μ of the polytene chromosomes implies an *overall* packing ratio of DNA in the chromosome of about 30; this is much less, of course, than that seen in metaphase chromosomes of somatic cells.

The idea that each polytene chromosome consists of many nucleoprotein fibres, concentrated in the bands and extended between them, is also supported by the observation of Swift (1964) that staining for histones (rather than for DNA) produces the usual banding pattern. The organization of the DNA into

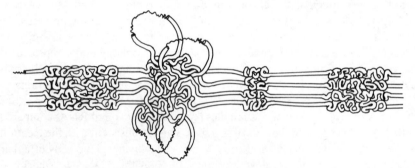

Figure 1.16: folded fibre model for polytene chromosomes. The interbands consist of extended DNA fibres; the bands consist of folded fibres. Puffs take place where the DNA is unwound from its supercoil and released to be available for transcription. After DuPraw (1970).

its fibre is therefore similar in both polytene and normal chromosomes. The arrangement of genes in a linear order along the polytene chromosome is probably a consequence of the extended state of the fibre, and this topography is unlikely to apply to normal metaphase chromosomes in which the fibre frequently folds back upon itself.

The genetic material in each band is proportional to its length; Rudkin (1965) found that the susceptibility to induction of mutations is proportional to band size and Lefevre (1971) found that the frequency of recombination follows the same rule. The bands seem to comprise units of organization for several chromosome functions. One pointer that a band may be replicated as a unit comes from a comparison of band sizes made by Keyl (1965) between two related subspecies of Chironomus thummi. Some parts of the homologous chromosomes can be distinguished from each other in the hybrid and the amounts of DNA in the corresponding bands measured.

The bands of C. th. thummi always contain exactly 2, 4, 8, or 16 times the amount of DNA in the homologous bands of C. thummi piger. Since the total ratio of DNA in the two species is the same in salivary glands and spermatocytes, this multiplication cannot be due to selective polytenization of some bands. Rather must duplication at some point in the evolutionary divergence of the two species have doubled the length of DNA in each haploid unit chromatid of these bands of C. th. thummi. Since unequal crossing over should produce triplet ratios in addition to doublings, recombination alone cannot explain the derivation of the longer bands. Each of these bands therefore seems to behave as an individual unit in evolution. This leaves as an explanation the idea that bands may be replicated as units and that aberrations in their replication may double the length of a band.

Under normal conditions, the entire DNA content of the euchromatic regions of the polytene chromosomes is replicated in several successive cycles. By incubating replicating polytene chromosomes of Chironomus with H^3-thymidine, Keyl and Pelling (1963) were able to show that replication of all parts of the chromosomes begins within about 30 minutes of the start of a cycle. This produces a distribution of label in which all parts of the chromosome are labelled in proportion to their DNA content. A second pattern is also observed, however, in which local labelling is restricted to individual bands. Keyl and Pelling suggested that this may result because the longer parts require more time to complete replication of their DNA, and therefore continue to incorporate label after the smaller bands have finished. This argues that the band can be equated with the unit of replication, since if each band contained many replicons, active simultaneously, all bands should achieve completion of their replication at about the same time. The interbands are probably replicated more rapidly than any of the bands. More recent studies by Plaut, Nash and Fanning (1966) and Plaut (1969) of the replication of polytene chromosomes of Drosophila have produced results consistent with this conclusion.

That one band might correspond to one gene was first suggested by Bridges (1935) because of the apparent equivalence of the cytological and genetic maps of Drosophila chromosomes. Cooper (1959) has estimated the number of bands in D. melanogaster as 5161, which is of the same order of magnitude as the expected number of genes. (Estimates for the number of genes may be derived by comparing the mutation rate per locus with the overall rate of deleterious mutations per generation as illustrated for man in chapter 4.) On the basis of a count of 1012 bands in the X chromosome, Rudkin (1965) estimated that the haploid content of each band is about 3×10^{-5} picograms; this is about 22×10^6 daltons, or 37,000 base pairs in length.

By constructing a detailed map with the electron microscope of part of the X chromosome, Berendes (1970) counted 116 bands; this compares with the 174 bands of the earlier light microscope map of Bridges (1938). The reduction in the number of bands is largely due to the elimination of doublets (doublet bands) which may therefore be an artefact of fixation for microscopy. If this reduction in band number applies to all chromosome regions, then the DNA content of a haploid chromatid of each band in the X chromosome would be about $4 \cdot 5 \times 10^{-5}$ picograms; this is about 34×10^6 daltons or 55,000 base pairs.

Another estimate for the average size of a band can be made by dividing the haploid DNA content of the Drosophila genome by the total number of bands. Rasch et al. (1971) found that Drosophila sperm contain about $0 \cdot 18$ picograms of DNA; the haploid genome is therefore about 11×10^{10} daltons. Diploid somatic cells contain about $0 \cdot 35$–$0 \cdot 36$ picograms. About 73% of the genome is found in the euchromatin of the bands; this is 8×10^{10} daltons of DNA per haploid genome. If the number of bands is two thirds of the light microscope estimate of 5161 (to allow for elimination of doublets), then the haploid DNA content of each of the 3420 bands is about $2 \cdot 3 \times 10^7$ daltons. This is about 4×10^4 base pairs, an estimate close to the 10^5 base pairs of the average band of Chironomus.

Even this approximate value for the DNA content of a band shows that it should contain far too much information to represent a single gene. This implies that either each band must contain many genes or that most of the DNA of the band does not code for protein but has some other function. However, that the genes are located in the bands (rather than in the interband regions) is implied by the equivalence of the cytological and genetic maps and confirmed by studies of gene action in salivary glands. We do not know whether this equivalence is precise enough to conclude that coding DNA is contained exclusively within the bands or whether it may extend also throughout the sequences of DNA present in the interbands. It is possible, however, that the structural arrangement of polytene chromosomes as extended interband and compact band regions in some way reflects the genetic organization of the DNA (see page 339).

In preparations of polytene chromosomes, some of the bands may appear in an expanded or *puffed* state in which chromosomal material is extruded for an appreciable distance out from the axis of the chromosome. A characteristic pattern of puffing is observed in different tissues and at different times of the development of the larva; a detailed series of studies of puffing in salivary glands of D. melanogaster has shown that particular bands always exhibit puffs at certain stages of development. Any given band may undergo a characteristic cycle of activity, in which it changes from its quiescent condition to display a puff during some developmental period until a specific time when it regresses.

Autoradiography of polytene chromosomes incubated with labelled precursors to RNA shows that the puffs are active in RNA synthesis. This suggests that a puff represents the site of transcription of messenger RNA needed at that particular stage of development; the control of puffing is discussed in chapter 6. Thin section studies of the fine structure of the puffs suggests that they represent sites where chromosome fibres unwind, although the fibres remain continuous with those in the axis of the chromosome (Beerman and Bahr, 1954). According to the folded fibre model of the chromosome, the puffs represent sites where the DNA tightly packed within the fibre is allowed to unwind, presumably so that it becomes accessible to RNA polymerase (for review see DuPraw, 1970). The characteristic puffing of particular bands therefore suggests that the band may be the unit of gene expression.

CHAPTER 2

Chromosome Functions During the Cell Cycle

Cell Division Cycle

Synthetic Periods of Interphase

Cytological observation of the cell divides its life into only two parts, interphase and mitosis. The cell starts its life cycle with the diploid set of chromosomes gained at telophase. For the duration of interphase, the chromosomes exist in the form of a network of interphase chromatin with few characteristic features. No cytological changes are observed until prophase of the next mitosis, when two diploid sets of chromosomes can be seen. The division of the chromosomes during mitosis then restores diploid status to the two daughter cells.

In contrast with the apparent cytological activity of chromosomes at mitosis and inactivity during interphase, it is between divisions that the synthetic state of the cell is determined and maintained. During interphase, the cell replicates its DNA and transcribes and translates RNA. Mitosis provides an interruption when virtually all synthetic activities are halted and the cell structure is reorganised to ensure an even segregation of chromosomes to the daughter cells.

Interphase can be divided into three parts according to the status of the chromosomes of the cell. The daughter cells generated by a mitosis enter the *G1 period*, during which they have a diploid set of chromosome material. Cells are said to leave G1 phase when they begin to replicate their chromosomes; the period during which DNA synthesis takes place is termed *S phase*. At the completion of S phase, the cells have replicated all their genetic material and therefore have two diploid sets of genetic information. Cells remain in this state for the *G2 period* until mitosis commences. S phase was so called as the synthetic period when DNA is replicated, G1 and G2 standing for the two gaps in the cell cycle when no DNA is synthesised.

The discovery which led to the idea that interphase can be divided in this way was the observation of Howard and Pelc (1953) that DNA synthesis is confined to a short period during the interphase of cells of Vicia faba. By using autoradiography to follow the incorporation of P^{32} into DNA, they were able to show that the label enters DNA for a period of only 6 hours during the 30 hours of the cell cycle. The S phase starts about 12 hours after the previous cell division, so that the cell cycle is divided into the sequence:

G1	12 hours
S	6 hours
G2	12 hours
M (mitosis)	

Growing interphase cells may therefore be classified as passing through G1, S or G2. Both the DNA and protein components of the chromosomes are synthesised during S phase, so that by its completion the amount of chromatin in the cell has doubled (although without visible change in its structure). The daughter chromosomes remain intimately associated with each other until they segregate at the subsequent mitosis. RNA and protein synthesis increase continuously throughout interphase, in contrast with the restriction of DNA synthesis to a limited part of the cell cycle. Somatic cells which have ceased to divide are usually found in the G1 state, with a diploid complement of chromosomes; for example, contact inhibited cells in culture halt in G1 phase (Todaro, Lazar and Green, 1965). Differentiated cells which have permanently ceased division are sometimes described as in the G0 state; although the status of their DNA is the same as that of G1 cells, they may undergo changes in other synthetic activities, related to their lack of ability to divide. It is a matter of controversy whether G1 and G0 represent genuinely different states of the cell, or whether G0 is a special example of the G1 state.

The amount of time spent in each phase of the cycle is a characteristic of any particular cell type. Defendi and Manson (1963) noted a general tendency for the length of S phase to be related to the DNA content of the cell, although this is not a precise rule. S phase is usually 6 hours or longer in duration in adult somatic cells, most commonly occupying 6–8 hours. However, it may sometimes take much longer. Restriction of the minimum length of S phase to 6 hours cannot be due simply to the large amount of DNA to be replicated, for embryonic cells may have much shorter S periods than the somatic cells of the adult. For example, Graham and Morgan (1966) found that S phase in early embryogenesis of Xenopus laevis may last for less than 15 minutes, although it occupies some 5 hours in adult somatic cells (see also page 385). A general tendency for S phase to be comparatively rapid in embryos, becoming longer throughout development, is also shown in mammals, but is much less pronounced, with the generation time of embryonic cells only a few hours short of that of adult cells. The more rapid replication of DNA in early embryogenesis (when cells can divide without appreciable growth) implies that the time taken to replicate DNA in adult somatic cells may in part reflect their particular organization of genetic material, but is likely largely to result from the different nature of the cell growth cycle as a whole.

The most variable period in the cell cycle is G1, which varies from its omission in some cells to its indefinite length in others. To a large extent, the length of the cell cycle in toto is therefore dictated by the duration of G1. This conclusion derives both from surveys of the length of the phases of the cell cycle in different mammalian cells (see Mitchison, 1971) and from experiments which have varied the length of the cycle of some particular cell grown in culture. For example, Tobey et al. (1967) found that the length of the cycle of Chinese hamster cells in culture can be changed by the details of the serum

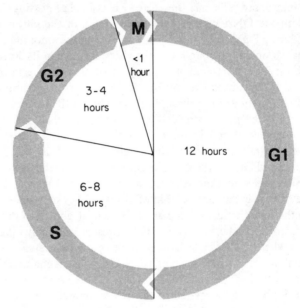

Figure 2.1: cell cycle of L cells growing in culture at 24 hours per doubling. Data of Stanners and Till (1960).

provided in their medium; the changes in the length of the cycle are confined largely to G1 phase.

The length of the G2 period may also vary in cells with different generation times, but in many respects this phase seems to comprise a pause after the completion of replication which provides time to enable the cell to prepare for mitosis. The period of G2 rarely occupies more than 6 hours and, in general, G2 is the most constant period when the cycles of different cells are compared. Experiments in which cells in different stages of the cell cycle are fused together to form a hybrid cell have suggested that a mitotic inducer may accumulate during G2, mitosis commencing when a critical level has been reached (see page 403).

For mammalian cells growing in culture, a typical cell cycle is that observed by Stanners and Till (1960) with mouse L cells. Figure 2.1 shows that the generation time is about 24 hours, some half of which is occupied by G1, with S phase lasting for 6–8 hours and G2 phase for 3–4 hours. Mitosis itself generally lasts for less than one hour. Variations in the cell cycle have been reviewed by Zetterberg (1970) and Mitchison (1971).

Assembly of organelles such as ribosomes and mitochondria takes place continuously during interphase. All components of the cell must double in the period between mitoses and it is generally assumed that both new and old components become mixed in the cytoplasm and are partitioned about equally between the two daughter cells at the ensuing division.

It is difficult to follow events in the cell cycle with much precision of timing. Although a single cell can be followed microscopically, this technique is of course limited in the parameters which can be measured. Most of the experiments discussed below which concern events at particular stages of the cell cycle have therefore utilised synchronised populations of cells growing in culture. One drawback inherent in the use of synchronised populations is that the cells generally remain synchronised for only a short period of time, usually about two generations. By the third generation, there is often an appreciable loss of synchrony, so that cyclical events can be followed through only two divisions of the cell.

Two general types of method are in common use to achieve synchrony (reviewed by Mitchison, 1971). One is to select cells at some particular stage of the cycle and subculture them as a synchronous population. Selection methods offer the important advantage that the cells are obtained in their "normal" state. One method used for selection is to isolate the new daughter cells which are preferentially released from a solid surface at mitosis. However, although this yields well synchronised cultures, it has the disadvantage that the cultures are limited to small numbers of cells. Much greater quantities of cells can be obtained by centrifuging a randomly grown culture so that cells in different stages of the cycle are separated according to their size and density. This does not yield such good synchrony, however, because the sizes of individual cells may vary considerably at any given stage of the cell cycle.

Induction methods rely on treatment of an asynchronous culture to delay cells more advanced in the cycle either in division or in replication so that the entire population accumulates at the same stage. One common method used to induce synchrony is to treat a culture with colchicine, or its derivative colcemid. These compounds interfere with the function of the spindle fibres and cause cells to accumulate in a metaphase-like state (see page 81). Transferring the cells to a medium lacking colchicine rapidly reverses the inhibition and allows all the inhibited cells to divide together. Because colchicine may damage the cells during long incubations, a brief period of inhibition is often used to achieve a partial synchrony, which is then succeeded by the application of some other method to fully synchronise the cells. The general disadvantage of using colchicine is that it is difficult to exclude the possibility that the drug may induce changes in cellular components which cause abnormal functions when the block is later removed.

Inhibition of DNA synthesis is commonly used to cause cells to accumulate at the beginning of S phase. High concentrations of thymidine halt DNA synthesis in mammalian cells, probably because of feedback effects on enzyme activities resulting from the increased size of the dTTP pool. This treatment is usually used in the form of a double thymidine block. Cells are treated with excess thymidine for a period equivalent to the sum of G2 + M + G1. Any cells which are in S phase when the thymidine is added halt DNA synthesis immediately. Cells in G2, M or G1 can continue through the subsequent stages of the cycle until they conclude G1 and enter S phase; they are halted at this point. This treatment therefore produces a population of cells some of which are halted in varying stages of progression through S phase, others of which are halted at the beginning of S phase. The blockade is then lifted for a period longer than S phase in duration; this allows all the cells to complete the S period, no matter at which point they are blocked. A second block of thymidine is then administered for long enough to allow the cells—most of which are now in G2—to continue through the cell cycle until they conclude the next G1. The entire population is therefore halted at the beginning of S phase.

Other inhibitors of DNA synthesis in common use include amithopterin (an antagonist of folic acid) and 5-fluoro-deoxyuridine (an analogue of the nucleotide precursors to DNA). Both these compounds act on the thymidine utilization pathway and their effects can be reversed by the addition of thymidine. Hydroxyurea is another effective inhibitor of eucaryotic DNA synthesis. All are often used to improve the synchrony achieved by a short incubation with colchicine. A disadvantage of these inhibitory treatments is that although they prevent S phase, they may allow the cells to continue with other synthetic activities. Mitchison (1971) has pointed out that this may have the result that the DNA synthetic cycle of the cell is halted whilst the growth cycle continues. This means that the relationship of DNA synthesis to the cell cycle may be changed; results obtained by induction synchrony must therefore be interpreted with some caution.

One of the most important parameters of the cell cycle is the length of S phase and the order with which different parts of the genome are replicated within it. Because individual chromosomes cannot be recognised during interphase, replication must be followed by labelling the cells with radioactive precursors—usually H^3-thymidine—during S phase, and then examining the labelled chromosomes at the subsequent metaphase. A short pulse dose of 10–20 minutes incubation with H^3 thymidine is usually allowed, after which metaphases are analysed for the presence of label at increasing intervals of time.

The first labelled metaphase plates to appear must represent division of cells which were at the very end of S phase when the label was added. The

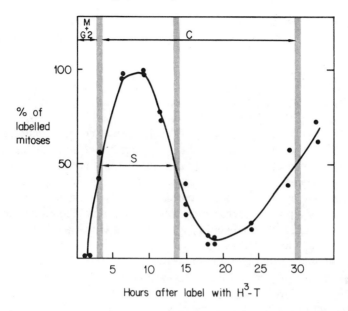

Hours after label with H³-T

Figure 2.2: measurement of the cell cycle. Hela cells were labelled with H³-thymidine for 30 minutes and the proportion of labelled metaphase plates among the mitotic cells was determined after various periods of incubation in unlabelled medium. The first labelled mitotic cells to appear must have been at the end of S phase when the label was applied; this interval therefore identifies G2 + M. Labelled cells increase in proportion as cells labelled during S phase pass through mitosis. The subsequent decrease represents cells that were in G1 during the labelling period and proceed through S and G2 to mitosis without being labelled. The increase after 20 hours represents the second division of the cells originally labelled during the earlier S phase. Measurements are usually made from the times when 50% of the cells become labelled or cease to become labelled (indicated by the shaded vertical lines). The interval to the first ascending 50% measure of 3·5 hours estimates M + G2; the period of 10·5 hours between the 50% ascending and 50% descending levels is a measure of S phase; the delay between subsequent ascending 50% measures of 26·5 hours estimates the cell cycle (C); by subtraction, G1 must be 26·5 − 10·5 − 3·5 = 12·5 hours. Data of Baserga and Wiebel (1966).

interval between addition of label and its appearance in metaphases thus measures G2 (+ part of M). The last labelled metaphase plates detected must represent cells which were at the very beginning of S phase when the label was added. The time taken for their appearance must therefore provide a measure of S + G2. As figure 2.2 shows, these periods are usually measured by taking the times when 50% of the cells first exhibit a label and when 50% of the cells fail to exhibit a label (reviewed by Baserga and Wiebel, 1969; Monesi, 1969; Mitchison, 1971). The interval between the appearance of two successive peaks of labelled metaphases measures the total length of the cell cycle. The appearance of label in individual chromosomes in the metaphase cells reveals the order of their replication.

Ordered Replication of DNA

Replication of eucaryotic DNA takes place by a semi-conservative mechanism in which the duplex DNA component of each chromosome directs formation of two new duplex DNA molecules, each of which forms the genetic material of one of the daughter chromosomes (see page 10). Replication does not proceed continuously along the DNA molecule, however, but instead is achieved by the simultaneous action of many replicating forks. In other words, each chromosome consists of many replicons.

The units of replicating DNA most frequently observed in autoradiography are 15–60 μ long. Huberman and Riggs (1968) and Huberman and Tsai (1973) found that these units consist of two replicating forks moving bidirectionally from a common origin (see chapter 10 of volume 1). That is, each replication fork starts from the origin at the same time, moving in opposite directions. Each fork therefore moves some 7–30 μ from origin to terminus. The rate of movement of the replicating forks seems to be about 0·5–1·2 μ per minute. By following the incorporation of a density label into the DNA of Chinese hamster cells, Taylor (1968) has noted growth rates of 1–2 μ per minute. After labelling Hela cells for 24 hours, Cairns (1966) found stretches of DNA at least 500 μ long; these presumably result from the joining together of adjacent replicating units. A 45 minute label gave lengths of 10–30 μ and a 180 minute label lengths of 50–100 μ, which suggests a rate of movement of the order of 0·5 μ per minute.

This rate of replication is very much slower than that of bacteria. The chromosome of E.coli, which is 1100 μ long, is replicated in 40 minutes by two growing points; this corresponds to a rate of movement of about 14 μ per minute. One reason for the ten fold slower rate of eucaryotic replication may be that the DNA is less accessible because of the structure of the chromosome. The enormous compression of eucaryotic DNA within the fibres of chromatin may impose restrictions upon the rate at which it can be unwound and replicated.

Numbers concerning the rate of replication of eucaryotic chromosomes are necessarily imprecise, because of the limitations of the techniques which can be used to follow the synthesis of eucaryotic DNA. However, from them a rough estimate can be made of the total number of replicons in the human genome and of the number which must be engaged in DNA synthesis at any particular time in S phase. A diploid human cell contains about $1·74 \times 10^6$ μ of DNA; if the average (bidirectional) replicon is 50 μ long, then there must be some 35,000 replicons in the nucleus. If replication were to proceed at a rate of 1 μ per minute, completion of DNA synthesis would require 30,000 hours. Since replication is in fact completed during the 6 hour period which defines S phase, some 5000 replicating forks must be active simultaneously in the cell at any moment. This corresponds to about 100 replicating forks (50 replicons) for each chromosome. Each replicon must be active for less than half an hour. These numbers are only very approximate and assume that replication takes place evenly throughout S phase (probably not strictly true). However, these calculations imply that the replication of eucaryotic DNA must be a complex process requiring coordination of the activities of very many replicons.

The large number of replicons in each chromosome implies that there must be many locations at which DNA unwinds. The isolation by Taylor (1973) and Taylor et al. (1973) of fragments of Chinese hamster DNA sedimenting at about 26S—this corresponds to a length of about 6 μ—representing recently replicated DNA suggests that nicks may be introduced in each replicon to serve this purpose.

How are the many replicons in the eucaryotic genome related to each other? Is DNA synthesis in individual replicons initiated at random, the process continuing until all the replicons have been duplicated, or is there an ordered pattern of replication of chromosomes and of replicons within each chromosome? That replication takes place in a definite order is suggested by experiments in which Mueller and Kajiwara (1966a) labelled synchronised Hela cells with C^{14}-thymidine at the beginning of S phase. The cells were then allowed to grow for several generations, resynchronised, and BUdR added during the early part of S phase. The C^{14}-label preferentially appears at the heavy density of the DNA segments containing BUdR which suggests that the same sequences are replicated at least at the beginning of each round of replication. Similar results were obtained in a comparable test for synchrony of regions replicating at the end of S phase.

These results imply that the temporal order of replication may remain the same in successive generations, although we do not know how precisely the order may be maintained (for review see Prescott, 1970). Because we lack the assays for particular genes available in bacteria (see Chapter 10 of volume 1), it is not in general possible to measure the time at which given chromosomal regions are reproduced. However, Giacomoni and Finkel (1972) have found that Chinese hamster and rat kangaroo cells grown in culture replicate their rRNA cistrons at a reproducible stage of the cell cycle. Hela cells do not show this specificity, but this may be because they are heteroploid and their chromosome complement has changed.

DNA synthesis starts simultaneously at many sites on each eucaryotic chromosome. In autoradiographic studies of Chinese hamster cells synchronised with FUdR, Hsu (1964) found that each chromosome seems to show its own characteristic pattern of labelling at each subsequent stage of S phase. One criterion for judging the precision of replication within a cell is to see whether homologous chromosomes show identical labelling patterns. In the Chinese hamster cells, homologous chromosomes show similar, but not identical patterns of replication. It remains possible that the two homologous copies of each autosome usually replicate at the same time and in the same sequence; there are insufficient data to come to a firm conclusion, but it is fair to say that there is no definitive evidence against this idea. It is important to remember also that synchronization techniques may disturb the usual pattern of replication of the cell.

The proportion of the genome under synthesis at any moment during S phase may vary. Remington and Klevecz (1973) found that Don C Chinese hamster cells display three maxima of synthetic activity during S phase; each peak of activity represents replicons whose activations are temporally related. Other cells, however, may display an increase in DNA content during S phase which is in general continuous.

A prominent feature in the order of replication is that certain chromosome regions are always replicated at the end of S phase. These regions include both the constitutive heterochromatin found at the centromeres of many chromosomes and the facultative heterochromatin produced by inactivation of complete chromosomes, such as one of the two X chromosomes of female mammals. Hsu (1964) noted that all Chinese hamster chromosomes have regions which characteristically replicate late in S phase; and Gavosto et al. (1968) found that many human chromosomes have late replicating regions at their centromeres or telomeres.

At least so far as the late replication of constitutive heterochromatin is concerned, both homologues appear to show the same behaviour. Pflueger and Yunis (1966) found that Chinese hamster cells taken from live animals or grown in culture show the same late replication pattern, which supports the idea that late replicating regions have some inherent property which distinguishes them from the sequences replicating earlier in S phase. Late replication has been reviewed by Lima-de-Faria and Jaworska (1968), who have pointed out that in a great variety of organisms heterochromatin both commences and completes its replication later than euchromatin.

Late replication of facultative heterochromatin is particularly striking because the active sex chromosome replicates during the same period as the autosomes, whereas its inactive homologue replicates much later. Working with the Syrian hamster, in which the X chromosomes are much larger than the autosomes and are therefore easy to recognise, Galton and Holt (1964) noted that homologous autosomes usually replicate together. In males, the Y chromosome and the long arm of the X chromosome behave like constitutive heterochromatin and replicate late; in females, the long arm of one X continues to behave in this way, but the entire length of the other X comprises late replicating facultative heterochromatin.

In human males, there is little difference in the time of replication of the autosomes and the X chromosome. Takagi and Sandberg (1968a, b) found that the chromosomes appear to replicate in an ordered sequence, with the Y chromosome forming the only large late replicating unit. However, in contrast with this and other similar experiments, Craig and Shaw (1971) reported that the human Y chromosome may complete its replication comparatively early in S phase. Knight and Luzzatti (1973) confirmed that its replication starts later than the autosomes and suggested that its replicons may function simultaneously instead of sequentially.

By using autoradiography to follow the synthesis of DNA in leucocytes cultured from human females, Morishima et al. (1962) observed that one of the X chromosomes is labelled at the same time as the autosomes, but the other is not labelled until a later time. Gilbert et al. (1965) used a protocol in which human red blood cells were labelled for a short time with H^3-thymidine, blocked in metaphase by the addition of colcemid 1·5 hours later, and metaphase plates examined at 3 hours after incorporation of the label. Only those cells which were between 90 and 180 minutes prior to mitosis in the cell cycle should be found in the form of metaphase plates. Only those completing S phase can have been able to incorporate the tritium into their chromosomes. These experiments showed that one X chromosome becomes intensely labelled compared with the other chromosomes. This implies that this X chromosome must be replicated at the end of S phase; the other must have been replicated earlier in the cycle.

The effect of labelling human cells for differing periods of time towards the end of S phase is shown in figure 2.3. Gavosto et al. (1968) found that a label administered only at the very end of S phase labels one X chromosome much more heavily than the other X chromosome and the autosomes. The centromeric regions are prominent amongst the other areas labelled at this time. When the labelling period covers only the last 11 or 21 minutes of S phase (parts a and b of the figure) the late labelling X chromosome accounts for a very large proportion of the incorporation of radioactive label. When the period of incubation with radioactive precursors is extended to cover the last 30 minutes of S phase (part c of the figure), the X chromosome is still heavily labelled, but other chromosomes can also be seen to suffer replication.

A single late replicating X chromosome is found in many species of mammal (see page 219). Results suggesting late replication of one X chromosome in female cells have also been obtained with the grasshopper Melanoplus by Lima-de-Faria (1959) and in the plant Vicia faba by Evans (1964).

The late replicating facultative heterochromatin both starts and completes its DNA synthesis later than euchromatin. Working with Chinese hamster cells, Hsu (1964) found that females do not start to replicate the inactive X chromosome until 4·5 hours after the beginning of S phase; males show a similar behaviour with the Y chromosome. Comings (1967a) found that the

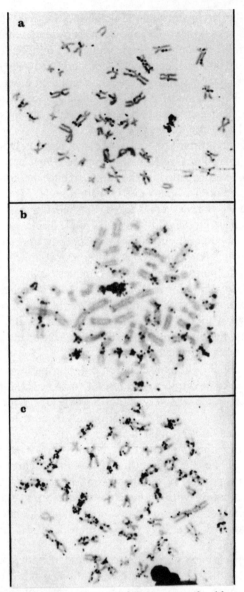

Figure 2.3: mitotic chromosomes of human erythroblasts which have incorporated labelled precursors to DNA at different times in the cycle. (a) labelling during the last 11 minutes of S phase identifies one X chromosome and a small number of other regions. (b) labelling for the last 21 minutes of S phase covers one X chromosome very heavily and an increased number of other regions, often centromeric. (c) labelling during the last 30 minutes of S phase shows that almost all chromosomes suffer some replication during this period. Data of Gavosto et al. (1968).

inactive X chromosome of cultured human fibroblasts commences replication some 2·5 hours after DNA synthesis has been initiated in euchromatin. Replication takes place at the same rate as in euchromatin for the next 3·5 hours (by which time most of the autosomes have completed replication) and then continues for a further 1·5 hours after they have ceased DNA synthesis. Gavosto et al. (1968) found that cells of all human tissues give the same pattern, in which total DNA synthesis falls towards the end of S phase, but the heterochromatic X chromosome continues to incorporate H³-thymidine after the other chromosomes have ceased to do so (see also Knight and Luzzatti, 1973).

The heterochromatic X chromosome appears to be replicated in its tightly coiled, dense state at a rate comparable to that prevailing in the less compact euchromatin. By measuring the extent of the area of the sex chromatin in sections of the nuclei of human cells, Mittwoch (1967) and Comings (1967b) showed that the heterochromatin does not disappear at any time—for example, because it has uncoiled for replication—but is present throughout the entire cell cycle. This implies that it must be replicated in its heterochromatic state and Comings (1967a) has in fact noted the incorporation of H³-thymidine into heterochromatic regions of the interphase nucleus.

There seems to be a shift during S phase in the nucleotide base composition of the replicating DNA. Flamm, Bernheim and Brubacker (1971) found that the DNA which is synthesised early in S phase in mouse lymphoma cells bands at a buoyant density greater than the average of the genome. Using Hela cells or mouse L cells, Tobia, Schildkraut and Maio (1970) found that DNA of high buoyant density is labelled if cells are incubated with H³-thymidine for one hour early in S phase, but the DNA replicated at the end of S phase has a lower buoyant density than the average. A similar shift has been found by Bostock and Prescott (1971b) in cultured CHO (Chinese hamster ovary) cells. These results imply that the sequences which are replicated at the beginning of S phase are rich in G–C base pairs, whereas those replicated at the end of the synthetic period are rich in A–T base pairs. (This conclusion applies to the main band of DNA; the replication of the repetitive sequences of satellite DNA is discussed in chapter 4.)

All these cell types grow in tissue culture, however, and may therefore have degenerated to heteroploidy from their diploid ancestors. Such a change in chromosome complement might alter the normal pattern of replication. Bostock and Prescott (1971c) therefore followed the buoyant density of DNA replicated in rabbit endometrium cells of defined chromosome complement; these cells are tetraploids which have precisely twice the diploid complement. A gradual shift takes place in the buoyant density of the replicating DNA as S phase proceeds. There are no discontinuous steps as the DNA of high buoyant density synthesized early in S phase is replaced first by DNA of average base composition and then, towards the end of S phase, by DNA of buoyant density lower than average. These results suggest that the density shift may be a general feature of the replication of mammalian DNA.

One model to account for this shift is to suppose that it represents a change from replication of euchromatic sequences early in S phase to heterochromatin later in the cycle. If DNA of high buoyant density is equated with euchromatin and DNA of low buoyant density represents heterochromatin, then the buoyant density of replicating DNA should decrease as heterochromatin begins to be replicated later in the cycle. However, when Bostock and Prescott (1972) prepared dense and light fractions of chromatin by a gentle centrifugation of cell extracts, they found no significant and constant variation in the buoyant density of the DNA contained in the two fractions prepared from various eucaryotic cells. The dense fraction may be approximately equated with heterochromatin and the light fraction with euchromatin (see page 190); in mouse cells the heterochromatic fraction contains DNA of slightly lower buoyant density than average, but in Chinese hamster cells the heterochromatic DNA is of slightly greater buoyant density than average. It is doubtful whether these differences are significant. This implies that the switch in base composition of replicating DNA during S phase does not represent a change from euchromatin to heterochromatin but must have some other cause (see also Comings and Mattoccia, 1972).

Association of Chromatin with the Nuclear Membrane

The genome of a bacterium appears to be attached to the membrane of the cell both at the origin where replication is initiated and also perhaps at the growing points. This attachment may represent the localization of the replication apparatus in the membrane; and it may have evolved in order to ensure an even segregation of duplicated chromosomes to daughter cells (see chapters 10 and 13 of volume 1).

Chromosome segregation in eucaryotic cells is of course accomplished by the mechanisms of mitosis and there is no evidence to suggest any association between polymerising enzymes and the nuclear membrane. However, it has been suggested that the replicating regions of the chromatin may be associated with the membrane of the nucleus; this might provide a more ordered structure for replication than the free association of enzymes with the network of chromatin fibres. Visualization of nuclei shows threads of chromatin attached to the membrane, especially at the clumps of heterochromatin. One of the implications of the great number of replicons in each chromosome is that each part of the DNA must unwind independently for replication; one way to provide a topological organization for this unwinding would be to anchor the DNA to the membrane.

Studies which support the idea of an association between replicating DNA and the nuclear membrane fall into two classes. Biochemical experiments analogous to those performed with bacteria show that when a pulse label is given to cells in S phase it first enters a cellular fraction containing the nuclear membrane but may later be chased out of it by unlabelled precursor.

After labelling the DNA of Hela cells with H^3-thymidine, O'Brien, Sanyal and Stanton (1972) sonicated isolated nuclei and fractionated the extract by centrifugation. The pellet from a low speed centrifugation contains membranes and attached fibres of nucleoprotein, comprising some 3-6% of the cellular DNA. Centrifugation of the supernatant at high speed gives a pellet of membranes, granules and fibrils, containing some 15-18% of the DNA. The supernatant of this centrifugation contains only fibrils and has about 80% of the cellular DNA. A pulse dose of H^3-thymidine labels the membrane associated DNA of the first fraction to a much greater extent than the DNA of the other two fractions. After a chase of 5 minutes, most of the DNA is found in the soluble fractions, the proportion increasing with the length of the chase. Newly replicated DNA is therefore preferentially associated with membrane fractions. Results suggesting a similar conclusion have been reported by Mizuno, Stoops and Peiffer (1971).

Extraction of cells may of course lead to a spurious association in vitro of cellular components. In bacteria, it is possible to show that specific genetic markers identifying the region of the origin are found in the membrane fraction. This argues that the association is genuine. In eucaryotic cells, this is not possible, and we must rely upon biochemical experiments alone. Although these experiments suggest that newly replicated DNA is found in different cellular fractions from the bulk of the DNA, they do not reveal whether this is because it is in some way linked to the membrane or because of some other property which influences its fractionation.

Newly synthesised DNA appears, indeed, to behave in a different manner from bulk DNA during extraction from cells. Friedman and Mueller (1968) and Fakan et al. (1972) found that upon extraction with reagents such as phenol or chloroform, newly synthesised DNA is found at the interphase, whereas bulk DNA—or a pulse label which has been followed by a chase— is found in the aqueous phase. The reason for this differential extraction probably lies with the single stranded nature of the new strands of DNA. These experiments therefore suggest that the presence of newly synthesised DNA in membrane fractions may be a consequence of its physico-chemical state and may not reflect attachment to the membrane.

Autoradiography of replicating cells has provided the second line of evidence which has been taken to suggest that replicating sequences may be attached to the membrane. After synchronising cells with amithopterin, Comings and Kakefuda (1968) used incubation with an H^3-thymidine label for 10 minutes to release the block and allow DNA synthesis, and then localised the grains produced by decay of the tritium through electron microscopy of the nucleus. This should identify the sites of the DNA replicated at the beginning of S phase. In more than 90 % of the cells examined, the label proved to be confined to the inner edge of the nuclear membrane.

There are three possible interpretations of this result. One possible artefact which could restrict labelling to the nucleus is that insufficient time might be allowed for the H^3-thymidine to diffuse into the nucleus. Against this interpretation are the experiments discussed below in which inner regions of the nucleus have been labelled. Another possible explanation is that the localization results from the synchronization procedure itself. A third interpretation is to suppose that replication takes place at the nuclear membrane and Comings (1968) has presented a model for the organization of interphase chromatin predicated on the concept that the replicating apparatus of each replicon is attached to the membrane. In support of this class of model, Alfert and Das (1969) have noted that the rate of DNA synthesis appears to be related to the surface area of the nucleus; this would be consistent with the idea that the availability of replication sites on the nuclear membrane controls the rate of replication.

The possibility that the synchronization technique may influence the apparent association of replicating DNA with the membrane has been investigated by Williams and Ockey (1970). CHO (Chinese hamster ovary) cells were blocked at metaphase with colchicine, released from the block, and then labelled with H^3-thymidine for 10 or 20 minutes after an interval corresponding to the length of G1 phase. This protocol labels the DNA replicated early in S phase; grains were found to cover the nucleus homogeneously, with no concentration at its periphery. If the metaphase cells produced by the colchicine treatment are blocked in early S phase by treatment with amithopterin or 5-fluoro-deoxyuridine, the same result is obtained. However, if the cells are labelled during late S phase, the grains tend to be located at the membrane, in particular in the condensed clumps of late replicating heterochromatin.

Further experiments have suggested that peripheral labelling can be produced in (apparent) early S phase by an artefact which results from the use of amithopterin. Ockey (1972) halted the Chinese hamster cells in metaphase by treatment with colchicine and then incubated them for varying lengths of time in amithopterin, after which the block to replication was released by the addition of H^3-thymidine. This treatment has been thought to label cells at the beginning of S phase. However, the cells undertake a synchronous division at between 19 and 21 hours after their release from colchicine (judged by a doubling in the cell number). The normal generation time of these cells is 12 hours. These results therefore imply that the cells are not prevented from replicating DNA and dividing by the presence of amithopterin, although the cycle is greatly slowed. One possible explanation is that their precursor pool is sufficient to allow them to complete one S phase, although at a slower rate than usual.

These cells therefore suffer an S phase which begins at about three hours after their release from colchicine. When the cells are released from inhibition by amithopterin, they can presumably continue replication under more normal conditions. However, the DNA which they synthesise upon release is not that usually replicated at the beginning of the S period, for by this time the cells are well into S phase. The DNA labelled by the H^3-thymidine thus represents that replicated at a later stage of S phase. This important conclusion means that cells which have seemed to be labelled at the beginning of S phase after treatment with amithopterin may in fact be labelled at a later point during the cell cycle.

Since after upto 18 hours in amithopterin the CHO cells exhibit no peripheral labelling, replication of DNA during the first S phase must take place at sites located throughout the nucleoplasm. Cell survival is high during this period and the effect of the amithopterin seems to be to reduce the speed of the cell cycle rather than to change its nature.

However, the second S phase which starts at about 24 hours after the cells have been released from metaphase has a different character. Cells which have remained in amithopterin upto this point exhibit an increasing amount of peripheral labelling with time of exposure to the inhibitor. By the time of 36 hours incubation, peripheral labelling occurs in virtually all the cells which incorporate H^3-thymidine into DNA when they are released from amithopterin. Increasing cell death accompanies the increase in peripheral labelling.

These results suggest that, at least for the first two thirds of S phase when euchromatin is replicated, DNA synthesis under normal conditions takes place at sites located throughout the nucleus. At the end of S phase, when heterochromatin is replicated, peripheral labelling is displayed by the replicons adjacent to the nuclear membrane. Amithopterin treatment does not change this pattern. However, when a thymineless state is maintained by extended incubation with amithopterin, an abnormal situation develops in which the second S phase shows replication at the membrane. Increasing cell death accompanies the abnormal replication. Ockey suggested that the cause of the peripheral replication and cell death may be a fragmentation of DNA into smaller units which aggregate on the inside of the nuclear membrane.

Experiments with several species of cells have confirmed that peripheral labelling is a consequence of the amithopterin treatment. Working with Chinese hamster cells, Wise and Prescott (1973) used mitotic selection followed by incubation with FUdR and Comings and Okada (1973) used mitotic selection followed by incubation with hydroxyurea to synchronize cells; both treatments inhibit DNA synthesis and so cause a selected G1 population to halt in the cycle at the G1/S interphase. Localization of an H^3-thymidine label by autoradiography suggested that replication takes place in the nucleoplasm.

One possible cause of homogeneous nuclear labelling is the length of the pulse doses of H^3-thymidine (in many experiments at least 5 minutes). A replicating fork moving at (say) 1 μ per minute could synthesize some 5 μ of DNA during this period. Even if replication were to take place at the nuclear membrane, this length of DNA might become stretched so that much of its label is detected within the nucleus, generating an apparently nucleoplasmic label.

This possible explanation has been excluded by Huberman et al. (1973), who labelled CHO cells with 30 second pulses of H^3-thymidine (see also Wise and Prescott, 1973). Similar experiments have been performed with mouse cells by Fakan et al. (1972). In an unsynchronized culture, both homogeneous and peripheral labelling are seen. With cells synchronized by colcemid, a pulse label during the first part of S phase generates a homogeneous labelling pattern; peripheral labelling becomes predominant later in S phase. The same results are found if a 10 minute label of H^3-thymidine is used, or if an 0·5 minute pulse label is followed by a chase of some hours with unlabelled thymidine. This implies that the location of DNA sequences in the nucleus is stable and does not change during replication; sequences located close to the membrane tend to be replicated towards the end of the synthetic period.

Initiation of Nuclear DNA Synthesis

The critical event in establishing the cycle of a cell appears to be the decision whether or not to enter S phase. A cell which replicates its DNA is almost inevitably committed to continue through G2 and mitosis into the next G1 phase. (An exception is the polyploid cells of tissues such as the liver.) A cell which does not commence replication may remain in the G1 (or the G0) state indefinitely. The cell cycle is therefore in effect controlled by the biochemical events in G1 which may lead to the initiation of DNA synthesis.

The control of DNA synthesis and its relationship to the cell division may be different in multicellular organisms and in unicellular organisms (in which we might expect to see some similarity with the bacterial control mechanisms described in chapter 13 of volume 1). Discussion here is therefore confined to mammalian cells, and in general concerns those grown in culture with which most experiments have been performed. Of course, these cells grow in an abnormal condition and their cycles may not be entirely typical of their counterparts within a multicellular organism. A more complete discussion of the cell cycle in various types of eucaryote has been presented by Mitchison (1971).

That signals from the cytoplasm control the initiation of DNA synthesis in the nucleus has been suggested by experiments utilising transplantation or cell fusion; for when a nucleus of an adult somatic cell which is inactive in replication is placed in a new cytoplasmic environment, it may be stimulated to synthesize DNA and divide (see chapter 7). These signals seem to be of low specificity since the cytoplasm of one species may activate the nucleus of another.

In a large number of mammalian cell types, RNA and protein synthesis must be allowed to take place late in G1 phase as a prerequisite for entry to S phase (reviewed by Baserga, 1968). This synthesis must take place in each G1 phase to allow the subsequent S phase to commence. In cells which have ceased division but may be stimulated to start again—such, for example, as liver cells after a partial hepatectomy—an early wave of RNA synthesis precedes DNA synthesis. These observations are consistent with the conclusion that the synthesis of protein factors is required in G1 (or in G0) phase before the cells can enter S phase. Of course, these experiments do not reveal whether the protein factors themselves interact directly to control replication, or whether they cause the production of other signals.

The effect on DNA replication of inhibiting RNA and/or protein synthesis appears to vary greatly in different cell types and even with different experimental growth conditions. Using Hela cells synchronised with amithopterin, Mueller and Kajiwara (1966b) found that the effect of inhibiting transcription with actinomycin depends upon the time at which the drug is given to the cells. Actinomycin exerts the greatest effect when added before S phase begins; the inhibitor continues to prevent DNA synthesis if added during the first two hours of S phase, but has little effect if added after this time. It usually takes about an hour before addition of the actinomycin exhibits any effect on DNA synthesis (see also Taylor, 1965).

Similar results have been obtained in many other experiments. For example, Fujiwara (1967) found that the addition of actinomycin to L cells during the G1 phase after a mitotic selection prevents initiation of DNA synthesis in 50% of the cells; addition of the drug during S phase appeared to have no effect. When Terasima and Yasukawa (1966) treated L cells with puromycin for 2 hours during G1, they found that the start of S phase is delayed for over an hour, no matter when the 2 hour inhibition period is timed within G1. This suggests that the whole G1 period is equally sensitive to inhibition, which would be consistent with the idea that there is a continual accumulation of proteins needed to induce DNA synthesis. These proteins must be coded by RNA molecules themselves synthesised during G1.

Many experiments have suggested that there is at least one discontinuous step essential for the initiation of DNA synthesis, located during G1 some 2–3 hours before the beginning of S phase. Whereas inhibition of transcription or translation before this moment inhibits the start of DNA synthesis, such treatments lose their effect after the step has taken place. Jakob (1972) found that when Vicia faba meristem cells are treated during G1 with actinomycin, a 60–70% inhibition of RNA synthesis within 3 hours or less of the start of S phase does not prevent the subsequent replication. But a similar treatment given 5 hours or more before the beginning of S phase reduces DNA synthesis by about 50%.

A further set of experiments, however, has suggested that the continuing synthesis of proteins during S phase may be essential to maintain replication. Baserga (1968) has noted that inhibitors of protein synthesis often prevent DNA synthesis in mammalian cells regardless of the part of the S phase at which they are added. For example, Highfield and Dewey (1972) found that puromycin and cycloheximide inhibit both initiation of DNA synthesis and its continuation in CHO cells. Weiss (1969) found that in Hela cells the effect of these antibiotics does not depend upon the time at which they are added, but DNA synthesis is inhibited whenever they are present. Hori and Lark (1973) observed that puromycin prevents new replicons from initiating DNA synthesis in Chinese hamster cells, but allows currently proceeding growing points to continue, supporting the concept that the critical steps in control of replication take place at initiation.

We do not know whether the effect of inhibiting protein synthesis during S phase differs in various cells because of genuine differences in the natures of the systems used to control replication or whether these effects depend upon the conditions in which tissue culture cells are grown. Initiation of DNA synthesis clearly depends upon the synthesis of RNA and its translation into protein at some time, or continuing period, during the preceding G1 phase. But we cannot yet define the nature of this step, although it is likely that inhibition of protein synthesis during G1 interferes with the control system which leads to the initiation of DNA synthesis rather than with enzyme functions needed for replication.

Studies of the proteins in mouse and Chinese hamster cells which can bind to DNA suggest that stationary phase and growth phase cells may be distinguished by their content of such proteins. Using columns of DNA-cellulose to bind proteins extracted from mouse 3T6 cells, Salas and Green (1971) isolated eight DNA binding proteins. One of these proteins, P6, is present in large amounts in cells in exponential growth but not in stationary phase cells; this protein appears to be synthesized during S phase. Two proteins present in greater amounts in stationary phase than in growth phase cells have since been shown by Tsai and Green (1972) to comprise precursors to collagen; these proteins are therefore not implicated in control of the cell cycle.

Using a similar technique, Fox and Pardee (1971) compared the DNA binding proteins of Chinese hamster ovary cells in various growth states. They found no differences in the proteins synthesized in G1 and S phase cells (apart from the histones comprising a low molecular weight fraction). But the proteins of cells deprived of serum differ from those grown on medium supplemented with serum. This suggests that the additional proteins synthesized in growth phase cells may distinguish the active cell cycle from the halted G0 state, rather than providing a control of S phase itself in the cycle of the growing cells.

Are the proteins which synthesize DNA produced only at some particular point(s) in the cell cycle or are they always present in the cell? Several technical problems make it difficult to follow the synthesis of DNA polymerase. A general problem in examining fluctuations in enzyme levels during the cell cycle is that synchronization of cells with inhibitors—which is often necessary to obtain enough cells to assay enzyme activities—may divorce the replication and cell growth cycles. Thus although cells may be forced to accumulate in metaphase or at the beginning of S phase, cellular activities other than DNA synthesis may not be inhibited in the same manner as replication. This may mean that any relationships between the synthesis of particular enzymes and the replication of DNA may be altered.

Another caution is that changes in enzyme activities may not reflect changes in their synthesis de novo, but may be caused by activation or inhibition of pre-existing enzyme molecules. In eucaryotic cells, it is rarely possible to demonstrate enzyme synthesis; fluctuations in the enzyme content of the cell can in general be followed only by an assay of activity in vitro. The activity of DNA polymerase is particularly difficult to assess, because in bacteria the presence of DNA polymerase activity in cell extracts has proved to be a misleading guide to the identity of the replicating enzyme (see chapter 10 of volume 1). Two DNA polymerase activities have been detected in extracts of mammalian cells, one located only in the nucleus and the other present in both nucleus and cytoplasm. We do not know which, if either, of these enzymes may replicate DNA in vivo.

There seems in general to be no simple correlation between the presence of DNA polymerase activity in cell extracts and the occurrence of replication. For example, with mouse tumour cells synchronized by selection after centrifugation through sucrose, Schindler et al. (1972) found that DNA polymerase activity fluctuates in the same way as total cellular protein and shows no sharp rise during S phase. These results are consistent with a continuous increase in polymerase activity during the cell cycle.

Although the total cellular content of DNA polymerase activity does not seem to display a peak during S phase, the localization of the enzyme within the cell may change during the cycle. One note of caution here is that the conditions of lysis may affect the apparent distribution of DNA polymerase. The DNA polymerase activity is generally found in the supernatant when cells are lysed in aqueous medium, but is distributed about equally between nucleus and cytoplasm after lysis in non aqueous medium. This may explain some of the different results reported for the presence of DNA polymerase in cell fractions prepared by different means.

An increase in the activity of nuclear DNA polymerase is often seen prior to S phase. Using mouse cells partially synchronised with FUdR and released from inhibition with thymidine, Littlefield, McGovern and Margeson (1963) found that the polymerase activity of the supernatant declines prior to DNA synthesis, whereas that of the nucleus increases. The overall cellular activity remains roughly constant. Comparable results for Hela cells have been reported by Friedman (1970). One interpretation is that DNA polymerase is synthesised in the cytoplasm and remains there until S phase, when it enters the nucleus. After S phase, the enzyme may decay or may return to the cytoplasm. In view of the lack of data on the stability of DNA polymerase and on the relationship between its activity and synthesis, no firm conclusion on its cycle is possible; it seems likely, however, that its synthesis is not restricted only to the part of G1 immediately prior to S phase and that its cellular function may be controlled more by its site within the cell than by a periodic synthesis.

The best evidence for a cycle of movement of the enzyme is derived from studies of sea urchin eggs. Unfertilised sea urchin eggs contain a high level of DNA polymerase activity, and the activity per embryo remains constant during early development, when the pre-existing enzyme is utilised without new synthesis. This situation may therefore be atypical so far as mammalian cells are concerned. Fansler and Loeb (1972) have isolated nuclear and cytoplasmic fractions from embryos at various times between fertilization and blastula; as development proceeds, the activity in the nuclear fraction increases at the expense of that of the cytoplasm. This suggests that as new cells form, the enzyme moves from cytoplasm to nucleus.

If eggs are fertilized simultaneously and allowed to develop, they can be assayed at varying times for their polymerase activity. When the embryos are grown in the presence of H^3-thymidine, successive waves of DNA synthesis are found, which shows that they develop synchronously. These cells have no G1 period, but commence DNA synthesis at the termination of mitosis (in telophase); the ensuing S period lasts for about 20 minutes, after which there is a 40 minute G2 until the next S phase commences.

DNA polymerase activity is found in the nucleus at the beginning of S phase. The activity increases 10 fold in nuclei when eggs pass from the 4 cell to the 8 cell stage and then declines as the cells complete DNA synthesis. This suggests that DNA polymerase may enter the cells at the beginning of S phase and then return to the cytoplasm at its completion. If chromatin is prepared from the nuclei, only 5% of the polymerase activity is lost; this implies that the nuclear sequestration of polymerase represents its association with the genetic material which it must replicate.

Although DNA polymerase itself does not seem to have a cycle of synthesis related to that of replication, some of the enzymes which are involved in the production of precursors for DNA synthesis appear to show an increase in activity during S phase. In particular, thymidine kinase activity tends to increase during DNA synthesis, returning to normal by the next G1 (see Baserga, 1968; Mitchison, 1971). Brent, Butler and Crathorn (1965) found that the activities of thymidine kinase and thymidylate phosphokinase, which between them synthesize dTTP from thymidine, fluctuate during the cell cycle in a manner correlated with DNA synthesis. Some of the increases in activity which are seen in S phase may be caused by substrate activation and not by synthesis of new enzymes (see Stubblefield and Murphree, 1967). However, Littlefield (1966) found that an increase in thymidine kinase activity in L cells can be prevented by puromycin and may therefore represent synthesis of new enzyme molecules. Bray and Brent (1972) measured the sizes of the deoxynucleotide precursor pools in synchronized Hela cells; but although there are fluctuations during the cell cycle, there is no precise correlation with DNA synthesis.

Although the enzymes of DNA synthesis do not appear to be produced solely during S phase in dividing cells but may be synthesized continuously, their production may be halted in non-dividing cells. Williams (1972a, b) found that when dividing erythroblasts mature to non dividing polychromatic erythrocytes in avian tissues, DNA polymerase activity declines, disappearing completely by the time mature erythrocytes are produced. The thymidine kinase activity of these cells is also closely correlated with their ability to replicate DNA.

The reverse series of changes is seen when human lymphocytes which have ceased division are stimulated to recommence growth by the addition of photohaemagglutinin (PHA); Loeb et al. (1970) found that DNA polymerase, thymidine kinase and TMP kinase activities all increase before replication of DNA starts. The increases are inhibited by addition of actinomycin with the PHA, which suggests that they result from de novo enzyme synthesis.

Analogous changes in DNA polymerase activity may take place when tissue culture cells pass from stationary to growth phase. Chang et al. (1973) reported that when L cells enter growth phase their cytoplasmic DNA polymerase activity increases, although the enzyme found only in the nucleus shows no increase.

Cells in which DNA synthesis has been suspended either permanently or indefinitely may therefore cease to synthesize the enzymes concerned with the reproduction of DNA. Reversal of this process is of course then essential if such cells are later stimulated to divide. But it seems probable that this change in activity comprises only an on/off switch; it is unlikely that much change in enzyme synthesis takes place during the cycle of dividing cells. Although the activities of some of these enzymes may be correlated with replication, it is unlikely that fluctuations in synthetic capacities or in the levels of the nucleotide precursors provides a control of DNA replication; there is no evidence to support the concept that the activities of the DNA synthetic activities may be limiting on replication. It is reasonable to suppose that these enzyme activities, if controlled, respond to systems related to those which control replication.

Synthesis of Histones

One notable exception to the general gradual increase observed in RNA and protein synthesis during the cell cycle is the synthesis of histones, which seems to be closely correlated with the synthesis of DNA in S phase. An early observation suggesting that the chromosome as a whole, and not just its genetic material, is reproduced during S phase was that of Alfert (1958), who found that the ratio of feulgen stain (which assays DNA) to alkaline fast green stain (which assays basic proteins such as histones) remains constant in onion tip roots. This implies that the content of histones must double when DNA doubles. Bloch and Goodman (1955) used a microspectrophotometric assay of fast green stain to show that an increase in the histones of rat liver nuclei accompanies the replication of DNA.

The duplication of particular chromosome regions has been followed in Euplotes, which in common with other ciliates has a densely staining macronucleus. Some hours before cell division the macronucleus suffers changes in its structure and staining. At the beginning of this reorganization, a narrow band appears at each end of the nucleus and the two bands progress towards each other until they meet several hours later. That these bands represent the sequential progress of replication from each end of the nucleus to the centre has been confirmed by autoradiography with H^3-thymidine and staining with feulgen dye.

By staining with alkaline fast green, Gall (1959) found that histone duplication accompanies the synthesis of DNA. Prescott (1966) found that incorporation of H^3-labelled amino acids into protein during G1 phase is linear, but there is a sharp increase in the content of labelled amino acids found in the nucleus during S phase. This does not show whether the histones are synthesised at the same time as the DNA, or whether they are assembled previously, but in either case they must enter the nucleus so that they associate with the newly synthesised DNA as the chromosome is replicated.

Several lines of evidence suggest that the histones are indeed synthesised de novo at the same time as DNA. By incubating onion root cells with an H^3-label of arginine or lysine, or of H^3-thymidine, for 30 minutes and then examining subsequent metaphases by autoradiography, Bloch et al. (1967) measured the lengths of G2 and S phase by the times when the label first appears and then disappears in the metaphase plates. The same parameters for the cell cycle can be derived by the use of either labelled amino acids or nucleotides, which implies that both components of the chromosome are synthesised at the same time.

Measurements of the cytoplasmic capacity of the cell to synthesize histones also suggest that histone and DNA synthesis are closely correlated. Robbins and Borun (1967) have followed the incorporation of labelled amino acids into Hela cells synchronised by selective detachment of mitotic cells. When the nuclei are isolated and their HCl-soluble proteins extracted—this fraction includes the histones—during G1 phase, incorporation into individual peaks increases uniformly and the relative ratios of the different proteins are maintained at a constant level. When DNA synthesis starts—measured by the incorporation of H^3-thymidine—there is an increase in the incorporation of labelled lysine into the two peaks of the HCl-soluble proteins which include the histones. Pulse chase experiments confirm that the histones are made in the cytoplasm and then transported to the nucleus. At the same time as the histones enter the nucleus, some non-histone proteins appear to be transported in the opposite direction, that is from nucleus to cytoplasm (see next section).

Synthesis of histones may be controlled at the level of transcription, for small polysomes containing the messenger RNA which codes for histones appear in Hela cells only during S phase (see page 264). However, although the synthesis of histones appears to be restricted to S phase and although this is the predominant time when they enter the chromosome, appreciable turnover of histone proteins occurs throughout the cell cycle. Histones already associated with DNA may therefore be replaced at times other than S phase (see page 144). This means that the structure of the chromosome is not static and its protein components do not remain inviolable between S phases.

Some of the amino acid residues of histones are specifically modified by the addition of methyl, acetyl, or phosphate groups after the protein has been synthesised. Modification seems to take place in the nucleus and at least some of the modifying enzyme activities appear to be associated with chromatin (see chapter 4). Although some of the modifications may take place during S phase at the period when histones are initially added to the chromosome, others appear to take place at a later stage in the cell cycle. Some of these modifications may be concerned with the turnover of histones in the chromosome. In this sense, the structure of the chromosome is not completely determined during S phase.

Although histones comprise the greater part of the proteins of the chromosome, non histone proteins also form an appreciable component of chromatin. These proteins, however, do not seem to be synthesised at the same time as the DNA and histones. Stein and Baserga (1970) found that synthesis of non histone proteins appears to continue at the same rate throughout the cycle of Hela cells. If this phenomenon is typical of other cells, it implies that newly synthesised DNA associates with the basic histone proteins soon after its replication, but that completion of the structure of chromatin by the addition of non-histone proteins need not necessarily take place at the same time. It is therefore possible that the structure of the chromosome is assembled more gradually than might be inferred from studies of histones and DNA alone.

Protein Synthesis and Cell Growth During Interphase

Cellular synthesis of nucleic acids and proteins is almost completely confined to interphase. The condensed chromosomes of mitotic cells are virtually inactive in RNA synthesis and the translation of previously synthesized messenger RNA is halted during the process of division. Replication of DNA, of course, is restricted to S phase. Synthesis of RNA and protein, by contrast, occurs continually during interphase and is halted only for the length of mitosis, resuming at the beginning of the next G1 (see page 329).

By following the growth of individual mouse fibroblasts with cytochemical assays, Killander and Zetterberg (1965a) measured the extent of DNA, RNA and protein synthesis as a function of the time elapsed since the last mitosis. The amount of protein is essentially measured by the cellular dry mass (which can be assayed by microinterferometry) and the amount of RNA and DNA is revealed by extinction at 265 mμ and 546 mμ after staining with feulgen.

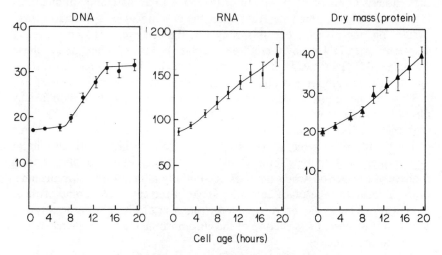

Cell age (hours)

Figure 2.4: accumulation of DNA, RNA and protein during cycle of mouse fibroblasts. G1 lasts for 8 hours; the S phase period is defined by the increase in DNA content which takes place from 8–14 hours after mitosis; and during G2 the cells possess double the G1 content of DNA. RNA and protein (given by dry mass) increase continuously during all three phases of the cycle. Data of Killander and Zetterberg (1965a).

The results of figure 2.4 show that the cycle of these cells comprises a G1 of 8 hours, an S phase of 6 hours, and G2 of 5 hours. (The amount of DNA fails to double exactly because the later age groups are contaminated by a few G1 cells). Although DNA synthesis is discontinuous, RNA and protein levels increase continuously. The technique does not allow measurements sufficiently precise to define the exact manner of growth; RNA levels may increase in either a linear or exponential manner. By following the increase in protein through the use of labelled amino acids, Zetterberg and Killander (1965) confirmed that the cellular content of protein doubles during the growth cycle.

These results show the growth of the cell as a whole. Zetterberg (1966a) extended these studies to follow the synthesis and accumulation of protein in the cytoplasm and nucleus of growing cells. The nuclear dry mass was determined by microinterferometry; subtraction from the total cell mass gives the mass of the cytoplasm. Figure 2.5 shows that the cytoplasmic mass increases predominantly during the first half of interphase. The predominant increase in nuclear mass takes place during the second part of interphase, its start coinciding roughly with S phase. Both nuclear and cytoplasmic dry mass double during the cycle, so that the relative proportions of the cell are the same at the end of G2 as at the beginning of G1, with about half the mass in the cytoplasm and half in the nucleus.

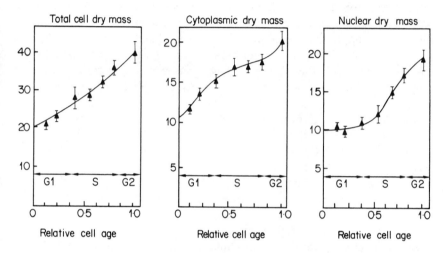

Figure 2.5: accumulation of protein during cycle of mouse fibroblasts. Total dry mass—that is protein content—increases continuously. Much of the increase in cytoplasmic mass takes place during G1 when the nuclear mass remains constant. Nuclear mass increases during S phase and G2, when cytoplasmic mass increases more slowly. Both nuclear and cytoplasmic mass double during the cycle. Data of Zetterberg (1966a).

The activity of cells in protein synthesis relative to the cytoplasmic mass is constant throughout the cell cycle, which implies that the total level of protein synthesis increases in proportion with the increase in content of the cytoplasm. During G1, most of the protein which is synthesised accumulates in the cytoplasm, generating the pattern of cytoplasmic growth during the first part of the cycle. During S phase, cytoplasmic protein synthesis continues at the same relative rate, but the mass accumulates in the nucleus instead of in the cytoplasm. This implies that much of the protein synthesised during the second part of the cell cycle must be transported to the nucleus. The growth cycle of the cell has been reviewed by Zetterberg (1970).

The transfer of proteins from cytoplasm to nucleus takes place throughout interphase and is not confined to the later part of the cycle when the nucleus grows in size. Zetterberg (1966c) gave cultures of L cells a pulse dose of H^3-leucine for 4 minutes and then followed the fate of the label by autoradiography. As time passes, more of the label is seen in the nucleus. This implies that newly synthesised proteins are transported from cytoplasm to nucleus. Since this transfer takes place at an even rate throughout interphase but no protein accumulates in the nucleus during G1, there must be a simultaneous movement of proteins from nucleus to cytoplasm in exchange during the early part of interphase.

This suggests a mobile view of the cell, in which proteins continually move in both directions between nucleus and cytoplasm. During G1, only the mass of the cytoplasm increases, for the numbers of proteins entering and leaving the nucleus appear to be in equilibrium. During S and G2, much of the new protein content of the cell accumulates within the nucleus, by transport from its site of synthesis in the cytoplasm. Analogous results have been obtained in other types of cell, including the lower eucaryotes (reviewed by Mitchison, 1971).

The accumulation of proteins which commences in the nucleus at S phase depends upon DNA synthesis. Auer et al. (1973) observed that when FUdR is used to inhibit DNA synthesis in L cells, total cellular mass continues to increase; synthesis of proteins is therefore independent of DNA synthesis. But the mass accumulates in the cytoplasm; transport of proteins from cytoplasm to nucleus to cause increase in nuclear mass during the latter part of the cell cycle therefore depends upon DNA synthesis.

A two way exchange of proteins between nucleus and cytoplasm has been observed in some detail in the unicellular organism Amoeba proteus. Byers, Platt and Goldstein (1963) first noted that if a cell is grown in H^3-leucine for long enough to label its proteins uniformly, transplantation of its radioactive nucleus to another, unlabelled cell is followed by a redistribution of the label. Within a matter of hours, the tritium can be seen by autoradiography to be concentrated over both the transplanted and the host nucleus; little label is found in the cytoplasm. This suggests that the cell nucleus contains a high concentration of proteins which migrate to the cytoplasm where they remain only briefly before returning to a nucleus; these species have been termed "rapidly migrating proteins".

There is a net transfer of labelled protein from donor (labelled) to host (unlabelled) nucleus until equilibrium is reached some hours after the transplantation. At this point, the host nucleus contains about 30% of the total radioactivity of the two nuclei. This inequality suggests that the nucleus may contain two classes of protein, those which migrate rapidly and those which remain in the nucleus (or at least turn over very much more slowly).

Further experiments have confirmed the conclusion that some nuclear proteins are in equilibrium with cytoplasmic pools and can return from cytoplasm to nucleus against a concentration gradient. Goldstein and Prescott

(1967, 1968) observed that the total amount of rapidly migrating protein in the nucleus and cytoplasm of Amoeba is about the same; since the nucleus comprises only about 2% of the cell volume, its concentration of the proteins must be some 50 times greater than that in the cytoplasm.

An important qualification to the interpretation of these experiments has been made by Legname and Goldstein (1972), who have observed that the transplantation technique introduces an artefact into protein migration. The maximum amount of protein to leave a transplanted nucleus migrates within 10 minutes; but equilibrium between the host nucleus and the cytoplasm is not reached until about 3 hours after the transplantation, when the concentration of label in the host nucleus reaches 25–50 times that acquired by the cytoplasm. This suggests that the rapid loss of proteins from the transplanted nucleus may be a consequence of damage caused during its micromanipulation.

In addition to the movement of these "injury proteins", however, some proteins do seem to suffer a genuine migration between nucleus and cytoplasm. If an H^3-labelled nucleus is transplanted into an unlabelled cell, left for one day, and then removed and replaced by an unlabelled nucleus, the second nucleus acquires a label about 5 times that of the cytoplasm. If the labelled nucleus is left in the cell, the final level of label in the second (previously unlabelled) nucleus is about 10 times that of the cytoplasm; it is presumably increased because of the presence of additional migrating proteins in the remaining originally labelled nucleus. These experiments show that labelled proteins may leave a nucleus, enter the cytoplasm, and then re-enter a nucleus.

Analysis of these results is complicated by the loss from the transplanted nucleus of the injury proteins. However, Legname and Goldstein suggested that the nucleus may contain three principal classes of protein. Some 40% of the total nuclear protein does not leave the nucleus during interphase, or at least turns over exceedingly slowly. About 45% of the nuclear protein is lost to the cytoplasm immediately after a transplantation, perhaps as the result of injury. Some of the injury proteins remain in the cytoplasm, but the other subclass returns to the nucleus. Some 15% of the nuclear protein seems to shuttle between nucleus and cytoplasm and may fall into two subclasses: molecules whose equilibrium lies so that they are concentrated within the nucleus, and those which undergo a similar migration but whose concentration is within the cytoplasm.

One important caution is that all these results rely upon experiments with transplanted nuclei, for these techniques do not allow the assay of the loss of protein from a nucleus within its original cytoplasm. Conclusions about the entry of proteins into the host nucleus against a concentration gradient are therefore more definite than those about the loss of protein from the donor nucleus. We can only speculate about the possible functions of rapidly migrating proteins; it is not possible to say whether they occur in all cell types or are a feature of the growth cycle of unicellular organisms.

The ability of proteins to enter nuclei rapidly is well documented, however, especially for situations which involve the reactivation of a previously inert nucleus. Transplantation of an adult somatic nucleus of Xenopus laevis into an enucleated egg is followed by the rapid entry of cytoplasmic proteins as it swells, and a similar situation is seen also when the inert nucleus of a chick erythrocyte is transferred by cell fusion into the cytoplasm of a Hela cell (see chapter 7). In the latter case, the reconstitution of an active chick nucleus is achieved by human proteins, so that at least some nucleocytoplasmic inter-actions are not species specific. These experiments also represent a situation in which a nucleus finds itself in an unusual situation, since it is persuaded to reverse its state of differentiation. However, the many different instances in which the accumulation of proteins in the nucleus is correlated with its activa-tion and DNA synthesis imply that this movement may reflect a physiological process which takes place during the normal cell cycle.

Although cells seem to vary somewhat in the age at which they start S phase (relative to the last mitosis), their mass may be more constant at this time. This observation suggests that DNA synthesis may be initiated when the cell mass passes some critical level, perhaps in a manner similar to the control of replica-tion in bacteria (see chapter 13 of volume 1). By analysing several different populations of mouse L cells during growth, Killander and Zetterberg (1965a, b) found that the mass of cells doubles each generation, but that the variation of initial mass—that is immediately succeeding each mitosis—is quite wide between the different populations. However, there is much less variation in mass at the beginning of S phase, which is consistent with the idea that replica-tion is linked to the accumulation of some critical cell mass. In agreement with this idea, cells of populations which enter G1 from mitosis with a low mass spend longer before initiating S phase than populations which start the cycle with a greater mass. Such a mechanism might evolve to ensure that the cell population retains the same mass through successive generations to counteract any uneven distributions of mass between daughter cells at mitosis.

Segregation of Chromosomes at Division

Reorganization of Cell Structure at Mitosis

The evolutionary importance of a mechanism to ensure that a cell distributes its replicated chromosomes evenly to the two daughter cells of a division needs no emphasis. Chromosome segregation at mitosis is an accurate process; Luykx (1970) has calculated that the frequency of mitotic non-disjunction, when homologues fail to separate, is about 0·03 % in the mouse. This figure is a maximum estimate for the frequency of this aberration, although it does not include any errors which may result from failures of other types, such as spindle malformation.

The reorganization of the cell which is required to achieve this accurate segregation is indeed drastic. Its usual compartments of nucleus and cytoplasm disappear, to be replaced by a spindle which consists of fibres running from the two poles of the cell. The attachment of chromosomes to the spindle fibres by their centromeres and their subsequent movement towards the poles of the cell achieves their separation into two diploid sets of homologous chromosomes. In reviewing mitosis, Mazia (1961) has suggested that genes may be organised into a small number of chromosomes precisely in order to ensure this even segregation; in this sense, chromosome organization is a device for attaching many genes to few centromeres.

Mitosis represents the culmination of events which have taken place during the preceding growth cycle of the cell. It is generally thought that the decision of a cell to divide is taken before it enters S phase. During interphase, the cell doubles in size and both reproduces its chromosomes and also synthesises any components needed to form the mitotic apparatus. Mitosis is therefore the final step in completing the division. Little is known about the molecular interactions which lead to the initiation of mitosis, but there is some evidence that a mitotic inducer may be produced during G2. Accumulation of a critical level of this inducer may act as a signal that the cell is ready to divide; if a nucleus is forced to enter mitosis prematurely, its division may be abnormal and its chromosomes are likely to suffer damage (see page 403).

In contrast to the emphasis placed on the mechanics of mitosis by cytological observations, it is in fact a short interruption—usually occupying less than 5% of the division cycle—in the normal functioning of the cell. Protein and RNA synthesis cease during the process of division and DNA has previously been synthesised; the rearrangement of cellular structures therefore takes place in the absence of any macromolecular synthesis. Mazia (1961) has noted that mitosis can usefully be considered in two stages: establishment of the mitotic apparatus, which is completed by metaphase; and transportation of the chromosomes to the poles by this apparatus.

Early observations of mitosis in the last century suggested the concept that each chromosome of the diploid set splits longitudinally into two sister chromatids, which are then segregated to the daughter cells by the ensuing process of division. The term *prophase* was originally coined to describe an early stage of mitosis before visible division of chromosomes into sister chromatids and *metaphase* to describe the succeeding stage after visible chromosome division. Of course, we now know that the chromosomes are reproduced well before the beginning of mitosis; indeed, they may now be visualized as double structures from the beginning of prophase, when two elongated threads appear to be coiled around each other. During prophase, the chromosomes contract and in metaphase the condensed chromosomes become aligned along the equatorial plane of the cell. Figure 2.6 shows a whole mount electron micrograph in which condensing human chromosomes enter prophase as clumps of fibres with many interconnections.

The end of prophase and the beginning of metaphase—sometimes described as *prometaphase*—is marked by dissolution of the nuclear membrane and the appearance of faint lines running between the two poles of the cell. These *spindle fibres* appear to terminate in regions with a star like structure of many

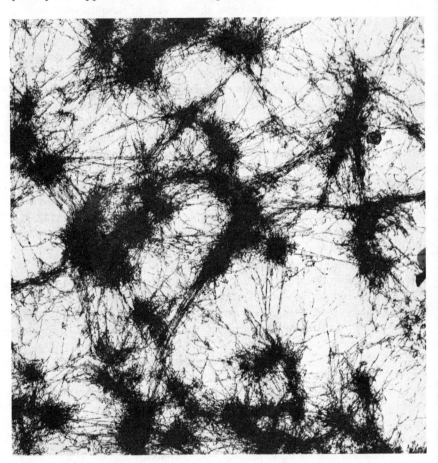

Figure 2.6: condensation of human chromosomes at prophase in a leucocyte, × 6700. Photograph kindly provided by Professor Ernest J. DuPraw; from *DNA and Chromosomes* (Holt, Rinehart and Winston, New York).

radiating lines (the asters). The structure of the cell at this time is said to comprise the *spindle*. *Anaphase* is defined by the period when the chromosomes move along the spindle fibres towards the two poles of the cell, one of each sister chromatid pair moving in each direction. *Telophase* describes the final stage when nuclear membranes form around the groups of chromosomes at each pole; this is a reversal of prophase, when the chromosomes return from the distinct individual structures to the network of interphase chromatin. After the separation of chromosomes has been completed, a progressive restriction takes place in the equatorial plane and this process, *cytokinesis*, completes the division of one cell into two. Cytokinesis seems to be functionally independent of the earlier mitotic events. Cell division has been reviewed in detail by Whitehouse (1969) and DuPraw (1968, 1970).

Formation of the spindle requires at least three different processes, which have been extensively reviewed by Luykx (1970). First, the macromolecular components which make up the spindle fibres must be synthesized and the centrioles must reproduce. The principal component of the spindle fibres appears to consist largely of one type of protein subunit. These protein chains must then be assembled into the larger asymmetrical structures of microtubules. Finally, the assembled structures must associate with each other and become orientated to form the spindle itself. These events take place successively and can be separated by inhibitors which influence particular parts of the cell cycle. For example, although dividing cells are inhibited in mitosis by the addition of colchicine or by a reduction in temperature, and consequently show no signs of orientated spindle components, their constituent subunits must have been synthesised; the spindle regenerates very rapidly when the inhibition is reversed, even though no RNA and protein synthesis is permitted.

A characteristic feature of the mitotic apparatus is the presence of the spindle fibres which run from pole to pole and attach to the chromosomes. These fibres were seen in stained sections of cells in early studies of mitosis and the controversy about their nature was settled only when Inoue (1953) observed the birefringence which they cause when living metaphase plates are viewed in polarized light. Using this technique, the birefringence pattern suggests that many fine filaments are organised parallel to the long axis of the cell (see Inoue, 1964).

The first suggestion that microtubules might be important in mitosis came from Ledbetter and Porter (1963); more recent work has been reviewed by Luykx (1970). Electron microscopy suggests that the spindle fibres are composed of microtubules and the microtubules correspond well in number and distribution with the intensity and pattern of birefringence seen in the polarizing microscope. Most of the microtubules are loosely organised in sheets and bundles—these are the spindle fibres observed by light microscopy. The microtubules located in different regions of the spindle are usually indistinguishable by the criteria which we can use at present (such as appearance in the electron microscope), but it is of course possible that there may be undetected chemical differences between them (for review see Margulis, 1973).

The mitotic apparatus can be extracted en masse from dividing sea urchin eggs; and asters, spindle fibres, centrioles are all visible in the preparation. Chromosomes are retained through at least the earlier stages of isolation. Mazia and Dan (1952) found that the mitotic apparatus behaves as a single unit in the isolation procedure and is relatively resistant to solution. The original methods of isolation relied upon extraction in either alcohol or detergents, but have been succeeded by protocols which include compounds with S–S bridges to protect the S–S bonds of the mitotic apparatus from reduction to –SH groups. Using these methods, it is necessary only to remove the membranes of the sea urchin eggs and to shake the cells in sucrose with dithioglycol; a structurally intact mitotic apparatus is then released (reviewed by Mazia, 1961).

The solubility of the mitotic apparatus depends upon the method used to isolate it. An estimate of its protein content has been made by Kane (1967), who found that it comprises about 0.5% of the total protein of the cell. By dispersing the apparatus in solution with compounds which reduce its S–S groups to free –SH groups, Sakai (1966) isolated a protein component which sediments at 2.5S and appears to have a molecular weight of about 34,700 daltons. This protein dimerises to a 3.5S particle of 67,000 daltons; the protein subunit has four free –SH groups and the dimer has 6 free –SH groups and one S–S bridge. One of the characteristics of this protein is that it is insoluble if Ca^{++} ions are present; this provides an assay to distinguish it from other components of the mitotic apparatus (see Bibring and Baxandall, 1968).

The 2.5S protein may be the basic unit which makes up the microtubules which in turn associate to form the spindle fibres. Microtubules of the mitotic apparatus are long, relatively straight cylindrical structures with a diameter which varies between 150Å and 250Å depending upon the method of fixation. When the microtubules are split open lengthways, their walls are seen to consist of 12–13 filaments. Using mitotic newt cells burst on a water surface, Barnicott (1966) showed that each filament consists of a series of globular particles of about 35Å in diameter. Kiefer et al. (1966) have suggested that each of these globular subunits may represent one of the 2.5S protein chains.

Spindle fibres may have a contractile action, so that it is interesting that the amino acid composition of the microtubule protein resembles that of actin. However, little is known about the interactions between protein subunits in the microtubule, apart from observations on the role of S–S bridges. The importance of S–S bridges in maintaining the structure of the mitotic apparatus is underlined by the effect of S–S compounds in ensuring isolation of an intact apparatus and of –SH compounds in making its components soluble.

That changes from –SH groups to S–S bridges are implicated in cell division is also suggested by the observation of Mazia, Harris and Bibring (1960) that the addition of mercaptoethanol blocks the division of sea urchin eggs. In a review of the structure of the mitotic apparatus, Sakai (1968) has noted that the total –SH group content of sea urchin eggs appears to remain constant throughout cell division. However, a protein of the cell cortex appears to be able to undertake an exchange reaction with the microtubule protein in which S–S bridges and –SH groups suffer an intermolecular rearrangement.

Sakai suggested that this interchange may be catalysed by an enzyme and may control the assembly of the apparatus.

Immunological assays have shown that the protein subunits of the mitotic apparatus are present in unfertilised sea urchin eggs, which supports the idea that mitosis depends upon the assembly of previously synthesised components into the apparatus and not upon their production at this time (see Mazia, 1961). Studies of the binding of colchicine to cells provide further evidence for this concept. The effect of colchicine in causing cells to accumulate in a metaphase-like state results from its interaction with microtubules (see Rizzoni and Palitti, 1973). Brinkley, Stubblefield and Hsu (1967) found that when colchicine is removed from inhibited cells, the spindle regenerates and spindle fibres appear between the separating centrioles. The components of the microtubules must therefore be present in the cell in an inert state.

By using H^3-colchicine, Borisy and Taylor (1967a, b) have shown that its binding to cells correlates with their production of microtubules. A protein component sedimenting at 6S in Hela cells binds the labelled inhibitor; a similar protein is found in sea urchin eggs. That this may be a component of microtubules in general, and not solely of those in the mitotic apparatus, is suggested by the isolation by Shelanski and Taylor (1967) of a 6S colchicine-binding protein from microtubules of sea urchin sperm tails. By incubating H^3-colchicine with Hela cells growing on C^{14}-leucine, Robbins and Shelanski (1969) have followed the time of synthesis of the protein activity which binds the inhibitor. There is no association between the tritium label and a C^{14}-leucine label given at metaphase. However, the extent of binding of H^3-colchicine to C^{14}-labelled proteins during interphase is proportional to the total protein synthesis of the cell. This implies that the basic protein component of the microtubules accumulates continuously throughout interphase, but is assembled into the mitotic apparatus only at the time appropriate for mitosis.

Establishment of the Spindle

The characteristics of the metaphase plate vary with the type of cell. But all mitotic cells contain two spindle poles, the regions towards which the anaphase chromosomes migrate. Each pole of the spindle in animal cells (and in some lower plant cells) typically is identified by the presence of a pair of centrioles, occasionally only a single centriole. We shall consider here only those poles organised around typical centrioles.

The centriole is a small hollow cylinder about 0·5–0·7 μ in length and some 0·25 μ in diameter. The wall of the cylinder consists of nine "triplet tubules" arranged in a dense matrix. As described by Stubblefield and Brinkley (1967), this gives a pinwheel appearance when viewed in cross section. Figure 2.7 shows that each of the nine triplets consists of three microtubules fused together. The three microtubules of each triplet lie almost in one plane, so that each triplet forms a blade which runs the full length of the centriole.

Ross (1968) has noted that only the inner microtubule of each triplet is strictly circular.

The centriole together with its surrounding material, which sometimes appears as a halo of dense granules, is known as the *centrosome*. The term *centrosphere* describes the region around the centrosome, including microtubules and membrane material. The *aster* includes the centrosphere and the

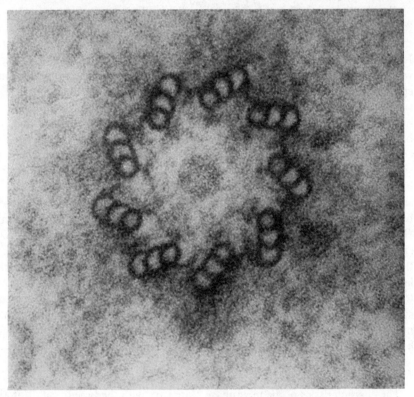

Figure 2.7: centriole of human lymphocyte × 240,000. Photograph of Ross (1968).

outlying radial arrangement of microtubules, membranes, rows of mitochondria and other components. The astral rays are often a predominant feature in light microscope studies of mitotic cells.

When two centrioles are present at a pole, they often lie in a plane approximately perpendicular to the spindle axis, although the axis of the centriole itself does not appear to be related to the axis of the spindle. During prophase, the spindle develops from the region of the centrosome, but it is likely that only one of the two centrioles is involved in its establishment. The position of the centrioles varies in cells of different species. In all cells with centrioles, however, the polarity of the spindle is determined either by them or by material associated with them. In cells which lack centrioles, the spindle microtubules must presumably interact with some other cellular component. It is probable that all eucaryotic cells use the same few basic devices for establishing the poles of the spindle; apparent variations in morphology may result from differences in cellular organization rather than reflecting fundamentally different mechanisms of division.

Centrioles appear to be capable of self replication; Brinkley and Stubblefield (1970) have observed that they reproduce by a mechanism in which a single daughter centriole is assembled alongside the parent centriole. One possible mechanism for this self replication is for each part of the parental structure to be duplicated and the copy then transported to the site nearby where the daughter centriole is under assembly. However, DuPraw (1970) has noted that centrioles may be assembled de novo in the absence of pre-existing centrioles in several cell types. For example, Turner (1968) has discussed the occurrence of centrioles in plant cells and has demonstrated de novo assembly in at least one instance.

The reproduction of centrioles must, of course, be linked to the cell cycle. Daughter centrioles first become visible about some 6 hours before mitosis as short cylinders orientated perpendicularly to the parental centrioles and close to their proximal ends. By the time of mitosis, most of the characteristic features of the centriole can be seen and the daughter assembly has reached about half its final length. It remains in this form during mitosis, so that each daughter cell usually receives two centrioles when the nucleus reforms about the two chromosome sets at telophase. One of these centrioles is the mature assembly which has provided the functions of the spindle pole; the other is the immature assembly which has played no role in the preceding mitosis. The immature centriole is presumably converted to its mature form at an early stage of the cell cycle, so that the cell has two mature centrioles, each of which may then begin to direct assembly of a daughter. Working with Hela cells, Robbins, Jentzsch and Micali (1968) have observed that the two centrioles commence their reproduction at about the same time that DNA synthesis begins, assembly ending by the completion of S phase. Working with L cells, Rattner and Phillips (1973) observed that centriole duplication does not depend upon DNA synthesis but may start during G1. Figure 2.8 shows the reproduction of centrioles during the cycle.

The idea that centrioles undertake self reproduction is difficult to assess. The principal component of their constituent microtubules is, of course, protein; but both RNA and DNA also seem to be associated with the centriole structure. It is possible that some structural information about their assembly is reproduced by a template like process which relies upon the structure of the pre-existing parental centriole. (This need not mean that the parental centriole itself divides, but simply that it is used to direct formation of the daughter centriole.) This would imply that some genetic information does not reside in DNA, but is contained in some manner in the structure of the centriole.

On the other hand, the process of centriole assembly may be more akin to the self assembly from macromolecular components which is displayed by ribosomes (see chapter 4 of volume 1). That this is so in at least some instances is argued by the de novo assembly of centrioles in some cells. The function of the DNA which is found in centrioles is not clear. However, the presence of

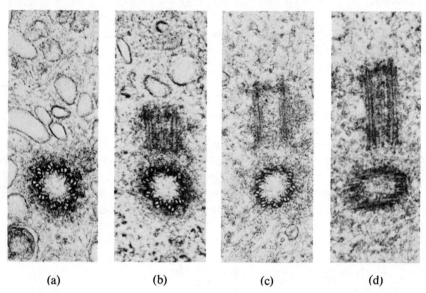

(a)	(b)	(c)	(d)

Figure 2.8: centriole reproduction during the cell cycle. (a) early G1; (b) S phase; (c) prophase; (d) metaphase. × 60,590. Data of Rattner and Phillips (1973).

DNA appears to be a feature common to both centrioles and structures related to them—kinetosomes—which are found as basal bodies in cells which have cilia (see Randall and Disbrey, 1965; DuPraw, 1970). The evolutionary and functional relationship between centrioles and kinetosomes is not known. However, one speculation has been that the DNA contains some information needed for reproduction of these assemblies, although there is at present no evidence to show whether it is transcribed and translated.

Function of the Centromere

Centromeres are not observed as visible structures in interphase nuclei, so they must develop as morphological entities during the early stages of chromosome preparation for mitosis and meiosis. The centromere region is one of the most conspicuous features of linear differentiation along the chromosome. Possession of a centromere is essential if a chromosome is to behave properly on the spindle; chromosome fragments which lack a centromere behave abnormally at cell division and may be lost from the spindle or left in the equatorial region. Mazia (1961) has observed that in fact the only essential part of the chromosome with regard to mitosis is the centromere.

The centromere of each chromosome attaches to the spindle fibres so that the chromosomes line up on the equatorial plane of the cell at metaphase and move towards the poles at anaphase. In some preparations, especially those fixed with OsO_4, the centromere region of each chromatid appears to contain a small intensely staining granule, sometimes termed the spindle spherule. Spindle fibres appear to attach to this granule rather than to the whole region included in the primary constriction. The term *kinetochore* is often used to describe the points on the chromosome to which the spindle fibres attach—loosely speaking, the terms centromere and kinetochore are used to describe the same region, but centromere is generally understood in a wider sense to include the whole area of primary constriction whereas the kinetochore comprises only the restricted part of it which attaches the chromosome to the spindle.

Whole mount preparations of chromosomes show that the number of chromatin fibres in the centromere region is much reduced, the reduction presumably generating the primary constriction. The model of DuPraw (1970) illustrated in figure 1.13 shows the kinetochore as a flat circular structure of diameter $0·2-0·25\ \mu$ and width $0·06-0·1\ \mu$ located within the primary constriction. In plant cells, and at early stages of mitosis in mammalian cells, the kinetochore may appear instead as a spherical structure of about this size. This model attributes the two functions of the centromeric region to different structural entities; maintaining the attachment of the two sister chromatids is mediated by chromatin fibres running between them, whereas attachment to the spindle fibres is the prerogative of the kinetochore alone.

The kinetochore first appears as a patch of condensed filaments packaged in irregular bundles at a small constriction on each side of the chromatid pair at early prophase. According to the studies reviewed by Brinkley and Stubblefield (1970), each patch is a spherical structure about $0·4\ \mu$ in diameter, apparently consisting of filaments 50–80Å in diameter. Even during early prophase the sister centromeres are probably structurally separate from each other, and they appear to develop synchronously during the subsequent stages of division. As prophase continues, there is a transition in centromere structure from these dense spherical masses to give less dense bands which extend for $0·1-0·5\ \mu$ one along each side of the duplicated chromosomes.

Studies of the ultrastructure of the centromeric region at this stage, such as those of Jokelainen (1967) with rat cells, reveal a dense inner layer some $0·15-0·30\ \mu$ thick of fibres which are continuous with the main body of the chromosome. Next there is a middle layer of about the same thickness but of low electron density. This in turn is surrounded by an outer layer, which may appear amorphous or may be seen to consist of fine filaments. This represents the kinetochore and is usually about 0·3–0·45 μ thick. Its underlying structure is controversial and has variously been interpreted as composed of discs or of coiled coils. Longitudinally, these layers appear as bands most commonly of length 0·2–0·25 μ at the surface of the chromosome.

The orientation of the kinetochores appears to be responsible for their characteristically different behaviour at mitosis and meiosis. In mitosis, sister kinetochores separate on the spindle so that homologous chromosomes move to different poles. This may result from the orientation of the kineto-chores on opposite sides of the sister chromatids so that one faces each pole. Figure 2.9 shows the attachment of spindle fibres in the kinetochores on either side of a pair of sister chromatids.

In the first division of meiosis, by contrast, the two sister kinetochores of each homologue pair of the bivalent do not separate but move together to one pole. The two sister kinetochores of the other homologue pair move together to the other pole. Then at the second division the sister kinetochores of each homologue pair separate as they do at mitosis. This behaviour may be the consequence of the orientation of the kinetochores, for at the first meiotic division the two kinetochores of each chromatid pair lie together on the same side of the homologue pair. By the second division, the sister kinetochores have separated so that they lie on opposite sides of the chromosome. The switch from adjacent to opposing orientations may be largely responsible for the difference in behaviour of the chromatids at the two divisions, although the molecular basis for this effect is unknown.

Connection of microtubules to the kinetochores usually commences at about the time of dissolution of the nuclear membrane. In micromanipulation experiments with spermatocytes of the grasshopper Melanoplus, Nicklas and Staehly (1967) and Nicklas (1967) found that the application of pressure with a needle tip can stretch the meiotic bivalent without detaching it from the spindle. This shows that the chromosome is firmly attached to the spindle and this firm attachment continues until the end of anaphase. Experiments in which a bivalent could be reversed in orientation by a 180° turn confirmed that chromosomes move towards the pole which the kinetochore faces.

Chromosomes usually move individually during mitosis so localised forces must act on individual spindle fibres and kinetochores. During the history of cytology, many types of long range or short distance forces have been postulated to act on chromosomes during mitosis. Mazia (1961) has dispelled this mystical view of chromosome movement and observed that the contractile forces which act through the microtubules attached to the kinetochore can provide an adequate explanation. A detailed study of chromosome movements on the spindle has been made by Nicklas and Koch (1972); for review see Nicklas (1970).

The force acting on the centromere appears to be a pulling one—force is

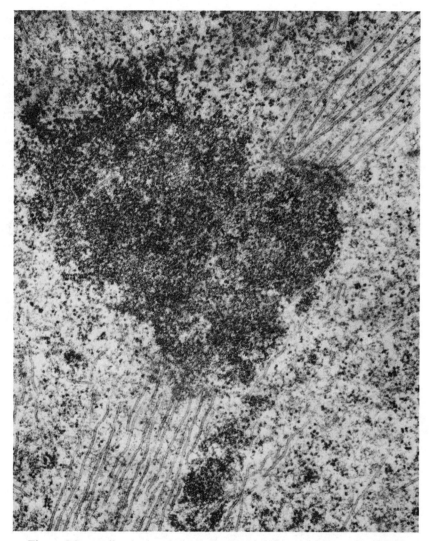

Figure 2.9: small subtelocentric chromosome of rat kangaroo fibroblast at metaphase. Microtubules are attached to both sister kinetochores. Magnification × 51,000. Data of Brinkley and Stubblefield (1970).

applied from the same direction in which the chromosomes are to move. This force varies in magnitude during division; in general, it seems to be large as metaphase starts, declines gradually during metaphase and then increases abruptly to mark the onset of anaphase. These same forces also exert some effect on other particles in the cell: chromosome fragments, mitochondria, persistent nucleoli may move towards the poles. The chromosomes follow an irregular path as they line up on the equatorial plane during metaphase; this alignment is generally thought to be the consequence of equal and opposite forces applied from the two poles on each chromatid pair (reviewed by DuPraw, 1968).

Anaphase is defined by the movement of the chromosomes towards the poles. The velocities of chromosome movement at this time are usually of the order of 0·2–5·0 μ per minute; the speed of movement seems to remain constant during anaphase. Since the distance between the poles of the mitotic cell is usually of the order of 10–30 μ, movement occupies between 2 and 60 minutes depending on the cell type (reviewed by DuPraw, 1968; Forer, 1969). Only the centromere shows active movement, the rest of the chromosome

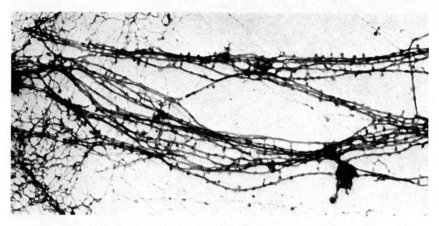

Figure 2.10: whole mount electron microscopy of mouse oocyte metaphase I chromosomes. Microtubules running horizontally insert into the chromatin fibres at the left. Magnification × 22,240. Photograph of Burkholder et al. (1973).

appearing to be dragged passively by virtue of its attachment to the centromere. This behaviour is readily deduced from the characteristic shapes which chromosomes acquire during mitosis and from the unusual configurations which are acquired by abnormal chromosomes, such as fusion products with two centromeres. In itself, this argues that the movement of chromosomes relies upon the properties of the centromere alone.

Spindle fibres fall into two general classes: those which are attached to various parts of the chromosomes or which run from pole to pole through the chromosomes; and those which are attached to the kinetochore. The former have been termed *continuous fibres* and the latter are described as *kinetochore* (or *chromosomal*) *fibres*. Figure 2.10 shows a whole mount electron micrograph of a mouse bivalent in the first metaphase of meiosis; the groups of microtubules insert into the chromosome at the left edge where they intermingle with fibres of chromatin. In addition to the kinetochore fibres, some fibres may attach to other parts of the chromosomes and these are known as *neocentric fibres*.

The kinetochore fibres do not appear to be contractile in nature, for they seem to move towards the poles at the same rate as the chromosomes themselves. The spindle microtubules therefore seem not to generate the mitotic force themselves, but serve rather to orientate the elements that do so—their role is to transmit the forces pulling the chromosomes to the poles. The poleward force acting on each centromere appears to be proportional to the distance from the pole, that is it depends upon the length of the kinetochore fibre.

Because cellular particles move towards the poles at metaphase, whereas the chromosomes remain in stationary alignment on the equatorial plane, it seems likely that the cellular component which generates the pulling force acts through all the microtubules extending along the spindle axis, including both kinetochore and continuous fibres. At this stage, the chromosomes may be immune from movement because of the equal and opposite attractive forces applied by each pole. Luykx (1970) has suggested that throughout metaphase spindle fibre protein is continually assembled and added to the microtubules at the kinetochore, to alleviate the pulling forces.

Since the first spindle fibres to appear in the cell at mitosis are associated with individual chromosomes, and bunches of fibres may appear to elongate from the region of the centromere, Luykx has proposed that this region may be regarded as a "kinetochore organiser", in some respects analogous to the nucleolar organiser. In their reviews of chromosome movements at mitosis, Bajer and Mole–Bajer (1970, 1972) have suggested that the kinetochore fibres stop growing at anaphase, with the result that the chromosomes move towards the poles. A prerequisite for this movement is also that the forces holding the two sister chromatids of each pair together cease to do so. Since the action of the spindle fibres is to transmit a pulling force rather than themselves to contract, the microtubules must shorten in length as the chromosomes approach the poles. This disassembly is presumably undertaken in the polar regions. And as a counterpart, any fibres which may happen to link a centromere to the more remote pole—that is the one from which it is moving away—must simultaneously be elongated. The assembly and breakdown of the mitotic apparatus in rat kangaroo cells has been reported by Roos (1973).

The function of the centriole in controlling the spindle fibres is unknown. It may itself direct the synthesis of spindle macromolecules, may be involved in the assembly of protein subunits into microtubules and their association into orientated spindle fibres, or may play only an indirect role in these processes. Microtubules of the mitotic apparatus appear to be attached to the base of the centriole, running perpendicular to its long axis. By contrast, the microtubules of the wall of a cilium appear to be direct extensions of those in the triplet of the kinetosome and therefore run parallel to the long axis.

Although all these microtubules appear to be of similar construction and chemical composition, there may therefore be several ways in which they can be assembled and attached to the centriole (or to the kinetosome). Brinkley and Stubblefield (1970) have presented a model for microtubule formation in which they are extended through assembly at the centrioles.

Chromosome Recognition in Meiosis

Segregation of Genetic Material in Gametes

Meiosis comprises a special form of nuclear division in which the progeny cells contain only a haploid complement of chromosomes instead of the diploid set found in the somatic cells of the adult. Each meiotic event produces four haploid germ cells. Prior to the meiosis, the diploid forerunner of the germ cells replicates its chromosomes, so that the cell enters division with the same two diploid sets with which it enters a mitosis. The four copies of each chromosome are distributed to different germ cells by two successive divisions, sometimes called *meiosis I* and *meiosis II*. The first division generates two diploid sets of chromosomes; the second division sees the production of two haploid genomes from each of these diploid sets.

The first division of meiosis is characterized by a very lengthy prophase, which may take upto several weeks in some organisms. At the beginning of meiosis, the chromosomes are seen as very fine, extended threads. During this stage, *leptotene*, they often have the appearance of beads on a string in which the chromomeres are clearly displayed. Leptotene cells appear under the light microscope to have two haploid sets of chromosomes, with only one thread corresponding to each homologue; although each chromosome has in fact replicated before the start of meiosis, their reproduction is therefore not apparent from their appearance at this early stage.

Zygotene describes the stage when homologous threads approach each other and pair side by side along some of their length. This *synapsis* increases until the chromosomes are paired along their entire length; the paired structures are called *bivalents*. Synapsis is completed by the stage of *pachytene* when the chromosomes appear as somewhat thicker threads.

The double nature of the two components of each homologous pair is clearly displayed at the subsequent stage of *diplotene*, when the homologues separate along much of their length but remain attached at certain restricted points. At this time, each of the homologues can be seen under the microscope to consist of two threads (chromatids); the overall structure of each bivalent therefore comprises four chromatids. Each chromatid represents one copy of the chromosome, containing the appropriate duplex of DNA.

The points where the homologues fail to separate from each other in the diplotene cell represent the *chiasmata* at which genetic exchange appears to take place. The appearance of the chiasmata suggests that they represent sites where the different chromatids are tightly associated with each other. The last stage of the extended prophase is *diakinesis*, when the tetrad structures coil up to form shorter and thicker chromosomes. Chiasmata which at diplotene are located at sites between the centromere and the telomere appear to move progressively towards the ends of the bivalent at this time in the process described as terminalization. Figure 2.11 shows two mouse homologues

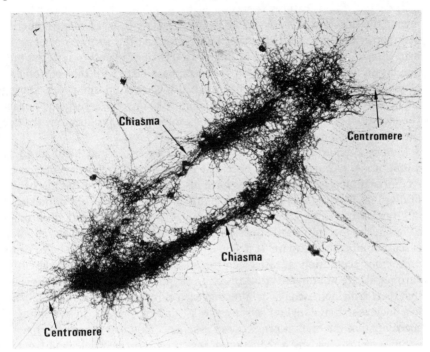

Figure 2.11: whole mount preparation of mouse oocyte metaphase I chromosome with complete terminalization of chiasmata. The positions of the centromeres may be inferred from the attachment of the micro-tubules to the chromosome; the positions of the chiasmata are marked by the passage between the non-sister chromatids of chromosome fibres. Magnification × 7,644. Photograph of Burkholder et al. (1973).

held together at the end of metaphase by terminal chiasmata. Chromatin fibres cross between non-sister chromatids at these locations (Burkholder et al., 1972).

During diakinesis any nucleoli disappear and the nuclear membrane disintegrates so that prophase I is succeeded by metaphase I. Each bivalent become aligned at the equatorial plane between the poles. A feature which marks a change in chromosome behaviour compared with mitosis is that the centromeres of sister chromatids lie adjacent to each other instead of opposed. Each bivalent thus consists of four chromatids organized in two chromatid pairs, the two centromeres of one pair facing one pole and the two centromeres of the other pair facing the opposing pole. During anaphase I, the two chromatid pairs of each bivalent therefore move towards the opposite poles of the spindle. At telophase I, nuclear membranes form around each of the diploid chromosome groups at the two poles.

A short interphase when the chromosomes temporarily appear to lose some of their individual features is succeeded by prophase II which leads to the second division. This in turn gives way to metaphase II, when the two nuclear membranes dissolve and two new spindles form. At this second division, the behaviour of the chromosomes is similar to their movements in mitosis: they line up on the equatorial plane of the metaphase plate with the centromeres of the sister chromatids of each pair aligned towards opposite poles. During anaphase II, the homologues separate and move to the poles; during telophase II a membrane surrounds each of the four haploid nuclei produced by the two successive divisions.

Meiosis is under genetic control and its characteristics may be changed by mutation. A parameter often observed is the overall frequency of genetic recombination. However, changes have also been detected in the mechanics of division. Davis (1971) isolated a meiotic mutant from natural populations of Drosophila which causes a high frequency of non-disjunction of all chromosomes at the second meiotic division of either male or female. The mutation is semi-dominant and maps at a single locus. Its action seems to be to cause precocious separation of sister centromeres, with the result that the sister chromatids are not directed towards opposite poles at metaphase. This means that there is an increased probability that each chromosome fails to be included in a nucleus so that daughter cells receive 0, 1 or 2 chromatids. It is tempting to speculate that the mutant gene codes for some component of the centromere.

Since meiosis leads to a reduction of the chromosome complement from its diploid state to a haploid condition, the germ cells contain only one allele representing each genetic locus. The segregation to different gametes of the homologous copies of each chromosome therefore means that if a parent is heterozygous, each germ cell has an equal chance of inheriting either version of the gene. A genetic reassortment of the genes of different chromosomes is provided by the independence of the movements of the copies of the different chromosomes of each set (which member of one bivalent enters any particular germ cell does not influence the behaviour of any other bivalent) and the later association of sperm and egg to regenerate the diploid cell.

The critical events in genetic recombination between genes carried on the same chromosome take place during the extended prophase of the first meiotic division. The result of a recombination event is to produce reciprocal copies of a chromosome, each of which has its genetic information derived in part from one of the parental homologues and in part from the other. That the formation of genetic recombinants is accompanied by a physical exchange of genetic material was suggested by early experiments using chromosome translocations; in crosses when the two homologues of one parent carry different genetic information and can also be distinguished from each other by cytological means, the formation of recombinant genetic classes is always correlated with the formation of new cytological classes.

The construction of linkage maps implies that each chromosome can be regarded as a linear array of genes in which those located close to each other tend to remain together through a mitosis, with the probability of recombination increasing with their distance apart. The characteristics of chiasma formation parallel those of genetic recombination; and this led to the suggestion that the chiasma represents the point where homologues suffer breakage and reunion. The likelihood of chiasma formation occurring between two chromosomal loci depends upon their distance apart. Chiasma formation exhibits the same positive interference shown by genetic recombination; the occurrence of one event reduces the probability of another in its immediate vicinity. Chiasma formation and genetic recombination are under genetic control and although characteristic for any species may be altered by mutation; the frequencies of chiasma formation and genetic recombination are generally correlated (reviewed by Henderson, 1969).

Genetic analysis shows that recombination must take place at the four strand stage of meiosis, with only two of the four chromatids involved in any given exchange event; the other two chromatids suffer no interruption in their integrity (extensively reviewed by Whitehouse, 1969; see also chapter 12 of volume 1). When more than one chiasma is formed in a chromosome bivalent, each event is independent in the sense that at each site any two of the four chromatids may be implicated in the breakage and reunion. The two strands which exchange genetic material at any chiasma may thus represent either those from the different homologues (causing genetic recombination) or those from sister chromatids (which does not of itself cause genetic recombination).

Formation of the Synaptonemal Complex

A crucial feature of genetic recombination is its precision. Any inaccuracy would change the nucleotide base pair sequence of DNA, but breakage and reunion takes place between two duplex molecules of DNA at exactly the same site in each. The molecular processes which allow two DNA molecules to break and rejoin at corresponding base pairs appear to make use of base pairing between complementary single strands of DNA and are discussed in chapter 12 of volume 1. In eucaryotic cells, the breakage and reunion event itself is preceded by synapsis of homologous regions and this pairing is a function of the chromosome structure as such rather than of its DNA content.

Chromosome pairing takes place in three stages. First, the homologous chromosomes (each of which has previously replicated) approach each other. We know virtually nothing about the attractive forces which must ensure that only corresponding chromosomes become associated. Then the two chromosomes become intimately associated with each other in the form of a *synaptonemal complex*, which is first generated at isolated points, usually including the telomeres, and then extends until the entire length of each chromosome is tightly synapsed. Formation of this complex brings the homologous regions of each chromosome into close apposition. However, the distances between corresponding DNA duplexes are still appreciable and the last stage prior to genetic exchange must be to bring the homologous polynucleotides into intimate contact; we know little about whatever mechanisms may be utilised to achieve this recognition.

The synaptonemal complex is visible only in the electron microscope; light microscopy reveals that the homologous chromosomes have become closely associated but lacks sufficient resolution to define their structure. Since the exact stage of meiosis is not apparent from electron micrographs of chromosomes in prophase I, cells must be followed by both light and electron microscopy to correlate the formation of the synaptonemal complex with progress through division. The complex was first observed by Fawcett (1956) and Moses (1956, 1958) as two dense parallel lines, *lateral elements*, separated from each other by a less dense central region containing a somewhat finer dense filament, the *central element*, running parallel to the two outside lateral elements. The triplet of parallel dense strands lies in a single plane that curves and twists along the axis of the chromosome. Moses (1968) observed that the two homologues of each bivalent lie one on each side of the complex, each associated with one lateral element. The complex often terminates on the inner membrane of the nuclear envelope. The synaptonemal structure can be found only in pachytene cells and disappears at the early stages of disjunction during diplotene.

As figure 2.12 shows, each lateral element has a diameter of about 400–500Å; the two elements are separated by some 1200Å. The central element in the middle of this region is usually about 200Å in diameter; however, it has a more tenuous appearance and is not visualized as clearly as the lateral elements. It seems to consist of fine filaments of a diameter less than 70Å. Moses and Coleman (1964) observed that each homologue consists of a network of chromatin fibres which are associated with the lateral elements. The sister chromatids of each homologue are not visible in the complex and each lateral element and its associated chromatin appears to comprise only a single structure. Moses (1968) noted that the synaptonemal complex joins homologous chromosomes over their whole length, including the centromeric regions, and extends without interruption from euchromatin into heterochromatin. The structure of the complex has been reviewed in detail by Westergaard and Wettstein (1972).

Although the triplet of parallel strands is the form typically taken by the complex in many types of cell, other classes of structure may be seen at synapsis. In some insects, the complex appears to consist of multiple elements, whose structure and function have been discussed by Moens (1969b). During the later stages of prophase in many mammals, the XY bivalent of male cells forms a structure sometimes termed the sex vesicle. Under the light microscope this appears as a dense body comparable to the nucleolus. Solari (1970a) has described the formation of the sex vesicle of the mouse, which under the electron microscope appears as two condensed homologous elements, larger in structure than the lateral elements of the synaptonemal complex, and in general not paired with each other. The different structures seen in meiotic cells may reflect a variation in the assembly of components similar to those which constitute the typical synaptonemal complex.

The components of the complex first become visible during leptotene when each unpaired chromosome—that is sister chromatid pair—takes the form of a network of chromatin fibres associated with a single axial element. Moens (1968) has described the approach and pairing of homologues in meiotic cells of the lilly, in which the leptotene chromosomes are loose bundles of fibres of diameter about 100–200Å. A longitudinal structure appears to develop in the centre of each homologue. Using whole mount preparations, Comings and Okada (1971b,c) have found that the chromatin fibres are attached to the axial elements as a series of lateral loops which occur in bunches; according to this interpretation, the chromomeres seen under the light microscope at leptotene represent the regions where the loops are concentrated on the axis.

Both ends of each unpaired leptotene chromosome are usually attached to the nuclear envelope. The telomeres of homologues may be attached at sites near to each other and their relationship may be a preliminary step which helps homologues to pair. The axial elements of homologous chromosomes begin to pair with each other during zygotene. Synapsis starts when some unknown mechanism brings about an approximate alignment of the two chromosomes to within some 3000Å of each other. Precise pairing to within 1000Å takes place as the synaptonemal complex forms, often by a zipper-like movement from the two telomeres. The formation of the complex of the mouse has been described by Solari (1970b) and extensively reviewed by Comings and Okada (1972) and Westergaard and Wettstein (1972).

The structure of the lateral and central elements is not well defined. However, Comings and Okada (1972) found that if leptotene cells are treated with DNAase the axial elements remain intact although the chromatid threads are digested. The lateral and central elements of the synaptonemal complex show the same resistance to DNAase; that they consist of protein is suggested by their rapid destruction with HCl, urea or trypsin. In contrast to earlier reports that these elements are resistant to RNAase, Westergaard and Wettstein (1972) observed that ribonuclease can degrade them; they therefore suggested that they may consist not only of protein but may also contain RNA. The demonstration that these elements do not contain DNA shows that they are

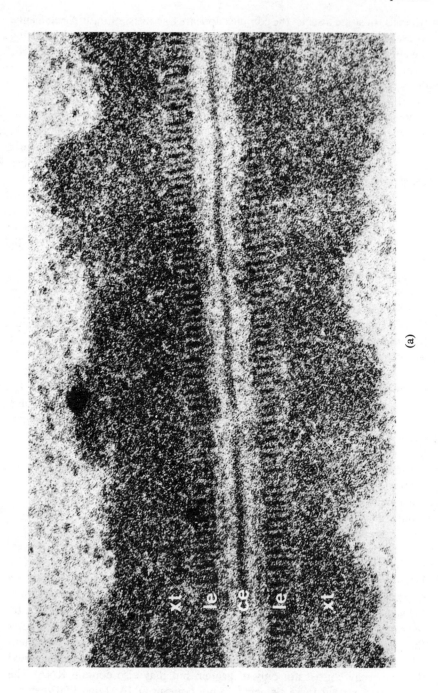

(a)

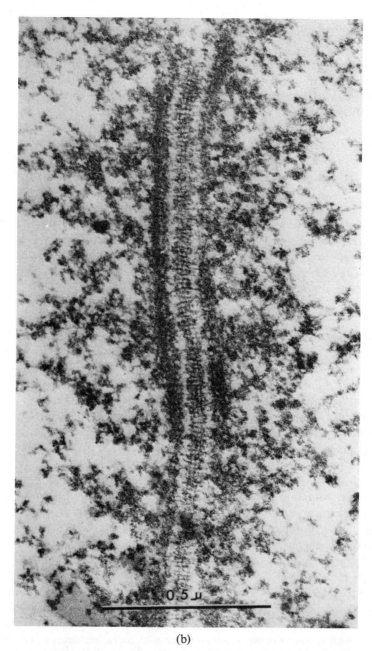

(b)

Figure 2.12: structure of the synaptonemal complex. (a) Section through a bivalent of Neottiella at pachytene. The lateral element (le) consists of alternating thick and thin bands embedded in the surface of the chromatin (xt). The diameter of the lateral element is about 500Å. The electron dense central element (ce) has a diameter of about 180Å. The distance between the two lateral elements is about 1200Å. Magnification × 61,200. Photograph of Westergaard and Von Wettstein (1970). (b) Electron micrograph of section through complex of Bombyx mori. Photograph of King and Akai (1971).

not constituted from the chromatin fibres and do not represent a rearrangement of the usual chromosome structure. Both lateral and central elements seem rather to be structures independent of the chromosome threads themselves, but associated with them in the synaptonemal complex.

The central element appears to be formed by filaments which extend from the two lateral elements to fuse in the central region (Moens, 1968; Solari and Moses, 1973). In their review of the structure of the synaptonemal complex, Comings and Okada (1972) have presented a model for its formation based on this interpretation. An alternative raised by Westergaard and Wettstein (1972) is that the central element may be synthesised elsewhere and subsequently assembled at the chromosomes to provide a structure which the lateral elements utilise to associate with each other. That there are some differences between the protein components of the lateral and central elements is suggested by the observation that only the former react with ammoniacal silver ions, and must therefore be more basic.

The precise appearance of the complex in whole mount preparations seems to depend upon the conditions of preparation. Solari (1972) observed that spreading on water causes a loss of the central element and much of the lateral elements, but leaves the chromatin fibres intact. This supports the concept that they consist of different components. By using appropriate salt conditions for spreading, Solari distinguished different types of material in the lateral element. The regions where chromatin fibres attach and the longitudinally running filaments of 65Å diameter are disrupted by DNAase. But the underlying bulk material, consisting of tightly packed fibrils, remains intact. The central element appears to consist of these fibrils alone. This suggests a model in which protein fibrils constitute the sole component of the central element, and form the major component of the lateral element in association with fibres which may contain a small amount of DNA. The lateral elements remain as single structures in different whole mount preparations, although Solari found that hypotonic treatment caused a loosening of the bulk material which could generate a spurious double appearance. This may explain observations suggesting that the lateral elements might be double in nature.

Some movement of the lateral components relative to the chromatin fibres seems to take place when the axial elements of unpaired chromosomes associate to become the lateral elements of the synaptonemal complex; Westergaard and Wettstein (1972) observed that the axial elements are located in the centre of the chromatin material, but the lateral elements lie to one side so that the chromatin of each homologue lies essentially on the outside of the synaptonemal complex.

The traditional view of chromosome pairing derived from light microscope studies is that only two homologues are paired at any point. Using triploid cells of the lily, Moens (1969a) found that the switch of pairing partners seen in light microscopy is accompanied at the level of the electron microscope by a switch of association of lateral elements. But in other instances, the axial elements of all three chromosomes appear to come into close contact. Comings and Okada (1971a) found that meiotic cells of triploid chickens may display synaptonemal complexes which have three lateral elements separated by two central elements. This shows that some variation in the normal structure of the complex is permitted when it is assembled. In general, synapsis takes place only between homologues and appears specific; for example, only a restricted length of the X and Y chromosomes may pair during mammalian spermatogenesis. On the other hand, Comings and Okada (1972) noted that the complexes can be found in haploid plants and in hybrids between two plants which are sufficiently unrelated to be unable to suffer genetic crossover. The specificity of formation of the complex may therefore vary and at present we do not know how appropriate chromosome sequences recognise their homologous regions.

Chiasma formation is not observed at pachytene in either the light or electron microscope. When the pairing of homologues is terminated and replaced by repulsion at early diplotene, the synaptonemal complex is shed from the bivalent, except at the chiasmata. Synaptonemal complexes generally disintegrate gradually during diplotene. First the central element is lost, then the fibres which apparently cross from the lateral element to the central element. In many organisms the separated axial elements disappear at once, although in some instances they may return to their previous position in the centre of the chromatin fibres before dissolution. At early diplotene, chiasmata seem to consist of short stretches of the synaptonemal complex which have been retained. However, we do not know whether the chiasma holds the bivalents together and thus protects the synaptonemal complex at these sites, or whether it is the retention of the synaptonemal complex which protects the chiasmata at this time. Whatever the order of events, however, it is clear that the homologous chromosomes remain intimately associated at these sites which presumably identify the locations where DNA has suffered breakage and reunion.

Although formation of a synaptonemal complex is essential before crossing over can take place, the complex represents an association of homologous regions of the chromosome and not of DNA itself. Meiotic chromosomes appear extended at prophase but, of course, are still many times shorter than the length of DNA which they contain. For example, Westergaard and Wettstein (1972) quoted a total length of 50 μ for the synaptonemal complexes of Neurospora, which must contain 16,000 μ of DNA. The packing ratio of DNA in the complex may therefore be several hundred. Even once appropriate regions of the chromosome have synapsed, therefore, it is still necessary for exactly corresponding sequences of DNA to match each other before genetic recombination can be initiated. We do not know how this recognition may take place. Meiotic chromosomes seem to have the general structure of a fibre folded both transversely and longitudinally and genetic exchange will therefore involve a breakage and reunion of the fibre at equivalent sites in each homologue. The location of a chiasma must therefore depend upon the folding of the fibre and DuPraw (1970) has provided an explanation for the mechanism which causes terminalization of chiasmata after diplotene by suggesting that this may represent a redistribution of the folded fibre within the chromosome.

Almost all the DNA of the genome is replicated before meiosis commences, but Hotta and Stern (1971) have found that a small amount of DNA synthesis takes place during two stages of meiosis. In cells of the lilly, about 0·3% of the genome seems to be replicated in zygotene; these sequences are rich in G–C content and suffer semi-conservative replication. One possible speculation is that this synthesis is concerned with pairing of DNA sequences. A repair-like synthesis of DNA seems to take place during pachytene, and according to current theories of genetic recombination this might be implicated in genetic exchange at a chiasma (see chapter 12 of volume 1). In a review of biochemical events during meiosis, Stern and Hotta (1969) noted that the DNA synthesis must be essential for if it is inhibited during early zygotene cells fails to proceed in meiosis. The protein synthesis which takes place during meiotic prophase is also essential, for its inhibition with cycloheximide prevents meiosis from proceeding into metaphase.

CHAPTER 3

Protein Components of the Chromosome

Sequences of Histone Proteins

Heterogeneity of Chromosomal Proteins

Chromosomes contain basic and other proteins and RNA as well as the DNA which comprises the genetic material itself. DNA and proteins associate to form the coiled coil of nucleoprotein constituting the tightly packed fibre which is in turn folded into the overall structure of the chromosome described in chapter 1. Structural studies reveal the general nature of this association, in which the DNA is largely covered in protein so that it is for the most part inaccessible to external agents. Local changes in the structure of the nucleoprotein may therefore be necessary before DNA can be replicated or transcribed.

Chromatin can be visualised as a nucleoprotein particle in which proteins interact both with DNA and with each other. The structure of the duplex DNA itself is well defined and the major proteins which are associated with it have been characterized. However, we cannot yet define the molecular structure of the chromosome; although chromatin seems to be homogeneous overall in its behaviour, we do not know how much precision exists in the interactions which generate the nucleoprotein structure. The fibre may vary in constitution from one stretch to the next; or it may have a regular, well defined organization of components.

The relative proportions of the components of chromatin vary according to the tissue, organism and method of preparation. However, the basic proteins always make up the greatest amount of the chromosomal protein; these are the *histones*. The *non histone* proteins are less basic in nature and have sometimes been described as *acidic proteins*, although this term is a misnomer in the sense that their isoelectric points are lower than those of the histones but need not in fact be acidic. In addition to the structural proteins, chromatin contains some enzyme activities, such for example as RNA polymerase. There is usually a little more histone by weight than DNA and somewhat less of the non histone proteins (whose reported proportion is more variable), so that the total mass of protein may approach twice that of the DNA. There is a much smaller amount of RNA, varying from 5–10 % of the DNA content; although some of this may be chromosomal in the sense that it is a structural component of chromatin, much may comprise nascent RNA chains which have not yet completed synthesis. According to Bonner et al. (1968a), calf thymus chromatin has the overall composition (relative to DNA content):

DNA	100
histones	114
non histones	33
RNA	7

Histones are rich in lysine and arginine and are small proteins, mostly of little more than 100 amino acids in length. The five classes of histone in the typical eucaryotic cell are usually classified according to their relative contents of lysine and arginine as lysine rich, slightly lysine rich, or arginine rich. The lysine rich histones of any cell occur in several closely related forms; but there are invariably only two unique proteins in the slightly lysine rich and two in the arginine rich fractions. The histone content of all eucaryotic cells therefore shows considerable homogeneity.

The function of histones has been a matter of some controversy. Stedman and Stedman (1950) made the first suggestion that histones might be specific genetic repressors which control gene activity. However, this would demand both tissue and species specificity in the histones but there is no such variation. Homologies of structure between corresponding histones of different species vary with the individual histone but at present appear to be considerable. This evolutionary conservation suggests that corresponding histones may play the same role in all organisms, which implies that their functions must be structural, probably involving comparable interactions with DNA.

The number of individual non histone proteins is less well defined, but also appears to be quite small; there are probably about 10 different non histone chromosomal proteins in a cell. These proteins are not yet sufficiently well characterized to define the relationship of the molecules of different tissues and organisms. However, it seems likely that any particular chromosome contains only a comparatively small number of protein components. The histones appear in general to be more implicated in binding to DNA; the non histone proteins probably bind to other proteins of chromatin rather than directly to DNA.

Separation of Histones

Two general types of method have been used to isolate histones (reviewed by Murray, 1965; Johns, 1971). Extraction with acid, usually sulfuric acid, of whole tissues, cell nuclei or washed DNP (deoxyribonucleoprotein) yields preparations of histone–salt such as histone–sulfate. These crude preparations

of histones can then be fractionated by ion-exchange chromatography to yield individual classes of protein (reviewed by Bonner et al., 1968a, b). An alternative method is to dissociate nuclei or nucleoprotein in solutions of high salt concentration and to use a series of extractions with salt followed by precipitations with acetone or ethanol (Johns and Butler, 1962; Johns, 1964).

The heterogeneity of histones varies with the method of preparation, for degradation occurs readily to generate additional protein species. By comparing the histones isolated by different methods of preparation, Rasmussen, Murray and Luck (1962) were able to show that only a comparatively small number of fractions represents genuine variations in protein sequences. These fractions were not themselves homogeneous, for some histones tend to aggregate and to be isolated together, and more recent work has shown that the number of histone fractions is restricted to five in virtually all eucaryotic cells.

Comparison of the histones prepared from different species shows that they are very similar. Although calf thymus has provided the classical tissue for experiments with chromatin, the results obtained with this tissue therefore appear to be typical of those obtained in analysis of cells of other organisms. The nomenclature used to describe histones is taken from the fractions of calf thymus. Fambrough and Bonner (1966) and Bonner et al. (1968) prepared chromatin from calf thymus or from pea seedlings by homogenising the tissue in a buffer, filtering to remove fragments of tissue and membranes, and collecting the precipitate from a low speed centrifugation. This comprises a crude chromatin pellet which can be purified by centrifugation through sucrose to give a gelatinous preparation. The content of protein is usually 1·3–2·0 times that of DNA by weight. Figure 3.1 shows the elution pattern obtained when a preparation of histone sulfate made from the chromatin of calf thymus is eluted gradually with increasing concentrations of guanidinium hydrochloride from ion exchange columns. The three fractions are: group I, lysine rich; group II, slightly lysine rich; group III–IV, arginine rich.

Although the histone I fraction is heterogeneous, all its major subfractions are closely related (see below). Fraction II consists of two peaks, IIa and IIb. Fraction IIa resembles an aggregate of IIb and III–IV, for gel electrophoresis suggests that it includes both these fractions. The formation of aggregates is a common occurrence in histone fractionations. Fraction IIb behaves in these preparations as though homogeneous, but other methods of isolation show that it consists of two unique proteins, IIb1 and IIb2. Fractions III and IV are not completely separated on these columns, but gel electrophoresis suggests that each comprises one unique protein. Sadgopal and Bonner (1970a) reported that Hela cell histones fall into the same five classes.

An analysis of pea seedlings—a tissue very different in evolutionary development from calf thymus—shows a very similar, although not identical, pattern of fractions. The figure shows that fraction I is more compact, fraction II

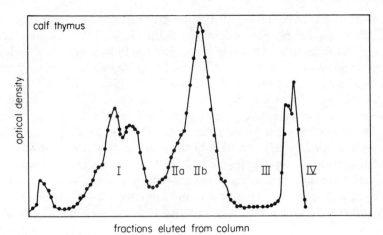

Figure 3.1: fractionation of histones on amberlite ion exchange columns by elution with increasing concentrations of guanidium hydrochloride. Data of Fambrough and Bonner (1966).

elutes with the main peak first and its shoulder second (the reverse of the order of calf thymus), and fractions III–IV are less well resolved. However, electrophoretic analysis confirms that the individual proteins of pea seedling correspond closely to those of calf thymus.

Stepwise release of histones from calf thymus chromatin with acid titration elutes the same fractions in the same order; Murray (1969) found that the lysine rich group I histones are completely eluted when the pH is reduced to 1·8, the slightly lysine rich group II histones by pH 1·2, and the arginine rich histones are released only when a pH of 0·6 is reached. Ohlenbusch et al. (1967) and Kleiman and Huang (1972) found that 0·6 M NaCl elutes the group I histones, 1·0 M NaCl elutes the group II proteins, and 3·0 M NaCl is needed to elute the fractions III–IV. These experiments suggest that the retention of histones in the chromatin structure depends to a large part on ionic forces (see also page 138).

In experiments using a series of fractionations, Johns (1971) prepared five fractions of histone from calf thymus. Johns (1964) had described these fractions as f1, f2, f3 according to their order of elution on carboxymethyl-cellulose columns in sodium acetate buffers of increasing strength. This order of elution is different from that displayed on ion-exchange chromatography of histone–sulfate preparations. The f1 group is the same as group I; these are the lysine rich histones. The lysine rich histones have the distinctive characteristic that they are extracted in 5 % trichloracetic acid or perchloric acid. The f2 fraction includes both the IIb and IV histones. Johns and Butler (1962) used extraction with ethanol to fractionate f2 into two species, f2a and f2b; Phillips and Johns (1965) then separated f2a into two fractions, f2a1 and f2a2, by precipitation with acetone from acid solutions. Fraction f2a1 corresponds to histone IV; fractions f2a2 and f2b together make up histone IIb and are usually described in this terminology as histones IIb1 and IIb2. Fraction f3 corresponds to the arginine rich peak III of ion exchange columns. The five histone species are therefore:

lysine rich:	f1	= I
slightly lysine rich:	f2b	= IIb2
	f2a2	= IIb1
arginine rich:	f3	= III
	f2a1	= IV

Correspondence between the different fractions obtained by the two isolation methods is not exact, for the peaks are not completely homogeneous (reviewed by Johns, 1971). However, in each case each peak consists very largely of one histone protein (with the exception of f1 = I which consists of several closely related proteins) and it is this one predominant component which is described by the number of the fraction. For example, Starbuck et al. (1968) have shown that each peak can be fractionated to yield one major histone contaminated to some small degree by most of the other histones when a method different from that of its preparation is used for analysis.

The general characteristics of the five histone fractions are given in table 3.1. All the histones are small and have about 25% basic amino acids, 13% acidic and 20% hydrophobic. The lysine rich histones are the largest, with a molecular weight of about 21,000; the other histones range in weight from 11,000 to 15,000 daltons. The lysine rich histones contain about 27% lysine, little arginine, and are also rich in alanine. The slightly lysine rich histones fall into two classes; f2b (IIb2) possesses 16% lysine and a moderate content of arginine, whereas f2a2 (IIb1) has 11% lysine, the same amount of glycine, and almost as much arginine so that it is sometimes called the arginine rich–lysine rich (AL) histone. Both the arginine rich histones, f3 (III) and f2a1 (IV) have about 10% lysine and 15% arginine; f3 is the only histone to contain

Table 3.1: Characterization of histones in five groups. The figures given apply to calf thymus, but overall amino acid compositions are similar in most organisms. About 25% of the amino acids are basic, about 13% are acidic and some 20% are hydrophobic. All the proteins are small. The lysine rich histones are heterogeneous, but the other groups each consist of a unique protein. Data of Fambrough and Bonner (1969), Phillips and Johns (1965), Yeoman et al. (1972), Iwai, Ishikawa and Hayashi (1970), Ogawa et al. (1969), DeLange et al. (1969b), Kinkade and Cole (1966).

histone	amino acid composition	number of amino acids (%)				molecular weight
		total	basic	acidic	hydrophobic	
lysine rich I = f1	27% lysine 2% arginine 24% alanine	207	31	5	10	~21,000
moderately lysine rich IIb2 = f2b	16% lysine 6% arginine	125	24	13	17	13,774
arginine rich, lysine rich (AL) IIb1 = f2a2	11% lysine 9% arginine 11% glycine	129	23	14	24	~15,000
arginine rich III = f3	10% lysine 15% arginine 2 cysteines	135	24	15	18	15,342
glycine rich, arginine rich (GAR) IV = f2a1	10% lysine 14% arginine 15% glycine	102	27	11	23	11,282

cysteine and f2a1 has a high content of glycine and is therefore sometimes known as the glycine rich–arginine rich (GAR) histone.

An electrophoretic separation developed by Panyim and Chalkley (1969a), using acrylamide gels in the presence of urea, shows that only five fractions of histone are present in virtually all eucaryotic cells. The lower photograph of figure 3.2 shows an analysis of calf thymus histones on short gels, on which the fractions run in the order: f1 (I) slowest, f3 (III), f2b (IIb2), f2a2 (IIb1), f2a1 (IV) fastest. Other mammals display the same order of separation;

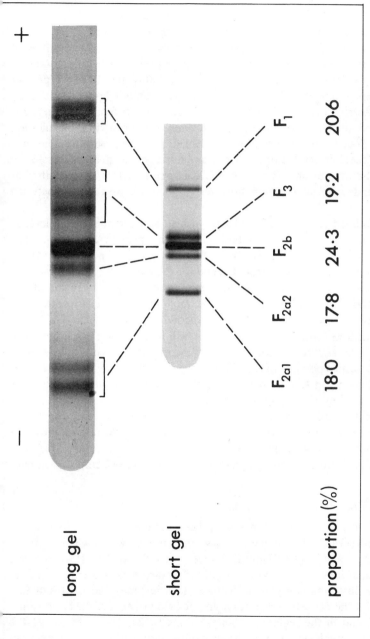

Figure 3.2: separation of calf thymus histones by acrylamide gel electrophoresis. The five bands of histones isolated on short gels (lower) can be further fractionated on long gels (upper). The proportions of each histone are similar in all calf tissues; the division of each fraction into further bands on the long gels shows more variation. Similar results are obtained with other eucaryotic cells. The bands of histone f1 on long gels represent molecules with closely related sequences; the multiple bands of other histones represent different extents of modification of a protein, the fastest running representing the least modified protein, each modification causing a reduction in mobility. Data obtained by the analysis of Panyim and Chalkley (1969a), kindly provided by Dr. R. Chalkley.

although mobilities are changed in other eucaryotic species, the same five fractions are always obtained.

The upper photograph of the figure shows that when longer gels are used to allow greater time for separation, most of the bands [usually all except f2b (IIb2)] may separate into further bands. The bands of f1 appear to represent different proteins of closely related amino acid sequences. The different bands of fractions f3 (III), f2a2 (IIb1) and f2a1 (IV) probably in each case represent modifications of one protein. For example, acetylation of a lysine residue may lower the electrophoretic mobility of a protein so that two bands are generated, one modified and one unmodified, if only some of the protein molecules are acetylated. Sequencing studies have confirmed this interpretation. The proportions of the different histone fractions are fairly similar in all calf tissues to those for the thymus shown in figure 3.2. The proportions of the different modified versions of each protein, however, may show more variation between tissues.

Many modified amino acids are found in histones; the more usual include α-N-acetyl derivatives at the N-terminus of the protein and ϵ-N-acetyl-monomethyl or dimethyl derivatives of lysine within the protein. Other residues which have been detected include α-N-trimethyl-lysine, 3-methyl-histidine, ω-N-monomethyl-arginine and o-phosphoserine (reviewed by DeLange and Smith, 1971). All of these modifications are introduced by specific enzymes after synthesis of the polypeptide chain.

In those histones which have been sequenced, modification is not complete at any particular site, so the histone molecules are heterogeneous in this sense. It is, of course, also possible that additional modifications may pass undetected because they exist at low frequency and the modifying groups are removed during preparation of the histone. Although the α-N-acetyl groups are stable, internal modifications appear to turn over at an appreciable rate; one reason for the apparent heterogeneity of histones may therefore be that modifications are introduced at specific times in the cell cycle and later reversed, so that unsynchronized cell populations yield histones in different states of modification.

Sequences of the Arginine Rich Histones

Both the arginine rich histones have been sequenced from calf thymus. Histone f2a1 (IV) has the sequence shown in figure 3.3 which was first determined by Ogawa et al. (1969) and DeLange et al. (1969a). This is the smallest histone and, as figure 3.5 shows, a prominent feature is its clustering of amino acids. This clustering is a general feature of all histones and is presumably important for their function in chromatin (see below). Most of the basic amino acids are found in the regions within positions 16–20, 35–40, 75–79, 91–95. The –NH₂ terminal sequence of the protein is almost completely lacking in hydrophobic residues and has a high concentration of basic amino acids. This has often been taken to suggest that this part of the molecule may be implicated in binding to DNA.

The amino terminus of histone f2a1 (IV) is blocked, consisting of N-acetyl serine. The lysine at position 16 bears an acetyl group and the lysine at position 20 may carry two methyl groups. Lysine 16 is acetylated in only 60% of the molecules; lysine 20 is dimethylated three times more frequently than it is monomethylated. This implies that there may be some variation in the f2a1 histone molecules in the cell with respect to their degree of modification.

Histone f2a1 (IV) of several other mammals and also of pea seedlings has been isolated and sequenced. The sequence in all those mammals investigated at present is identical with that of calf thymus. The sequence of pea seedling

$$
\begin{array}{cc}
\text{Ac} & \text{Me}_2 \\
\end{array}
$$

Ac–Ser–Gly–Arg–Gly–Lys–Gly–Gly–Lys–Gly–Leu–Gly–Lys–Gly–Gly–Ala–Lys–Arg–His–Arg–Lys–
          10   20

–Val–Leu–Arg–Asp-Asn-Ile – Gln – Gly – Ile –Thr–Lys–Pro–Ala–Ile – Arg–Arg–Leu–Ala–Arg –Arg–
          30   40

–Gly–Gly–Val – Lys–Arg–Ile – Ser–Gly–Leu–Ile – Tyr –Glu–Glu – Thr–Arg–Gly–Val – Leu–Lys–Val–
          50   60

–Phe–Leu–Glu–Asn-Val –Ile – Arg –Asp-Ala-Val –Thr–Tyr–Thr–Glu–His–Ala–Lys– Arg–Lys–Thr–
          70   80

–Val– Thr–Ala–Met-Asp-Val-Val– Tyr– Ala– Leu–Lys –Arg–Gln–Gly–Arg– Thr–Leu –Tyr-Gly– Phe–
          90   100

–Gly – Gly– COOH

Figure 3.3: sequence of histone f2a1 (IV) of calf thymus, pig thymus, rat tumour, Novikoff hepatoma. Histone f2a1 of pea seedling differs at only four sites: isoleucine occupies position 60, arginine occupies position 77, the lysine at position 20 is not modified and there is more complete modification of the acetyl-lysine at position 16. Data of DeLange et al. (1969a, b), Ogawa et al. (1969), Sautiere et al. (1970, 1971) and Wilson et al. (1970).

histone f2a1 differs in only four positions from that of calf thymus. Two of these differences concern modifications rather than the amino acids themselves. The lysine at position 20 is not methylated or given any other modification in the pea; and the lysine at position 16, by contrast, is more completely modified. This shows that the position and even the extent of modification is under a precise genetic control which is different in different organisms. The only amino acid substitutions involve the replacement of valine at position 60 of the calf histone by isoleucine in the pea; and the replacement of lysine at position 77 by arginine. Both these changes are conservative since the amino acid substituted has the same general character and the replacement demands a genetic change of only one base pair.

This stringent conservation of sequence argues that histone f2a1 must play the same role in both pea and calf. Its extent of evolutionary preservation appears to be much greater than that of any other protein; DeLange and

Smith (1971) calculated that its rate of evolution must be about 0·06/100 residues/100 million years, which is many times more stable than any other protein known at present. The most likely explanation for this surprising result is that virtually any mutation to this histone is lethal. This implies that the entire sequence of the histone must be important for its function.

There are few sequences in common in the DNA of organisms so distantly related as the plant pea and the mammal calf. We should therefore expect

$$
\begin{array}{c}
\text{Me} \qquad\qquad\qquad\qquad\qquad \text{Ac} \\
\text{\textbar}\text{O-3} \qquad\qquad\qquad\qquad \text{\textbar}
\end{array}
$$

NH$_2$ -Ala- Arg-Thr-Lys-Gln-Thr-Ala-Arg-Lys-Ser-Thr-Gly-Gly-Lys-Ala-Pro-Arg-Lys-Gln-Leu-
                                            10   20

$$
\begin{array}{c}
\text{Ac} \qquad\qquad\quad \text{Me} \\
\text{\textbar} \qquad\qquad\quad \text{\textbar}\text{O-3}
\end{array}
$$

-Ala - Thr - Lys-Ala-Ala-Arg-Lys-Ser-Ala- Pro-Ala-Thr-Gly-Gly- Val- Lys- Lys-Pro-His-Arg-
                                            30   40

-Tyr- Arg- Pro-Gly - Thr-Val -Ala- Leu-Arg- Glu - Ile -Arg-Arg-Tyr- Gln- Lys- Ser-Thr- Glu - Leu-
                                            50   60

- Leu- Ile - Arg- Lys- Leu-Pro- Phe-Gln- Arg- Leu-Val-Arg-Glu- Ile - Ala-Gln- Asn- Phe-Lys-Thr-
                                            70   80

$$
\text{SH}
$$
$$
\text{\textbar}
$$

-Asp-Leu-Arg-Phe-Gln-Ser-Ser-Ala-Val-Met-Ala-Leu-Gln-Glu-Ala- Cys- Glu- Ala - Tyr - Leu-
                                            90   100

$$
\text{SH}
$$
$$
\text{\textbar}
$$

-Val - Gly- Leu-Phe-Glu-Asp-Thr-Asn-Leu-Cys-Ala-Ile- His-Ala-Lys- Arg -Val -Thr- Ile - Met-
                                            110   120

-Pro-Lys- Asp-Ile - Gln-Leu-Ala-Arg-Arg -Ile- Arg-Gly-Glu-Arg-Ala - COOH
                                            130

Figure 3.4: sequence of histone f3 (III) of calf thymus. Histone f3 of the carp has the same except that the cysteine at position 96 is replaced by serine and there is no acetylation of the lysines at positions 14 and 23. Data of DeLange, Hooper and Smith (1972, 1973) and Hooper et al. (1973).

there to be few, if any, features of control of gene activity common to both organisms. The conservation of histone sequences thus argues strongly that histone f2a1 at least must play some structural role in the chromosome and cannot be concerned with any specific control of gene activity. One possible implication of this conclusion is that DNA may be packed into the chromosome in a manner which is independent of its particular sequences of base pairs; this offers the prospect that there may be common features in the structures of all eucaryotic chromosomes. Although conservation of sequence has been demonstrated only for the arginine rich histones f2a1 (IV) and f3 (III) and for parts of some of the lysine rich fractions, the general similarities in the electrophoretic mobilities and amino acid compositions of corresponding histones in different species suggest that these results may be of general application.

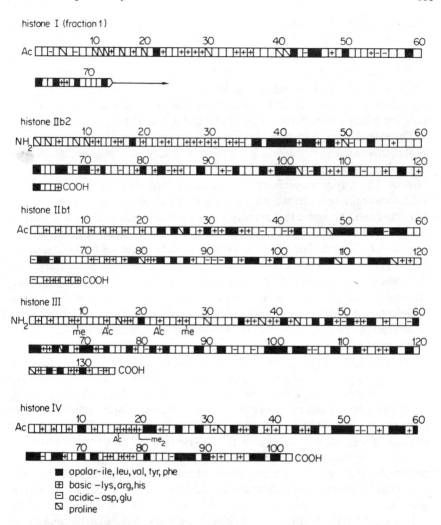

Figure 3.5: distribution of amino acids in calf thymus histones. There is always a concentration of basic groups at the NH_2-terminal end of the protein. Hydrophobic groups are usually clustered in restricted regions which may encourage helix formation. Based on data of figures of this chapter.

The complete sequence of histone f3 (III) of calf thymus has been obtained by DeLange, Hooper and Smith (1972, 1973) and shows some of the same features as histone f2a1. As figures 3.4 and 3.5 show, histone f3 also has clusters of amino acids which generate basic and hydrophobic regions in the molecule. The $-NH_2$ terminal sequence of the protein again lacks hydrophobic residues and carries a high basic charge; it may be involved in binding to DNA.

As is the case with histone f2a1, the sites of modification are all located in the –NH$_2$ terminal region; histone f3 also shares the characteristic that modification is not complete at any site. Some 25% of the molecules of f3 carry an acetyl group on the lysine at position 14; 29% are acetylated at position 23. At position 27, 40% of the molecules carry dimethyl lysine and 34% have monomethyl lysine. Trimethyl lysine has also been detected. The lysine at position 9 may be similarly methylated, although usually less extensively.

The extent of modification varies with the preparation of the histone and it is of course possible that low degrees of modification may pass undetected, for example because the modifying groups are removed during extraction of the histone. The only common feature of the methylated sequences in both histones is that an arginine is located on the NH$_2$ terminal side of the modified lysine; but other residues share this arrangement of amino acids and are not methylated. In addition to the methylations and acetylations, serine residues may be modified, for Marzluff and McCarty (1972) found that two of the seryl residues in the NH$_2$–terminal half of the protein appear to be partially phosphorylated. The sites of modification of both histones f2a1 and f3 are located in the –NH$_2$ terminal regions of the molecule which are highly basic and have few hydrophobic groups; this has prompted the suggestion that the function of the modifications may be to control the binding of the histone to DNA.

Histone f3 is the only one which contains cysteine. Fambrough and Bonner (1968) noted that pea histone f3 has only one cysteine residue, whereas calf thymus f3 has two cysteines. Panyim, Sommer and Chalkley (1971) observed that this difference is a general one; the f3 histone of most organisms, including plants, vertebrates and the lower mammals, contains a single cysteine, but two cysteines are present in higher mammals. The two cysteines of calf thymus f3 can form either intra- or intermolecular S–S bridges and their formation during extraction of the protein generates dimers and higher aggregates, which were responsible for some of the early reports of histone heterogeneity. However, it seems likely that in vivo the cysteines remain in the reduced state with free –SH groups.

Histone f3 (III) of the carp, Letiobus bubalus, has been sequenced by Hooper et al. (1973) and is almost identical with the protein of calf thymus. The sole difference in the sequence of amino acids is that the cysteine at position 96 of calf thymus is replaced by serine in the carp; the single cysteine remaining in the carp f3 is therefore that located at residue 110. The same two lysine residues which are methylated in calf thymus—at positions 9 and 27—are also methylated in the carp, although with reversed intensity so that carp lysine 9 is more greatly methylated than lysine 27. A difference in modification is that there is no acetylation in the carp histone f3, whereas calf thymus may be acetylated at the lysine residues at positions 14 and 23.

The very stringent conservation of sequence in the evolution of the arginine rich histones f2a1 (IV) and f3 (III) implies that the complete sequence of each

NH_2- Pro- Gln- Pro-Ala-Lys- Ser-Ala-Pro-Ala-Pro-Lys-Lys- Gly-Ser- Lys-Lys-Ala-Val-Thr-Lys -
                              10   20

-Ala - Gln - Lys -Lys-Asp-Gly-Lys- Lys -Arg-Lys-Arg-Ser -Arg-Lys- Glu - Ser-Tyr - Ser-Val- Tyr -
                              30   40

-Val - Tyr- Lys- Val- Leu- Lys-Gln- Val - His- Pro- Asp-Thr-Gly - Ile - Ser-Ser- Lys-Ala-Met-Gly-
                              50   60

-Ile - Met-Asn-Ser- Phe-Val-Asn-Asp-Ile- Phe-Glu-Arg-Ile - Ala-Gly- Glu-Ala-Ser-Arg- Leu-
                              70   80

-Ala - His -Tyr- Asn- Lys-Arg-Ser-Thr- Ile - Thr-Ser-Arg-Glu -Ile - Gln - Thr-Ala-Val-Arg- Leu-
                              90   100

-Leu- Leu- Pro-Gly-Glu- Leu-Ala-Lys- His-Ala-Val-Ser- Glu-Gly-Thr- Lys-Ala-Val-Thr- Lys-
                              110   120

-Tyr - Thr- Ser-Ser-Lys-COOH

Figure 3.6: sequence of histone f2b (IIb2) of calf thymus. Data of Iwai, Ishikawa and Hayashi (1970).

of these histone proteins is implicated in a common function in species as far apart as mammals, fish and plants. However, we cannot at present define whatever fundamental property of these histones depends upon this evolutionary conservation.

Sequences of the Slightly Lysine Rich Histones of Calf Thymus

The slightly lysine rich histones display the same general properties which characterize the arginine rich histones; basic groups are clustered at the N-terminal ends of the proteins, which may therefore constitute the DNA binding sites. Histone f2b (IIb2) has only one hydrophobic group in the first 35 residues; histone f2a2 (IIb1) has only two in the first 30 amino acids. Figure 3.6 shows the sequence of f2b determined by Iwai, Ishikawa and

Ac - Ser-Gly- Arg-Gly-Lys-Gln-Gly-Gly-Lys- Ala- Arg- Ala- Lys-Ala- Lys- Thr- Arg-Ser-Ser- Arg-
                              10   20

-Ala - Gly - Leu - Gln - Phe-Pro -Val - Gly - Arg- Val- His -Arg-Leu- Leu- Arg-Lys-Gly - Asp-Tyr- Ala -
                              30   40

-Glu - Arg- Val - Gly - Ala- Gly-Ala-Pro -Val - Tyr -Leu-Ala- Ala- Val -Leu-Glu - Tyr-Leu-Thr-Ala-
                              50   60

-Glu - Ile - Leu-Glu- Leu-Ala-Gly -Asn-Ala- Ala-Arg-Asp-Asn-Lys-Lys -Thr-Arg-Ile - Ile - Pro-
                              70   80

-Arg- His-Leu-Gln- Leu-Ala-Ile -Arg -Asn-Asp-Glu -Glu -Leu-Asn-Lys-Leu- Leu-Gly- Lys-Val -
                              90   100

-Thr- Ile -Ala - Gln- Gly-Gly-Val-Leu-Pro-Asn- Ile-Gln- Ala-Val -Leu-Leu-Pro-Lys-Lys-Thr-
                              110

-Glu - Ser-His- His- Lys-Ala -Lys-Gly -Lys-COOH

Figure 3.7: sequence of histone f2a2 (IIb1) of calf thymus. Data of Yeoman et al. (1972).

Hayashi (1970). Histone f2a2 has been sequenced by Yeoman et al. (1972) and as figure 3.7 shows has basic residues in clusters at positions 29–36, 71–77, 123–129. Sections 43–70 and 100–117 are largely hydrophobic and have no basic amino acids.

These histones appear to suffer less modification than the arginine rich histones. With the exception of the N-terminal acetyl group, no methyl, acetyl or phosphate groups have been detected in f2a2 of calf thymus (although modification may take place in other organisms—see below). Although Marzluff, Miller and McCarty (1972) have detected an N-acetyl-lysine in the N-terminal half of calf thymus f2b, acetylation is only partly complete and the modification has not been placed within the molecule. These results are consistent with the data of figure 3.2 which show that the arginine rich histones f3 (III) and f2a1 (IV) each display more than one electrophoretic band on long gels; but the slightly lysine rich histones f2a2 (IIb1) and f2b (IIb2) show less heterogeneity, with usually only one band each.

Microheterogeneity of Lysine Rich Histones

Lysine rich histones alone can be extracted from chromatin by 5% perchloric acid. Their lysine/arginine ratio is very high, for lysine constitutes some 27% of their amino acid content and arginine only 2%; they also have a high content of alanine, about 24%. Kinkade and Cole (1966) found that the lysine rich histones of calf thymus (the group I or f1 fraction) can be fractionated into four main components, each of which has a similar amino acid composition and about the same molecular weight of 21,000 daltons. These histones therefore comprise the longest class.

Each of the three peaks (two are analysed together because of their incomplete separation from each other) gives 50–55 tryptic peptides; 45 of these peptides are common to all three fractions. This suggests that the lysine rich histones may form a closely related group of proteins in which there is microheterogeneity; each protein may have most of its sequence in common with the other histones of the group, but should display some differences. One possible model for their structure is to suppose that all the lysine rich histones have a common DNA binding region but differ elsewhere in the molecule.

A similar situation prevails in rabbit thymus, in which Bustin and Cole (1966a) found that the lysine rich histones give five overlapping peaks upon fractionation on amberlite ion exchange columns with narrow gradients of guanidium chloride. All the fractions have about the same size, charge and amino acid composition. According to their elution from sephadex columns, the lysine rich histones seem to have an extended rather than globular structure. Bustin and Cole (1969b) found that peak 3 of these histones can be cleaved with N-bromo-succinimide into two main components. (This reagent breaks a polypeptide chain at the sites of any methionine residues.) The carboxy terminal fragment of the histone is more basic than the amino terminal fragment; this contrasts with the situation prevailing in the other classes of histones.

By examining these two fragments, Bustin and Cole (1970) have shown that the molecule seems to have three regions. The N-terminal region contains a large proportion of lysine and all of the proline present in the N-terminal half of the protein. The centre of the molecule is depleted in lysines and is devoid of prolines. And the C-terminal part of the molecule is rich in proline and in lysine. This raises the possibility that the two ends of the molecule, especially the C-terminal, may interact with DNA, the central region remaining free to interact with other components of chromatin.

By sequencing the N-terminal regions of two of the peaks of histone I from rabbit thymus and one of the peaks from calf thymus, Rall and Cole (1971)

```
Rabbit-3  Ac-Ser-Glu-Ala-Pro-Ala-Glu-Thr-Ala-Ala-Pro-Ala-Pro-Ala-Glu-Lys-Ser-Pro-Ala-Lys-Lys-
Rabbit-4  Ac-Ser-Glu-Ala-Pro-Ala-Glu-Thr-Ala-Ala-Pro-Ala-Pro-Ala----Lys-Ser-Pro-Ala-Lys-Thr-
Calf  - 1  Ac-Ser-Glu-Ala-Pro-Ala-Glu-Thr-Ala-Ala Pro-Ala-Pro-Ala-Pro-Lys-Ser-Pro-Ala-Lys-Thr-
                                             10                                    20

          Lys----Lys-Ala-Ala-Lys-Lys-Pro-Gly-Ala-Gly-Ala-Ala-Lys-Arg-Lys-Ala-Ala-Gly-Pro-
          Pro-Val-Lys-Ala-Arg-Lys-Lys-Lys-Ser-Ala-Gly-Ala-Ala-Lys-Arg-Lys-Ala-Ser-Gly-Pro-
          Pro-Val-Lys-Ala-Ala-Lys-Lys-Lys-Lys-Pro-Ala-Gly-Ala-Arg-Arg-Lys-Ala-Ser-Gly-Pro-
                                             30                                    40

          Pro-Val-Ser-Glu-Leu-Ile-Thr-Lys-Ala-Val-Ala-Ala-Ser-Lys-Glu-Arg-Asn-Gly-Leu-Ser-
          Pro-Val-Ser-Glu-Leu-Ile-Thr-Lys-Ala-Val-Ala-Ala-Ser-Lys-Glu-Arg-Ser-Gly-Val-Ser-
          Pro-Val-Ser-Glu-Leu-Ile-Thr-Lys-Ala-Val-Ala-Ala-Ser-Lys-Glu-Arg-Ser-Gly-Val-Ser-
                                             50                                    60

          Leu-Ala-Ala-Leu-Lys-Lys-Ala-Leu-Ala-Ala-Gly-Gly-Tyr-
          Leu-Ala-Ala-Leu-Lys-Lys-Ala-Leu-Ala-Ala-Ala-Gly-Tyr-
          Leu-Ala-Ala-Leu-Lys-Lys-Ala-Leu-Ala-Ala-Ala-Gly-Tyr-
                                             70
```

Figure 3.8: sequences of lysine rich histones (f1 = I group) aligned to show the similarities between fractions 3 and 4 of the rabbit and fraction 1 of the calf. Data of Rall and Cole (1971).

have shown that the amino acid sequences of this part of each molecule are extensively related, although not identical. The peaks RTL-3, RTL-4 and CTL-1 have been compared; CTL-1 is the first peak eluted from the calf thymus preparation, RTL-3 and RTL-4 the last two of rabbit thymus. Although their positions of elution do not correspond, all peaks show the very similar N-terminal sequences of figure 3.8.

Lysine, alanine and proline comprise 75% of the first 40 amino acids; this overall composition is the same as that of the last 110 residues of the C-terminal part of the molecule (that is the C-terminal fragment released by N-bromosuccinimide). This supports the idea that both terminal regions may be implicated in binding to DNA. One unusual feature of these molecules is that there are 7 prolines in the first 40 amino acids, making helix formation much less likely in this region. The next proline residue does not occur until position 105, so that there must be 65 residues lacking proline. This distribution implies that helix formation may take place in the central region of the molecule but not at the N-terminal region; if the N-terminus is involved in binding to DNA, it must therefore do so in some form other than a helix.

Synthesis of f2c (V) in Nucleated Erythrocytes

Some species—such as fish, birds and reptiles—possess nucleated erythrocytes which are completely inactive in transcription and replication of DNA; the sole function of these cells is to manufacture haemoglobin, as it is in organisms in which the erythrocyte has lost its nucleus. The chromatin of the inert erythrocyte nucleus suffers a loss of the f1 lysine rich histones, which are

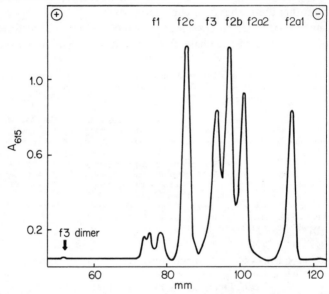

Figure 3.9: electrophoretic analysis of avian erythrocyte histones. Histone f1 has largely been replaced by the tissue specific histone f2c (V). Data of Sanders and McCarty (1972).

replaced by a new histone molecule. Edwards and Hnilica (1968) found that this unusual histone, described as f2c or V depending upon the method used for its preparation, is present in the nucleated erythrocytes of chicken, fish and frogs and is absent from all other tissues. Figure 3.9 shows the electrophoretic analysis of Sanders and McCarty (1972) of the total histone of avian erythrocytes in which almost all the f1 has been replaced by f2c.

As avian erythroblasts mature to erythrocytes, their nuclei condense dramatically, their chromatin becomes more compact and their capacity for transcription is suppressed. Dick and Johns (1969) found that the maturation of duck erythrocytes is accompanied by a decrease in the amount of f1 and a corresponding increase in f2c, so that the total content of f1 and f2c together is about the same as that of f1 in other tissues.

The lysine rich f1 fraction begins to be replaced by f2c early in the maturation of the erythrocyte. Sotirov and Johns (1972) found that chicken erythroblasts contain about 20–25% of the level of f2c finally achieved in the erythrocyte; Billeter and Hindley (1972) found that at a subsequent stage of maturation the early polychromatic erythrocyte contains 30–40% of its final level of f2c. By using density gradient centrifugation to separate populations of erythroid cells of differing degrees of maturity, Appels, Wells and Williams (1972) have also shown that the amount of f2c increases gradually during the maturation of the cells. Of the total histone present, erythroblasts contain 16% of f1 and 17% of f2c; polychromatic erythrocytes contain 12% f1 and 24% f2c. All the other histones are found in the same amounts at each stage of maturation.

A common idea about the role of f2c is to speculate that its accumulation might be related to the inactivation of the erythrocyte genome; the function of this histone might be to cause DNA to take up a condensed structure in which it is completely inaccessible for transcription. Erythroblasts are the last cells in the maturation pathway to divide; polychromatic erythrocytes no longer do so. That appreciable amounts of f2c are present in the erythroblasts even before they have ceased synthetic activity therefore implies that the presence of f2c is not in itself sufficient to inactivate the genome.

All the histones are synthesised in dividing erythroblasts since they all become labelled when the cells are incubated with C^{14}-labelled amino acids. Appels and Wells (1972) have shown that f2c is the only histone to become labelled in the non dividing polychromatic erythrocytes. When a pulse label is chased out of f2c it decays at about the same rate as its incorporation; this shows that f2c is turning over rapidly and therefore does not accumulate in the cell. This observation thus prompts the speculation that if f2c is implicated in the inactivation of the genome its loss of ability to turn over might be critical rather than its presence as such; the dynamic status of f2c molecules in earlier erythrocyte cells might mean that they do not block transcription, whereas in more mature cells the histone may become irreversibly fixed to DNA. However, production of f2c is not the only change which takes place in avian erythrocyte chromatin, for Vidali et al. (1973) observed that the complement of non histone proteins in the nucleus changes during maturation of the erythrocyte from its precursor cells.

Histone f2c behaves as a lysine rich histone in extraction, for Johns and Diggle (1969) found that it can be isolated from chromatin by using 5% perchloric acid. Greenaway and Murray (1971) found in amino acid composition and sequence studies that it is rich in some of the amino acids found in high proportion in the lysine rich f1 histones; f2c contains 24% lysine, 11% arginine, 16% alanine and 13% serine. Its lysine content is therefore nearly as great as that of f1, although its arginine content is appreciably greater and its alanine content somewhat lower. Analyses of fish, avian and amphibian erythrocytes show variations in composition, so this histone fraction may be species specific.

Chicken histone f2c can be separated on amberlite columns into two peaks (Va and Vb in the alternate nomenclature). The amino acid compositions and chymotryptic maps of both fractions appear identical. The protein contains only one methionine residue, so that cleavage with ethidium bromide gives two fractions; the smaller fraction includes the N-terminus and has the sequence shown in figure 3.10. This end of the molecule therefore contains more hydrophobic residues than are usually seen at the N-terminus.

The difference between Va and Vb appears to reside in the amino acid located at position 15, which is glutamine in Va and arginine in Vb. This charge difference may be sufficient to explain the difference in separation of the two proteins. An examination of the histones present in individual chickens shows that they contain either Vb alone (arg 15) or Va and Vb (gln 15 + arg 15). This suggests that the heterogeneity may be under genetic control, with Va and Vb representing two different alleles for the histone. In this case, it should also

NH_2-Thr- Glu - Ser- Leu - Val - Leu - Ser - Pro - Ala - Pro - Ala - Lys - Pro - Lys - $\overset{\text{Arg}}{\underset{\text{Gln}}{}}$ - Val - Lys - Ala - Ser -
                                            10
- Arg - Ser - Ala - Ser - His - Pro - Thr - Tyr - Ser - Glu - Hsr -
                          30

Figure 3.10: sequence of the N-terminal region of histone f2c (V) of chicken erythrocytes. Position 15 contains arginine in some molecules and glutamine in others. Hsr is homoserine. Data of Greenaway and Murray (1971).

prove possible to find chickens which have only histone Va. If this explanation for the heterogeneity of f2c is upheld, this is the first example of a genetic polymorphism allowing variation in the histone sequence of an organism; and this in turn would imply that there is only one gene locus concerned with coding for histone f2c.

Evolution of Histone Sequences

The arginine rich histones appear in general to be the most conserved in evolution and the lysine rich comprise those most open to variation. Using an electrophoretic separation in which histones are fractionated into five bands on short gels, Panyim, Bilek and Chalkley (1971) found that histones f3 (III) and f2a1 (IV) have the same mobilities in all organisms and tissues. Some heterogeneity results from modification, of course, but the mobilities of the different bands remain the same although their relative proportions may vary. The slightly lysine rich histones f2a2 (IIb1) and f2b (IIb2) show small changes in mobility in different classes of vertebrate.

Comparisons of the histones of different species by immunological cross reactivity also suggest that similarities in structure have been retained in evolution. Stollar and Ward (1970) prepared antisera to histones f2a2 (IIb1), f3 (III) and f2a1 (IV) of calf thymus; they found that each antiserum shows a good cross reaction with the corresponding histones of species as diverse as human, chicken, frog and lobster. (The antiserum to f2b showed cross reaction with the other histone fractions and so could not be tested.)

The lysine rich histones appear to be the only class with any tissue specificity; in any given organism all of the other classes appear to remain constant in character in all cells. The f1 histones also show the greatest changes in amino acid composition and electrophoretic mobility through the species; their electrophoretic behaviour on the gels of Panyim et al. (1971) is almost species specific. Using antisera prepared against the complete f1 fraction, Bustin and Stollar (1972) found that these fractions exhibit both tissue and species specificity—a marked contrast with the responses to antisera corresponding to the other histone fractions.

The basis for these differences in the content of lysine rich histones is not yet clear. Kinkade (1969) noted that the complement of lysine rich histones appears to remain the same in any particular organism, but the proportions of each histone in the f1 fraction vary in the different calf tissues. Pipkin and Larson (1972) have noted that changes seem to occur in f1 histones when dividing cells are compared with those of quiescent tissues. A small amount of what appears to be an extra lysine rich histone, which moves more slowly than the other f1 histones, has been observed in non dividing tissues. A difference in the f1 histone content, although not in the other histones, has been noted by Oliver and Chalkley (1972a, b) in a comparison of the histone proteins of the larval and adult forms of D. melanogaster and D. virilis.

With the exception of any possible differences in the content of lysine rich histones, the only differences in histone content between the tissues of an organism lies in their extent of modification. In particular, there appears to be an approximate correlation between modification and the rate of division; presumably this relates to the reproduction of the genetic material rather than to the growth of the cell itself. For example, Panyim and Chalkley (1969b) found that rapidly dividing calf tissues have equal amounts of acetylated and non-acetylated histone f2a1 (IV); slowly dividing tissues have a greater concentration of the non-acetylated form relative to the acetylated molecule. An analogous difference has been detected between Drosophila larvae and adults, where Oliver and Chalkley (1972a, b) found that acetylation of f3 (III) is more common in the larva than in the adult.

The conservation during evolution of general similarities between the histones argues against suggestions that their role might be to act as genetic repressors and implies rather that their function in chromatin may be structural. Similarities between the different histone fractions have been discussed by Phillips (1971). We do not know how great the conservation of sequence is in most instances, although the remarkable similarities of calf, pig, rat and pea histone f2a1 (IV) and between calf and carp histone f3 (III) shows that in at least these species the complete sequences of the arginine rich histones are demanded for some function which must presumably be identical in both organisms. Since each histone may interact both with DNA and with other proteins of chromatin, this implies that the general features of chromatin structure in different organisms may be closely related and may even result from identical interactions between components of the chromosome. Modifications of histones may be related to their association with DNA after synthesis and not to any specific functional role in chromatin. This leaves only the limited number of f1 proteins and the non histone proteins to provide heterogeneity in the proteins of the chromosome.

Non Histone Proteins of Chromatin

Although the histones form the greater part of the proteins of chromatin, non histone proteins which are less basic in nature comprise an important component of the eucaryotic chromosome. The non histone proteins display more diversity than the histones but they too have only a limited heterogeneity. Although this fraction includes enzyme activities such as RNA polymerase, the enzymes provide only a small amount of the total protein, so that the principal non histone proteins probably have some other functional or structural role in chromatin.

When histones are removed from chromatin by extraction with sulfuric acid, the remaining complex of DNA and protein can be dissolved in a solution containing 1% SDS and the DNA removed by centrifugation to leave a supernatant fraction of non histone proteins. Elgin and Bonner (1970) found that the proteins isolated by this method separate into about 13 bands on acrylamide gels. Fractions of rat liver, rat kidney and chicken liver display almost identical electrophoretic bands, which suggests that their non histone proteins may be similar and may therefore play a structural role rather than that of controlling gene activity. However, the proteins of pea bud have electrophoretic mobilities different from those of the vertebrates, so non histone proteins do not seem to have been as well conserved in evolution as the histones.

Although the number of non histone proteins varies in different preparations from about 12 to a little more than 20, limited heterogeneity has been noted in all tissues and organisms at present studied. (Far more than 20 proteins would be needed to act as specific gene repressors.) Shaw and Huang (1970) analysed non histone protein content by dissolving all the proteins of chromatin in 7M urea–3M NaCl; gel electrophoresis shows that in addition to the five histones there are upto about 20 bands of non histone proteins. The amino acid composition of the non histone protein fraction usually shows an acidic/basic ratio of about 1·4. Using SDS gels, Bhorjee and Pederson (1972) found 22 bands in Hela cell extracts, ranging from 15,000 to 180,000 daltons with an average weight of 45,000 daltons.

When chromatin proteins are extracted in urea–NaCl the basic histones are not retained by hydroxyapatite columns, but the more acidic non histone proteins are retained and can be eluted by increasing concentrations of phosphate salt solution. McGillivray, Caroll and Paul (1971) found that according to this separation chromatin contains about 60% histone and 40% non histone proteins. About 13 bands are found on electrophoresis in SDS; a similar pattern is produced by the proteins of kidney, liver and spleen of mouse. These bands are the same as those found when the non histones are extracted with acid. McGillivray et al. (1972) found that the overall ratio of non histone to histone proteins varies with the method of preparation, but is usually between one third and two thirds.

Analysis on SDS gels separates proteins according to their molecular weights alone; it is therefore possible that a single electrophoretic band may contain more than one protein and that the apparently similar patterns of different tissues result from the presence of different proteins of the same molecular weight. However, Elgin and Bonner (1972) have identified a similar number of proteins by using a different method of isolation and this supports the idea that the heterogeneity of the non histone proteins is limited to about twelve proteins overall. They fractionated the chromosomal proteins of rat liver into six groups on sephadex ion exchange columns; two of the groups contain histones and four contain non histone proteins. Further fractionation shows that two of the non histone fractions probably comprise single polypeptide chains and two are heterogeneous, each containing 4–6 proteins. The non histone proteins are larger than the histones, with molecular weights from 15,000 to 100,000 in these preparations and have acid/base ratios (without allowance for amides) of from 2·7 to 1·2. Analyses of this kind detect only proteins which comprise more than about 1% of the total non histone protein, so that any enzyme activities or regulator proteins which are present in lesser amount are not included in the non histone proteins identified at present.

Non histone proteins of rat liver have been extracted by Teng, Teng and Allfrey (1971) by a procedure which involves precipitating chromatin in salt, extracting the histones and any lipids, and then extracting non histone proteins from the residue with phenol. After this treatment the proteins can be dialysed back into the aqueous phase. Separation of the proteins in SDS shows about 20 fractions, this analysis generating different patterns from liver and kidney chromatins.

All analyses of non histone proteins identify fractions in the same range of molecular weights and acidic/basic ratios, although the total number of proteins varies. It is difficult to come to any firm conclusion about the heterogeneity of non histones because the proteins of chromatin itself may become contaminated with other nuclear proteins and because additional heterogeneity may arise from degradation during preparation. Resolution of the structures and functions of the non histones therefore requires a more detailed characterization—in which the main problem is lack of sufficient protein for analysis—of the individual proteins.

Some of the non histone proteins can be distinguished by their modification. Kleinsmith and Allfrey (1969) found that calf thymus nuclei possess a phosphoprotein fraction which has about 4–5 phosphate residues in 100 amino acid residues. The phosphate turns over rapidly in vivo; the protein fraction contains a phosphokinase activity which can add phosphate groups in vitro, in which case they are stable; their removal must be due to some other protein which is absent from the extract. Teng, Teng and Allfrey (1970) found that many of the rat liver non histone proteins can be labelled with P^{32}, most of the phosphate modifying serine and some threonine. The pattern of phosphorylation appears to be tissue specific and most of the proteins are labelled if rats are injected with P^{32}.

A small proportion of the non histone proteins seems to bind specifically to DNA and others bind in a less specific manner. Kleinsmith, Heidema and Carroll (1970) have chromatographed a preparation of rat liver non histone proteins labelled with I^{131} through columns of DNA adsorbed to a cellulose matrix. Although there is appreciable non-specific binding to DNA, only a small number of proteins appears to bind specifically, according to their retention by rat DNA but not that of other species. When proteins are first adsorbed to salmon sperm DNA, to reduce any non-specific binding activity, some of the eluted proteins can then bind to rat DNA; this binding is probably specific for sequences of the rat DNA.

Using a similar protocol, Van den Broek et al. (1973) chromatographed non histone proteins through two successive columns, the first of E.coli DNA-cellulose and the second of rat liver DNA-cellulose. About one third of the protein fraction binds to the E.coli DNA and about 4% to rat liver. Calf thymus DNA can be substituted for rat liver DNA, so that although the binding is specific for eucaryotic DNA it is not specific for the same species.

Some of the non histone proteins therefore have a relatively non-specific affinity for DNA and a much smaller number binds only to homologous or to related DNA; of course, this latter fraction provides a minimum estimate of the specific binding proteins, for proteins which bind specifically to rat DNA may also have a less specific affinity which would allow them to bind to columns of heterologous DNA.

Modification of Histones

Histone Modification during the Cell Cycle

The selection of a small number of lysine residues for modification indicates that the reaction is highly specific and that different enzymes may act on different histones. The modifying groups are not stably incorporated into the protein chains, but turn over; both the modification and demodification activities are associated with chromatin, so that histones are modified at internal residues after they enter the nucleus and not at the time of their synthesis on polysomes. N-terminal acetylation, however, may take place at an earlier stage before the histones leave the cytoplasm.

When Hela cells are labelled with C^{14}-acetate the arginine rich histones achieve the highest label during a 90 minute incubation. This is consistent with the results of sequencing studies. Wilhelm and McCarty (1970) found that when a 90 minute pulse is followed by a 90 minute chase, the label in f3 (III) is replaced entirely and only 25–30 % of the label remains in f2a1 (IV).

The complete loss of label in f3 results from the rapid turnover of unstable internal acetyl groups; the intermediate kinetics of f2a1 result from the loss of internal acetyl groups but the retention of N-terminal acetate, which is stable and does not turn over. Histone f2a2 (IIb1) also possesses both external and internal acetyl groups and therefore displays intermediate kinetics of loss. The f1 histone fraction which suffers only N-terminal acetylation accordingly retains its acetate label during a chase.

The same histones are labelled in calf thymus, in which Vidali, Gershey and Allfrey (1968) found that C^{14}-acetate largely enters histones f3 (III) and f2a1 (IV), with histone f2a2 (IIb1) gaining a lesser extent of label. (Contrasting results have been obtained with Chinese hamster cells, in which Shepherd, Noland and Hardin [1972] found acetyl groups of histones to be stable; instability of histone modifications is therefore not an invariant feature of their structure in all cell types.)

The time of modification varies but much of it appears to be correlated with the time of histone synthesis and addition to chromatin (reviewed by Allfrey, 1971). Wilhelm and McCarty (1970) found that cycloheximide does not inhibit histone acetylation in Hela cells, which implies that the reaction does not take place on newly synthesised proteins alone but may modify both old and new histones together. In general, N-terminal acetyl groups are added to histone molecules in the cytoplasm soon after their synthesis and are stable; internal acetyl groups are added to the ε-position of lysine within the nucleus, largely during S phase. These groups are often unstable and may be lost later during the cell cycle (in which case the old histones as well as the new may be acetylated at the next S phase).

Similar temporal restrictions are shown by other modifications. Shepherd, Hardin and Noland (1971) found that in CHO cells the f1 histone group is not labelled with methyl groups, f2b (IIb2) gains methyl groups only during S phase, and the f2a (IIb1 and IV) histones and f3 (III) start to take up methyl groups during S phase but do not complete methylation until 4–5 hours after the completion of DNA synthesis. All the methylated histones show turnover of their methyl groups.

Acetylation and other modifications seem to be implicated in the association of histones with DNA rather than with their synthesis as such. The addition of internal acetyl and methyl groups at the time of DNA synthesis, followed by their loss afterwards, suggests that the presence of these groups may be necessary before and whilst histone is binding to DNA; but that they become unnecessary and are therefore removed once the histones have become part of the structure of chromatin. The function of these modifications thus seems to be concerned in some way with ensuring that chromatin gains its proper structure during its reproduction; this concept also explains why both newly synthesised and pre-existing histones are modified during S phase, for at this time rearrangements in the chromatin may take place to permit replication, so that old as well as new histones may have to associate with DNA.

Changes in histone modification occur when a hepatectomy is performed on rats to cause the remaining lobes of the liver to regenerate. Pogo et al. (1968) found that in normal rat liver there is a rapid turnover of H^3-acetate incorporated into the arginine rich histones; in regenerating liver the turnover becomes much slower. Since the acetylation does not seem in this case to be coupled to histone synthesis itself one possible speculation is that it is implicated in changes in the genome which take place to change the pattern of gene expression.

Histones are also phosphorylated during regeneration of rat liver; Sung, Dixon and Smithies (1971) found that histone f2a2 (IIb1) is phosphorylated on its N-terminal serine. Old histone is phosphorylated as well as the newly synthesised molecules. In normal rat liver, the f1 fraction consists of one major (more than 90%) band and 3 minor, slower running bands. These are separated by irregular spacings and are probably histones of different sequences. Balhorn, Chalkley and Grenner (1972) found that regenerating liver contains an additional two bands; the fastest band contains only 50% of the total f1 histone but is augmented by two new bands which run more slowly. Treatment with alkaline phosphatase restores the one major and three minor bands characteristic of normal liver; this suggests that the splitting of the fastest band into three bands is caused by the addition of phosphate groups. The f1 histones do not seem to be acetylated or methylated; and we do not know how many phosphorylation events take place and how many of the f1 histone molecules are modified in this way.

Histone phosphorylation also seems to be linked to the cell cycle. Balhorn et al. (1972) found that histones f1 and f2a2 (IIb1) of cultured HTC cells start to gain phosphate groups at the end of G1 phase; phosphorylation declines at the end of S phase. No phosphate is incorporated into the other histones and there is no phosphorylation in the histones of cells in stationery phase. When HTC cells are transferred from stationary to exponential phase, phosphorylation increases rapidly and some 70% of the pre-existing f1 histones become phosphorylated. When the cells return to stationary phase, phosphorylation declines to zero. Phosphorylation therefore depends upon the growth and division either of normal liver cells (that is when regeneration is stimulated by partial hepatectomy) or of liver tumor cells (shown by the coordination of phosphate incorporation and cell growth in HTC cells).

In exponentially growing HTC cells, phosphorylation is a unique event which occurs only once to each f1 histone molecule in each round of cell division. Both newly synthesised and pre-existing histones are phosphorylated. Oliver et al. (1972) found that if cells are grown in H^3-lysine, the H^3-label is found in the f1 electrophoretic bands corresponding to the phosphorylated molecules; this shows that newly synthesised histones are phosphorylated. If cells are grown in H^3-lysine for three generations and grown to stationary phase, the isotope removed, and the cells grown in normal lysine to exponential phase, the label is once again found in the phosphorylated bands; pre-existing histones must therefore incorporate phosphate groups.

If a 20 minute pulse label with H^3-lysine is given to HTC cells in exponential growth and the cells are allowed to grow for one complete division cycle, the entry of the label into the electrophoretic bands containing phosphorylated histones can be followed through the cell cycle. Immediately after the pulse, most of the label is found in parental, unmodified f1. By 40 minutes later, 75% of the label is found in phosphorylated molecules. During the next 12 hours, the level drops with a half life of about 5 hours. At 24 hours, one cell cycle later, the lysine label enters the phosphorylated electrophoretic bands again. This suggests that one burst of phosphorylation takes place at the same time in each cell cycle, transferring both newly synthesized and pre-existing f1 histones to the phosphorylated bands. There is a time lag of 30–60 minutes after the synthesis of f1 proteins before they are phosphorylated, which suggests that the histones may be associated with chromatin when the phosphate groups are added.

Phosphorylation seems to comprise the only modification to f1 in several species of cell, although the other classes of histones which also gain phosphate groups may differ. Gurley, Walters and Tobey (1973) found that f1 (I) and f2a2 (IIb1) are phosphorylated during S phase in CHO cells. Working with the same cells, Shepherd, Hardin and Noland (1972) found that f2b (IIb2) seems to be phosphorylated largely during S phase in the same cells, but that f1 is phosphorylated throughout the cell cycle. The reason for this discrepancy is not clear.

Phosphorylation of f1 takes place only in growing and not in stationary phase cells. However, in contradiction to observations that phosphorylation takes place during S phase, Lake and Salzman (1972) reported that in Chinese hamster cells growing in culture the stage of the cycle when f1 is phosphorylated is mitosis. The reaction is catalysed by an f1 phosphokinase activity associated with chromatin. We may speculate that such a reaction might be implicated in the condensation of chromosomes.

The metabolism of f1 also appears to differ from that of the other histone fractions. Gurley, Engers and Walters (1973) observed that there appears to be a pool of f1 proteins in the cytoplasm which does not enter chromatin and may be extracted from cells in association with the polysomes. The function of this pool is not known, but its existence implies that the cellular population of f1 histone molecules may fall into more than one compartmentalized class.

It is difficult to determine the proportion of histone molecules which suffers each particular modification because of the transient association of the modifying group with the protein. Extraction of histones for sequencing studies utilises a tissue of cells in different states of the cell cycle and therefore bearing different numbers of modifying groups. This may explain why the histones of calf thymus which have been sequenced always show incomplete modification at any particular site. Definition of the degree of modification would therefore require extraction and sequencing of histones from synchronized cells.

The only general conclusion at which we can arrive at present is therefore that N-terminal acetylation takes place in the cytoplasm shortly after histone synthesis; these acetyl groups are stable and are thus always present in histones extracted for sequencing. Acetyl, methyl and phosphate groups are in general added to positions within the histone, often to both newly synthesised and pre-existing protein molecules, at S phase when chromatin is reproduced; most if not all of these internal modifications are not stable and it is possible that not all of the sites which are modified have been detected in sequencing studies. The general function of these modifications appears to be concerned with the assembly of histones into chromatin.

Changes in Histone Modification in Trout Testis Cells

Trout testis is a particularly interesting tissue in which to study the modification of histones because the set of basic proteins associated with DNA is completely replaced during spermatogenesis. In trout and in salmon, the histones are removed from DNA during maturation of the spermatid and are replaced by protamines. The protamines are small proteins, with an average length of 33 amino acids, and are very rich in arginine. They are heterogeneous and Dixon and Smith (1968) noted that they too are modified during the maturation process. Lam and Bruce (1971) found that protamines also replace histones during spermatogenesis in the mouse. One possible reason for the substitution is that the presence of the protamines distinguishes the maternal and paternal genomes after fertilization of the egg.

The stem cells of trout testis divide to give two cells which undergo a further mitosis to generate primary spermatocytes. Louie and Dixon (1972b) observed that these cells then divide to give secondary spermatocytes; and it is these cells which suffer meiosis to generate spermatids. Maturation of the spermatids generates the spermatozoa. These classes of trout testis cell can be separated because they become progressively smaller during maturation. The testis contains an asynchronous selection of cells at any time, but centrifugation through density gradients separates several classes of cell according to their sizes.

The last synthesis of DNA in these cells is detected before the secondary spermatocytes enter meiosis and the rate of histone synthesis (and modification) parallels the rate of DNA synthesis during development from spermatocyte to spermatid. Protamines appear in middle stage spermatids; after this stage histones are progressively reduced until finally there is none remaining in the mature sperm. Little histone synthesis then takes place and by the late spermatid stage histone synthesis has declined to 1 % of the level of the earlier cell class. In trout testis cells which have just begun to synthesize protamines there is extensive phosphorylation of seryl residues; in mature spermatozoa the protamines are found in a dephosphorylated form. Template activity declines as the protamines replace the histones and the genetic material coils up more tightly.

All four of the lysyl residues at positions 5, 8, 12, 16 of histone f2a1 (IV)— the N-terminal sequence is identical in trout testis and calf thymus—can be acetylated in developing trout testis cells. Candido and Dixon (1971) found that five bands of histone f2a1 are generated on ion exchange columns; these bands contain 0, 1, 2, 3 or 4 acetyl residues. The lysine at position 5 is on average acetylated to only about one third of the level found at positions 8, 12 and 16. Louie and Dixon (1972a) followed the kinetics of labelling of histone f2a1 with C^{14}-acetate groups and found that very soon after its synthesis the histone molecule is acetylated twice. Acetylation then continues at a slower rate to give forms of the protein with 3 and 4 acetyl groups. It is then slowly deacetylated to give forms which have 1 and 0 acetyl groups. The acetylation and deacetylation of a newly synthesised histone f2a1 (IV) takes about one day.

The other histones are also modified, although the kinetics of modification are not so well established. In trout testis histone f3 (III)—which has the same sequence of 25 amino acids from the N-terminus as that of calf thymus— acetylation is also confined to this region of the molecule. Candido and Dixon (1972b) found that the two main sites of acetylation are at positions 14 and 23 (which are also acetylated in calf thymus); the lysine residues in positions 9 (methylated in calf thymus) and 18 (not modified in calf thymus) are also acetylated. The lysine at position 27, which is methylated in calf thymus, does not seem to be modified in trout testis.

Histone f2b (IIb2) also contains four lysine residues which can be acetylated. The sequence of this histone is similar to that of calf thymus upto position 18, the only difference being the deletion of an ala-pro dipeptide in the sequence ala-pro-ala-pro between positions 7 and 10. The lysines at positions 10 and 13 are acetylated; a smaller amount of acetylation is also found at position 5 and 18. All four sites of modification follow the general pattern and are therefore located in the N-terminal region of the protein. Histone f2a1 (IIb1) has the same 17 N-terminal amino acids as calf thymus, with the exception that the gln at position 6 is replaced by thr; and Candido and Dixon (1972a) found that only the lysine at position 5 is acetylated. The f1 histones do not seem to be modified at all by acetylation. Since sequences of amino acids identical to those on each side of the acetylated lysine residues are found elsewhere in the histone molecules but are not modified, acetylation may depend not only upon recognition of the appropriate amino acid sequences by modifying enzymes but also upon the accessibility of these sites in the structure of the histone.

The cycle of acetylation and deacetylation is accompanied by a cycle of phosphorylation in all histones except f3. Phosphorylation takes place independently of acetylation; Dixon (1972) reported that histones f2a2 (IIb1) and f2a1 (IV) are phosphorylated on their N-terminal serine residues, histone f2b (IIb2) is phosphorylated on the serine at position 6; and the major site in the phosphorylation of f1 histone, by contrast, lies in the C-terminal region of the protein.

The time of histone phosphorylation depends upon the class of histone. Louie, Sung and Dixon (1973) reported that histone f2a2 (IIb1) is phosphorylated soon after its synthesis; the proportion of newly synthesized histone in the phosphorylated band upon gel electrophoresis rises to about 25% during a 4 hour period and then declines as the phosphate groups are removed. Histone f2a1 (IV), by contrast, suffers an appreciable lag period after its synthesis before phosphate groups are added. These two phosphorylation events may therefore serve different functions. About 5% of the histone in trout testis usually carries phosphate groups at any particular time.

Protamine is also phosphorylated on seryl groups; Louie and Dixon (1972c) reported that one or two phosphate groups are added to each protamine molecule within 5–10 minutes of its synthesis; the protamine is then converted to a form carrying three phosphate groups over the next 5–10 hours. Phosphorylation is succeeded by a dephosphorylation reaction which occupies several days. In spermatids which have just begun to synthesize protamines, most of these proteins are found in their phosphorylated state but as the spermatids mature the protamines become progressively dephosphorylated.

What is the purpose of these modifications to histones and protamines? Sung and Dixon (1970) first proposed that modification of the DNA binding regions of histones might serve to control their affinity for DNA. The electrostatic attraction for DNA of the unmodified N-terminal histone regions may be great enough to allow the proteins to bind to negatively charged nucleic acid without specificity for particular sequences of DNA. But modifications, especially of lysine residues, by reducing the positive charge in the protein could lower its ionic attraction for DNA, so that binding takes place only to appropriate base pair sequences which have a particularly high affinity for the protein.

Introduction of a phosphate group alters the local charge structure of the protein because it bears two negative charges. Acetylation reduces the local positive charge by eliminating the contribution of a lysine residue. Another possible effect of acetylation has been proposed by Lewin (1973), who noted on the basis of molecular model building that the steric changes caused by introduction of acetyl groups may be important. He suggested further that methylation may influence hydrophobic forces in the histone in addition to its effect upon the charge carried by lysine. The overall effect of the various modifications may be to change the conformation of the histone, perhaps by helping to promote the formation of helical regions from random coils.

Once the histone is bound to its appropriate site on DNA, removal of the modifying groups may increase its ionic attraction for the duplex and thus make its association at this position more stable. When it becomes necessary to release the histone from DNA, modification may be a prerequisite for reducing attraction to the point where DNA and histone are allowed to part. This may explain why modifications of histones in trout testis cells appear to occur at two periods in the life of the histones.

The acetylations or phosphorylations which take place soon after synthesis may be concerned with ensuring that histones bind correctly to DNA; the modifying groups are removed when the binding has been achieved. Most of the phosphorylation events take place at an early time and probably have this function. The modifications which take place later in the development of the cell and influence preformed rather than newly synthesized histones may be needed to release histones from DNA so that they can be replaced by protamines. Candido and Dixon (1972c) found that C^{14}-acetate is incorporated into the internal lysines of histones both in early cells which are still synthesising histones and in late cells which are replacing the histones with protamines. This suggests that acetylation is implicated in both reactions. The phosphorylation of protamines is presumably needed to ensure that they bind correctly to DNA when the histones are replaced. Louie and Dixon (1972c) observed that the removal of phosphate groups from the protamines seems to correlate in time with the condensation of chromatin; it may perhaps play some role in assisting the DNA to take up its final compact conformation within the sperm.

Structure of Chromatin

Accessibility of DNA in Chromatin

Our general view of the state of DNA in chromatin is of a nucleic acid duplex extensively covered in protein. The histones possess the greatest affinity for DNA and all classes of histone can bind directly to the nucleic acid; the association between DNA and histones appears to be largely responsible for ionic neutralization of the negative charges of DNA. The affinity for DNA of the non histone proteins is much lower; they probably interact largely with each other and with histones and may bind directly to DNA to a lesser extent. Whether the RNA present in chromatin has any structural role is not known at present.

Several classes of experiment suggest that the DNA of chromatin is probably not in general accessible to macromolecules. The extensive nature of its association with chromosomal proteins is revealed by the elevation of the melting temperature of almost all the DNA of chromatin compared with free DNA in solution. The DNA of chromatin is protected against degradation by the enzyme DNAase in a manner suggesting that most of it is bound to proteins. (An approach to measure free DNA directly has relied upon titration with polylysine to assay the number of free phosphate groups in DNA of chromatin; however, this has proved misleading because the reaction is not restricted to free phosphate groups.) When chromatin is used as a template for directing RNA synthesis in vitro its activity is very much less than that of pure DNA; the reduction in transcription is caused by the histones associated with DNA (see page 354). Functional inactivity therefore accompanies the inaccessibility of bulk DNA; this implies that gene expression must demand the activation of particular regions of DNA.

One of the most controversial properties of chromatin has been its solubility in aqueous solution. In early experiments, Zubay and Doty (1962) found that when calf thymus chromatin is homogenised with saline citrate it generates a sediment which becomes gelatinous in water as the chromosomes swell. Chromatin can be precipitated from this viscous gel with 0·15 M NaCl. Soluble preparations can be derived by changing the conditions of incubation to include reagents such as capryl alcohol, which suppresses surface denaturation. An alternative is to dialyse the viscous preparation against a phosphate buffer; centrifugation yields a supernatant which contains a solution of chromatin.

The state of the chromatin is critical in assessing its response to added macromolecules, for one explanation for the apparent inaccessibility of DNA is to suppose that it is precipitated in a form which enzymes such as RNA polymerase cannot penetrate. Sonnenberg and Zubay (1965) and Roy and Zubay (1966) have suggested that chromatin may be inactivated by aggregation into gels; Barr and Butler (1963) also experienced difficulty in obtaining soluble complexes of calf thymus chromatin, but Butler and Chipperfield (1967) found that the decrease in DNA template activity found in the presence of histones does not parallel the precipitation of DNA, which can therefore be no more than partly responsible for its inactivity. Bonner and Huang (1966) have pointed out that the conditions of incubation are critical and that gel formation can be prevented only when Mg^{++} and Mn^{++} ions and nucleoside triphosphates are present. Another possible cause of precipitation is contamination with nuclear proteins during extraction and these must be removed by centrifugation through sucrose. Chalkley and Jensen (1968) found that calf thymus chromatin can aggregate by forming cross links between histone molecules, but this interaction may be reduced by the use of urea.

Digestion of calf thymus chromatin by DNAase proceeds much more slowly than digestion of free DNA. Mirsky (1971) followed the progress of digestion by measuring the release of nucleotides which are soluble in perchloric acid. (Other assays which have been used, such as measuring the nucleotide content remaining in the supernatant after the undigested residue has been precipitated by centrifugation, underestimate the extent of digestion.) The general manner of degradation of DNA proved to be the same with two different types of preparation of chromatin, although somewhat faster with one than the other. This confirms the idea that different methods for preparing chromatin yield products which have different, although related, structures.

The extent of the digestion reaction depends upon both the time for which chromatin is exposed to DNAase and upon the concentration of the enzyme. Little DNA in the chromatin preparation is free of protein and immediately available to the enzyme, for in conditions in which virtually all of a pure DNA preparation is digested much of a chromatin preparation remains protected. Figure 3.11 shows the effects of time and enzyme concentration for substrates of free DNA, isolated calf thymus nuclei, a preparation of chromatin made

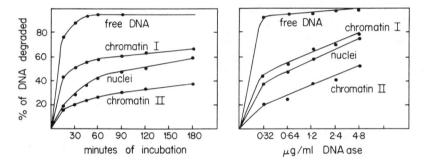

Figure 3.11: degradation of DNA, nuclei and chromatin preparations by DNAase. Left: the time course of reaction with an enzyme concentration of 1·2 μg/ml. Right: the degradation achieved after incubation for 180 minutes with different concentrations of enzyme. Chromatin I was prepared by the method of Paul and Gilmour (1968) and chromatin II by the method of Clark and Felsenfeld (1971). Although the extent of protection varies with the preparation, the DNA of nuclei or of chromatin is degraded by DNAase progressively as a product of time and concentration. Data of Mirsky (1971).

according to the method of Clark and Felsenfeld (1971), and chromatin prepared by the method of Paul and Gilmour (1968).

As the left panel of the figure shows, with an enzyme concentration of 1·27 μg/ml, almost 75% of free DNA is destroyed within 15 minutes; less than 20% of the DNA in nuclei is degraded. One preparation of chromatin is less sensitive than free nuclei and suffers loss of only 15% of its DNA in this period; the other preparation is more sensitive and 40% is degraded. Similar results of the time course of action of other concentrations of the enzyme show that in all cases free DNA is degraded rapidly; and the nuclear and chromatin preparations are degraded more slowly but progressively. With longer incubations, an appreciable proportion of the DNA in chromatin is lost; the right panel of the figure shows that the proportion depends upon the enzyme concentration. With the lowest enzyme concentration, 92% of free DNA is digested, but only 37% of DNA in nuclei and 43% and 20% of the DNA of the two chromatin preparations. At the highest concentration of enzyme, however, 75% of the DNA of the nuclear and the more sensitive chromatin preparations are lost.

These results suggest that although free DNA can be attacked immediately by the enzyme, little of the DNA of nuclear and chromatin preparations is available in a form unprotected by proteins; most of this DNA must be covered by proteins which prevent immediate access of the enzyme and is therefore attacked only with higher enzyme concentrations and longer incubations. The DNA of chromatin seems to be largely homogeneous in its response to degradation; if it falls into two classes, one free (unprotected by proteins) to immediate attack and one inaccessible (protected by proteins), the amount of DNA in the free class must be small.

The greatest amount of free DNA can be estimated by the proportion of DNA of chromatin degraded in conditions in which almost all of a free DNA preparation is attacked; this suggests that less than 20% of the DNA is unprotected. These results contradict the conclusions drawn by Clark and Felsenfeld (1971), who found that 5 μg/ml of DNAase degraded 50% of the DNA of chromatin in a lengthy incubation; they suggested that this might mean that half of the DNA is free of protein, but more extensive experiments suggest that the final amount of DNA degraded depends upon the product of enzyme concentration and time of incubation so that ultimately all the DNA of chromatin would be degraded.

In experiments using the enzymes DNA polymerase and RNA polymerase as probes for the accessibility of DNA, Silverman and Mirsky (1973) found that chromatin provides a very poor template compared with that of purified DNA. Of the order of only 5% of the DNA in chromatin appears to be available for use by these enzymes. They observed that errors in scintillation counting related to a concentration-dependent interaction of DNA with filters may have been responsible for earlier suggestions that the chromatin template may be relatively more accessible.

Are all histones equally effective in protecting DNA from degradation? Mirsky and Silverman (1972) extracted either the lysine rich histones (with 0·1 M citric acid) or the arginine rich histones (with ethanol-HCl solutions) from calf thymus nuclei. The rate of digestion of the DNA of the treated nuclei by DNAase is increased by removal of either type of histone; but weight for weight the removal of lysine rich histones renders the DNA most sensitive to DNAase. In analogous experiments in which histones were added to the calf thymus nuclei before degradation with DNAase, Mirsky, Silverman and Panda (1972) found that addition of either lysine rich or arginine rich histones inhibits the reaction, with lysine rich histones the more effective.

That the lysine rich histone f1 group may be concerned with the overall structure of chromatin is suggested by the properties of chromatin from which the lysine rich histones have been preferentially extracted. Removal of f1 histones by 0·6 M NaCl renders the remaining preparation soluble in 0·3 M NaCl, in which complete chromatin is insoluble. Tuan and Bonner (1969) noted that the removal of histone f1 by salt does not change the optical absorbance or ORD (optical rotatory dispersion) of chromatin; Bradbury et al. (1972a) observed that it does not change the X-ray diffraction pattern. These results imply that the lysine rich histones may be less intimately associated with DNA itself than the other histones.

By preferentially extracting the arginine rich histones from calf thymus nuclei with solutions of ethanol-HCl, Littau et al. (1965) found that upto 50% of the total histone can be extracted without causing any marked change in the appearance of chromatin. When 60% of the histone has been extracted, the clumps of chromatin in interphase nuclei are replaced by a fibrous network. Extraction with citric acid, which removes the lysine rich histones, is more disruptive; when 20% of the total histone—that is the f1 group—has been extracted, the network of fibrils replaces the usual clumps of chromatin. The restoration of lysine rich histones restores the clumped appearance. So far as can be judged from the appearance of partially dehistonized chromatin, therefore, the loss of the lysine rich histones has a much greater effect upon the structure of chromatin than the loss of arginine rich histones.

One possible explanation is that the lysine rich histones play an important role in determining the surface structure of chromatin so that their loss causes a general loosening of the structure (perhaps because they cross link different regions of the chromatin fibre). This interpretation is supported by the studies of Brasch, Setterfield and Neelin (1972), who found that the extraction of lysine rich f1 histones from chicken liver cells causes the chromatin fibres of 200Å diameter to be replaced by tightly packed fibres of only 100Å diameter; this may mean that the removal of lysine rich histones allows the supercoiled chromatin fibre to unwind. However, they also observed that the general distinction in the nucleus between euchromatin and heterochromatin is maintained so that the overall structure of interphase chromatin does not depend on the lysine rich histones alone in these cells. The idea that lysine rich histones are particularly important in the overall packing of the chromatin fibre is consistent with the demonstration that their removal has the greatest effect in making the DNA available to DNAase.

Stabilization of DNA by Histones

When the DNA of chromatin is melted by an increase in temperature, the thermal denaturation profile is very different from that of DNA itself (the melting of DNA is discussed in chapter 4). Figure 3.12 shows that only a small amount of the DNA of chromatin melts at the Tm value characteristic of free

DNA; a much larger proportion melts gradually over a somewhat higher temperature range. This biphasic melting curve also suggests that little of the DNA is free of protein; most of the regions of the duplex are stabilised by interaction with proteins, so that they do not melt until a higher temperature than that of free DNA.

By measuring the rate of change of hyperchromicity with temperature, a derivative curve of the melting profile can be drawn; this reveals how much of the DNA melts at any given temperature. Li and Bonner (1971) found that pea bud chromatin shows a broad melting profile with at least three phases. A small amount (about 4%) of DNA melts with a Tm of 42°C, which is that of free

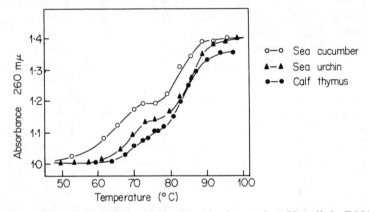

Figure 3.12: melting curves of eucaryotic chromatins. Very little DNA melts at the Tm characteristic of free DNA under these conditions; virtually all the DNA is protected by proteins and thus displays an elevated Tm. The biphasic melting profiles of these chromatin preparations, in which some of the DNA melts between 60–70°C and some in the range 80–90°C, are typical of eucaryotic chromatin; although the Tm values depend upon the species of chromatin, DNA usually falls into stabilized classes. Data of Subirana (1973).

DNA under the conditions of incubation. Some of the stabilised DNA melts at 60°C and the remainder at 81°C. The proportions of DNA melting at the two higher temperatures are decreased and the amount melting at the Tm of free DNA is increased when histones are successively dissociated from chromatin.

This shows that neither of the stabilised Tm values results from the inter-action of one histone alone with DNA, but each must be caused by the melting of regions of DNA duplex which are stabilised by interactions with all the histones. In addition to the major Tm transition, there is a shoulder on the 66°C peak, corresponding to a small peak at 55°C. This transition decreases only after most of the histone has been dissociated from the DNA; it may represent DNA which is not stabilised by direct binding of histone and possible explanations are that it results from interactions with non histone proteins or represents regions of DNA between two adjacent histone molecules.

The melting curves of chromatin derived from other sources also display more than one phase of stabilised DNA. Subirana (1973) found that under conditions in which free DNA melts at 55°C, calf thymus chromatin shows two transitions, at 69°C and 83°C; there is virtually no melting due to free DNA. Similar biphasic melting curves are exhibited by the chromatin of sea urchins and sea cucumbers. Li, Chang and Weiskopf (1972, 1973) obtained the derivative curve of calf thymus chromatin shown in the upper panel of figure 3.13; this has two stabilised transition bands, one at 71°C (corresponding to the 66°C band of pea bud chromatin) and one at 83°C (corresponding to the 81°C band of pea bud). In addition, the calf thymus curve shows a tail at 50–60°C which corresponds to the 55°C shoulder of pea bud. There is no peak at the 42°C region in which free DNA would melt under their conditions, so that there must be virtually no regions of extensive free DNA. (The absolute Tm values depend upon the conditions used for melting, including factors such as ionic strength, so that the stabilised transition values must in each case be compared with the melting point of free DNA under the same conditions.)

By contrast with the multicomponent melting curve of chromatin, the complex of polylysine with DNA displays the simple curve of the centre panel of figure 3.13, with the two components melting at 42°C (free DNA) and 98°C (DNA bound to the polypeptide). The different degrees of stabilization of DNA in chromatin must therefore represent a characteristic of its reaction with chromosomal proteins.

When polylysine is bound to chromatin, the hyperchromicity of the regions melting between 50–70°C disappears and is replaced by a new transition band at 98°C which corresponds to the melting of DNA-polylysine complexes. The lower panel of figure 3.13 shows that there is only a slight decrease in the transition band at 83°C. When sufficient polylysine has been added to bind 60% of the DNA, the area of the 98°C band shows that only about 30% of the DNA melts at this temperature.

Polylysine must therefore bind preferentially to those regions of chromatin originally stabilized to a lesser extent, presumably displacing the less basic regions of histones and/or the non histone proteins. Polylysine then binds to the better stabilized regions, but must add to the proteins covering the DNA which melts at 83°C instead of replacing them, since there is little diminution in the 83°C transition peak. These experiments thus suggest that the experiments of Clark and Felsenfeld (1971), in which DNA was titrated with polylysine, overestimated the number of free phosphate groups; the polylysine binds to chromatin proteins as well as to DNA and has only a 50% efficiency of binding so that the reaction of chromatin with polylysine does not depend upon the amount of free DNA.

Binding to DNA of individual histones also stabilises the duplex structure

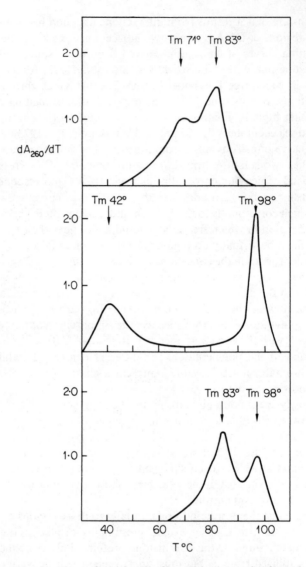

Figure 3.13: derivatives of melting profiles. This shows the dependence of
the rate of change of optical density during denaturation plotted against
temperature. Upper: calf thymus chromatin displays a biphasic melting
curve comprising two protected components, one with a Tm of 71°C, the
other with a Tm of 83°C. Centre: DNA–polylysine complex possesses free
DNA not bound by the polypeptide with a Tm of 42°C and bound DNA
with an elevated Tm of 98°C. Lower: addition of polylysine to chromatin
replaces the 71°C transition peak with a 98°C melting band but does not
alter the 83°C band. Data of Li, Chang and Weiskopf (1972).

against denaturation. Shih and Bonner (1970a) reported biphasic melting curves in which one component is free DNA melting at a low Tm and the other is stabilised DNA melting at a higher Tm. The proportion of the two bands in a derivative curve profile depends upon the ratio of histone to DNA; as the concentration of histone increases, more DNA is stabilised. Similar curves are shown by all histone fractions, although the extent to which each histone stabilises DNA varies. Under the conditions used, free DNA melts at 47°C and chromatin at 74°C; complexes of DNA-histone f1 (I) melt at 76°C, a complex of DNA-histones f2b and f2a2 (IIb2 and IIb1) melts at 81°C; and a complex of DNA-histone f2a1 (IV) melts at 84°C. DNA is even more strongly stabilised by other basic proteins; protamine raises the Tm to 92°C, poly-arginine to 98°C and polylysine to 99°C.

Derivative melting profiles more complex than these, however, have been reported by Li and Bonner (1971), who found that calf thymus histone f2b (IIb2) generates two peaks of stabilised DNA (that is in addition to the peak of free DNA not bound by the histone); they suggested that these peaks might correspond to the two principal stabilised peaks of chromatin. Each peak appears to be caused by a different part of the f2b molecule. This histone can be cleaved into two parts by reaction of its methionine residues with cyanogen bromide; Li and Bonner found that the N-terminal half stabilises DNA to a Tm of 70°C and the C-terminal half—which is less basic—stabilises DNA to a Tm of 57°C. These results suggest that both parts of the histone molecule bind to DNA.

The obvious explanation that each of the stabilised peaks of chromatin corresponds to one of the two levels of stabilization seen in these experiments is probably too simple. Experiments with partially dehistonised chromatin show that each histone contributes overall to each peak. Because the histones are small proteins and each molecule or half molecule can cover only a short length of DNA, each individual melting event may implicate a length of DNA associated with more than one histone molecule; as the different classes of histone seem to be intermingled in DNA rather than segregated, melting at the molecular level probably involves lengths of DNA bound to different classes of histone. To explain the two stabilised peaks of chromatin by different affinities for DNA within regions of histone molecules therefore requires a model in which all histones have stronger and weaker DNA binding regions; and the histones must be orientated together so that each sequence of nucleotide base pairs is associated either with the more strongly binding parts of several different histones or with their more weakly DNA-binding regions.

That all histones contribute to the stabilization of DNA in chromatin is suggested by experiments in which different classes of histone are successively dissociated from chromatin. Ionic extraction removes the lysine rich histones first and the arginine rich histones last. Ohlenbusch et al. (1967) and Kleiman and Huang (1972) found that increasing concentrations of NaCl elute the histones in the groups: f1 (I) at about 0·6 M NaCl; f2b and f2a2 (IIb2 and IIb1) at about 1·0 M NaCl; and f3 and f2a1 (III and IV) at 3·0 M NaCl. Smart and Bonner (1971a) found that deoxycholate elutes histones from chromatin in a different order; as its concentration is increased from 0·015 to 0·1 M, histones are removed in the order: f2b and f2a2 (group IIb); f3 and f2a1 (III and IV); and histone f1 (I).

The melting profiles of the partially dehistonized complexes broaden as increasing amounts of histone are removed but do not divide into regions of free DNA and protected DNA. Even when all of histones f2b and f2a2 (group II) have been removed by deoxycholate, there are no regions melting at the low Tm of free DNA. Smart and Bonner (1971b) obtained similar results when histones were instead extracted sequentially with NaCl. The stabilization of DNA is therefore reduced but not abolished as histones are removed in either of the orders of extraction achieved by deoxycholate or NaCl. Chromatin thus cannot consist of stretches of DNA bound to one histone alone; but all classes of histone must contribute to the stabilization of each region of the duplex.

The decrease in Tm as histones are extracted is not simply a function of the total weight of histone removed, for the release of a given amount of lysine rich f1 histones by NaCl causes a greater decrease than removal of the same weight of slightly lysine rich (f2b and f2a2) group II histones by deoxycholate. However, if the fraction of histone removed is plotted in terms of positive charge content—the lysine rich histones are the most positively charged, the slightly lysine rich are the least charged—against the decrease in Tm, a similar curve is obtained for either order of dissociation of histones. This suggests that the number of positive groups in each histone is an important parameter in determining the stability of DNA associated with it.

That high concentrations of salt abolish DNA-histone interactions reveals their ionic nature. However, hydrophobic forces also play a role in maintaining the structure of chromatin and stabilising the DNA duplex. According to electrostatic considerations alone, the lysine rich f1 histones should bind to DNA the most strongly since they have the greatest basic/acidic ratios. But in low salt the f1 histones are extracted first; and in 0·6 M NaCl, when most of the ionic interactions between histones and DNA should be abolished, almost all of the histones except f1 remain bound to DNA in the structure of chromatin. This implies that forces other than ionic neutralization of charge must play some role in the binding of histones to DNA (for discussion see Lewin, 1973).

Urea reduces the affinity of histones for DNA and lowers the concentration of salt which dissociates them from chromatin. Kleiman and Huang (1972) found that histones do not bind to free DNA in the presence of urea until salt concentrations are reached well below those needed to remove histones from chromatin in its absence. The ready elution from chromatin of histone f1 with NaCl suggests that it binds the most weakly to DNA; but when histones are allowed to associate with DNA and the salt concentration of a mixture is reduced in the presence of urea, histone I is the first to bind, at 0·2 M NaCl. As more salt is eliminated, the other histones also bind.

The presence of urea inhibits hydrophobic interactions; the reversal by the addition of urea of the affinity of f1 histones for DNA relative to the affinities of the other histones implies that abolition of hydrophobic forces most greatly reduces the interactions of these latter histones with DNA. In the presence of urea, histones are removed from DNA at lower concentrations of NaCl, that is their binding is less secure. In solutions of 6 M urea, histones f2b and f2a2 (IIb2 and IIb1) dissociate at 0·15 M NaCl, compared with 1·0 M NaCl in the absence of urea; extraction of the arginine rich histones f3 (III) and f2a1 (IV) depends strongly upon conditions such as the time of incubation and the concentration of urea, but is achieved within a range of about 0·6 M NaCl if urea is present, compared with 3·0 M NaCl in its absence. Extraction of f1 becomes partly easier in the presence of urea, with a dissociation level of 0·3 M instead of 0·6 M NaCl.

Experiments upon the dissociation of histones from calf thymus chromatin in phosphate buffers suggest a similar conclusion. Bartley and Chalkley (1972) found that histone f1 (I) is not affected appreciably by an increase in urea concentration and is eluted from chromatin by a low salt concentration of from 0·4–0·6 M. But the slightly lysine rich histones f2b and f2a2 (IIb2 and IIb1) are released together with the lysine rich histones if urea is added, instead of at higher ionic levels; in the presence of 3·0 M urea, these histones dissociate in 0·4 M phosphate buffer compared with an ionic level of the order of 1·0 M in the absence of urea. The concentration of urea has a pronounced effect upon the dissociation of the arginine rich histones f3 and f2a1 (III and IV); in 2 M urea there is no dissociation by 1·0 M phosphate, but in 6 M urea dissociation takes place in 0·6 M phosphate.

Propyl urea, an effective reagent in disrupting hydrophobic bonds, is more effective in dissociating histones than urea itself; the substitution of propyl urea for urea allows the dissociation of a greater amount of histone by any particular salt concentration. The effect of temperature on dissociation also confirms the importance of hydrophobic forces. The strength of hydrophobic interactions increases with temperature; histone preparations are usually incubated in conditions of low temperature (2°C) to avoid degradation, but in phosphate buffer and urea there is a shift in the affinity of histones for DNA when the temperature is increased to 20°C. All the histones except the f1 class bind more strongly to DNA at 20°C than at 2°C; the f1 histones bind less strongly.

Results obtained in phosphate buffer may be more precise than those obtained by using NaCl, for Bartley and Chalkley have observed that degradation takes place when histones are released from chromatin in NaCl and this may obscure quantitation of the release. But the same relative changes in histone dissociation take place in either ionic condition when urea is added;

these imply that hydrophobic forces play the most role in the binding to DNA of the arginine rich histones f3 (III) and f2a1 (IV), play some role in the interaction of the slightly lysine rich histones f2b (IIb2) and f2a2 (IIb1), but play little role in the binding of f1, which must be largely ionic in nature.

Structures of Histones in Chromatin

Histones in general display a low content of helix and in solution appear for the most part to exist in random coil conformations. By using NMR (nuclear magnetic resonance) to assess the proportion of the protein chain in the form of a helix, Bradbury et al. (1967) showed that the extent of helix formation increases with ionic strength (for review see Lewin, 1973). With individual histones in solution, histone f1 (I) increases from no helix in water to 10% helix in M NaCl; and histones f2b (IIb2), f2a2 (IIb1) and f3 (III) all increase from 5–10% in water to 25–30% in M NaCl. Histone f2a1 (IV) comes out of solution at M NaCl and so cannot be determined.

The helical content of total histone in chromatin appears to be of the order of 20%, so that it is within the same range displayed by the isolated histones in solution. However, measurements of the helical content of histones in chromatin are complicated by the presence of many components and Bradbury and Crane-Robinson (1971) have noted that such experiments in general tend to be unreliable and of poor reproducibility. Although the overall structure of histones may be very different when free in solution and part of chromatin, we may speculate that some of the regions of helix formation may be the same in both situations.

Helical regions of proteins tend to contain a high proportion of apolar residues—leucine, isoleucine, valine, alanine, methionine, phenylalanine, tyrosine, tryptophan—whereas the non helical regions have a greater proportion of polar side chains—lysine, arginine, histidine, aspartic acid, glutamic acid—which tend to engage in electrostatic repulsions that prevent helix formation. Proline also disrupts the formation of helix, by virtue of its stereochemical structure.

If these general rules are followed by histones, the clustering of each class of amino acid within the protein must mean that only certain local regions are likely to take up a helical structure. These regions appear to be those in the central parts of all histones; the polar content of the N-terminal and (albeit to a lesser degree) the C-terminal parts of the protein suggests that they are unlikely to display helix formation of their own accord. Considering the histone structure in isolation, then, the most basic sequences which probably bind DNA should not favour helix formation; but the central regions of the molecule, which may interact with other chromatin proteins, may do so. (Figure 3.5 shows the clustering of residues in histones.)

The potential of a protein for helix formation may be inferred by drawing a helical wheel of the form shown in figure 3.14; helical segments are usually

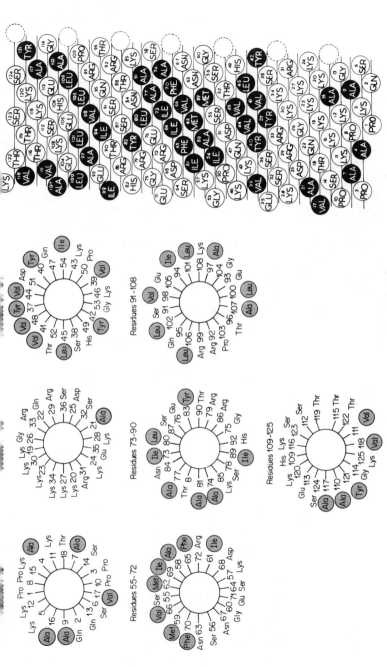

Figure 3.14: structure of calf thymus histone f2b (IIb2). Left: sequence written in the form of helical wheels, with the apolar residues which contribute to helix formation identified by circles. This suggests that helical structures are most likely to form in the centre of the molecule. Right: a representation of the two dimensional surface of a helical sequence, showing that the apolar residues (shaded) tend to be concentrated in restricted regions. After Boublik et al. (1970b).

characterized by the clustering of apolar residues on one side of the helix to form a stabilised arc. Helix formation in histone f2b (IIb2) of calf thymus is therefore most likely to take place in the four central wheels representing residues 37–54, 55–72, 73–90, 91–108. The probability of helix formation in the two wheels representing the N-terminal sequence and the wheel representing the C-terminal sequence is very much less. A representation of the two dimensional surface which would be displayed if the histone took up a helical structure illustrates the extent of the concentration of the apolar residues in restricted regions.

A similar model applies to the other histones. As the sequence of histone f2a1 (IV) of figure 3.3 shows, for example, the high number of charged groups and proline residues in the N-terminal region confers a very low potential for helix formation. The immediate C-terminal region of the protein is also unlikely to form a helix. The clustering of amino acids in helical wheels suggest that the central regions are likely to form helices, residues 55–72 having the greatest potential and residues 37–54 and 73–90 also showing high probability. Comparable structures are generated by drawing helical wheels for histones f2a2 (IIb1) and f3 (III); there is at present no reason to suppose that the sequence of f1 will display a different type of structure.

NMR studies have suggested that helix formation is indeed restricted in histones in solution to the predicted regions (Bradbury et al., 1967; Boublik et al., 1970a, b). Bradbury et al. (1972b) have made a more detailed study of histone f2b (IIb2), which can be cleaved into parts by the reaction of cyanogen bromide with its two methionine residues; the N-terminal fragment consists of amino acids 1–58 and the C-terminal fragment comprises residues 63–125. Except for residues 40–47, all the potential for helix formation is found in the carboxyl part of the molecule. Little change in conformation takes place in the N-terminal half when the salt concentration is raised, but there is an appreciable increase in the secondary structure of the C-terminal fragment. The average behaviour of the two fragments together is that of the complete molecule; this suggests that two halves of the intact molecule behave independently and may have different functions in chromatin.

By using a computer programmed analysis of the NMR spectra of the histones, Bradbury and Rattle (1972) have followed the changes of conformation in salt solution more precisely. In general, the regions of the molecule which take up a helical structure correspond to those predicted to do so by their content of apolar residues. Salt induced changes in the conformation of histone f1 take place in residues 50–106, the central part of the molecule, whereas there is little helix formation in residues 1–40 and 107–216. In histone f2a1 (IV), the entire molecule apart from the N-terminal region seems to be involved both in the development of intrachain secondary structure and in interactions between different polypeptide chains. In histone f2b (IIb2), sequences 30–50 and 70–102 are principally involved in conformational changes when the salt concentration is raised; these changes include the acquisition of secondary structure by the chain and its interaction with other chains.

Changes occur in the NMR spectra when f2b is added to DNA (in M NaCl); the regions of the histone stabilised by DNA correspond to residues 1–30 and 103–125. This implies that the structure of histone molecules is altered by their interaction with DNA; it is possible to construct models in which the histone takes up a helical form when it is bound to DNA by ionic neutralization of charge. The most likely model for the assembly of chromatin is therefore to suppose that histones possess helical regions in the central parts of the polypeptide chain; and it is these regions which give each histone its characteristic structure in chromatin and which interact with the other histones and non histone proteins. The more basic N-terminal and C-terminal ends of each histone do not spontaneously generate helical structures but lose their random coil conformation to acquire secondary structures when they interact with DNA. It is possible that each end of the histone may independently interact with DNA, leaving the central portion free to interact with other components of chromatin.

Studies of chromatin using circular dichroism suggest that the structure of DNA is altered by its association with histones; it is packaged into a coiled coil. Partially dissociating some histones from the chromatin in part releases the conformation of DNA, so that it has a structure closer to that of free DNA. Although there is at present no agreement about which histones make the greatest contribution to maintaining the supercoil of DNA, Bartley and Chalkley (1973) have found that the dissociation of f1 has the least effect; this is consistent with proposals that f1 may be implicated more in maintaining the coiled structure of the chromatin fibre itself rather than in binding to DNA. Inhibition of hydrophobic forces by the addition of urea also causes the DNA to be released into a conformation more like that of free DNA; the supercoiling of DNA may therefore depend upon hydrophobic interactions as well as upon its ionic binding to histones.

In the overall structure of chromatin, DNA forms a regular supercoil, bound to histones by both ionic and hydrophobic forces. In addition to their affinity for DNA, the histones have considerable ability to aggregate with each other; and these attractive forces may play an important role in maintaining the structure of chromatin. The importance of ionic bonds is emphasized by the effect of salt on chromatin and by observations such as those of Miller, Berlowitz and Regelson (1971) that addition of the polyanion polystyrene sulfate to mealy bug cells reverses the condensation of chromatin. The importance of hydrophobic bonds both between histones and DNA and between histones themselves is revealed by the effect of urea upon the stabilization of DNA structure and upon the dissociation of histones from chromatin.

A model for the coiling of DNA in chromatin has been presented by Bartley and Chalkley (1973). They suggested that the major destabilising force is the rigidity of duplex DNA—the DNA prefers a rod like conformation to a supercoil—and the repulsion of electrostatic charges in chromatin (chromatin has about one quarter of the negative charge of DNA itself, distributed evenly along its structure). The major stabilising force to maintain the structure might be a torque generated by the twisting of the DNA molecule upon its axis. This torque could be maintained if the histones are bound firmly to sites on DNA and if the secondary structure of the histones pulls these sites towards each other. The model suggests that the torque is relieved by the formation of supercoils, which forces chromatin into its compact form.

Stability of Chromatin Structure

The genetic material itself is of course a stable component of chromatin; a label in DNA is diluted only by replication, when it segregates semi-conservatively between the daughter chromosomes. A label incorporated into protein does not behave in this manner, but is randomly segregated to daughter chromosomes and is gradually lost from them (see chapter 1). Too little is known about the non histone proteins to draw any conclusions about their stability in the chromosome. Studies of histone turnover have produced different results in different systems, but some experiments suggest that they may be stable components of the chromosome which remain associated with DNA for at least several generations of the cell (although of course any one histone molecule may not remain bound to the same stretch of DNA but may suffer a redistribution during S phase). [One possible criticism of experiments which have seemed to show that histones are unstable components of chromatin is that contamination with proteins which do turnover rapidly could suggest a spurious apparent instability.]

Histones appear to comprise a stable component of chromatin of HTC cells in both stationary and exponential growth; Balhorn et al. (1972) found that when a radioactive label is diluted by the synthesis of new molecules, both histone and DNA display the same half life of 24 hours. This shows that the histones remain in the chromosomes throughout the cell cycle; they do not turnover and are diluted only by the association with chromatin of an equal amount of histone at replication. Working with Hela cells, Hancock (1969) showed that histones are stable chromosome components for at least eight generations. This metabolic stability reinforces the idea that histones are structural components of the chromosome.

Little is known about the synthesis of non histone proteins and their addition to chromatin. Bhorjee and Pederson (1972) have followed the non histone content of Hela cells during the growth cycle; the same electrophoretic bands are present at all stages of interphase. However, there are differences in the amounts of each band; in particular, a small number of bands is reduced by up to half in protein content during S phase. This suggests that some of the non histone proteins may behave in the same way as histone and associate with chromatin when DNA is reproduced, so that its overall protein content remains the same; a smaller number of proteins may be added at a later stage.

One time at which the protein components of chromatin might change is the transition from interphase to mitosis when the chromosomes must become much more tightly folded. Some analyses have shown that isolated metaphase chromosomes appear to have about twice the protein content of interphase chromatin. Huberman and Attardi (1966) and Maio and Schildkraut (1969) found that about 15 % of the metaphase chromosome is DNA, compared with 30 % in interphase chromatin. Since the DNA content must be the same, the protein content must have doubled at metaphase.

This change results from the association of additional non histone proteins with chromatin. Sadgopal and Bonner (1970b) found that Hela cell interphase chromatin has the composition:

DNA—100; histone—115; non histones—86; RNA—12.

In metaphase chromosomes the histone content remains the same but the non histone protein increases to a value of about 355 relative to DNA. The extra non histone protein is therefore sufficient to account for the increase in total protein of chromatin during mitosis. Hancock (1969) has also shown that there is no change in the histones present in mitotic chromosomes but that non histone protein content increases. In quantitative studies, Skinner and Ockey (1971) observed that metaphase chromosomes of M. agrestis possess only a half to a third of the f1 histone content of preparations of interphase chromatin, in contrast to the lack of change in the histones of the chromosomes of other organisms.

About half of the non histone protein added to the chromosomes at mitosis shares with histones the property of extraction by HCl (although it is not extracted by H_2SO_4); but although it is acid soluble, this fraction is distinct from the histones (Sadgopal and Bonner, 1970b). Some of the additional proteins therefore have properties which may be intermediate between those of the histones and the non histone proteins of interphase chromatin.

One of the changes in cellular structure at mitosis is the dissolution of the nucleolus, after which many of the nucleolar fragments associate with the chromosomes; these fragments consist largely of ribonucleoprotein particles which may provide at least an appreciable proportion of the non histone protein added to the chromosomes. This may also account for the large increase in RNA content at metaphase—to a relative value of 65—which is large ribosomal.

Using a different method of preparation, Stein and Farber (1972) reported a failure to detect any changes in chromatin proteins at mitosis. We do not know which preparations resemble the true state of the chromosomes within the cell, but it is possible that much of the additional non histone proteins represent extraneous material—from the nucleolus or elsewhere—which associates with the chromosomes for the duration of mitosis but is not part of its structure. It is presumably the protein components of chromatin which are responsible for mitotic coiling, but at present we have little information on whether this is achieved by changes in the interactions of pre-existing proteins or by addition of new proteins.

Calculations of the extent of packing of DNA into the chromatin fibre show that it must have a more compact structure at metaphase (see chapter 1). In spite of the differences in composition of interphase chromatin and metaphase chromosomes, Shih and Lake (1972) observed that circular dichroism and thermal denaturation profiles of both types of preparation remain the same (although of course differing from pure DNA). Pardon et al. (1973) reported similar X-ray diffraction patterns for interphase and metaphase preparations. These results imply that the additional proteins associated with metaphase chromosomes do not change their fundamental structure; mitotic condensation must therefore take place without changing the manner of supercoiling or the degree of stabilization of DNA to an extent detectable by these techniques.

Assays of the availability of DNA of interphase chromatin to exogenous reagents support the concept originally suggested by Mazia (1961, 1963) that the chromosomes may undergo a continuous cycle of coiling and uncoiling, instead of discontinuous steps just prior to and after mitosis. Pederson (1972) found that the chromatin of early S phase cells is more sensitive to DNAase than the chromatin of late S phase or G2 or G1 cells; half the concentration of DNAase is required in early S phase to achieve a given amount of degradation. This change reflects a property of bulk DNA, not simply that of newly synthesised DNA, for if cells are labelled with H^3-thymidine for 15 minutes before the chromatin is extracted and digested, the pulse label suffers degradation to the same extent as the bulk of DNA. A comparison of African green monkey tissue culture cells grown in stationary and exponential phase showed that the contact inhibited cells are less sensitive to DNAase; this agrees with the concept that they remain in a G1-like phase, whereas the growing cells must pass through the more sensitive S phase.

The reactivation of an inert nucleus is accompanied by a visible increase in the volume of its chromatin; the formation of more diffuse chromatin is accompanied by a substantial increase in the binding of dyes by DNA (see page 398). Association with small molecules can therefore be used as an assay for the accessibility of DNA in chromatin. Pederson and Robbins (1972) obtained Hela cells in mitosis and G1 by selective detachment and in S and G2 phases by a double thymidine block; the cells were then exposed to H^3-actinomycin, which binds specifically to the guanine bases of DNA. Figure 3.15 shows that there is a continuous variation in the relative binding of actinomycin to DNA; it is at a minimum at the time of mitosis and increases during G1 to reach a maximum just around the beginning of S phase, after which it declines until the next mitosis.

These results suggest that the condensed chromatin fibres of mitotic cells gradually uncoil, thus making their DNA more accessible to actinomycin and to DNAase, until they reach their most relaxed structure in preparation for the

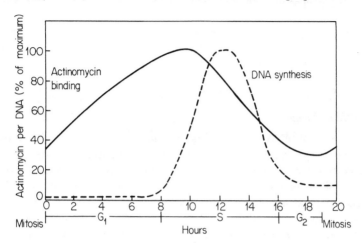

Figure 3.15: binding of actinomycin to DNA during the cycle of Hela cells. Chromatin suffers a continuous cycle of condensation and unfolding; it appears most unwound just prior to S phase and most compact at mitosis. Data of Pederson (1972).

replication of DNA. After its reproduction, chromatin gradually becomes more condensed in preparation for mitosis. The state of the mitotic chromosomes therefore represents one extreme of a spectrum of chromosome coiling; there seem to be no points at which sudden changes take place in the chromatin fibres. Of course, this may make it more difficult to resolve the interactions which are responsible for chromatin coiling, since interphase chromatin may contain a spectrum of structures undergoing continuous change.

CHAPTER 4

Sequences of Eucaryotic DNA

Repetitive Complexity of the Genome

The C Value Paradox

The amount of DNA in the haploid genome of a eucaryotic cell—known as its *C value*—appears to be greater by an order of magnitude or so than would be predicted from the number of genes which we expect to code for proteins. Calculations of the number of genes in a eucaryote can be only rather crude since they depend upon assumptions about recent evolution; however, such calculations suggest that if all, or even most, of the DNA of a eucaryote such as man were devoted to coding for proteins, the number of deleterious mutations occurring each generation would be great enough to kill the species.

The frequency of deleterious mutation at each genetic locus is usually taken to be about 10^{-5}–10^{-6} in each generation (see Muller, 1967). Ohta and Kimura (1971) have supported this estimate by comparing the amino acid sequences of corresponding proteins in different species; they derived a frequency of mutation of about 8×10^{-9} for each amino acid per year. This is consistent with a rate per locus of about 2.5×10^{-6} per generation if proteins are on average coded by genes of about 1000 base pairs (0.67×10^{6}) daltons and if the generation time is one year; these parameters are usually taken as typical of those which may have prevailed during evolution of the species used in these estimates.

The human genome is fairly typical of the mammals and consists of some 3×10^{9} base pairs, or 10^{9} potential amino acid coding sites. If the genome is devoted entirely to coding for proteins, a mutation rate of 8×10^{-9} per amino acid per generation would imply that about 8 replacements occur in each generation. Since many of these changes are likely to be deleterious, any species with a genome of this order of size would suffer a heavy genetic load from mutation. In another calculation, taking the rate of deleterious mutation per locus to be 10^{-5} per generation, Ohno (1971) has pointed out that a human genome of 3×10^{6} genes would accumulate 30 deleterious replacements each generation—again constituting an unbearable genetic load. Consistent with these estimates are the calculations derived by Kohne (1970) from comparisons of the rates and extents of renaturation together of the DNAs of different species. The differences between man and monkey DNAs suggest that about 5 base changes occur in each gamete.

In contrast with these values, Muller (1967) suggested that the overall rate of deleterious mutation prevailing during the evolution of man may have been about 0·5. If the frequency of mutation per locus is 10^{-5}–10^{-6} per generation and if the overall rate of mutation is 0·5 per generation, then the number of genes should be of the order of 5×10^4–5×10^5. An estimate of the order of 10^5 genes in the human genome therefore seems reasonable; Ohno (1971) has suggested 4×10^4.

The number of genes in the human genome consistent with mutation rates is therefore very much less than the coding potential; if these estimates are correct, less than 5% of human DNA should have the function of specifying proteins. Even allowing for the uncertain values of some of the parameters used in these calculations, the discrepancy between a prediction of 10^5 genes from mutation rates and of 3×10^6 genes from the genome size, is so large as to suggest that at least a large proportion of the human genome does not code for proteins.

A problem which is associated with the apparent excess of DNA in the eucaryotic genome is that C values do not display any correlation with evolutionary complexity. We should expect the amount of DNA in the genome of a species to be roughly proportional to its complexity; but this does not seem to be so. As figure 4.1 shows, the *minimum* size of genome reported for each class of eucaryote increases in a manner depending upon evolutionary development. This shows that a certain increase in genetic information must accompany evolution; however, many species have a much greater content of DNA than the minimum in any class.

Although our data are obviously incomplete and include only a small proportion of the eucaryotes, as figure 4.2 shows most classes of eucaryote comprise a range of species with a variation in genome size of about 10 fold. Mammals have genomes which fall into a particularly small range of DNA contents; the C value is usually 2–3 picograms (2–3×10^9 base pairs). Amphibia, by contrast, vary much more widely, from less than 1 picogram to almost 100 picograms; even closely related amphibia may have greatly different contents of DNA in their nuclei. Taking the members of each class as a whole, there is therefore no relationship between position on the evolutionary scale and the C value of a species. The lack of correlation between relative genetic complexity and DNA content is sometimes referred to as the C value paradox.

Closely related species must presumably utilise about the same number of genes in their development; their differences in C value may therefore be explained in either of two ways. One class of model supposes that differences in DNA content reflect differences in the degree of amplification of genes within the haploid genome. According to such models, species with a low C value should have a small number of copies of each gene, but related species with high C values should have a much greater number of copies in each haploid genome. This would mean that although the absolute amount of DNA might vary, the genetic complexity of the species would remain the same. Genetic considerations argue against models which suggest that there is more than one copy of each gene (see chapter 1); analyses of DNA sequences also refute this idea, at least on a basis widespread enough to explain the C value paradox, although we cannot at present exclude the possibility that some individual genes exist in more than one copy.

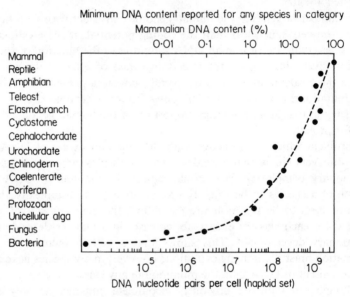

Figure 4.1: minimum content of haploid genome found in each class of species. DNA content increases with complexity; but other members of each class may have appreciably greater DNA contents as shown in figure 4.2. Data of Britten and Davidson (1968).

The apparent excess of DNA in the eucaryotic genome, coupled with wide variations in C value, can instead be interpreted to mean that only a small proportion of the DNA codes for proteins, the rest having some other function. According to this class of model, related species should have the same amount of DNA coding for protein, but might differ widely in the amount of DNA serving other functions. The idea that the excess DNA is concerned with control functions, perhaps acting as recognition sites instead of elaborating proteins, is again difficult to reconcile with variations in C value between related species; we might expect similar species to have similar control networks as well as genes coding proteins. One possible explanation for the excess DNA is that it has some structural significance in the chromosome; this might even to some extent be independent of its sequence. However, these models also

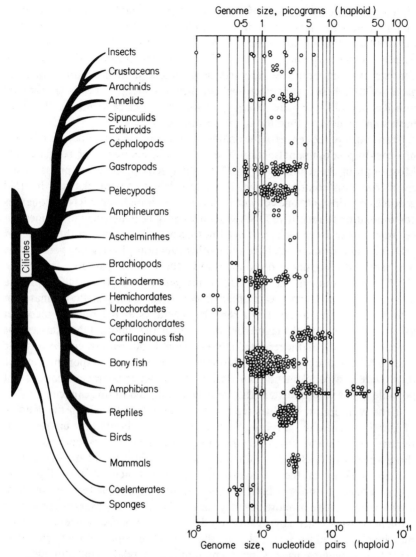

Figure 4.2: Genome sizes of the eucaryotes. Data of Britten and Davidson (1968).

encounter serious problems; it is difficult to visualise structural roles for an amount of DNA varying from most to virtually all of the genome.

There is at present therefore no satisfactory explanation for the large contents of DNA of eucaryotes and for their wide variations. Some models, however, do not appear to be consistent with the parameters which define the division of sequences of eucaryotic DNA into three classes. Some sequences are unique; there is only one copy of each sequence in the haploid genome. Another major component includes sequences which are repeated many times; this intermediate fraction comprises a spectrum of varying degrees of repetition, usually greater than one thousand fold. The smallest class of eucaryotic DNA consists of short sequences repeated an enormous number of times, often in excess of 10^6; these highly repetitive sequences do not appear to be transcribed into RNA, and as they are in large part located at the centromeres, they may have some structural function in the chromosome.

Transcripts of sequences of the unique (or non repetitive) component of DNA provide the predominant fraction of sequences represented in messenger RNA. Most or all of the sequences of mRNA that specify proteins are probably derived from this component, which is equivalent to saying that the haploid genome contains only one copy of each gene. (Although some RNA sequences corresponding to repetitive DNA sequences have been detected in the messenger fractions of some cells, this need not necessarily mean that they code for proteins and they may well have some other function.) Restricting genes to the non repeated component does not resolve the C-value paradox, for unique DNA may account for around half of the genome and this itself displays a coding potential much greater than is consistent with genetic estimates of gene number. This means that either the genetic estimates must be incorrect by at least an order of magnitude, or much of the unique DNA (as well as probably virtually all the repeated DNA) does not code for proteins. A possible role for the sequences of the intermediately repetitive component may be to provide control elements, but until they are better characterized it is not possible to say to what extent they may be implicated in this capacity.

If the variation in C values represents an amplification within the genome of identical genes, those species with high C values should lack unique components and should have only repeated sequences; correspondingly, only those species of each class with the lowest C value should display unique sequences. However, there is no evidence to support such a relationship; most species appear to possess both unique and repeated sequences, although of course their relative proportions may vary. In fact, the very existence of unique sequences argues strongly against models which suppose that excess DNA may be accounted for by amplifying the number of copies of each gene. The repetitive sequences usually represent families of related but not identical sequences, so that it is unlikely that they include any substantial portion of amplified genes (which would all be of identical sequence).

One exception to the limited number of copies of each gene is of course the situation of the ribosomal RNA genes. All eucaryotes appear to possess a large number—of the order of 100 or more—genes coding for rRNA, which are usually clustered on the chromosome associated with the nucleolus. (Of course, the total contribution of these sequences to the DNA of the genome is very small, less than one per cent.) So far as we know, all these genes code for ribosomal RNA of identical sequence; this implies that the cell possesses some mechanism which prevents or selects against mutations which change the sequence of the rRNA coded by each gene. Since these genes do not code for protein and are expressed by the transcription apparatus of the nucleolus, their degree of repetition and identity of sequence may represent a special case. We do not know whether there may be any repetition and clustering of the genes which code for ribosomal proteins.

Denaturation and Renaturation of DNA

When a solution of DNA is heated, very striking changes occur in many of its physical properties, such as viscosity, light scattering and optical density. This occurs in a narrow temperature range which depends upon the pH and type of buffer; at physiological ionic strengths it is usually in the region of 80–90°C, but may be lowered to as little as 40°C by the presence of reagents which assist denaturation, such as formaldehyde. This *denaturation* or *melting* of DNA is usually characterized by the temperature of the mid point of transition, Tm, sometimes referred to as the *melting temperature*; the process represents the disruption of the double helix of DNA into separated single strands. The physical nature of denaturation has been discussed by Marmur, Rownd and Schildkraut (1963).

The denaturation of DNA can be followed in several ways. The heterocyclic rings of the nucleotides absorb ultraviolet light very strongly, with a maximum around 260 mμ; but the absorption of DNA itself is some 40% less than would be expected from a mixture of free nucleotides of the same composition. This is described as the hypochromic effect; although the cause is not completely understood, it appears to result from mutual interactions of the electron systems of the bases. It is thus at its greatest when they are stacked in the parallel array of the double helix. The degree of hypochromicity is a sensitive measure of the physical state of DNA, since any departure from the ordered configuration of the double helix is reflected by a loss of hypochromicity— or increase of hyperchromicity—that is an increase in optical density. Figure 4.3 shows the melting curves of various DNAs followed by increase in optical density. All the DNAs display the same general melting profile, although each has a characteristic Tm (see later).

When DNA which has been denatured by heating is cooled, the ultraviolet absorbance decreases again, typically to a level some 12% above that of the original solution. Doty et al. (1960) found that when this renatured solution is in turn reheated, the optical density increases again but in a manner which depends upon the speed with which the denatured DNA has been cooled. After a quick cooling, reheating results in a gradual increase in optical density without a sudden transition. This suggests that short regions of DNA which happen to be complementary have associated, but without reformation of the original duplex, so that the short renatured regions have varying stabilities and melt over a wide range of individual transition temperatures, generating a gradual increase in optical density. But when the denatured solution is cooled more slowly, on reheating there is a characteristic transition close to the usual Tm. This suggests that slow cooling allows a substantial proportion of the original duplex molecules to reform to give a DNA which more closely resembles native DNA; it is this process which is usually described as *renaturation*.

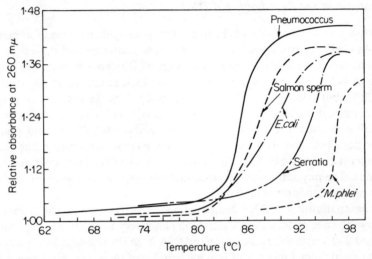

Figure 4.3: Melting curves of DNA from different species. Although the Tm values of particular DNAs are different, the general curves of their melting profiles are similar. Data of Marmur and Doty (1959).

Another technique which can be used to determine the state of DNA is density gradient equilibrium centrifugation. Meselson, Stahl and Vinograd (1957) observed that when a solution of some dense low molecular weight solute—such as sucrose or caesium chloride—is centrifuged at high speed, an equilibrium is reached between the opposing tendencies of sedimentation and diffusion. This produces a continuously increasing density of solution along the direction of the centrifugal force. When a macromolecule such as a nucleic acid is placed in this solution, it moves to an appropriate equilibrium position where its density equals that of the solute. This is known as its *buoyant density*. It stabilises in a band around this density corresponding to a Gaussian distribution.

Double stranded and single stranded DNA have different characteristic buoyant densities. In either case, the density of the DNA depends upon its base composition. Molecules of duplex DNA which are rich in G and C have a greater density than those which are rich in A and T; buoyant density increases more or less in proportion to the increase in G–C content, from 1·700 gms/cm³ at 42% G–C to 1·718 gms/cm³ at 59% G–C. The empirical formula:

$$\rho = 1\cdot660 + 0\cdot00098(G + C)$$

is often used to deduce the base composition of duplex DNA from its buoyant density.

The separated strands of denatured DNA can be centrifuged to density equilibrium in alkaline solution (the alkali maintains the DNA in its denatured state); single strands have a greater buoyant density than duplex molecules and denaturation of E.coli DNA changes its buoyant density from 1·710 gms/cm³ to 1·725 gms/cm³. The buoyant density of single strands of DNA is determined by their relative proportions of pyrimidines and purines; buoyant density increases with pyrimidine content. In duplex DNA, of course, the pyrimidine and purine contents are equal; but they may be quite different in the single strands, in which case the two strands of a duplex may be separated by alkaline density gradient centrifugation.

The buoyant densities of DNAs produced by renaturation experiments support the idea that fast cooling produces a random aggregation whereas slow cooling permits formation of longer duplex regions. Molecules renatured slowly have a buoyant density close to that of the original native DNA before its denaturation. But the rapidly cooled molecules have a higher density, which might arise from the random reaggregation of complementary parts of several different strands along one given strand.

Columns of hydroxyapatite have more recently been used to separate single stranded and double stranded molecules of DNA (see below). At low ionic strength (0·12 M phosphate buffer), hydroxyapatite retains duplex DNA but single strands are eluted. The denaturation of DNA can therefore be followed by applying duplex molecules to the column and following the release of single strands as the temperature is increased. Renaturation of a solution of complementary single strands of DNA can be assayed by the amount of DNA retained by the column. Hydroxyapatite columns offer the advantage that both the single stranded and duplex molecules of DNA may be isolated; although single strands alone are directly eluted when DNA is applied to the column, the duplex molecules which are retained can be eluted by increasing the salt concentration to 0·27 M phosphate buffer.

Reformation of duplex molecules from denatured single strands is specific for complementary sequences of DNA; strand reunion does not occur between DNA molecules of different sequences. Whereas the denatured single strands of DNA of any one species renature under the appropriate conditions, therefore, single strands derived from unrelated species fail to associate. But when two species have some DNA sequences in common, these sequences can renature when two preparations of denatured DNA are mixed. The extent of renaturation of the DNA from one species with that of another therefore depends upon the degree of similarity between their genomes, so that it provides a criterion of their evolutionary relationship. Laird and McCarthy (1968a) have compared the relationships between different species of Drosophila according to their inter-DNA renaturation; Bendich and McCarthy (1970) have measured the thermal stability and extent of duplex formation between the DNAs of barley, oats, rye and wheat. The extent of heterologous duplex formation in general appears closely related to the predicted evolutionary relationship.

Hybridization between DNA and RNA

Since renaturation depends upon the pairing reaction between complementary bases, denatured DNA should be able to anneal with complementary sequences of RNA as well as DNA. The product of this reaction is a hybrid with one strand of DNA and one strand of RNA. The original experimental technique of hybridization in solution is at a disadvantage in examining the formation of hybrids between DNA and RNA, for hybrid formation must in this case compete with renaturation between the original parent strands. This difficulty has been overcome by the development of methods for immobilizing the DNA single strands so that they cannot renature with each other, although they remain free to hybridize with other nucleic acids which may later be added.

In a technique developed by Bolton and McCarthy (1962), denatured single strands of DNA are immobilised in a gel of agar; the preparation of DNA-agar is then incubated with the molecules which are to be hybridized, usually by using the agar as a column to which the molecules to be hybridized are added. Complementary sequences of RNA or DNA are retained on the column but sequences which are not complementary to the immobilised DNA are washed through. By adding a radioactive preparation of RNA or DNA to the column, the retention of radioactive label can be used to measure the extent of hybridization.

The use of agar columns has now given way to a more convenient method for immobilising the DNA, developed by Gillespie and Spiegelman (1965) and Nygaard and Hall (1963, 1964), which allows a faster and more quantitative assay of the hybrid. In moderately concentrated salt solution, nitrocellulose filters adsorb single strands of DNA. Molecules of RNA are not adsorbed, but if a DNA-filter preparation is incubated with a solution of RNA, the complementary sequences of RNA hybridize with the DNA and are therefore retained by the filter. As figure 4.4, shows by using labelled RNA the filter can be counted for retention of radioactivity to measure the degree of hybridization. This technique can also be used for DNA–DNA hybridization by treating the DNA-filter with a solution of albumin; Denhardt (1966) found that this prevents any further single strands of DNA from adhering to the membrane (although those already adsorbed are retained). If a preparation of single strands of DNA is added, only those molecules which can form duplexes with the DNA of the filter are retained.

The absolute efficiencies of hybridization of DNA with RNA depend on several factors in addition to the degree of complementarity (see later). A more powerful technique is therefore to use competition assays to test the

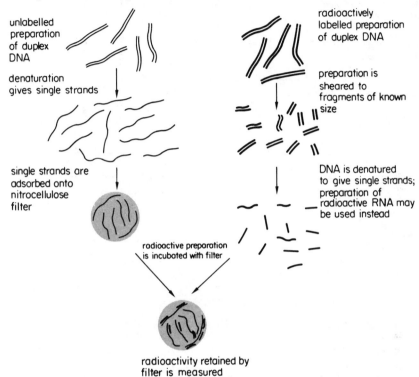

unlabelled preparation of duplex DNA

denaturation gives single strands

single strands are adsorbed onto nitrocellulose filter

radioactive preparation is incubated with filter

radioactivity retained by filter is measured

radioactively labelled preparation of duplex DNA

preparation is sheared to fragments of known size

DNA is denatured to give single strands; preparation of radioactive RNA may be used instead

Figure 4.4: hybridization of denatured DNAs by the membrane filter technique. Labelled RNA may be used instead of DNA. The conditions of incubation of the filter with the labelled preparation may be varied to control the rate by changes in temperature, concentration of labelled DNA in solution, time of incubation, ionic strength of solution.

homology of two RNA (or DNA) preparations. As figure 4.5 shows, several filters are obtained from a preparation of denatured DNA and are tested in a series of incubations with RNA. Each incubate contains a constant amount of a radioactively labelled RNA (or single stranded DNA) preparation, but increasing amounts of a different, non-labelled preparation. The assay measures the degree of competition between the labelled and unlabelled RNAs for DNA; the influence of the increasing concentration of the unlabelled RNAs on the binding of the labelled RNA molecules depends on their homology.

If the two preparations of RNA share no similarity of sequence, they must

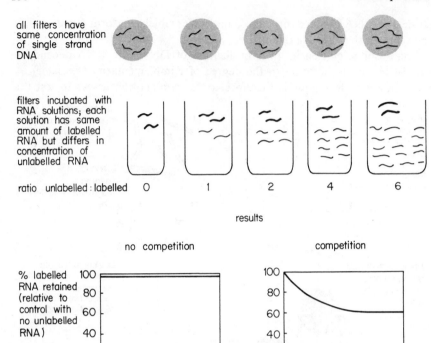

all filters have same concentration of single strand DNA

filters incubated with RNA solutions; each solution has same amount of labelled RNA but differs in concentration of unlabelled RNA

ratio unlabelled : labelled   0   1   2   4   6

results

no competition   competition

% labelled RNA retained (relative to control with no unlabelled RNA)   100   80   60   40   20

1 2 3 4 5 6
ratio unlabelled / labelled RNA

Figure 4.5: competition assay for RNA sequences. Filters of single stranded DNA are incubated with a series of preparations of RNA; each preparation contains the same amount of labelled RNA but a different amount of unlabelled competitor. Left: if the two RNA preparations have no similarity, the same amount of labelled RNA is retained on all the filters, irrespective of the concentration of unlabelled RNA. Right: the binding of the labelled RNA is reduced to a plateau of 60% by increasing concentrations of the unlabelled preparation. This shows that 40% of the sequences of the labelled RNA preparation are also found in the unlabelled preparation.

hybridize with different sequences of the immobilised DNA. In this case there is no competition between them, so that increasing the concentration of the unlabelled preparation has no effect on the binding of the radioactive molecules; the same amount of radioactivity is retained by all the filters. But if the two preparations are homologous in base sequence they must compete with each other for the same sites on DNA. As the concentration of the unlabelled preparation is increased, it is increasingly successful in binding to DNA so that less of the radioactive preparation can hybridize. The greater the similarity between the two preparations, the more pronounced is the displacement of the radioactive molecules by increasing concentrations of unlabelled species. The retention of radioactivity declines to a plateau as the concentration of the unlabelled species increases; the level of this decline measures the extent of homology between the preparations.

Influence of Temperature on Renaturation

Three structural parameters govern the stability of a DNA duplex and thus the temperature at which it dissociates into complementary single strands; the relative contents of the two types of base pair, the proportion of properly

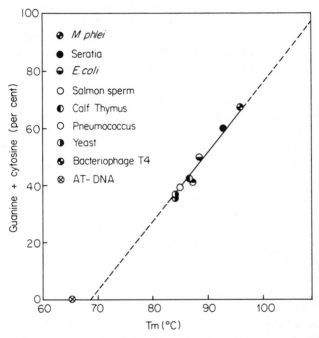

Figure 4.6: dependence of melting temperature on G–C content of DNA. Duplexes of DNA are more stable with increasing G–C content. Data of Marmur and Doty (1959).

paired bases, and the length of the duplex (see McCarthy and McConaughy, 1968). Figure 4.6 shows that the T_m increases with the proportion of G–C base pairs. This results from both intra- and inter- base pair effects. Since each G–C pair has three hydrogen bonds compared with the two of each A–T pair, greater energy is required to disrupt a G–C pair than an A–T pair. It has also been calculated that G–C pairs make a greater contribution than A–T pairs to base stacking interactions which help to stabilise the duplex. The energy required to separate individual base pairs of a duplex therefore increases with its content of G–C pairs. When the denaturation and renaturation of DNAs of different base compositions are compared, it is thus necessary to make a correction to compensate for their relative G–C contents.

Any mispairing between bases reduces the stability of the duplex. In a duplex of DNA comprising two strands which are perfectly paired with each other, every base pair must be disrupted before the strands can separate. This is the condition which we may expect to prevail when native DNA is isolated. However, the presence of any non-matched pairs of bases which are not held together by hydrogen bonds reduces the energy required for strand separation, since these pairs make no contribution to the stability of the duplex. As an approximate correlation, 1·5% non-matching lowers the Tm by about 1°C. This situation may be encountered in studies of the thermal stability of DNA which has previously been denatured and renatured. The reduction in the Tm of such preparations from the value characteristic of native DNA therefore provides a measure of the extent of mispairing which has taken place during renaturation.

The length of the duplex also influences its denaturation and renaturation. Variations due to length are usually eliminated by shearing preparations of DNA to a standard size, often about 500 nucleotides long. Under the conditions which are usually used for denaturation, this treatment reduces the stability of the duplex by about 1–2°C.

In addition to these factors intrinsic to the duplex itself, the conditions of incubation greatly change the Tm at which DNA melts. These parameters include pH and ionic strength; under physiological conditions a Tm of 80–90°C is usually observed. It is sometimes advantageous to utilise a much lower temperature of reaction, in which case reagents such as formamide may be used to lower the Tm to within a range of about 40°C.

Renaturation of separated complementary strands of DNA depends upon the same parameters. In a detailed discussion of the mechanism of renaturation, Wetmur and Davidson (1968) reported that when DNA is allowed to renature in solution at low concentrations (less than 10 μg/ml) the reaction follows second order kinetics with respect to DNA phosphate concentration; this would be expected of a reaction depending upon random collision between DNA strands. At higher concentrations of DNA, however, the kinetics become more complicated due to the aggregation of renatured DNA molecules.

Renaturation between heterologous eucaryotic DNAs or hybridization between DNA and RNA is commonly performed by incubating an immobilised preparation of single DNA strands with a solution of denatured DNA or RNA. McCarthy (1967) observed that the rate of renaturation is proportional to the amount of DNA in solution at low concentrations, when the soluble strands do not interact with each other. The reaction rate is proportional to the amount of DNA on the filter, but when the concentration of the immobilised DNA is increased so that it is in a great excess over that in solution (more than 5–10 fold) the reaction rate ceases to be limited by the amount of immobilised DNA and depends only upon the concentration of the nucleic acid in solution.

Renaturation of two complementary single strands to give a duplex of DNA takes place in two stages. Wetmur and Davidson (1968) suggested that the first stage is rate limiting and consists of a nucleation reaction in which a comparatively short length of duplex is formed. This is succeeded by a more rapid zipper like reaction which brings the rest of the complementary strands into their base paired duplex conformation.

Renaturation depends heavily upon the level of ionic strength below 0·4 M NaCl, but is independent of it above this value; 0·18 M NaCl is commonly used as a standard condition for renaturation. The length of the fragments of single stranded DNA is important; the rate of renaturation is proportional to the square root of length. When a solution of DNA strands is renatured with an immobilised preparation, the size of the immobilised species does not control the reaction, but the length of the fragments in solution remains important. Preparations to be tested for complementarity with immobilised strands of DNA are therefore usually first degraded to a standard size.

The temperature of renaturation controls the rate and character of duplex formation of eucaryotic DNA. The maximum rate of proper renaturation of DNA is achieved at a temperature 20–30°C below the Tm. McCarthy (1967) observed that the rate of renaturation follows the two step curve with temperature shown in figure 4.7. At low temperatures (20–50°C) loose aggregates of unrelated strands may form rather rapidly; this does not represent renaturation. At higher temperatures (50–80°C) renaturation forms duplex molecules of DNA, with an optimum rate of renaturation at about 60–65°C, compared with a Tm for native DNA of 85°C under these conditions.

The probability of two complementary strands meeting and rejoining after their separation depends upon their relative concentrations. If only one type of DNA molecule is present, the chance of renaturation taking place is much greater than if there are, say, ten or one hundred different sequences present. In accordance with this prediction, Marmur and Doty (1961) found that the extent of renaturation of the DNAs of various species depends upon the size of the genome; under given conditions of incubation, the greatest renaturation is shown by phage DNA and the smallest by the eucaryotes.

At the optimum temperature for renaturation, however, the rate of reaction of mouse DNA to form duplexes is similar to that of B. subtilis DNA; figure 4.7 shows similar curves for the bacterial and eucaryotic DNAs between 60°C and 80°C. Since the haploid genome of the mouse has 500 times more DNA than that of the bacterium, we should expect the eucaryotic DNA to renature much more slowly; if it consists of non-identical sequences, each individual sequence should be present in a concentration 500 times lower than that characteristic of the bacterial DNA. One explanation for the relatively rapid renaturation of mouse DNA is therefore that there is considerable duplication of sequences in the genome, so that the concentration of each individual sequence is increased and any particular single stranded region is not restricted to renaturation with its original partner, but may associate with any one of a number of other sequences. A further point is to suppose that renaturation of mouse DNA is not completely specific under these conditions

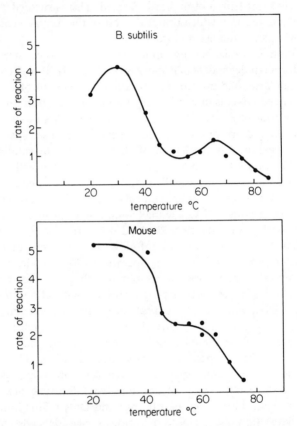

Figure 4.7: reassociation of sheared denatured DNA in solution with denatured DNA bound to membrane filter. Associations of single strands in the temperature range 20–50°C represent non-specific aggregation of unrelated sequences; renaturation to form duplex molecules takes place in the range from 50–80°C. Data of McCarthy (1967).

(whereas that of bacterial DNA is), so that single strands of mouse DNA which are related but not identical in sequence may anneal together.

The nature of the eucaryotic genome can in part be deduced from the characteristics of the duplex molecules formed by renaturation (that is by incubation in the range 50–80°C); these can be assayed by their thermal stabilities. Duplex molecules of high stability—that is with accurate base pairing—have a high Tm; duplex molecules formed under conditions of lower stringency are less stable and have a correspondingly lower Tm. There are in principle two ways to measure the Tm. One is to follow the increase in optical density as the strands of DNA separate. Another is to follow the release of single strands from a membrane filter or column of hydroxyapatite as the temperature is raised; this demands complete separation of complementary strands before release, so that separation may be retarded by the existence of

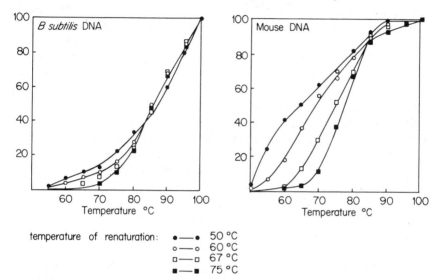

temperature of renaturation: ●——● 50 °C
                              ○——○ 60 °C
                              □——□ 67 °C
                              ■——■ 75 °C

Figure 4.8: Thermal dissociation profiles of different preparations of renatured DNA. Denatured preparations of single strand DNA were renatured on membrane filters at 50°C, 60°C, 67°C, 75°C. The figure shows the melting curves of each of the double stranded DNAs produced by each of these incubations. Left: DNA of B. subtilis renatures into duplexes of about the same stability at all incubation temperatures. Right: Mouse DNAs renatured at higher temperatures are more stable than those renatured at lower temperatures; the higher the temperature, the more stringent are the conditions for base pairing so that more accurately matched duplexes are formed. Data of McCarthy (1967).

short regions rich in G–C base pairs which are more stable. The Tm measured in this way is therefore denoted as Tmi, the midpoint of irreversible strand separation. This is a few degrees higher than the Tm measured by increase in optical density.

Different relative stabilities are displayed by bacterial and eucaryotic DNAs renatured at increasing temperatures. Figure 4.8 shows the thermal stabilities of the B. subtilis and mouse DNAs renatured at different temperatures under the conditions of figure 4.7. Bacterial DNA forms only one kind of duplex, irrespective of the temperature (within the 50°–80°C renaturation range) at which the single strands are allowed to associate. The temperature therefore controls the rate of renaturation only; the same duplex molecules are formed at all temperatures.

The stability of renatured eucaryotic DNA, however, depends upon the temperature of renaturation. Early results pointing to the importance of temperature were obtained by Martin and Hoyer (1966), who found that when mouse DNA is renatured using the agar technique, the duplex molecules formed at 65°C have a higher Tm than those renatured at 50°C. The results of McCarthy (1967) and McCarthy and McConaughy (1968) showed a continuous increase in duplex stability with the temperature of renaturation. Figure 4.8 shows that the mouse DNA renatured on filters at 50°C has a low stability and melts gradually over a range between 50–80°C. Duplex molecules renatured at 60°C, 67°C and 75°C show increasing stabilities, the last displaying a Tm of about 78°C which is some 7°C below the Tm of native DNA.

The accuracy of base pairing when single strands of eucaryotic DNA associate therefore depends upon the stringency of the conditions used during the hybridization. When renaturation takes place at high temperature, duplex formation must be very accurate to offer enough energy of reaction to overcome the temperature promoted dissociation of duplex molecules. Lowering the temperature provides less stringent conditions for strand association and a certain degree of mismatching is permitted in duplex formation; the two strands need not exhibit perfect complementarity.

The differences in the thermal stabilities of bacterial and eucaryotic DNAs formed at increasing temperatures show that their relative complexities of sequence are very different. Under conditions of renaturation, bacterial DNA can form only duplex molecules of a stability close to that of native DNA; figure 4.8 shows that the Tm of duplexes renatured from 50°C to 75°C is about 85°C. This means that any given denatured single strand must anneal with its original complementary sequence or it must remain single stranded.

But with mouse DNA the predominant product of reaction represents the association of single strands which are only partially complementary. That is, they represent different regions of the genome which have related but not identical sequences. The rate of reassociation at different temperatures therefore provides a measure of the number of sequences in the mouse genome which are similar enough to form duplex molecules. There are many degrees of similarity in mouse DNA and the precision of the relationship demanded between two renaturing single strands declines as the temperature is lowered.

At low temperatures the conditions for renaturation are not stringent and sequences of only low similarity may anneal with each other; the duplex products therefore contain appreciable mispairing and melt at low Tm values when the temperature is later raised. Higher temperatures demand a better fit between the renaturing complementary strands, so that more stable duplexes are produced. This suggests the concept that mouse DNA includes families of related base sequences in which the resemblance between the members of each family varies; renaturation at low temperatures allows the less related members of each family to renature, whereas higher temperatures restrict recognition to more closely related sequences.

Renaturation of Unique Sequences

Reassociation of two complementary strands of DNA depends upon random collision and is therefore a function of their concentration and the time available for reaction. Britten and Kohne (1968) have shown that the

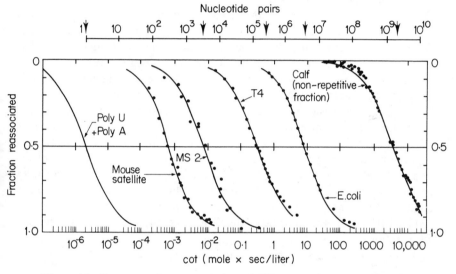

Figure 4.9: Cot curves for reassociation of different DNAs. Increasing Cot values are required by DNAs of greater sequence complexity; the Cot$_{1/2}$ value of renaturation can be used to estimate the length of the basic repeating unit of the (haploid) genome. These estimates are indicated on the axis for nucleotide pairs. Data of Britten and Kohne (1968), obtained by following decrease in optical absorbance as renaturation proceeds in 0·12 M phosphate buffer.

parameter which controls the extent of completion of renaturation is the product of DNA concentration and the time of incubation. This is usually expressed as the *cot*:

$$cot = \text{moles of nucleotides} \times \text{seconds/litre}$$

A cot of 1 mole-second/litre is produced when DNA is incubated for one hour at a concentration of 83 μg/ml (this corresponds to a solution with an optical density of about 2·0 at 260 mμ).

Cot curves take the form shown in figure 4.9. As the genome becomes more complex, the concentration of any particular sequence within it is reduced, so that a greater cot value is needed for its renaturation. The range of cot curves from phages to eucaryotes varies over eight orders of magnitude and a logarithmic scale is generally used. The complexity of any particular DNA component may be characterized by its $cot_{1/2}$, the cot value which is needed to allow its renaturation to proceed halfway to completion. The $cot_{1/2}$ is inversely proportional to the concentration of complementary DNA sequences; this means that it is proportional to the number of *different* unique sequences present in the genome.

If a DNA consists of, say, a sequence 100 nucleotides long which is repeated many times, then the concentration of this sequence in any given amount of DNA is very high; complementary lengths can find each other rapidly and renaturation requires only a low cot. On the other hand, if a genome consists of one long sequence of DNA which is unique, then each single strand has only one complement with which it can pair in this length; the concentration of any particular sequence is therefore much lower, so that renaturation is slow and a high cot is needed before the reaction can proceed.

The longer the basic unique sequence in the genome, the harder it becomes for complementary strands to find each other; the $cot_{1/2}$ therefore measures the length of the basic repeating unit in a DNA preparation. Figure 4.9 shows that there is a close correlation between $cot_{1/2}$ values and the lengths of phage and bacterial DNAs. For example, MS2 has a length of about 4000 nucleotides and a $cot_{1/2}$ close to 10^{-2}; the E.coli genome of $4 \cdot 6 \times 10^6$ base pairs is therefore some 10^3 times longer, so that the concentration of any particular sequence is reduced 1000 fold, and a correspondingly greater $cot_{1/2}$ of almost 10 is needed for reaction.

An ideal cot curve is symmetrical and follows the equation:

$$\frac{c}{c_0} = \frac{1}{1 + kc_0 t}$$

where c_0 is the initial concentration of DNA single strands and c is the concentration remaining single stranded after time t; k is a constant. Britten and Kohne (1969a) pointed out that the cot values representing 0% and 100% renaturation should have a ratio of about 100 for any single preparation; if renaturation requires a range of cot values much greater than this, the DNA preparation must consist of more than one component.

The DNAs of viruses and procaryotes therefore consist of single components, as figure 4.9 shows, with $cot_{1/2}$ values proportional to the content of DNA in the genome. The renaturation curves of eucaryotic DNAs, however, are more complex. Figure 4.10 compares the cot curve of calf thymus DNA with that of E.coli. Whereas the bacterial DNA comprises a single component which can be defined by its $cot_{1/2}$ value of about 6, the eucaryotic DNA possesses two separate components. The first 40% of the DNA reassociates before a cot of $2 \cdot 0$ is reached and has a $cot_{1/2}$ of $0 \cdot 03$. Little if any reaction occurs during the next one hundred fold increase of cot. Further renaturation then takes place as the cot is increased from 10^2 to 10^4; the $cot_{1/2}$ for this component is about 3000. This implies that calf thymus DNA consists of two types of sequence. The slowly renaturing component of $cot_{1/2} = 3000$ must consist of a very long unique sequence which is therefore present in the genome at low concentration; the more rapidly renaturing component of $cot_{1/2} = 0.03$ must be present in the form of many copies of much shorter repeating units whose concentration in the genome is correspondingly greater.

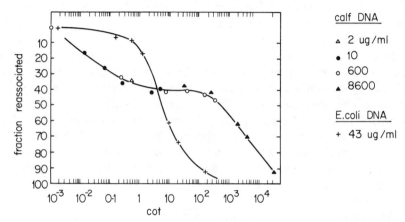

Figure 4.10: reassociation of denatured calf thymus DNA measured by retention of duplex molecules on hydroxyapatite. Forty per cent of the DNA renatures rapidly with a cot of less than 2·0. The remaining DNA sequences renature slowly, requiring cot values of 10^2–10^4. Although the controlling parameter is the product of DNA concentration and time of incubation, the most convenient conditions are obtained by using low concentrations of DNA for low cot values (so that renaturation is not too rapid) and using higher concentrations for the greater cot values (to avoid very prolonged incubations). The different symbols on the curve show the concentrations used to determine the renaturation at each cot value. Data of Britten and Kohne (1968).

The correlation between renaturation kinetics and nucleotide sequence complexity shown in figure 4.9 demonstrates the concept that the $cot_{1/2}$ of a DNA of unique sequence is directly proportional to its length. We may therefore rely upon bacterial DNA as a standard and write the equation:

$$\frac{cot_{1/2} \text{ (unknown preparation)}}{cot_{1/2} \text{ (bacterial DNA)}} = \frac{\text{repeating length of preparation}}{\text{length of bacterial DNA}}$$

Under the conditions used for renaturation in figure 4.10, E.coli DNA which has a length of $4·6 \times 10^6$ base pairs has a $cot_{1/2}$ of about 6·0. Applying the equation to calf thymus DNA therefore shows that the slow component has a basic repeating unit of about $2·3 \times 10^9$ base pairs. The total DNA content of bull sperm is about $3 \cdot 2 \times 10^9$ base pairs; since the slow fraction represents 60% of the total calf DNA, it must have a length of some $1 \cdot 9 \times 10^9$ base pairs in each haploid genome. The coincidence of the length of the renaturing unit and the length of the slow fraction of the genome of calf DNA shows that this component consists of unique DNA sequences; there is no repetition of sequences within this part of the genome so its $cot_{1/2}$ remains proportional to its length. (The properties of the more rapidly renaturing component of low cot are discussed in the next section.)

For the cot values of different preparations of DNA to be compared they must of course be measured under the same conditions of ionic strength, temperature and fragment size. If two preparations are renatured under different conditions, their $cot_{1/2}$ values can be compared only after a suitable correction to allow for any differences in renaturation rate due to their conditions of incubation. One condition which has commonly been used is $0 \cdot 18$ M ionic strength at $60°C$ with a fragment length of 500 nucleotides. Recent experiments have used SSC solution, which consists of $0 \cdot 15$ M NaCl and $0 \cdot 015$ M sodium citrate. A temperature of $60°C$ is $25°C$ below the Tm of B. subtilis DNA under these conditions and therefore lies within the optimum range for renaturation (see figure 4.7). B. subtilis is often used as a bacterial standard because its G–C content of 40% closely resembles that of many higher plants and animals.

When renaturation requires incubation for extensive periods of time, high temperatures tend to cause chain breakages and depurination reactions in DNA. It is therefore advantageous to perform the renaturation reaction at a lower temperature. In order to remain within the temperature for optimum renaturation, the Tm of DNA may be lowered by reagents such as formamide. McConaughy, Laird and McCarthy (1969) found that a 48% solution of formamide lowers the Tm of B. subtilis DNA to $60°C$; renaturation at $37°C$ therefore lies within the optimum temperature range. Figure 4.11 illustrates the effect which these different conditions of incubation have upon the cot values of renaturation. The conditions most usually employed are in SSC at $60°C$, when B. subtilis DNA has a $cot_{1/2}$ of $5 \cdot 0$; or in 48% formamide in $5 \times SSC$ (that is 5 fold increase in ionic strength) at $37°C$, when B. subtilis DNA has a $cot_{1/2}$ of $1 \cdot 6$. Since the reaction in formamide under these conditions is three times faster than in SSC at $60°C$, $cot_{1/2}$ values obtained in 48% formamide at $37°C$ must be corrected by a factor of three for comparison with those obtained at $60°C$ in $5 \times SSC$.

The renaturation reaction may be followed either by the decrease in optical density as duplex regions reform or by using hydroxyapatite. If DNA is renatured to a given cot and passed through a column of hydroxyapatite, molecules which contain duplex regions are retained, even though only part of their length may be double stranded; the reaction therefore appears to proceed more rapidly. Cot values thus depend also on the method used to follow renaturation; if all other conditions are equal, a correction of 2 may be used to convert those obtained on hydroxyapatite to values which would be obtained by following the decrease in optical density. But although absolute $cot_{1/2}$ values may vary with the conditions of renaturation, the ratios between the $cot_{1/2}$ values of two different DNA preparations should remain constant.

To avoid the need to make corrections for $cot_{1/2}$ values measured under different conditions, a bacterial standard DNA may be included in the mixture used to renature a eucaryotic DNA. A small amount of E.coli or B. subtilis DNA, distinguished by a radioactive label, does not interfere with the renaturation of a much larger amount of eucaryotic DNA; but it provides an internal standard against which the eucaryotic renaturation curve can be compared.

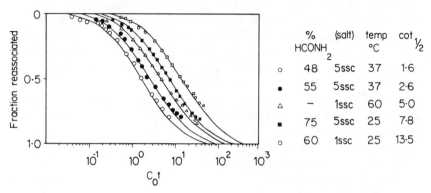

	% HCONH$_2$	(salt)	temp °C	cot$_{1/2}$
○	48	5ssc	37	1·6
●	55	5ssc	37	2·6
△	—	1ssc	60	5·0
▲	75	5ssc	25	7·8
□	60	1ssc	25	13·5

Figure 4.11: renaturation of B. subtilis DNA (2×10^6 base pairs) under different conditions. Formamide lowers the Tm of DNA so that the reaction may be carried out at lower temperature; the reduction in Tm is proportional to the concentration of formamide. Ionic strength increases the rate of renaturation. The table shows that $cot_{1/2}$ values may vary over a ten fold range with these variations in renaturation conditions. Data of McConaughy, Laird and McCarthy (1968).

This automatically compensates for any variation in incubation conditions in individual experiments. Using this method, Laird (1971) has confirmed that the slow fractions of eucaryotic DNAs are indeed unique. Compared with bacterial DNA, the $cot_{1/2}$ values of the slow components of Drosophila melanogaster and mouse are in almost exact proportion to their content of DNA.

The unique sequences of the eucaryotic genome in general account for more than half of the total DNA. There appears to be no repetition within these sequences, which presumably include the many individual genes coding for different proteins; we do not know what proportion of the unique sequences codes for protein and whether it has any other functions. But within the limitations of the renaturation technique (see later) we can at least exclude the possibility that this part of the genome contains many identical copies of each gene. Nor does this seem a likely structure for the more rapidly renaturing components, which seem to consist of many related but not identical sequences repeated a very large number of times. Models which suggest that the genome includes extensive duplications of each gene are therefore inconsistent with the kinetics of renaturation.

Comparison of the renaturation kinetics of the DNAs of various amphibians shows that their large genomes do not evolve by the development of polynemic chromosomes containing several copies of the haploid genome of their ancestors. Straus (1971) compared the cot curves of three amphibia whose haploid genomes have DNA contents in the ratios 1:3·5:7. If DNA has been laterally copied to produce multistranded chromosomes in the larger genomes, all three DNAs should renature at the same cot values; the addition of further copies of the complete genome should not alter the proportion of any renaturing component. The slowest renaturing component should therefore have the same $cot_{1/2}$ in all three DNAs in spite of the 3·5 and 7 fold increases in genome size. According to this model, therefore, only the amphibian with the smallest genome should possess unique DNA.

But the $cot_{1/2}$ of the slowest component increases with the genome size; all three species possess single copy DNA. All the species display characteristic cot curves and the differences in genome size are also accompanied by changes in the relative proportions of the different components; the larger size appears to possess more repetitive sequences, which suggests that many duplications of small sequences may accompany the evolution of larger genomes. Data of this kind therefore exclude polynemic models for the amphibian chromosome in which excess DNA is explained by repetition of all their sequences.

DNA Sequences of Intermediate Repetition

Repetitious DNA occurs in all eucaryotes, with the possible exception of yeast. All these species therefore have two principal components of DNA sequence: a slowly renaturing fraction of high cot which consists of unique sequences; and a fraction renaturing at an intermediate cot which consists of sequences in general repeated some 10^3–10^5 times. In addition to these two components there is often a very rapidly renaturing component of extremely low cot—sometimes known as satellite DNA—which consists of sequences repeated an enormous number of times, of the order of 10^6.

The degree of repetition of the intermediate fraction can be calculated by comparing its length in the haploid genome with the repeating unit corresponding to its $cot_{1/2}$. The calf intermediate fraction shown in figure 4.10 has a $cot_{1/2}$ of 0·03; comparison with the cot curve of E.coli DNA suggests that the corresponding repeating length should be 2×10^4 base pairs. But the total amount of DNA in this fraction is 40% of the haploid genome, that is about $1·3 \times 10^9$ base pairs. This means that the fast component behaves as though consisting of sequences repeated about $0·65 \times 10^5$ times. This calculation does not reveal how many functional sequences there are within the repeating length of 20,000 base pairs; on average, however, there must be many sequences whose combined lengths total 20,000 base pairs, each repeated 65,000 times in the genome.

Reiterated sequences occupy varying proportions of different eucaryotic genomes and the intermediate fractions usually renature over a wide range

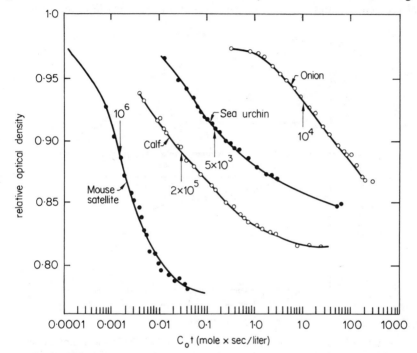

Figure 4.12: renaturation kinetics of repeated fractions of DNA. The arrows indicate the $cot_{1/2}$ points of each curve and the degree of repetition predicted from the sequence length which corresponds to this $cot_{1/2}$. The mouse fraction is a very rapidly renaturing (fast) component; the calf, sea urchin and onion DNAs represent the intermediate fractions. Data of Britten and Kohne (1968).

of cot values which correspond to several components representing varying degrees of repetition. Figure 4.12 shows the kinetics of renaturation of isolated intermediate fractions; this demonstrates the extensive variation in the components of each fraction. The average numbers of repetitions suggested by these curves are 2×10^5 for the calf, 5×10^3 for the sea urchin and 10^4 for onion DNA. It is important to note that these figures represent only *average* values; each intermediate fraction consists of sequences of various lengths repeated different numbers of times, but corresponding overall to the total repeating length and number of repetitions inferred from the $cot_{1/2}$ value.

The reassociation curve of mouse DNA shown in figure 4.13 displays slow, intermediate and fast components. McConaughy and McCarthy (1970b) observed that the slow component comprises about 75% of the DNA and has a $cot_{1/2}$ value corresponding to a unique sequence. The intermediate component corresponds to about 15% of the genome and appears to contain sequences repeated from 10^3–10^5 times. The fast component constitutes 10% of the genome and seems to consist of a short sequence repeated some 10^6

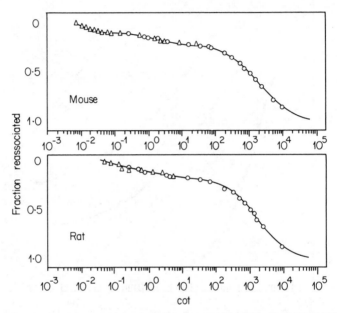

Figure 4.13: renaturation kinetics of rodent DNA. This shows cot values obtained in 48% formamide–5 × SSC at 37°C. Mouse DNA contains fast (10%), intermediate (15%) and slow (75%) components; rat DNA appears to lack or to have a much smaller quantity of the fast component. Data of McConaughy and McCarthy (1970b).

times (see next section). Figure 4.14 shows the proportion of the genome devoted to different degrees of repetition in the mouse. Rat DNA displays similar renaturation kinetics but has a much smaller proportion of fast component.

The nature of the sequences of the intermediate component is not entirely clear. However, the thermal properties of renatured intermediate DNA molecules suggest that they consist of families of related sequences. The different components of eucaryotic DNA can be isolated by using hydroxyapatite columns to separate the duplex molecules renatured at any cot from those remaining single stranded. A cycle of renaturations followed by isolation of the duplex molecules can therefore be used to separate the DNA components which renature at different cot values. Renaturation to a very low cot—of the order of 10^{-2}—allows only the fast fraction to form duplexes. When the single strands recovered from a hydroxyapatite column are then renatured to a cot of about 10, the intermediate fraction reassociates. Only the unique fraction should then remain single stranded. Of course, the DNA renaturing in each cycle is to some extent contaminated by the other components and more than three successive renaturations and isolations on hydroxyapatite may be needed to isolate reasonably pure components.

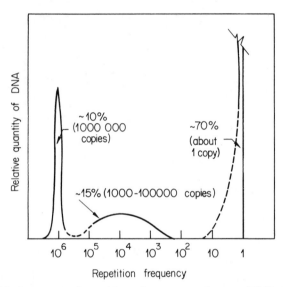

Figure 4.14: frequency of repetition of sequences of mouse DNA. Regions of the curve which have been determined are shown by solid line; those which are uncertain by the dotted line. No sequences of between 100 and 1 times repetition have been detected by renaturation experiments. Data of Britten and Kohne (1968).

Renatured slow components generally have a thermal profile resembling that of native DNA; they melt sharply at a Tm only a little below that of native DNA. This shows that strand association is accurate; each unique sequence anneals with its exact complement. As figure 4.15 shows, however, the intermediate repetitive fraction of bovine DNA melts gradually over a wide temperature range; similar melting curves are seen for other intermediate fractions (Britten and Kohne, 1968; McConaughy and McCarthy, 1970b).

The low stability of renatured intermediate DNA shows that reformation of this component does not represent perfect matching between one strand and its original complement. Rather does duplex reformation take place by the association of strands which are derived from related but not identical sequences. Intermediate components may therefore consist of families of related sequences; the many members of each family therefore comprise a set of nucleotide sequences which have sufficient similarity with each other to renature but which are not identical. Renaturation is thus accomplished by the association of sequences comprising different members of a family.

The total length of the repeating unit deduced from the $cot_{1/2}$ value, however, depends upon the assumption that renaturation is perfect. The general effect of mismatching is to reduce the rate of renaturation between two sequences so that a higher cot is required for strand association; this means that the

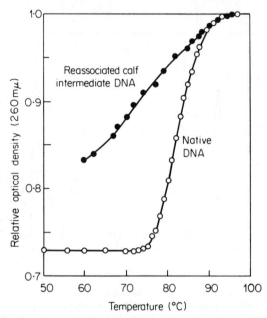

Figure 4.15: comparison of melting profile of renatured calf intermediate DNA and native DNA. Whereas native DNA has a sharp Tm, renatured intermediate DNA melts over a wide range, suggesting that strand association took place by binding between non exactly matching sequences. Data of Britten and Kohne (1968).

repetitious fractions renaturing at any particular cot value have a shorter repeating unit than inferred from a comparison with bacterial DNA. The extent of repetition is therefore correspondingly greater. Estimates of the length of the repeating unit and its degree of repetition are therefore only approximate; the corrections for cot values which can be made to allow for renaturation of mismatched sequences are discussed below (page 183).

One interpretation of the wide melting range of renatured intermediate DNA is to suppose that there is a wide spectrum of relationships between the individual members of a family. Renaturation between the closely related members of a family produces duplex molecules whose base pairing is almost perfect and which therefore melt at a temperature close to that of native DNA. Reassociation of the least related members of the family generates the duplex molecules which melt at the lowest temperature of the wide melting curve.

Fractionation of renatured intermediate DNA on hydroxyapatite columns supports the idea that sequences of increasingly close relationship form the duplex molecules which melt at increasing temperatures. When renatured intermediate DNA is bound to hydroxyapatite columns and then incubated at increasing temperatures, the single strands eluted at each temperature can be recovered. When these single strands are then renatured, they form duplex molecules which, although not those originally eluted from the first columns, represent strand associations amongst the same set of sequences. These duplex molecules can in turn be adsorbed to hydroxyapatite and their melting followed when the temperature is again increased. Britten and Kohne (1968) reported that the same average relationship is retained—the fraction derived by renaturing the single strands eluted at, say, 65°C, in turn melts with an average Tm of 65°C; the fraction collected when a temperature of 70°C was reached can be renatured to give duplexes which melt at 70°C, etc.

This suggests that each set of fragments is derived by renaturation between DNAs with characteristic divergences of sequence. The strands first eluted from the original hydroxyapatite columns of intermediate DNA consist of the members of a family with only low resemblance to each other; when isolated they can therefore on average renature only poorly once again. The last strands to be eluted comprise a fraction of sequences more closely related to each other, so that when isolated they tend to renature more accurately. Thermal chromatography thus separates members of a family according to their degree of resemblance.

Cot values describe the renaturation of DNA in solution. However, the kinetics of renaturation may also be followed when one DNA preparation is in solution and the other is immobilised on agar, a membrane filter or a hydroxyapatite column. Because the renaturation reaction takes place much more slowly when one set of strands is immobilised, renaturation under these conditions is confined to low cot values (1–100) relative to those which may be obtained in solution. This restricts renaturation performed with immobilised DNA to repeated sequences. The DNAs renatured in the experiments of figures 4.7 and 4.8 therefore represent the repeated sequences alone and do not include the slowly renaturing unique fraction. The results of figure 4.8 thus demonstrate that the conditions of incubation greatly influence the quality of the duplex molecules reformed when intermediate sequences renature; only closely related sequences of each family associate at high temperature, but under less stringent conditions the less closely related sequences of each family may associate to form duplexes of low stability.

The proportion of the genome which appears to consist of repetitive DNA in any experiment depends upon the stringency of the criteria used to allow base pairing. At one extreme, conditions of low stringency may allow random reassociation of unrelated sequences; at the other, conditions of high stringency may virtually prevent the detection of repetitive DNA. As the stringency of renaturation is increased, only more closely related members of a family may renature and ultimately each sequence must find its original complement. Britten and Davidson (1971) have observed that when calf DNA is renatured at a temperature only 8°C below the Tm of native DNA, only precisely complementary sequences can renature and less than 10% of the genome appears to be repetitive. But at 36°C below the native Tm, 55% of the genome renatures at low cot values which imply that it must be repetitive.

Similar results have been obtained with other organisms. Bendich and McCarthy (1970) noted that barley, oats, rye and wheat appear to have a greater proportion of redundant DNA than animals in the conditions usually used for renaturation; but although most or all of the DNA appears to be repetitive at 61°C, under more stringent conditions only 50–55% is found in this fraction. The DNA composition of D. melanogaster appears to be less complex than that of the mammals, for Laird and McCarthy (1968b, 1969) found that the duplex molecules renatured at 60°C or 67°C have Tm values of 77°C and 78°C, only 4°C less than that of the Tm expected for perfect duplexes with 43% G–C. This suggests that only unique sequences reform under these conditions and that most if not all of the genome is unique. But when renaturation takes place at 50°C, two separate melting components are formed, one with a Tm of 55°C (representing repeated sequences) and the other with a Tm of 78–80°C (the unique component).

The effect of temperature on the proportions of the different components of eucaryotic DNA supports the interpretation that the intermediate fraction consists of families each comprising many sequences which are related to varying extents, but with little if any repetition of identical sequences. The proportions of the genome devoted to slow, intermediate and fast components are generally taken as those detected at 25°C below the Tm, at the optimum rate for renaturation; but, of course, these values are relative to the conditions of incubation and do not represent any absolute partition of the genome (for review see McCarthy and Church, 1970).

Organization of Repetitive Sequences

Fractionation of DNA by Buoyant Density

When mouse DNA is sedimented to equilibrium in a CsCl density gradient, two peaks are separated. Kit (1961) first observed that in addition to the main band—which sediments at about 1·701 g/ml, corresponding to its G + C content of 42%—about 8% of the DNA is found as a *satellite* band of buoyant density about 1·691 g/ml. Satellite DNAs have since been separated from the main band of DNA in a variety of eucaryotes; their proportion of the genome

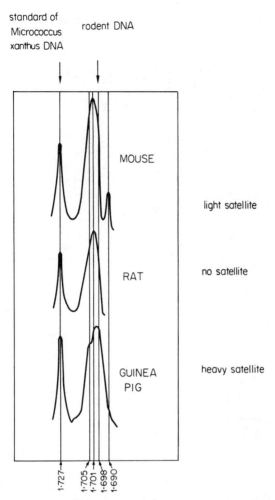

standard of
Micrococcus
xanthus DNA

rodent DNA

MOUSE

light satellite

RAT

no satellite

GUINEA
PIG

heavy satellite

1·727 1·705 1·701 1·698 1·690

Figure 4.16: buoyant density of rodent DNAs in CsCl gradients. Data of
McConaughy and McCarthy (1970b).

ranges upto about 10% and they may be either heavier or lighter than the
main band DNA (or may be absent altogether). Figure 4.16 shows the fractionations of the DNAs of mouse, rat and guinea pig on CsCl; the mouse has a
satellite lighter than the main band, the rat possesses no satellite, and the
guinea pig genome displays a satellite denser than the main band.

The usual basis for separation of DNAs by buoyant density is their content
of guanine plus cytosine—the greater the $(G + C)/(A + T)$ ratio, the higher
the buoyant density (see page 155). Although this relationship is observed by
the main band DNAs and by some of the satellites which have been found,
it is not an invariable guide to the base composition of satellites, many of

Table 4.1: properties of satellite DNAs of mammals. Only one satellite in each of these mammals has been resolved by analysis in neutral CsCl (left). Analysis in gradients using Ag^+—or sometimes Hg^{2+}—may resolve other satellites in addition to this one (centre); when rerun on neutral CsCl these satellites usually have buoyant densities close to that of the main band. The $G + C$ proportions shown are those determined by direct analysis; they may differ appreciably from those expected from the neutral density—satellites often have abnormal sedimentation properties. The single strands of many satellites can be separated in alkaline CsCl (right); the strands are complementary so that the base compositions of the heavy strands are complementary to those for the light strands given in the table. Data of Corneo et al. (1968, 1970a, 1971) and Yasmineh and Yunis (1971); see also Walker (1971) and Yunis and Yasmineh (1972).

| | analysis in neutral CsCl | | | | | analysis in Ag^+–Cs_2SO_4 gradients | | | analysis in alkaline CsCl | | | | |
| | main band DNA | | satellite DNA | | | | | | | light strand base composition | | | |
organism	density (g/ml)	G + C (%)	density (g/ml)	% total DNA	no.	rerun in CsCl	G + C (%)	density in Ag^+	density (pH 12) single strands	A	T	G	C
mouse	1·701	42	1·691	8·0			34		1·725/1·752	44	22	20	14
guinea pig	1·699	39	1·705	5·5	I	1·705	39	1·493					
								1·535	1·692/1·778	37	24	3	36
			—	2·5	II	1·704	44	1·463	1·738/1·769				
			—	2·5	III	1·704	—	1·456	1·740/1·762				
cow	1·699	41	1·713	7·0	II	1·713	55	1·510	1·750/1·769				
								1·570					
				3·0	I	1·706	46	1·440	1·773/1·769				
man	1·700		1·687	0·5	I	1·687		1·481	1·707/1·783				
								1·444					
				2·0	II	1·693		1·509	1·740/1·750				

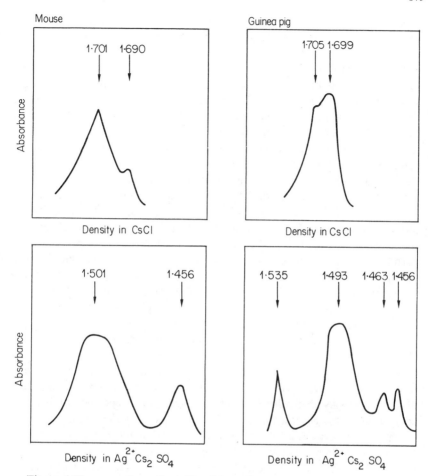

Figure 4.17: separation of satellite DNAs from main band DNA. The presence of silver ions (lower gradients) improves the resolution of satellites compared with their separation in CsCl gradients (upper). With mouse DNA (left) the satellite is completely resolved by the addition of silver ions. With guinea pig DNA (right) the satellite present on CsCl is resolved as a separate heavy band on Cs_2SO_4–Ag^+ and two additional satellites are resolved on the light side of the main band. Data of Corneo et al. (1968, 1970).

which show anomolous sedimentation; table 4.1 shows that the $G + C$ contents of satellites whose composition has been determined directly is often appreciably different from the value which would be predicted from their buoyant density. Satellite DNAs may have G–C/A–T ratios either greater or smaller than the main band.

Centrifugation through gradients of Cs_2SO_4 which contain silver ions improves the resolution of satellite DNAs. As figure 4.17 shows, the mouse satellite is more clearly separated from the main band under these conditions. In some organisms, centrifugation through $Ag^+-Cs_2SO_4$ gradients reveals additional satellites which are not separated by centrifugation in CsCl. In the guinea pig, for example, the single satellite band found on CsCl is also resolved as a satellite in $Ag^+-Cs_2SO_4$ (at 1·535 g/ml instead of the 1·493 g/ml of the main band), but two further satellites are also separated (at 1·463 and 1·456 g/ml). As the data of table 4.1 show, the satellites found when calf and human DNAs are centrifuged through CsCl are also resolved at a greater distance from the main band by sedimentation in $Ag^+-Cs_2SO_4$; a second satellite is also found in each instance. The bands which are separated on $Ag^+-Cs_2SO_4$ can then be recovered and the table shows the buoyant densities which the isolated satellites display when rerun on neutral CsCl gradients.

The basis for the improved resolution of satellites in the presence of silver ions is not clear. It was at one time thought that this separation depends on the A–T/G–C ratio; but as the data of table 4.1 show there seems to be no correlation between the G + C content of satellite DNAs and their buoyant densities in silver ions. Satellite I of the guinea pig is heavier in $Ag^+-Cs_2SO_4$ than satellite II, which has the greater G–C content; but satellite I of the cow is lighter than satellite II, which has the greater G–C content. This implies that satellite DNAs are abnormal both in their buoyant densities in CsCl and in the extent to which they bind silver ions when centrifuged through $Ag^+-Cs_2SO_4$. Although satellites which can be isolated on neutral CsCl have a base composition distinct from that of the main band, this is not necessarily true of satellites separated on $Ag^+-Cs_2SO_4$ gradients. The atypical behaviour of satellite DNAs on density gradients appears to be independent of their base compositions and is shown by both G–C rich and A–T rich satellites.

On alkaline gradients the strands of mouse satellite DNA separate into heavy (H) and light (L) strands. As table 4.1 shows, sequence analysis of the separated single strands reveals pronounced asymmetries; the light strand contains 45% A, 20% T and 22% G, 13% C; the proportions of the complementary bases are reversed in the heavy strand. Strand separation in alkali is displayed by many, although not all, satellite DNAs. An extreme case of asymmetry in composition is displayed by satellite I of the guinea pig, which contains only 3% G (and 36% C) in its light strand; these compositions are reversed in the heavy strand, so that one strand contains virtually all the cytosine of the satellite and the other contains almost all its guanine.

Sequence Analysis of Satellite DNA

When mouse satellite DNA is denatured into single strands, the separated strands renature with each other very rapidly to give a duplex close in structure to the original native DNA. Waring and Britten (1966) found that the cot curves of satellite renaturation (see figures 4.12 and 4.18) correspond to a sequence length of some 300–400 base pairs repeated some 10^6 times.

Although satellite DNAs differ widely in base composition, buoyant density, and silver binding capacity, they all appear to share one common characteristic; their sequence complexities are very low. Figure 4.18 shows the cot curves for the renaturation of several isolated satellites; although the length of the repeating unit varies, all satellites renature very rapidly and appear to consist of a short nucleotide sequence repeated very many times.

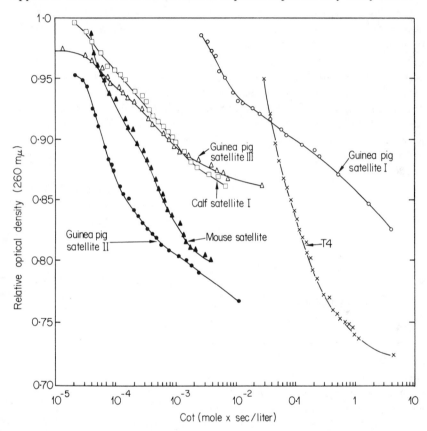

Figure 4.18: renaturation kinetics of satellite DNAs performed in 0.18 M NaCl at 60°C. Data of Corneo et al. (1970).

Low sequence complexity and abnormal sedimentation properties in CsCl or in the presence of silver ions seem to be correlated. As figure 4.18 shows, all three of the guinea pig satellites—only one of which can be isolated in neutral CsCl, the other two separating from the main band only in Ag^+–Cs_2SO_4—have low $cot_{1/2}$ values; rapid renaturation is thus a characteristic of both classes of satellite. The fast components of eucaryotic DNAs seem in general to have unusual silver binding characteristics, whether or not they are found as satellites on neutral CsCl. Maio (1971) found that the DNA of cultured kidney cells of the African green monkey forms a single, apparently homogeneous, band at 1·699 g/ml in neutral CsCl gradients. But about 20% of the DNA comprises a rapidly renaturing fraction corresponding to about 10^6 repeats of a sequence of about 350 base pairs. This DNA appears as a light fraction in Ag^+–Cs_2SO_4 gradients. Binding silver ions abnormally and/or displaying unusual buoyant density values in CsCl thus seem to be general characteristics of fast components and may be related to their repetitious nature rather than to their base compositions.

The repeating lengths inferred from $cot_{1/2}$ values rely upon a comparison between the renaturation of eucaryotic DNA and that of bacterial DNA. However, mismatching takes place when the related but non identical members of a eucaryotic DNA family renature, whereas bacterial DNA reforms only well matched duplex molecules. When the effect of mismatching on the reassociation of related eucaryotic DNA sequences is taken into account, the estimates for repeating length suggested by the $cot_{1/2}$ values of satellite DNAs appear to be too large.

The first step in renaturation is formation of a few base pairs in the correct phase; further base pairs are then formed rapidly by a zipper-like propagation from this site. The formation of mismatched pairs of bases reduces the rate of association, so that repeated sequences which are similar but not identical renature at a reduced rate which depends upon the probability that a mismatched pair will be encountered before a stable duplex has been formed. Southern (1971) has calculated that this reduces the rate of association according to the equation:

$$\log(R_m/R_0) = n \log(1 - p)$$

where R_m is the rate of reassociation of the mismatched duplex, R_0 is the rate of reassociation if there is no mismatching, n is the nucleus of base pairs needed for stable duplex formation (about 10–20) and p is the fraction of mismatched base pairs in the final duplex. The rate of renaturation which is measured by experiment is R_m.

The correlation between cot curves and nucleotide sequence complexity, which supposes perfect formation of the double helix, therefore depends upon R_0. If n and p are known, R_0 can be calculated and this corrected rate of reassociation can be used to derive a corrected cot value. In fact, since R_m is given by $1/cot_{1/2}$—for the cot is inversely proportional to the rate of renaturation—we can rewrite the relationship as:

$$\log \frac{cot_{1/2} \text{ (perfect duplex)}}{cot_{1/2} \text{ (experimental value)}} = n \log(1 - p)$$

Values for n and p at different temperatures have been deduced by Sutton and McCallum (1971) from studies of the thermal dissociation profiles of renatured mouse satellite DNA. The Tm of renatured mouse satellite DNA is about 5°C below that of the native DNA, which suggests that about 7% mispairing takes place during strand reassociation. However, this is an average value for there are different degrees of mispairing within the reformed duplex molecules. Different classes of duplex can be separated by thermal fractionation; after renatured satellite DNA has been bound to hydroxyapatite, single strands are released by progressively raising the temperature. Sutton and McCallum divided the eluted single strands into four fractions, corresponding to denaturation temperatures of 80°, 82·5°, 85° and 90°C. Each of these fractions was then renatured under standard conditions to determine its cot curve. The fractions eluted at lower temperatures require higher cot values for renaturation. This implies that the fractions eluted at lower temperatures are less stable because they represent those duplex molecules with a high degree of mismatching; it is therefore more difficult to renature them. More accurately paired duplex molecules are eluted at higher temperatures and, because of their greater strand complementarity, can be more readily reassociated and therefore require lower cot values.

The degree of mismatching—that is a value for p—can be calculated from the Tm values of the strands eluted from hydroxyapatite at each temperature step. By measuring the $cot_{1/2}$ of each of these fractions it is possible to make a plot of $\log(1/cot_{1/2})$ against $\log(1 - p)$; this gives a slope of $-n$. Values for n derived from this correlation curve should be valid for all DNAs of similar base composition and may therefore be used in correcting cot values obtained in other experiments; n is 13 ± 2 base pairs at 60°C and is 18 ± 3 base pairs at 70°C.

A value for p can be derived from the thermal stability of any renatured satellite; the satellite DNA is denatured, renatured, and the decrease in Tm compared with that of the native DNA provides an estimate for p. A value for n is taken from the correlation curve obtained with mouse satellite DNA. Correction of $cot_{1/2}$ values by applying a factor of $n \log(1 - p)$ to the observed values gives the data of table 4.2. The corrected $cot_{1/2}$ values are those which

Table 4.2: Correction of Cot values for renaturation of satellite DNAs. When the extent of mismatching is taken into account, a reduced Cot value can be calculated for the renaturation rate which would be shown by a perfectly renaturing duplex. This value is used to calculate the length of the repeating unit—in base pairs—in the satellite. Data of Sutton and McCallum (1971).

DNA	ΔTm °C	observed $Cot_{1/2}$	corrected $Cot_{1/2}$	corrected repeating unit
mouse satellite	5	$6\cdot6 : 10^{-4}$	$2\cdot6 \times 10^{-4}$	140
guinea pig satellite I	3	$4\cdot1 \times 10^{-4}$	$2\cdot4 \times 10^{-4}$	130
guinea pig satellite III	8	$9\cdot5 \times 10^{-4}$	$2\cdot1 \times 10^{-4}$	110
calf satellite	10	$5\cdot0 \times 10^{-3}$	$7\cdot2 \times 10^{-4}$	390

would be shown by a perfectly renaturing duplex; they can therefore be compared with standards of bacterial DNA to calculate the length of the repeating unit. This suggests a length of the order of 110–140 base pairs for mouse satellite (140) and the guinea pig satellites I (130) and III (110) and a length of almost 400 base pairs for the calf satellite. The degree of repetition of the satellite sequences must therefore be correspondingly greater than that estimated from the repeating length suggested by the uncorrected $cot_{1/2}$ values.

The sequence from which guinea pig satellite DNA I has evolved appears to be much simpler than even the corrected short repeating unit. Southern (1970) separated the complementary strands of guinea pig satellite DNA I in alkaline CsCl. Degrading the isolated single strands with diphenylamine— which interacts specifically with purines—leaves sequences of pyrimidines. These fragments are short and can be separated by two dimensional chromatography; they can then be sequenced.

The most frequently occurring fragment in the L strand is CCCT. And all except 10% of the total sequences found can be derived from CCCT by a single mutation, for example CCTT, CTCT etc. In the H strand the most frequent sequence is TT; T, TTT, T_2C are also common. This is consistent with a basic repeating structure:

L	5'	C C C T A A	3'
H	3'	G G G A T T	5'

which must have been multiplied many times during evolution. This basic sequence must have been modified by the accumulation of mutations during evolution; the introduction of sufficient mutations at an early stage of amplification of the basic sequence—say at 10–20 copies—could produce a repeating unit of the order of 100 base pairs.

These mutations cannot have taken place at random with respect to the present sequence, for the expected derivatives of the basic sequence are not found at the predicted frequencies. One possible explanation is that the basic sequence was multiplied many times, after which mutations modified it in a non random manner because of selection pressures for and against certain sequences. However, a more likely explanation is that the satellite sequences have been produced by an amplification process in which the periods between successive amplifications were long enough to allow further mutations to be introduced at each stage. The predominance of any particular mutation in the final sequence may therefore depend upon how early in the amplification it was introduced—the earlier it occurred, the more times it will have been multiplied and the greater its representation today. According to this model, it is not necessary to invoke selection pressures to explain the non random divergence of the present sequence from its ancestral sequence.

The presumed ancestral structure of six nucleotides contains one CCCT in the L strand and one TT in the H strand. The frequencies with which these two sequences are found today are 1/23 nucleotides and 1/11 nucleotides, instead of the ancestral 1/6 nucleotides. This demands a total of 23 % mutation in the L strand and 20% mutation in the H strand during divergence. At a rate of 4×10^{-9} changes/base pair/year—shown by some proteins—this level of mutation would take about 50 million years to achieve. This is consistent with estimates for the length of time since the guinea pig diverged from other rodents.

The concept that satellite DNAs contain short sequences which are very highly repeated is supported by the experiments of Kurnit, Schildkraut and Maio (1972) in which denatured strands of satellite DNAs were allowed to renature in high salt; under these conditions, only the nucleation reaction of renaturation takes place so that a short duplex sequence is formed but there is no subsequent rewinding of the remaining single strands. This reaction therefore measures the complementarity of rather short sequences. In 7·9 M CsCl in tris-EDTA buffer at 25°C, satellite DNAs of mouse, human I, guinea pig I, II and III, and calf I and II nucleate in 20 hours with less than 5% subsequent rewinding according to their buoyant densities. Human main band and bacterial DNAs, however, show no reaction under these conditions. These results thus imply that all satellite DNAs may share the same general type of nucleotide sequence repetition in their structure.

The structure of satellite DNA does not consist entirely of an L strand which comprises a repetition with some divergence of a basic repeating unit, matching an H strand which has a repetition with divergence of the complementary repeating unit; this would demand that all the sequences of either separated strand are related to the same single strand nucleotide sequence. However, Flamm, Walker and McCallum (1969) found that H strands or L strands of mouse satellite DNA reassociate with themselves to a limited but reproducible extent in the absence of the other strand—about 15% of either strand self anneals. This implies that each strand must therefore contain short sequences which are complementary (that is rather than identical).

Cytological Hybridization of Satellite DNA

That fractions of satellite DNA consist of rather short sequences of nucleotides repeated many times in related but non identical copies is shown by their $cot_{1/2}$ values and by sequence analysis. This conclusion makes no implication about the arrangement of the repeated sequences within the genome. Are sequences of very repetitive DNA organised in clusters consisting of many adjacent copies of satellite DNA or are satellite sequences interspersed amongst the unique and intermediate sequences? We may ask also whether the sequences of satellite DNA are found in all regions of all chromosomes or whether there is some restriction upon the loci containing them.

An extension of hybridization techniques allows the hybridization reaction to be performed with DNA intact in chromosomes; DNA is denatured in situ and then hybridized with a radioactively labelled preparation of DNA or RNA in solution. Binding of the labelled preparation to the denatured DNA of the chromosomes is detected autoradiographically and can be used to locate the regions of chromosomes to which the added radioactive DNA or RNA binds.

In the technique for cytological hybridization developed by Gall and Pardue (1969) and John, Birnsteil and Jones (1969), cells are squashed beneath a cover slip, frozen on dry ice, washed with ethanol and air dried. The slides are then dipped in an agar solution, the agar is allowed to gel, and the chromosomal DNA is denatured by placing the slides in NaOH for 5 minutes. The treated slide is then incubated at 60°C for 12–15 hours with a saline solution of radioactive DNA or RNA to allow hybridization to take place. After the hybridization reaction has been completed, any unbound radioactive DNA or RNA is removed by washing. The slides are then dipped into an emulsion and stored to allow grains to develop from decay of the radioactive label. The film is later developed and the slide is stained with a chromosomal dye to locate the genetic material.

Oocytes of Xenopus amplify their copies of the genes coding for rRNA; each pachytene nucleus produces extra-chromosomal rDNA, which can be detected cytologically as a densely staining cap on one side of the nucleus. During diplotene, this DNA spreads over the inner surface of the nuclear envelope and produces multiple nucleoli. Amplified rDNA constitutes about 70% of the total DNA of these cells. Figure 4.19 shows that when a labelled rRNA is added to these cells the label becomes concentrated over the cap.

Radioactive DNA of Xenopus which lacks the sequences of rDNA can be prepared by centrifuging DNA through CsCl gradients and discarding the rDNA, which bands at a high buoyant density corresponding to its G–C rich sequences. When Pardue and Gall (1969) incubated this DNA with cytological preparations, they found that the labelled DNA binds to the chromosomes but not to the cap. The difference of the locations identified by rDNA and by DNA lacking the rDNA sequences shows that the reaction is specific. Other controls include the failure of mouse satellite DNA to bind at all to the Xenopus cells and the abolition of binding by treatment of the cells on the slide with DNAase. Cytological hybridization therefore provides a method for localising the regions to which particular added DNA or RNA sequences hybridize.

The conditions of cytological hybridization allow only low cot values to be achieved; with somatic cells of adult tissues—that is those which contain only the usual diploid complement and have no gene amplification—added sequences of the fast or intermediate repetitive components may bind to the chromosomal DNA, but unique sequences should not react. Cytological hybridization is thus a suitable technique for identifying the location of satellite DNAs in the ultrastructure of the chromosome. Two techniques have been used to obtain a radioactively labelled preparation of the satellite sequences. Cells may be grown in culture in H^3-thymidine and the labelled satellite DNA fractions isolated by their buoyant density. As an alternative, the satellite DNA may be transcribed in vitro into RNA sequences by the RNA polymerase of E.coli, using an incubation mixture which contains

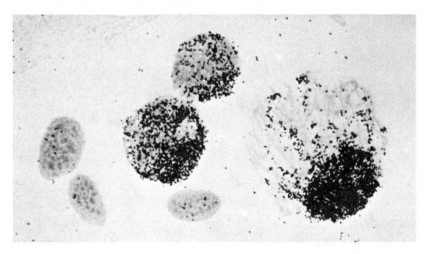

Figure 4.19: cytological hybridization between labelled rRNA and pachytene nuclei of Xenopus oocytes. The label is concentrated over the cap where the DNA containing the genes for ribosomal RNA is located. Data of Gall and Pardue (1969).

labelled precursors to RNA. The radioactive RNA is then hybridized with the chromosomal DNA.

Using labelled mouse satellite DNA, Pardue and Gall (1970) found that the satellite sequences are confined to the heterochromatic regions at the centromeres of mitotic cells. Figure 4.20 shows that the centromeric regions coincide with the area of silver grains developed by the action of the hybridized satellite DNA. Metaphase plates are usually used for cytological hybridization so that the individual chromosomes can be identified. However, when interphase nuclei are used the satellite sequences bind preferentially to the chromocentres which are thought to be formed by the aggregation of centromeric heterochromatin. Similar results have been obtained by Jones (1970) with an RNA transcript instead of satellite DNA itself. If main band DNA or its transcript is used instead of the satellite fraction, the grains detected by autoradiography are located over all regions of the genome of either mitotic or interphase cells; this shows that the sequences of intermediate repetition are not confined to any particular region of the chromosomes.

The correlation between satellite DNA and centromeric heterochromatin appears to be a general one (reviewed by Walker, 1971a, b; Rae, 1972b). Jones and Corneo (1971), for example, found that the RNA transcribed from human satellite DNA anneals to the centromeres of mitotic chromosomes. Jones and Robertson (1970) have tested the RNA transcribed in vitro from D. melanogaster DNA under renaturation conditions which permit only the rapidly renaturing sequences to reform duplexes. Their results suggest that highly reiterated sequences in general—that is both those located within

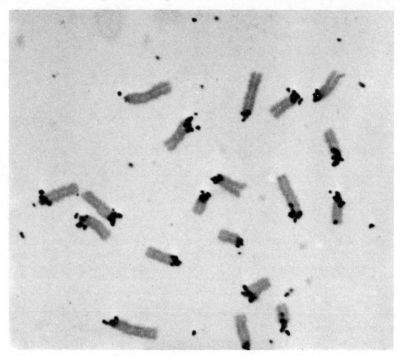

Figure 4.20: cytological hybridization between labelled mouse satellite DNA and mitotic cells. The label is located at the centromeres of the mouse chromosomes, identifying these regions as the location of satellite DNA. Data of Pardue and Gall.

satellite DNA fractions and any others located within the main band but also included in the fast component—anneal to centromeric heterochromatin. Using interphase cells of D. melanogaster and of mouse, Rae (1970, 1972a) has shown that rapidly annealing sequences—generated by transcription of the DNA fractions which renature rapidly—label the chromocentres heavily. It seems likely that the fast component which represents the highly repeated DNA consists of sequences which are located largely at the centromeres of all eucaryotic chromosomes. Cytological hybridization does not usually identify fast components at the telomeres, although staining may show that these regions contain heterochromatin.

Comparison of the sequence compositions of diploid and polytene cells also implies that the centromeric heterochromatin must be rich in satellite DNA sequences. Cytological observations of the increase in DNA content as salivary gland cells increase their degree of polytenization suggest that euchromatin may undertake some eight or nine doublings of DNA, whereas the DNA contained within the chromocentral aggregate of heterochromatin may double only two or three times (see chapter 1). Gall, Cohen and Polan (1971) reported that the DNA of D. melanogaster consists of a main band of density 1·702 g/ml in CsCl, with a single satellite comprising 8% of the DNA which bands at 1·689 g/ml; D. virilis DNA comprises four fractions, the main band sedimenting at 1·700 g/ml and satellites sedimenting at 1·692 (I is 25% of total DNA), 1·688 (II is 8% of DNA), 1·671 (III is 8% of DNA). The satellites renature very rapidly and have asymmetrical single strands. These proportions are found in DNA extracted from diploid tissues; however, if DNA is extracted from polytene tissues the satellite fractions are undetectable or are present in very much reduced amounts (see also Dickson et al., 1971).

That the absence of the satellites in the DNA of polytene tissues corresponds to the failure of the centromeric heterochromatin to replicate has been shown by cytological hybridization. If RNA bearing a radioactive label is transcribed from the D. melanogaster satellite or the D. virilis satellite I by E.coli RNA polymerase, it binds to the heterochromatin of the chromocentre of salivary gland cells. But diploid nuclei of the cells of imaginal discs—larval tissues located near the salivary glands—bind the same amount of this RNA as the polytene chromosomes. This implies that these satellite sequences must be replicated at most a very small number of times in the formation of polytene chromosomes.

Quite different results are obtained with RNA transcribed from the intermediate sequences. If Drosophila DNA is denatured, the fast component—that is satellites and any highly repetitive main band sequences—can be removed by isolating the duplex molecules which renature very rapidly; the remaining sequences can then be transcribed into RNA by the E.coli enzyme. Only the intermediate and not the unique transcripts hybridize under cytological conditions. Labelled RNA is found at many loci along the polytene chromosomes; in diploid nuclei the label is restricted to euchromatin. However, the level of labelling is very much lower in diploid than in polytene nuclei, confirming that the sequences of intermediate repetition located throughout the DNA of euchromatin are replicated during polytenization.

Since the satellite of D. melanogaster comprises only 8% of the total DNA but some 25% of the genome is contained in the chromocentre and fails to replicate in correlation with euchromatin, the centromeric heterochromatin must include repetitive sequences of the main peak as well as those of the satellite fraction. This conclusion is supported by experiments in which Rae (1970) found that RNA transcribed from rapidly renaturing components of the main band also hybridizes at the centromeric regions. Centromeric heterochromatin in D. melanogaster is sometimes considered to comprise two regions; α-heterochromatin which comprises the chromocentre itself and the surrounding granular regions of β-heterochromatin. The β-heterochromatin does seem to undergo replication and contains a large proportion of the repeated sequences remaining in the polytene chromosomes.

Fractionation of chromatin according to its density also suggests that satellite DNAs are located within heterochromatin. A sonicated suspension of chromatin—sonication reduces the size of the chromatin fragments—can be separated by comparatively gentle centrifugation into three fractions. Yasmineh and Yunis (1971) reported these as:

1. heterochromatin associated with the nucleolus (25% of total DNA);
2. an intermediate fraction containing heterochromatin, euchromatin and some nucleoli (20% of total DNA);
3. euchromatin (55% of total DNA).

The terms euchromatin and heterochromatin are used in this context simply to denote the lighter and heavier components of chromatin according to their rates of centrifugation; since euchromatin and heterochromatin are defined cytologically by their relatively diffuse and relatively dense appearances, this correlation is reasonable although it is of course imprecise.

When the DNA from each of these fractions derived from guinea pig chromatin is sedimented through gradients of Ag^+–Cs_2SO_4, the DNA of the first fraction shows a four fold enrichment of the satellite DNA fractions; the DNA of the second fraction displays the same relative proportions of each component as that of total DNA; and the DNA of the third fraction has only a single peak of main band DNA. Corneo, Ginelli and Polli (1971) obtained similar results with a fractionation of human chromatin; and Yasmineh and Yunis (1971) reported that fraction 1 of calf chromatin contains 36% of the total DNA, 16% of which comprises the heavy satellite and 7% of which represents the light satellite; fraction 2, which also contains 36% of the total DNA, has little satellite and fraction 3 (28% of total DNA) contains no satellite fraction. These results imply that about 85% of the total satellite DNA content of the calf is contained in the heterochromatic fraction. Yasmineh and Yunis (1969) found that sonication of mouse liver nuclei releases a fraction 1 which consists of 70% satellite DNA.

The fractionation of satellite DNA into these different chromatin components supports the concept that heterochromatin has a structure which is more dense than that of euchromatin and which contains much if not all of the satellite DNA (and presumably also any fast component which is highly repetitive but is located within the main band fraction). The nucleolar organiser usually appears to be close to centromeric heterochromatin in the cell; we do not know whether there is any functional significance in their apparent association. (However, the sedimentation together of much of the heterochromatin and the nucleolar material probably explains results such as those of Schild-kraut and Maio (1968) in which some satellite DNA appears to be present in nucleoli.)

Although satellite DNA is located at the centromeric heterochromatin, this does not imply that the highly repeated sequences are organised in a continuous array on DNA; the folded fibre model for the chromosome discussed in chapter one implies that sequences of DNA which are distant along the length of the fibre may be located at the same point in the chromosome. Because DNA is sheared to a size of the order of 500 nucleotides (300,000 daltons of duplex DNA) before hybridization, the cot curves described above do not distinguish possible arrangements of individual satellite sequences, for both continuous arrays of satellite sequences or satellite sequences interspersed with unique sequences would both be broken to the size of the individual member of the satellite family.

Renaturation with longer sequences of denatured DNA should throw some light on the organization of the satellite sequences in the genome. However, interpretation of these experiments must take into account a possible change in kinetics of renaturation of repeated sequences with a decline in fragment size. When DNA is sheared to a very small size, apparently unique sequences might be produced by fragments containing satellite sequences which have suffered more extensive changes in their sequence than the average member of the repeated family. Southern (1971) has pointed out that although the proportion of DNA behaving as repeated sequences increases with the length of the fragment at a low size range, this does not necessarily mean that short segments of repeated sequences are scattered amongst unique sequences; those sequences which appear to be unique at very low size values may simply comprise the repeated sequences containing an unusually high proportion of base changes.

When D. melanogaster DNA is sheared to a length of 300–600 nucleotides and renatured at a cot of 0·1, the very rapidly renaturing sequences can be isolated by chromatography on hydroxyapatite columns. Kram, Botchan and Hearst (1972) reported that the renatured duplex molecules band at 1·691 g/ml and seem to have a lower G–C content than the average of the genome. However, they cannot be isolated as a satellite when native DNA of molecular weight greater than 8×10^6 daltons (12,500 base pairs) is centrifuged through CsCl gradients. This implies that repetitive sequences low in G–C content are interspersed in the genome with regions which are richer in G–C base pairs.

A model of this nature predicts that the buoyant density of renatured duplex molecules depends upon the size of the fragments. The duplex molecules renatured at low cot generate two peaks when sedimented through CsCl; one at 1·691–1·694 g/ml and another at 1·702–1·706 g/ml. The proportion of the two peaks depends upon the size to which the Drosophila DNA is sheared before denaturation and renaturation. At a length of 600 nucleotides, all the rapidly renaturing DNA forms duplex molecules of buoyant density 1·691 g/ml. As the length of DNA is increased to 25,000 nucleotides the proportion of this peak declines and that of the heavier peak increases; at the same time, the total amount of DNA found in the fast component increases from 8% to 16%.

The density of the heavy peak suggests that it may consist of renatured satellite duplex molecules with tails of single stranded DNA which are more complex in sequence and so have not renatured. The lighter peak may comprise more completely renatured sequences of satellite alone. Kram et al. compared these results with a computer simulated programme designed to find arrangements of satellite sequences which would generate the observed buoyant density peaks by random breakage of the DNA. They suggested that the rapidly renaturing sequences may consist of lengths of some 2500–10,000 base pairs separated from each other by less rapidly renaturing sequences of 3000–7000 base pairs. The more slowly renaturing sequences might belong to a different component of the genome—either the intermediate or unique fractions—or might include variants of the satellite sequence which are more distant from the ancestral sequence and so renature more slowly. There may thus be serial repetition of satellite sequences, presumably in the regions of the chromosome fibre concentrated at the centromere (see below).

Since heterochromatin in general appears to be replicated later during S phase than euchromatin (see page 56), the sequences of satellite DNA located at the centromeres should be replicated towards the end of S phase. By labelling synchronised L cells at different times during S phase, Bostock and Prescott (1971a) found that the buoyant density on CsCl of the main band DNA decreases as replication proceeds, apparently representing the shift from sequences richer in G–C content synthesised at first to those poorer in G–C synthesised later. The reason for this shift is not known but it appears to be a feature of replication in several mammalian cell types (see page 59). Satellite DNA seems to be preferentially synthesised during the third quarter of S phase. The satellite fraction is not labelled by a pulse given during the last quarter of S phase, but the DNA labelled during the second half of S phase includes satellite sequences. A small amount of satellite is labelled during the second quarter of S phase.

A similar order of replication has been observed in mouse lymphoma cells by Flamm, Bernheim and Brubacker (1971), who found that satellite sequences become labelled during middle and late S phase. Of course, it is not possible to completely equate satellite fractions and centromeric heterochromatin; although much of the satellite DNA always appears to be located at the centromeres, other sequences may also be present within them, including further members of the fast fraction which are located in main band DNA as well as any intermediate or unique sequences. Definition of the relationship between replication of euchromatin and of heterochromatin thus requires resolution of the other components which may be present in the heterochromatin.

A striking temporal division of replication has been observed by Bostock, Prescott and Hatch (1972) in cells of the kangaroo rat, Dipodomys ordii, transformed with SV40 and grown in culture. Satellite DNAs comprise about 60% of the total DNA and are present in enriched amounts in heterochromatic fractions; although occupying the centromeres they extend beyond them in an asymmetrical manner and may fill entire chromosome arms. Synthesis of main band DNA takes place during the first half of S phase; satellite DNA is synthesised during the second half. The switch from main band to satellite replication may be abrupt, for intermediate patterns of labelling are rare in synchronised cells. One of the two satellites (banding at 1·714 g/ml) shows a behaviour very similar to that of the mouse satellite; label is incorporated during the third quarter of S phase and ceases well before its end. The other satellite, banding at 1·707 g/ml, is labelled close to the end of S phase. Autoradiography shows that labelling of the centromere ceases 1–2 hours before the end of S phase.

Function and Evolution of Repetitive Sequences

The genomes of all eucaryotes appear to comprise three types of sequence: unique lengths of DNA present in only one copy per haploid genome; intermediately repetitive sequences present in from 10^2 to 10^5 copies, the range of frequencies depending upon the organism; and highly repetitive sequences, often included in a satellite fraction of distinct buoyant density, consisting of more than 10^6 repetitions of rather short sequences of the order of 100 nucleotides long. Cytological hybridization shows that the sequences of intermediate fractions are located in all regions of the chromosome, but most if not all of the highly repetitive fraction is found at the centromeres of mitotic chromosomes and is present in the heterochromatic chromocentres of interphase cells.

The sequences of RNA present in eucaryotic cells appear to correspond only to the unique and intermediate sequences, for Flamm, Walker and McCallum (1969) found that in no case can an excess of RNA extracted from the liver, spleen or kidney of the mouse hybridize to the separated strands of satellite DNA. Another experiment implying that the highly repeated sequences are not transcribed is that of Sieger, Pera and Schwarzacher (1970), who found that the chromocentres of interphase cells of M. agrestis do not incorporate a label of H^3-uridine.

Three general classes of model have been proposed for the functions of highly repeated sequences. Britten and Davidson (1971) have suggested that they may represent pools of material which is not in genetic use but which may be utilised in further evolution. Another possible implication of the failure of these sequences to be transcribed is that they represent recognition sites in the chromosomal DNA; one early model was to postulate that they might be elements recognised by the molecules which control gene expression. Highly repeated sequences might also provide sites at which transcription or replication is initiated. However, these models would imply that satellite sequences should be found at all chromosome locations.

The concentration of satellite sequences at the centromeres suggests a third model; they may play some structural role in the chromosome, perhaps one for which their sequence is of comparatively small importance. Walker (1971a, b) has suggested that the presence of a bulk of DNA at the centromere may confer an advantage on the chromosome which carries it; the quantity of DNA at the centromere might be important in helping the chromosomes to pass successfully through the movements demanded of them at cell division. For example, increased bulk of centromeric DNA might be better able to resist breakage when chromosomes are pulled to the poles by the spindle fibres. Other possible functions of this class suggested for satellite sequences by Walker, Flamm and McClaren (1969) include assistance in meiotic pairing, recognition of homologous centromeres, and folding of the chromosome fibre. Satellite sequences might be present at different parts of the chromosome fibre which are brought together at the centromere.

The small repeating unit and the high repetition of satellite sequences implies that the present content of satellite DNA must have evolved by a many fold replication of the original sequence, integration into the genome at the centromeric regions, and spreading through the population. It seems likely that the same satellite sequences are found at all the centromeres of a chromosome set. One possible mechanism which has been suggested to explain the widespread occurrence of satellite DNA in a species is that its original integration into the genomes of the population may have been assisted by a virus. Britten and Kohne (1969b) suggested that repetitive sequences may have originated in a saltatory replication—a sudden event in which particular sequences suffered an excessive replication of tens or hundreds of thousands of copies of one sequence. Walker (1971) has pointed out that because base substitutions of the ancestral sequence do not seem to have occurred with equal frequencies at all positions of the hexamer of guinea pig satellite I, there may have been a series of successive replication events separated in time by the introduction of mutations.

That satellite sequences have developed recently in the evolution of rodents is shown by the lack of common sequences between the satellites of different rodent species. Hennig and Walker (1970) have shown that the proportions and characteristics of satellite sequences vary even within a closely related group of rodent species. The proportion of DNA in the satellite varies from about 2% (the limit of resolution in buoyant density experiments) to about

10%; some rodents have a satellite which is distinct from the main band of DNA—by either greater or lower buoyant densities—whereas others lack such a component in CsCl gradients, but contain a rapidly renaturing component within the main band of DNA. A similar lack of relationship between the satellites of related species has been found by Mazrimas and Hatch (1972) amongst the kangaroo rats, in which the proportion of DNA in the satellites of different species varies from none to half of the genome in neutral CsCl.

Three related species of Drosophila, D. hydei, D. neohydei and D. pseudo-neohydei, all have satellite DNAs which band on CsCl at positions equivalent to a G–C content higher than that of main band: D. hydei has a satellite of 55% G–C which occupies 3% of the genome; D. neohydei has a component of 62% G–C corresponding to 3·4% of the genome; and D. pseudoneohydei has a fraction of 59% G–C comprising 6·5% of the genome. Hennig, Hennig and Stein (1970) transcribed the satellites of D. neohydei and D. pseudo-neohydei into labelled RNA; each RNA hybridizes well with its parent DNA but poorly (5–10 times less well) with the DNA of the other Drosophila species. Although these species can interbreed, similarities between their satellites therefore appear to be limited.

Wide variations in the properties of satellite DNAs in a group of species implies that the satellites may have evolved since the separation of the species. In many instances, satellite DNAs appear to share in common only the characteristic of very low complexity of sequence. [The extreme case of sequence simplicity in a satellite DNA has been reported in the crab, which has a satellite of alternating A and T bases (reviewed by Laskowski, 1972). However, Blumenfeld and Forest (1971) found that eggs of D. melanogaster also contain a poly-dAT complex, much of which may be coded by the Y chromosome. We do not know whether extremely simple satellites of this nature serve the same function as those of greater sequence complexity.]

Although few similarities have been reported between the satellite DNAs of groups of related species, Sutton and McCallum (1972) have shown that the satellites of four mouse species are related to each other and appear to have diverged from a common ancestor. The Thailand mice M. caroli, M. famalus and M. cervicolor lack the satellite band at 1·691 g/ml which is characteristic both of the European M. Musculus and of another Thailand mouse, M. castaneus. (The presence of the satellite in the latter two species confirms their predicted close evolutionary relationship since they are inter-fertile.) On Ag^+–Cs_2SO_4 gradients, M. caroli has one heavy and two light satellites which band at 1·695 and 1·967 g/ml on neutral CsCl. Reassociation curves show that these satellites are of about the same complexity as that of M. musculus.

Appreciable cross reaction takes place between the satellite of M. musculus and the DNAs of the Thailand mice to form duplex molecules of low stability with a Tm of about 60°C. This implies that sequences related to the members of the highly repeated family of M. musculus must be present in the Thailand mice. Allowing for the mismatching which takes place when these hybrid DNAs are formed, the satellites of M. musculus and M. caroli appear to have a common repeating unit of 10 nucleotides at 60°C and of 19 nucleotides at the less specific temperature of 50°C. The cross reaction therefore detects a shorter repeating unit than the accurate repeating unit of M. musculus of 120 nucleotides. Sutton and McCallum pointed out that the short internal repeating length of 10 nucleotides is not detected when the M. musculus satellite is reassociated because it has diverged so far from its ancestral sequence that associations in badly matched registers are substantially slower than those in the register of the longer, more accurate repeat.

The cross association of the light satellites of M. musculus and M. caroli implies that they are derived from the same general family of sequences; sequences of M. famalus and M. cervicolor also appear to be derived from this family. The extent of divergence between the satellites of the different species is greater than the divergence within each satellite, for whereas the unique DNA sequences of the different mice reassociate to form hybrids with a ΔTm of only 5°C, hybrids between the satellite sequences are much less stable.

If unique and satellite sequences have suffered mutation at the same rate, this result must imply that divergence between satellites of the mice of an ancestor population must have preceded its evolution into separate species. However, if the satellite of each species was produced by a saltatory replication of a sequence selected at random, there should be no relationship between them. Sutton and McCallum therefore suggested that the Mus satellites are descended from a common ancestral sequence which suffered duplication and divergence to generate a family of repeated sequences; different members of this family must then have been the substrates for saltatory replications in the populations which evolved into the present species.

The low stability of the hybrid satellites might be produced by either of the two relationships shown in figure 4.21. If the satellites of individual Mus species were generated by different saltatory replications of related members of a highly diverged family of repetitive sequences, then the hybrid duplex satellite molecules should be held together by a consistent but imperfect pairing along their length which represents the relationship between the two ancestral sequences from which they are derived. This model predicts that the structure of the self-reassociated satellites should be different from that of the hybrid products of a cross reaction, since each satellite should consist of a more accurate repetition selected by saltatory replication of one of the ancestor sequences. An alternative model is to suppose that individual Mus satellites may have been generated by saltatory replications of different sequences sharing only a short common element. In this case, each hybrid duplex should consist of short well matched sequences representing the

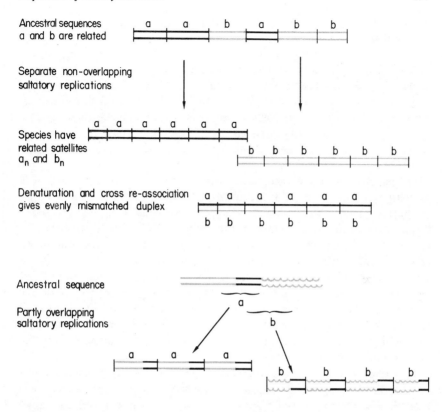

Figure 4.21: models for evolution of satellite DNA. Upper: the common ancestor contains a repetition of the related but non-identical sequences *a* and *b*. Sequence *a* is utilised for one saltatory replication whilst sequence *b* is utilised in the other. Further divergence is later imposed on these sequences in the separated species. Because *a* and *b* are related, cross-reassociation experiments produce a duplex which is evenly mispaired. Lower: two different sequences are utilised for the saltatory replications, but have in common the dark sequence. Although further divergence takes place during evolution of the separated species, cross reassociation experiments allow the descendents of the common sequence to renature; the descendents of the remaining part of each sequence are not complementary and remain single stranded. This hybrid, although also uniform, therefore has a structure in which well matched duplex sequences alternate with non-complementary single strands. Data of Sutton and McCallum (1972).

common element, separated by dissimilar single strand regions which are not complementary.

Both models are consistent with the uniform nature of the hybrid product, which in spite of its low stability melts at a sharp Tm. But the first model predicts a consistent degree of mispairing whereas the second model predicts the presence of many single strand regions which would remain unpaired. Reassociated molecules with these classes of structure can be distinguished by their response to the enzyme nuclease S1; this is specific for denatured DNA and can degrade internal single strand sequences within a duplex structure.

The activity of the nuclease depends upon the temperature at which it is incubated with its substrate. If the duplex is formed by accurate reassociation of short sequences, there should be appreciable lengths of single stranded regions even at low temperature; if the duplex is formed by consistent mispairing the activity of the enzyme should increase as temperature rises and thus encourages strand separation to generate single stranded regions. The hybrid satellites are not digested by nuclease S1 at low temperature, but increase in instability with temperature; this result suggests a consistent degree of mispairing in strand reassociation. The differences observed in Tm values of the separately reassociated and cross-reassociated satellites would be consistent with a 13% random mutation of sequence in the members of the ancestral family before the saltatory replication event.

Two important corollories follow from the influence of mismatching in slowing the kinetics of renaturation. The first is that the cross reaction between heterologous DNAs derived from different species may demand cot values which are higher than those required for self-renaturation; sequences which are related but have evolved so that they are no longer identical may have a greatly reduced rate of reassociation so that their apparent reaction is less than would accord with their relationship. Cot values for cross-reassociation of repeated sequences can be corrected for mismatching by characterising the thermal profile and hence the stability of the hybrid duplex; unique sequences may of course fail to anneal under normal cot conditions so that the level of reaction is too low to follow.

Another important implication is that the reduced rate of renaturation between related sequences, compared with that between perfectly complementary sequences, may explain the apparent absence in the eucaryotic genome of sequences repeated from 5–100 times. If a sequence is repeated five times, for example, but each of the five copies has diverged in evolution by 10% from its ancestor, then each copy will renature with its true complement some five times more rapidly than with one of the other, related copies, with which its reaction rate is reduced by mismatching. This implies that on average only the precisely complementary copies would be detected during renaturation. It is therefore possible that the eucaryotic genome contains families of a very low degree of repetition which have not been detected in renaturation experiments.

According to the kinetics of renaturation, the intermediate fraction consists of sequences repeated from 100 to 10^5 times; some at least of these sequences are transcribed into RNA. However, we do not know whether they are translated into proteins. One interpretation of the sequence complexity suggested by the renaturation experiments is that genes coding for proteins comprise the unique fractions and that control signals of some nature—whether recognised as such in the genome itself or active when transcribed into RNA—constitute the intermediate sequences. The satellite sequences, of course, are not transcribed and must play some structural role. We have no evidence on the possible origin of the intermediate fraction, but its sequences are species-specific and in cross-association experiments show extents of reaction which correspond to the evolutionary relationships of species.

Distribution of Repetitive Sequences in DNA

The kinetics of renaturation reveal only the total length of sequences in each component and the average repetitive frequency. But this makes no implication about the lengths or locations of individual repeated sequences. Cytological hybridization shows that intermediately repetitive sequences are located in all regions and that highly repetitive sequences are largely confined to the centromeres. Because of the folding of DNA in the chromosome, however, this does not reveal the organization of sequences in DNA.

That there is some serial repetition of sequences within the genome is suggested by the experiments of Thomas et al. (1970). When fragments of eucaryotic DNA are treated with exonuclease to digest their ends, duplex molecules with tails of single stranded DNA are produced. These tails can then self anneal to form circular structures of various sizes. The tails must therefore comprise complementary sequences which are repeated a short distance farther along the genome. *Folded circles* of this type are found both in main band and in satellite DNAs. Another technique for preparing circles is to expose the fragments of DNA to high temperature or pH, separate the chains, and then anneal at a low concentration. This produces *slipped circles*.

Formation of circles may be followed in preparations of isolated main band or satellite DNA. Hennig, Hennig and Stein (1970) found that satellite circles as small as 300 bases can be formed from some Drosophila satellites; the circles formed by the main band are usually larger. About 20% of the satellite DNA and about 14% of the main band DNA forms circles. However, highly repeated sequences of DNA may remain in the main buoyant density band after removal of the apparent satellites and such sequences may be responsible for much of its circle formation.

By using DNA extracted from polytene chromosomes of Drosophila species, Lee and Thomas (1973) attempted to examine the circles formed in fragments of DNA lacking the highly repetitive satellite sequences (which are under represented in polytene cells). About 15% of the DNA of D. melanogaster, D. hydei or D. virilis forms circles. The stability of the renaturation between complementary tail sequences which generates the circles can be assessed from their melting profiles; this criterion suggests that circles are formed by annealing between complementary sequences of a length in excess of 100–200 nucleotides.

The optimum fragment length for ring formation is 1–2 μ; below and above this length fragments are formed increasingly rarely. The dependence of circle formation upon fragment length suggests that the serially repeated units which anneal under these conditions are confined to regions of DNA about 5 μ long; that is they are organized in blocks of upto about 6000 nucleotides in length. In a study of circle formation in mouse and in Necturus, Pyeritz and Thomas (1973) obtained results similar to those displayed by Drosophila DNA.

Two models to explain these results have been considered by Thomas, Zimm and Dancis (1973) in their theoretical analysis of ring formation. A tandem repeat model supposes that the same sequence is repeated many times in immediately adjacent copies; an intermittent repeat model postulates that the repeated sequences which form the circles are separated from each other by unique sequences. Bick, Huang and Thomas (1973) suggested that these models might be distinguished by the properties of the rings at their points of closure, since a prediction made by the intermittent repeat but not by the tandem repeat model is that single stranded sequences of unique DNA should be present which find no complements. Their failure to detect such sequences therefore tends to support the tandem repeat model.

These results were originally interpreted to show that sequences of the intermediate fraction are organized in such a way that all the members of one family are located in one region of the chromosome. More recent results, however, have demonstrated that in Drosophila circles are formed largely, although not entirely, from the satellite DNA fraction. By analysing the Thomas circles isolated from D. melanogaster, Schachat and Hogness (1973) showed that some two thirds of their DNA is derived from the highly repetitive fraction, the remainder corresponding to sequences of the intermediate fraction. The buoyant densities of isolated Thomas circles also indicate that most are derived from the satellite DNAs. Since the circles hybridize under cytological conditions with regions of centromeric heterochromatin, they seem for the most part to represent fairly large blocks of satellite sequences derived from the DNA of the centromeres.

This conclusion is supported by the analysis of satellite DNAs made by Peacock et al. (1973), which showed that these highly repeated sequences can be isolated as high molecular weight DNA; more than 80% of the rapidly renaturing fraction exists as sequences organized in rather long blocks. After removal of these satellites from Drosophila DNA, the remaining fraction displays a greatly reduced ability to form circles; under these conditions, some three quarters of the total circle formation appears to have been due to satellite sequences.

The members of each highly repeated satellite family in Drosophila therefore appear to be organized in the form of contiguous blocks of DNA, each block residing within the centromeric heterochromatin. Although these sequences account for most of the circle formation, a small proportion of the circles appear to be derived rather from the fraction of intermediate repetition. Since most intermediate sequences appear to be interspersed in the genome with non-repeated sequences (see below), these circles may represent a minor proportion of intermediate DNA which is organized in tandem repeats. Its function and relationship to the other intermediate sequences is not known.

A visual technique to follow the distribution of intermediate repeated sequences in the genome has been utilized by Wu, Hurn and Bonner (1972). After removing the fast component which occupies 6–8% of the genome of Drosophila melanogaster, by extracting the duplex molecules renatured at very low cot values, DNA sheared to a length of 800 base pairs may be renatured under conditions when only the intermediate sequences providing 15% of the genome can form duplex molecules. The DNA of the duplex molecules can then be recovered as a sample of the intermediate component and used in further renaturations.

The distribution of the repetitive segments in the genome can be determined by denaturing these short fragments and then renaturing them with denatured DNA of much greater molecular weight; this forms short duplex regions along a backbone of single stranded DNA. The lengths of the reformed duplex regions can be measured with the electron microscope. The lengths of duplex formed most frequently are 150–200 base pairs (about 10^5 daltons or 0·05 μ), the distance between them varying from 100–3000 base pairs and most commonly about 750 (corresponding to 5×10^5 daltons or 0·25 μ). This technique does not allow an accurate estimate to be made of the lengths of the repeating units because bushes of single strand DNA are found at the ends of each duplex, but suggests that they are probably rather short, of the order of 125 nucleotides long on average. This would be consistent with some regulatory rather than protein coding role.

A model for the Drosophila genome on the basis of these results has been proposed by Bonner and Wu (1973), who suggested that sequences of the intermediate fraction are interspersed amongst unique sequences. This is therefore a model of intermittent repeat. The kinetics of renaturation suggest that the intermediate fraction of D. melanogaster consists of sequences repeated on average about 30–35 times, comprising about 15% of the genome, that is about $1·7 \times 10^7$ base pairs. The total length of the repeating fraction— that is the combined lengths of one member of each family—is about 5×10^5

nucleotides. If the average length of the individual repeating unit is 125 bases, then there must be about 4000 families. According to this estimate of the length of the individual repeating unit, therefore, the intermediate fraction of D. melanogaster DNA consists of 4000 different families, each family comprising about 35 copies on average of a sequence of about 125 nucleotides in length.

Results which suggest the interspersion of repetitive with non repetitive sequences of the genome in Xenopus have been obtained in an analysis of renaturation by Davidson et al. (1973). In these experiments, labelled Xenopus DNA was sheared to give preparations of defined lengths varying from 450 to 3700 base pairs. After denaturation, these preparations of DNA were renatured with an excess of a standard preparation of unlabelled Xenopus DNA of length 450 nucleotides.

Binding of the labelled DNA to hydroxyapatite after renaturation increases with the fragment length from 450 to 3700 nucleotides. Almost 80% of the 3700 nucleotide long fragments bind when renatured to a cot of 50, at which only the repetitive sequences can anneal to form the duplex molecules retained on the hydroxyapatite columns. But only some 45% of the Xenopus genome is repetitive under these conditions. This implies that the fragments of 3700 nucleotides contain both repeated sequences (responsible for their renaturation and retention on hydroxyapatite) and non repeated sequences which do not react.

This interpretation is supported by the results of experiments in which the 3700 nucleotide fragments retained on the hydroxyapatite columns were recovered, denatured, sheared to a length of 450 nucleotides, and then renatured with the standard unlabelled preparation. Only 46% of the now shorter labelled fragments form duplex molecules, demonstrating that the repeated and non repeated sequences originally present on the same molecules of 3700 nucleotides may be separated by shearing to a length of 450 nucleotides.

Because DNA molecules sheared to lengths of either 1400 or 450 nucleotides and then renatured display similar Tm values when subsequently melted, renaturation of fragments of different lengths appears to rely upon reaction between the same classes of sequence. But the extent of duplex formation—measured by hyperchromicity—is less with the 1400 nucleotide long fragments. This suggests that both 450 nucleotide long and 1400 nucleotide long fragments possess a repeated element which renatures and a non repeated element which remains single stranded. The hyperchromicity of the 450 nucleotide long fragments suggests that they do not form complete duplexes but renature only in part, the average length of the repeating unit corresponding to about 300 ± 100 nucleotides. The fragments of length 3700 nucleotides react at about the same rate as fragments of 450 nucleotides in length when renatured with a standard preparation of length 450 nucleotides. This implies that each 3700 fragment possesses only one or two repetitive elements.

About half of the Xenopus genome appears to comprise alternating sequence elements of the repetitive and non repetitive classes, the repeating elements varying in length from 200–400 nucleotides and the non repeating elements varying on average from 650–900 nucleotides. Analysis of the renaturation kinetics suggests that much of the remaining DNA comprises repeating units whose length cannot at present be defined separated by unique sequences of from 4000–8000 nucleotides.

Analysis of the DNA of the sea urchin suggests that its genome displays a sequence interspersion similar to that of Xenopus; by measuring the renaturation of DNA sheared to various lengths with an excess of 450 nucleotide long fragments, Graham et al. (1974) showed that about half the sea urchin genome comprises 300–400 average length repetitive segments interspersed with somewhat longer non-repetitive segments, of average length 1000 nucleotides. A further 20% of the genome displays a similar pattern of alternation, but with the repetitive sequences separated by rather longer non-repeated sequences, of about 4000 base pairs. About 6% of the genome consists of clustered repetitive sequences; and about 22% comprises what may be solely non-repeated sequences, or at least has only a few repetitive sequences. As Davidson et al. (1973) have pointed out in their review of these results, the similarities between the Xenopus and sea urchin genomes, whose sequence arrangements differ only in the exact values of the critical parameters in spite of their distance apart on the evolutionary scale, imply that interspersion may be a general phenomenon in the eucaryotes.

Although the non-repeated component of the genome almost certainly includes the structural genes, the function of the intermediate repetitive component cannot at present be defined. The interspersion of repetitive with non-repetitive sequences is consistent with models which suppose that the repeated sequences may provide elements which control the structural genes. One particularly critical problem is to define the relationship between the unit of transcription and the organization of the genome revealed by hybridization experiments; it is important to resolve whether these rather short repeated sequences are transcribed and if so whether they enter the polysome fraction in the cytoplasm or are degraded in the nucleus (see chapter five). Such problems have been discussed by Davidson and Britten (1973) in their review of the organization of the eucaryotic genome and its implications for the control of transcription (see also chapter six).

Organization of Heterochromatin

Banding of Chromosomes with Giemsa and Quinacrine

That chromosomes have a characteristic overall ultrastructure in which a quaternary coiling is imposed upon the folded fibre is shown by their characteristic appearances during mitosis. The specificity of this structure is indicated also by the bands seen in the polytene chromosomes of Dipteran salivary glands and in the appearance of loops in the lampbrush chromosomes of amphibian oocytes. More recently, the interactions of chromosomes with fluorescent stains and chemical dyes has revealed that each individual chromosome in the haploid set of any organism reacts in a different manner; this indicates that each must have its own characteristic structure.

That there are local variations in the molecular organization of DNA and proteins in the ultrastructure of the eucaryotic chromosome is therefore implied by the banding patterns generated by staining with Giemsa or quinacrine dyes. Under mild conditions of pre-incubation, a characteristic array of bands is produced by either reagent in the arms of each chromosome; since the two reagents generate essentially the same pattern, the reaction must depend upon some underlying structural arrangement of the chromosome components. The heterochromatic X chromosome usually displays the same banding pattern as its euchromatic homologue; this implies that its inactivation to yield facultative heterochromatin does not alter at least those features of its ultrastructure which interact with the dyes. Under more stringent conditions of pre-incubation, reaction in the chromosome arms is abolished and dense staining takes place at the regions of centromeric heterochromatin; this implies that there may be some substantial difference between the organization of euchromatin and facultative heterochromatin on the one hand and that of constitutive heterochromatin on the other.

The dense staining with Giemsa which takes place at the centromeres of mitotic chromosomes subjected to cytological hybridization is also produced when the hybridization step with added radioactive DNA or RNA is omitted. Arrhigi and Hsu (1971) and Yunis et al. (1971) found that treatment with alkali of cells on slides, followed by incubation in the absence of exogenous nucleic acid, leads to the characteristic dense staining around the centromere. In this technique, known as *C-banding*, fixed chromosomes are exposed to upto 0·07 M NaOH for 60–120 seconds and then incubated in 2–6 × SSC (a standard saline citrate solution) at 65°C overnight. Comings et al. (1973) noted that either of these treatments alone can itself produce centromeric staining when the treated cells are later incubated with Giemsa, but that the combination of the alkaline treatment and the lengthy incubation in SSC produces a clearer definition.

Untreated chromosomes stain more or less uniformly with Giemsa along the arms, although centromeres show a light stain. Sumner, Evans and Buckland (1971) found that the reactions of human chromosomes to Giemsa stain depend upon the severity of the treatment to which they are subjected before staining. Treatment with alkali produces the centromeric staining of the C-banding technique shown in figure 4.22. However, when chromosomes are first subjected only to milder conditions of incubation when they are

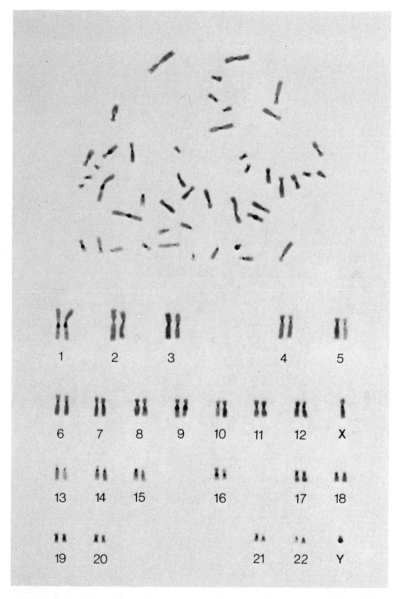

Figure 4.22: C-banding of human chromosomes. Upper: metaphase plate subjected to banding. Lower: chromosome set arranged to show individual members. Photograph kindly provided by Mr. Norman Davidson, MRC Clinical and Population Cytogenetics Unit, Western General Hospital, Edinburgh.

Gene Expression

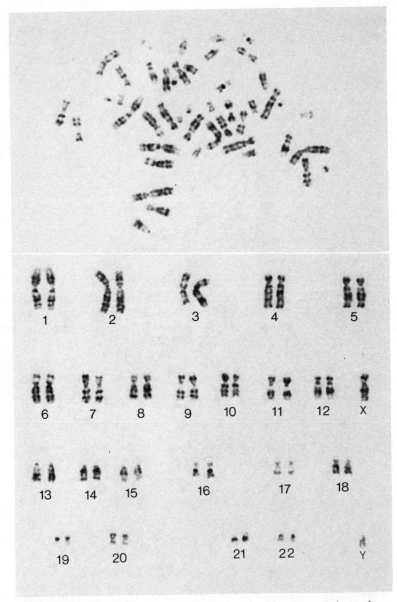

Figure 4.23: G-banding of human chromosomes. Upper: metaphase plate subjected to banding. Lower: chromosome set arranged to show individual members. Photograph kindly provided by Mr. Norman Davidson, MRC Clinical and Population Cytogenetics Unit, Western General Hospital, Edinburgh.

incubated for 60 minutes at 60°C in the 2 × SSC, Giemsa staining generates a pattern of bands along the arms of each chromosome, with little or no reaction at the centromere (see also Schendl, 1971). This technique is known as *G-banding*; the G bands of the human complement are shown in figure 4.23 (see also figure 4.24).

The fluorochromes quinacrine and quinacrine mustard interact with the chromosomes of many species. In particular, the Y chromosome of man fluoresces very vividly. Chromosomes cannot usually be identified in interphase cells, but after quinacrine staining the human (and some monkey) Y chromosome can be visualised as an intense blob of fluorescence. Pearson, Bobrow and Vosa (1970) showed that this reaction in effect identifies a chromatin body for the male (corresponding to the single inactivated X chromosome which forms the Barr body of the female). Bobrow, Pearson and Collacott (1971) were able to show by this technique that the human Y chromosome is associated with the nucleolus. It is the distal part of the long arm of the Y which fluoresces and this region appears to comprise constitutive heterochromatin.

The fluorochromes interact less intensely with the other chromosomes of the cell, but give *Q-banding* patterns in which fluorescent regions are located in characteristic bands along the length of each chromosome. Using this technique, Caspersson, Zech and Johansson (1970) have shown that each human metaphase chromosome can be identified by its characteristic banding pattern. Caspersson et al. (1972) reported that the same Q-banding pattern is found in metaphase chromosomes of all human tissues—there is no variation in structure with differentiation. The natures of structural abnormalities involving rearrangements of chromosomal material have also been identified by Q-banding (Miller et al., 1971a). Differential banding patterns may be a general characteristic of mammalian chromosome sets; although all the mouse chromosomes are acrocentric and therefore indistinguishable on metaphase plates, Miller et al. (1971b, e) showed that the different chromosomes can be identified by their characteristic Q-banding patterns.

The bands produced in human chromosomes by application of either G-banding or Q-banding appear to be almost identical; Sumner, Evans and Buckland (1971) observed that the only difference is that the Giemsa stain generates two bands on the Y chromosome instead of the single fluorescent spot. A similar situation is found in the mouse; Rowley and Bodmer (1971) reported that Giemsa and quinacrine staining produce the same banding patterns in all mouse chromosomes (for a detailed comparison see Nesbitt and Francke, 1973). The centromeres of the mouse chromosomes show the least fluorescence of any region; and the amount of centromeric heterochromatin revealed by the C-banding technique seems to be directly proportional to the reduction in fluorescence of the centromeres upon Q-banding.

Aula and Saksela (1972) also noted that the centromeres of human chromosomes fail to interact with quinacrine.

Two types of chromosome region may therefore be delineated by these banding techniques. Centromeric heterochromatin reacts especially poorly with either Giemsa or quinacrine under the mild conditions used for G-banding and Q-banding; however, it reacts intensely with Giemsa stain under the more stringent conditions of alkaline denaturation used in the C-banding technique. The regions which form bands in the chromosome arms react only in the mild conditions of G-banding and Q-banding. The basis for the reaction of Giemsa stain is unknown, but quinacrine appears to bind directly to DNA. The identity of the G-banding and Q-banding patterns implies that both may depend upon the same structural features in the organization of the chromosome, rather than representing some particular sequence of DNA. Banding patterns may thus identify chromosome regions which have a different manner of association of DNA and protein from the non-banding regions, although the reaction is not understood in any detail.

There is in general assumed to be a correlation between the location of constitutive heterochromatin, late replicating DNA, and the regions which respond to chromosomal dyes. Constitutive heterochromatin is clearly distinct from facultative heterochromatin, since organisms in which one X chromosome is active and the other is inactivated show identical C-banding, G-banding and Q-banding of both their X chromosomes. One interpretation of banding patterns is to suppose that the constitutive heterochromatin may fall into two classes: centromeric heterochromatin (identified by C-banding) and constitutive heterochromatin at other locations (identified by G-banding and Q-banding).

Although the centromeric regions stained by C-banding can be identified as constitutive heterochromatin by both structural and functional features, however, the nature of the regions of the chromosome arms which react to G-banding and Q-banding is less clear. That the centromeres are surrounded by constitutive heterochromatin is well established in many species; these regions appear condensed when stained, aggregate to form chromocentres during interphase, contain satellite DNA sequences which are not transcribed into RNA, and replicate towards the end of S phase. But the identification as heterochromatin of the G-bands and Q-bands is less well defined. Since these regions are defined not by function but by visible staining characteristics, any correlation between apparent heterochromatin in the arms and banding patterns relies at least in part upon a circular argument.

In the human complement, these regions appear to be late replicating; Ganner and Evans (1971) found that virtually all the late replicating material can be accounted for by the facultative heterochromatin of the inactive X chromosome, the centromeric regions stained by C-banding, and the sites in the chromosome arms stained by G-banding and Q-banding. This provides some evidence that the G-bands and Q-bands may represent heterochromatin. However, no distinct sequences of DNA have been identified in these regions and we do not know whether or not they are active in transcription. We can at present therefore say only that the bands appear to represent chromosome segments whose DNA-protein interactions differ from that of other regions in the euchromatin of the arms; it is possible, but not proven, that the bands represent inactive regions of constitutive heterochromatin.

Some organisms contain chromosome regions which stain very intensely with quinacrine, such as the Y chromosome of the human complement. Vosa (1970) observed that several species of Drosophila possess chromosome regions which show an analogous intense fluorescence, more like that of the human Y than of the typical chromosome band. In metaphase cells of D. melanogaster, all of the intensely staining regions correspond to sites identified as heterochromatin (although not all heterochromatin stains intensely). There is a general correlation between constitutive heterochromatin, late replication and fluorescence, although it is not absolute.

In Samoaia leonensis, the metaphase chromosomes—which are about twice the size of those of Drosophila—contain large regions of heterochromatin around the centromeres which fluoresce very intensely after Q-banding. Because these regions are extensive, they can be identified with some precision and Ellison and Barr (1972) found that there is an exact correlation between intense fluorescence and late replication.

In M. agrestis, much of the material of the two sex chromosomes is constitutive heterochromatin; Arrhigi et al. (1970) found that it hybridizes with the 35% of the genome which consists of repeated sequences and stains intensely with Giemsa in the C-banding technique. Mukherjee and Nitowsky (1972) showed that the same regions fluoresce intensely upon Q-banding. This material therefore appears to comprise a third class in its staining response; it reacts intensely to both C-banding and Q-banding. These regions of constitutive heterochromatin are identified as such by their content of repetitive sequences and inactivity in transcription; but they share with the G-bands and Q-bands of chromosome arms a response to quinacrine (although a much more intense one than the bands) unlike the failure of reaction of the centromeric heterochromatin.

The mammalian genome may therefore contain three classes of constitutive heterochromatin. Centromeric heterochromatin contains repetitive sequences of inactive DNA and responds only to C-banding. Segments within the arms of the chromosomes respond to both G-banding and Q-banding; but their identification as constitutive heterochromatin is more tentative. Extensive regions of heterochromatin occupying upto whole chromosome arms share with centromeric heterochromatin the inclusion of repetitive sequences of inactive DNA; they also respond to the C-band technique and in addition fluoresce intensely upon Q-banding.

Mechanisms of Band Formation

Treatment with DNAase abolishes the fluorescence of chromatin, but RNAase and proteases fail to have this effect; this confirms that it is the DNA component with which quinacrine interacts. At first, it was thought that the mustard group of quinacrine might bind to the N-7 of guanine. But quinacrine itself has the same specificity as quinacrine mustard, although it is less active. Weisblum and de Haseth (1972) have found that the fluorescence of a solution of quinacrine mustard is reduced as increasing amounts of DNA are added. The reduction is proportional to the G–C content of the DNA; poly-dG alone is active in quenching fluorescence, which implies that this effect residues in the guanine base. Although poly-dA and poly-dT are themselves inactive, a mixture to give poly-dAT strikingly enhances fluorescence. This suggests that fluorescent banding is produced by reaction with A–T base pairs of duplex DNA. Pachmann and Rigler (1972) found that acridines show a reaction with DNA which is similar to that of the quinacrines and it too is reduced by poly-dG and enhanced, although to a less degree, by poly-dAT. However, the range of values of fluorescence of different chromosome regions appears to be wider than might be explained by variations in base composition; the identity of Q-bands and G-bands also suggests that the reaction must in part at least be controlled by the availability of DNA within chromatin rather than by its sequences.

That the patterns produced by G-banding and Q-banding are the result of the state of DNA in its local interaction with chromosomal proteins is suggested by the generation by other techniques of similar banding patterns. In general, treatments which influence chromosomal proteins and their association with DNA may under appropriate conditions produce bands when the chromosomes are later stained with Giemsa. Seabright (1971) found that incubation with proteolytic enzymes yields a characteristic banding pattern with human chromosomes. Kato and Yoshida (1972) found that treating chromosomes with 5 M urea or with 2 M urea and 0·05% SDS—both of these solutions can be used to extract proteins from chromatin—induces banding patterns similar to those of G-banding. The reaction takes place very rapidly, within 2–7 seconds at 0°C or room temperature. Figure 4.24 shows an electron micrograph of a Chinese hamster chromosome banded with trypsin; the compact nature of the bands is clear.

By testing a variety of reagents for their effect in including bands, Kato and Morikawa (1972) showed that acids are inactive, salts are inactive when neutral although sodium ions may be effective, but immersion of slides in alkali for a few seconds is sufficient to produce a banding pattern. Chelating agents are not active, but protein denaturing agents such as SDS generate bands. Salts and denaturing agents require 15–20 minutes for reaction, but alkalis and surface agents act more rapidly. Since untreated chromosomes

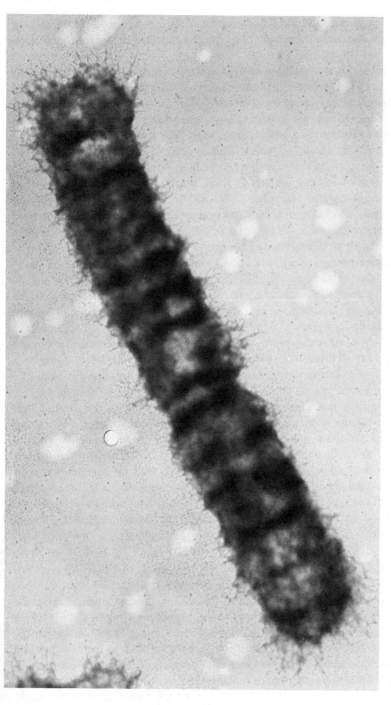

Figure 4.24: Electron micrograph of a Chinese hamster chromosome banded with trypsin. Fine details of the band and interband regions can be distinguished and there is a slight dispersion of the chromatin fibres at the edge of the chromosome. Magnification × 14,040. Photograph kindly provided from the unpublished data of Dr. G. D. Burkholder.

bind Giemsa intensely in all regions, these results support the idea that G-banding is caused by the removal of proteins from those regions of the chromosome which do not bind the stain.

The interaction of proteins with DNA may control C-banding as well as G-banding. The presence at the centromere of satellite DNA was at first thought to be responsible for the staining produced by C-banding, perhaps because the DNA might preferentially renature after the denaturation step with alkali. However, the relative states of the DNA sequences of euchromatin and heterochromatin do not seem to influence C-banding. Acridine orange is a fluorochrome which under appropriate conditions can discriminate between single stranded and duplex DNA; single strands of DNA fluoresce orange whereas double strands fluoresce green. Stockert and Lisanti (1972) followed the progress of renaturation by performing the cytological denaturation used in C-banding and then allowing renaturation to take place for periods from 1–2 seconds to 10 minutes before staining with acridine orange.

When the preparation is allowed to renature only very briefly, for 10–30 seconds, the centromeric regions of all mouse chromosomes (except the Y) fluoresce green; the remaining regions are orange. This confirms that the centromeric regions contain repeated DNA sequences which can renature rapidly. But after as little as 2 minutes of renaturation, the difference in colour between the centromeric regions and the rest of the chromosomes decreases until they are all green. When interphase nuclei are used instead of metaphase chromosomes, the chromocentres become green more rapidly than the remaining areas.

Renaturation times which allow all the DNA to renature do not change the relative proportions of the regions stained in C-banding; this implies that the single stranded or duplex state of the DNA does not determine the reaction with the stain. That the differential extraction of proteins from the regions which do not stain may be important is suggested by experiments with formaldehyde. The addition of formaldehyde fixes proteins in the chromosomes so that they cannot be extracted during the denaturation–renaturation reaction. Preparations fixed with formaldehyde show a deep and uniform stain with Giemsa over all chromosome regions, as do untreated chromosomes.

That C-banding is independent of RNA or histones of chromatin is shown by the failure of RNAase or acid to influence the reaction. One possible explanation is that the concentration of DNA at the centromeres might be greater than elsewhere in the chromosome; however, Comings et al. (1973) showed by a feulgen reaction that the concentration of DNA within centromeric heterochromatin is if anything slightly lower than in the chromosome arms. They suggested that the packing of DNA in constitutive heterochromatin may be relatively fixed, so that during interphase (when euchromatin unwinds) the heterochromatin is more compact, but the condensation of euchromatin for mitosis generates a structure in which the heterochromatin is no more condensed than euchromatin.

By following the state of DNA with acridine orange, Comings et al. showed that all DNA is denatured by the alkaline treatment of the C-banding technique; upon removal of the alkali all the DNA rapidly renatures. Treatment with $2 \times$ SSC at 65°C overnight then denatures the DNA again. However, the same C-bands are seen no matter whether the DNA at the centromeres is single stranded or duplex and irrespective also of the state of the DNA of the chromosome arms; intense C-banding can be seen in chromosomes in which DNA has renatured or in which it has been maintained in a denatured state by fixation with formaldehyde. This confirms that C-banding is unrelated to the state of the DNA itself.

But measurements of the amount of DNA after treatment shows that both the alkaline step and the incubation in SSC overnight remove significant quantities of DNA from chromatin. Feulgen or radiolabelling assays of the DNA remaining after treatment show that incubation with 0·07 M NaOH for 30, 60, 90 or 180 seconds removes 21%, 25%, 31% and 49% of the total DNA. Treatment with $2 \times$ SSC alone for 24 hours at 65°C removes 28% of the DNA. A cytological denaturation procedure in which alkali is used for 30 seconds and is followed by incubation with SSC removes more than half of the chromosomal DNA.

Whole mount electron microscopy and centrifugation analysis of the DNA remaining shows that the sequences extracted are derived preferentially from the chromosome arms, leaving the repetitive sequences in the centromeres. This implies that the DNA of centromeric heterochromatin is maintained in a type of structure different from that of other chromosome regions. Its resistance to extraction may in part explain the localization of C-bands at the centromeres, but cannot provide a complete explanation because C-bands can also be produced by proteolytic enzymes; mild treatment with these enzymes generates G-bands in the chromosome arms, but more extensive treatment abolishes reaction in the arms and instead produces centromeric regions sensitive to staining.

The model proposed by Comings et al. to account for these results postulates that C-banding results from the relative resistance to disruption of the centromeric regions and is caused whenever the structures of the arms are disrupted, either by removal of their DNA or by extraction or degradation of their proteins. This would also be consistent with the observation that the cytological denaturation procedure renders the centromeres sensitive to quinacrine so that they fluoresce brightly instead of failing to react.

The cause of C-banding therefore seems to be extensive damage to all chromosome regions except the centromeres. That there are variations within the structures of the regions of the chromosome arms is of course shown by the patterns produced by G-banding and Q-banding. No DNA is extracted from the chromosome during these treatments and reaction seems to be a consequence of the selective removal of only some proteins. Giemsa stains untreated chromosomes intensely, although the centromeres are the most lightly reacting components. This suggests that removal of some proteins from the arms during the G-banding technique may prevent Giemsa from reacting at the non-staining regions, leaving only the bands. Further extraction must abolish the structure of the bands so that they no longer react with Giemsa; however, it is not clear why this should then confer the ability to stain upon the centromere.

That the sequence of DNA does not determine the reaction with dyes is suggested by the observation of Rodman and Tahilani (1973) that feulgen—which reacts with both purines A and G—can be used as a reagent to produce banding patterns of the G-type. This argues that band formation depends upon differential access to DNA. But on the other hand, Dev et al. (1972) observed that fluorescent antibodies against adenosine, which react only with denatured DNA, may be used to produce G-type bands. To achieve these results, it is necessary to heat the slides for one hour at 65°C in 95% formamide-SSC; pre-incubations which are sufficient to achieve G-bands do not support anti-A fluorescence. Consistent with the concept that anti-A fluorescence results from the reaction of the antibody with A–T rich sequences of denatured DNA is the further observation of Schrek et al. (1973) of the response of human metaphase chromosomes to fluorescent anti-C. Cytosine bases can be maintained in a state suitable for reaction by first destroying their guanine partners by photo-oxidation in the presence of methylene blue. The banding pattern which results is the reverse of that produced by Q-banding or by anti-A antibodies.

However, because the conditions of pre-incubation are critical in determining the responses of chromosomes to these antibodies, it is possible that differential access to DNA may be important instead of or as well as the nucleotide content. Indeed, the differentiation between bands and non-bands by all these techniques is much greater than might be expected to prevail on the basis of changes in nucleotide content. That these reactions do not depend on base contents alone is suggested also by the observation that regions of centromeric heterochromatin are exempt from the relationship which otherwise prevails between anti-A and anti-C bands, presumably because their structure is different from that of the chromosome arms.

Different protocols are used for G-banding and Q-banding; in both techniques the chromosomes are first fixed on a slide, but for G-banding the fixed cells must be incubated in SSC before staining with Giemsa, whereas quinacrine may be applied directly to the slide to give the Q-banding patterns. However, in spite of the difference in pre-incubation procedures the banding patterns are identical. Quinacrine must presumably react to give bands in regions in which the DNA is accessible to the reagent; these appear therefore to be the same regions which upon gentle disruption of protein structure fail to lose the ability to bind Giemsa. Until the nature of the reaction with Giemsa is better characterized it is not possible to define more precisely how these regions may differ from those which do not form bands.

The bands may possess sequences of DNA which are different from those of the non-banding regions; the destruction of DNA in the chromosome arms during the procedure for cytological hybridization means that although hybridization experiments show the presence of satellite DNA at the centromere, they must be of much lower precision within the chromosome arms. Further experiments are therefore required before we can come to any conclusion about the nature of the sequences of the bands. It is possible that the regions which form bands differ in their non-histone protein content from the non-banding regions; and both these segments of the chromosome arms may be different from the centromere. However, we cannot yet equate the variation in the distribution of chromosomal components in the arms with any function of the chromosome.

Inactivity of Facultative Heterochromatin

In eucaryotes in which the female has two X chromosomes, one of the two homologues appears condensed throughout the cell cycle in the form of heterochromatin—this is the Barr body or sex chromatin often noted in human female cells in interphase. [Female sex chromatin was for many years thought to consist of both X chromosomes until Ohno, Kaplan and Kinosita (1959) showed that it comprises only one X chromosome in the rat.] Maintaining one of the two X chromosomes in the inactive conditions of facultative heterochromatin ensures that both (XX) females and (XY) males have the same number of X-carried genes in an active state; single X inactivation therefore provides a dosage compensation for the inequality of representation of X-carried genes in males and females.

The facultative heterochromatin of the inactive X chromosome therefore represents a condition different from that of constitutive heterochromatin. Constitutive heterochromatin contains highly repetitive sequences of DNA and also differs in its content of chromosome proteins. The genetic information contained in the inactive X chromosome, however, is homologous with that which is expressed in its active euchromatic counterpart. The inactive X chromosome must presumably differ from its active homologue in its content of chromosomal proteins. But the identical response to banding techniques of the heterochromatic and euchromatic X chromosomes implies that the difference between them does not reside in any change of organization comparable to that distinguishing euchromatin and constitutive heterochromatin. Facultative heterochromatin must therefore be formed by the imposition of a more condensed structure upon the usual organization of the X chromosome.

The behaviour of facultative heterochromatin has been followed in some detail in the mealy bug. In this unusual organism, the chromosomes which a male inherits from his mother are euchromatic, but the entire chromosome set inherited from his father becomes heterochromatic (although they were euchromatic in the father). The male therefore has a maternal euchromatic set of chromosomes, which remain euchromatic when passed on to a daughter but which become heterochromatic when passed to a son. The paternal heterochromatic set of chromosomes of a male is discarded when gametes are formed. In the female, both sets remain euchromatic.

The formation of heterochromatin develops gradually during the early stages of embryogenesis; the decision to make this set heterochromatic is probably taken before the male chromosomes enter the egg. As Brown (1966) has noted, the formation of heterochromatin is correlated with genetic inactivity. The pattern of inheritance in mealy bugs shows that males express and transmit only those genes received from their mother. The condensed heterochromatic set of chromosomes of the male fails to incorporate H^3-uridine into RNA (Berlowitz, 1965) and is replicated later during the division cycle. Both genetic and biochemical analyses thus show that the formation of facultative heterochromatin is associated with an inactivation of its genes.

Inactivation of one of the two sex chromosomes of female mammals often takes place at an early stage of development. Lyon (1961) observed that mice heterozygous for X-linked colour mutants have a variegated phenotype in which some areas of coat are normal in colour but others are mutant. This suggests that one of the two X chromosomes is inactivated at random in every cell of the embryo at an early stage of development. If the paternal X chromosome is inactivated in some cells and the maternal X chromosome is inactivated in the others, and if this state is perpetuated in their descendents, then some of the cells of the coat should have an active paternal X chromosome whereas others have an active maternal X chromosome.

Figure 4.25 shows that if one of these X chromosomes is wild type for some coat colour gene and the other carries a recessive mutant allele, those cells with the active wild type X chromosome are normal. But those cells in which the wild type X chromosome has been inactivated, so that the X chromosome carrying the mutant gene is active, express the mutant characteristic. Coexistence of the descendents of the two types of inactivated cell generates the variegated phenotype.

That the variegated phenotype results from inactivation of the genes of one X chromosome is confirmed by the observation that coat colour alleles which are carried on autosomes do not behave in this manner; heterozygotes carrying one wild type and one mutant gene are normal in appearance. But these same alleles show a variegated pattern if they are translocated from their normal autosomal position to the X chromosome. In any particular cell, the choice of which X chromosome is inactivated appears to be random, so that on average half the cells have an active maternal X chromosome and half

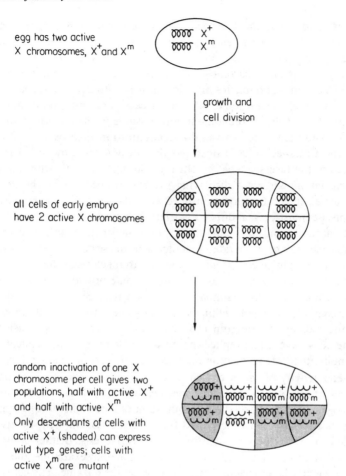

egg has two active
X chromosomes, X⁺and Xᵐ

growth and
cell division

all cells of early embryo
have 2 active X chromosomes

random inactivation of one X
chromosome per cell gives two
populations, half with active X⁺
and half with active Xᵐ
Only descendants of cells with
active X⁺ (shaded) can express
wild type genes; cells with
active Xᵐ are mutant

Figure 4.25: variegation in mice heterozygous for X-linked coat colour genes. Chromosome X^+ carries a wild type gene; its homologue X^m carries a mutant allele. Some of the coat cells are derived from ancestors with an active wild type chromosome X^+ and an inactive mutant chromosome X^m; others are derived from ancestors in which X^+ was inactivated so that the mutant gene of the active X^m chromosome is expressed. Patches of coat colour may represent descendents of a common ancestor cell.

have an active paternal X chromosome. Russell (1961) proposed that in female mammals one X chromosome is active but all the others are inactivated; so in abnormal organisms carrying, for example, three X chromosomes, two are inactivated by the formation of facultative heterochromatin. Lyon (1962) noted that mice of the type XO show normal development, which implies that one X chromosome is sufficient; they do not show variegation but express only the gene carried on the single X chromosome, which remains active.

Confirmation that the heterochromatic X chromosome carries inactive alleles whilst those of the euchromatic X are active has been provided by correlating the phenotypes and genetic states of individual cells of mice heterozygous for an X chromosome carrying a translocation. The chromosome X^t carrying Cattanach's translocation is cytologically distinguishable from the normal chromosome, X^n since it carries an insertion of autosomal material which makes it the longest chromosome in the mouse complement.

The translocated region carries dominant autosomal genes for coat colour. Ohno and Cattanach (1962) observed that in females showing variegation, the areas of the recessive coat colour are composed of cells with an inactive heterochromatic X^t; areas displaying the colour specified by the dominant genes carried in the translocation have an active X^t and a heterochromatic X^n. This demonstrates a direct association between heterochromatin and genetic inactivity and also shows that autosomal regions attached to the X chromosome may be inactivated together with the sex chromosome itself.

Autoradiography of cells from mice with Cattanach's translocation suggests that the inactive heterochromatic X chromosome provides the late labelling material found in many mammalian cells (see page 56). Evans et al. (1965) showed that a single late labelling X chromosome is found in all mouse cells; when the cells are taken from heterozygotes for Cattanach's translocation, half the cells have a late replicating X^t and half have a late replicating X^n. This implies that late replication may accompany genetic inactivation.

Genetic activity of the X chromosome can be inferred from the presence in the cell of gene products which it specifies. The enzyme glucose-6-phosphate dehydrogenase is carried on the X chromosome of both eutherian and marsupial mammals and has been widely used for such studies; different forms of the enzyme can readily be separated by gel electrophoresis. Beutler et al. (1962) showed that single X inactivation takes place in man by examining the erythrocytes of females heterozygous for a disorder in the enzyme. The cells consist of two populations; cloning shows that some cells display one phenotype whereas others display the alternative coded by its allele. By examining the cells of females heterozygous for two inherited metabolic diseases, Migeon et al. (1969) and Romeo and Migeon (1970) demonstrated a similar clonal heterogeneity for two further enzymes, HGPRT (hypoxanthine guanine phosphoribosyl transferase) and α-galactosidase; both are coded by genes carried on the X chromosome. These results confirm that one X chromosome is inactivated at random in each cell of the early embryo.

The random inactivation of one of the two X chromosomes of females which is found in eutherian mammals such as mouse and man is replaced in marsupial mammals, such as the kangaroo, by a directed inactivation of the paternal X chromosome. Euros, wallaroos and kangaroos are marsupials which can interbreed to give sterile hybrids and which are distinguished by the possession of morphologically distinct X chromosomes. The euro has an X chromosome about one and a half times that of the wallaroo, although the other chromosomes are the same size; the red kangaroo has an X chromosome which is about twice the size of that of the euro. Each of these species has a G-6P DH specified by the X chromosome and all three forms of the enzyme are electrophoretically distinct.

In crosses between female euros and male wallaroos and between female wallaroos and male red kangaroos, Richardson, Czuppon and Sharman (1971) found that the electrophoretic patterns of the enzyme show the usual X-linked inheritance. But female progeny show only the electrophoretic band corresponding to that of the mother; this suggests that the paternal X chromosome is preferentially inactivated.

As table 4.3 shows, there is an inverse correlation between the derivation of the enzyme and the late replicating X chromosome. Using these crosses

Table 4.3: inactivation of X chromosome in crosses between marsupials. Data of Richardson et al. (1971) and Sharman (1971).

parents		hybrid progeny	
female	male	G-6-P DH	late labelled X
euro	wallaroo	euro	wallaroo
wallaroo	kangaroo	wallaroo	kangaroo
kangaroo	euro	—	euro

and also that between the female red kangaroo and the male euro, Sharman (1971) incubated leucocytes of the hybrid progeny with H^3-thymidine and identified the late replicating X chromosome by autoradiography. In almost every case, the X chromosome of paternal origin shows late replication. This supports the concept that the paternal X-chromosome is inactivated and therefore replicates late, so that only the genes of the maternal X chromosome are expressed.

A useful cross which can be made between eutherian mammals uses donkey and horse as the parents. When the donkey is the male parent, the hybrid progeny are sterile mules; performing the cross in the opposite orientation yields hinnies. The X chromosome of the donkey is almost acrocentric and that of the horse is sub-metacentric; the two X chromosomes code for G-6-P DH enzymes of different electrophoretic mobilities. In early experiments with cultured leucocytes from female mules, Mukherjee and Sinha (1964) found that in half the metaphase cells one parental X chromosome is late labelling and in the other cells the reverse parent labels late.

Other experiments, however, have suggested that a non-random situation may prevail in the hybrid animals. Hamerton et al. (1971) examined the expression of G-6-P DH in female mules and hinnies by cloning fibroblasts from the hybrid animals. Any particular cell line shows the presence of only one of the two types of G-6-P DH. But more cells are found with the horse enzyme—this suggests that the donkey X chromosome may be more prone to inactivation. Further support for this idea has been lent by the demonstration that in lymphocyte and fibroblast cultures the late labelling X chromosome is more often that of the donkey than that of the horse. The expression of the donkey gene for G-6-P DH is in almost exact inverse proportion in frequency to the late replicating donkey X chromosome. Using fibroblast strains established from female mules, Cohen and Ratazzi (1971) have obtained similar results.

That this preferential expression of the horse genes is a tendency and is not absolute is emphasized by the results of Hook and Brustman (1971), obtained by analysis of 54 female mules; in two organs the horse G-6-P DH enzyme is predominant in some samples, but in others the results accord with the predictions of random inactivation. One explanation for these non random results is that the formation of heterochromatin may usually start at an earlier stage in the donkey than in the horse, so that the donkey chromosome is inactivated in more cells. Another possibility is that inactivation is random, but that some selection of cells prevails so that those with an inactivated donkey chromosome have a selective advantage and provide the ancestors for the cells of the adult. (We should note that these experiments do not reflect a situation comparable to the non-random inactivation of the marsupials, for it is always the donkey chromosome which is preferentially inactivated, no matter whether it is derived from the male or female parent.)

That some cell selection may operate during the embryogenesis of normal mammals has been suggested by Mukherjee and Milet (1972), who have made use of the observation that chimeras can be produced by fusing two eight-cell mouse embryos in vitro; the fused embryo then develops into a single mouse, even when the two embryos it is derived from have different sexes. When XX and XY embryos are fused, the two cell types are found in different proportions in the developing tissues; this supports the idea that mechanisms such as unequal segregation and cell selection may help to determine the final distribution of cells. Random inactivation of one X chromosome may therefore take place early in embryogenesis, but non random treatment of these cells and their descendents might influence the constitution of the adult. However, in at least some instances adult mammals appear to suffer random single X inactivation. An important note of caution is also that the behaviour of interspecific hybrids may be different from that of the parents with respect to single X inactivation. The importance of results obtained in mules and hinnies is therefore that they show a good correlation between the formation of heterochromatin, genetic inactivation and late replication.

Formation of Facultative Heterochromatin

Several models have been proposed to explain how one of the X chromosomes of female mammalian cells is inactivated (reviewed by Lyon, 1971,

1972). There is little consistency in the time at which inactivation takes place in different species, but it is in general early in embryogenesis. Gardner and Lyon (1971) utilised a variation of the chimera technique to define the time of inactivation in the mouse; instead of fusing two embryos to give a single product which develops into a mouse, they injected a host blastocyst of one genotype with a single cell taken from a donor blastocyst of another genotype. By taking the female donor cell from the blastocyst of an embryo heterozygous for X-linked coat colour genes they were able to distinguish its descendents from those of the cells of the host blastocyst, which carried autosomal genes for a third, distinguishable coat colour.

If single X inactivation takes place before the blastocyst stage, the chimeric mouse should have cells expressing only one of the two coat colour genes carried on the X chromosome; for one of the two X chromosomes of the donor cell should have been inactivated before its injection and all its descendents will be of the same class. But if inactivation takes place after the blastocyst stage, patches of each X-linked coat colour mutant should be produced in the adult mouse; for the injected cell must have two active X chromosomes, can divide to give descendents of the same type, and only then suffers random inactivation of one X chromosome in each of the descendent cells.

Since the donor coat colours were widely distributed over the chimeric mice, single X inactivation must take place after the blastocyst stage (that is 3·5 days of embryogenesis). Inactivation probably takes place very soon after because biased animals are often found in which one of the two X-linked coat colours predominates; this suggests that there has been chance selection from amongst a rather small number of cells.

Once inactivation has taken place, it is perpetuated in all the descendent somatic cells; the state of inactivation is therefore maintained through many cell divisions and chromosomal replications. Lyon (1972) has pointed out that formation of facultative heterochromatin falls into two stages; first one of the X chromosomes is inactivated, and then this state must be reproduced at every cell division. The mechanisms involved in the two stages may be different.

By fusing mouse A9 fibroblasts with human diploid cells of a female heterozygous for alleles of the enzyme G-6-P DH, Migeon (1972) showed that the inactivated X chromosome remains in this state and does not express its gene for G-6-P DH. Cell fusion of this nature often reactivates genes which have been turned off in the differentiated cell (see chapter 7); the stability of the inactive X chromosome therefore implies that the mechanism which is responsible for maintaining facultative heterochromatin in its inert state must be different in nature from whatever mechanisms are responsible for switching off genes which are not utilised in differentiated cells. The formation of facultative heterochromatin thus appears irreversible in somatic cells.

Inactivation of genetic material is not confined to the X chromosome itself but can include autosomal material translocated from its usual location to the X chromosome. Inactivation of euchromatin transferred to regions of heterochromatin appears to be a general phenomenon in many species. Lewis (1950) noted that somatic mosaicism is produced in Drosophila when a gene is translocated from euchromatin to heterochromatin; heterozygotes in which the gene for white eyes is carried close to heterochromatin show a variegation effect in which some of the facets of the eye are white instead of red. Inactivation seems to spread along the translocated autosomal segment, proving most effective at the region closest to the heterochromatin and less effective farther away.

Similar effects are found in the mouse. Russell (1963) and Russell and Montgomery (1970) showed that when an autosomal gene is transferred to the X chromosome its action is suppressed in some cells; when the wild type allele carried on the X is inactivated, its recessive counterpart can be expressed to give variegation effects. When an autosomal segment becomes attached to a part of the X which causes inactivation, the effect appears to spread along the autosomal genes; the more distant autosomal regions may therefore fail to be inactivated although those closer to the heterochromatin are inert. Inactivation also seems to depend upon which part of the X chromosome carries the autosomal segment—some parts of the X chromosome do not appear to inactivate attached autosomal translocations. Inactivation in autosome-X translocations has been reviewed in detail by Eicher (1970).

Although autosomal regions translocated to certain parts of the X chromosome may fail to be inactivated, Lyon (1964) has noted that there is no evidence to support the idea that any part of the X chromosome itself fails to be inactivated. The entire X chromosome appears to be inactivated, and the inactivation may spread into attached autosomal regions, although with an efficacy which depends both upon the length of the autosomal region and on its site of attachment to the X chromosome. These results led Russell (1964) to suggest that single X inactivation may be initiated at some centre or centres and then spreads throughout the length of the X chromosome; autosomal regions attached to sites on the X close to the centres of inactivation are likely themselves to be inactivated, but when the site of translocation lies at a more distant point on the gradient of inactivation the effect does not spread into the autosome.

In this sense inactivation may rely on a mechanical spread from the inactivating centres; the effect is strong enough to be perpetuated throughout the nucleohistone fibre of the X chromosome itself, but may fail to spread into autosome regions attached to parts of the X chromosome more distant from the inactivating centre and may also be perpetuated inefficiently along the length of the autosome regions. According to this model, the decision to inactivate the chromosome involves events at a limited number of specific centres; and the formation of facultative heterochromatin along the length of the chromosome then depends upon a sequential inactivation of chromosome regions.

That inactivation is not an autonomic property of regions of the X chromosome but takes place only in genetic material attached to an inactivating centre is suggested by an analysis of mice carrying Searle's translocation. This translocation comprises a reciprocal exchange between the X chromosome and an autosome; translocation heterozygotes contain the four types of chromosome:

> X, A—derived from normal parent
>
> X–A, A–X—derived from Searle's translocation parent

where X–A contains most of the X-chromosome (the breakage point is not symmetrical) and A–X contains the remaining regions.

The reciprocal exchange transfers the $+^{Ta}$ coat colour gene from the X chromosome to the autosome; it is carried on the A–X translocation chromosome. Lyon et al. (1964) found that when mice carry Searle's translocation, all the heterozygous $+^{Ta}$/Ta mice—the $+^{Ta}$ is on A–X and the Ta is on X—are wild type. This shows that the $+^{Ta}$ allele is not inactivated in its new position on the A–X chromosome, whereas it can be inactivated when it is carried by an X-chromosome which is converted to facultative heterochromatin. Genetic inactivation therefore takes place only when a gene is attached to an inactivating centre of an X chromosome.

Inactivation of the X chromosome in the translocation heterozygote is not random; genes carried on the normal X chromosome are always inactive and the segments of the translocated X–A chromosome are always active. There are two possible explanations for this effect. One is that inactivation is random, but that cell selection takes place because cells with an inactivated X–A chromosome are for some reason inviable. Another is that only the normal X chromosome can be inactivated because the break in the Searle's translocation in some way prevents its inactivation; this might happen if the site of breakage coincides with the inactivating centre so that its structure is disrupted.

The position effect variegation found in the inactivation of autosomal genes attached to Cattanach's translocation—those genes more remote from the point of breakage are less likely to be inactivated—is itself under genetic control. The X^t chromosome produced by Cattanach's insertion has a length of autosome inserted about two thirds of the distance along it; the translocated autosomal segment carries the wild type genes $+^P$ and $+^{Cch}$, whose recessive alleles code for pink eye (*P*) and chinchilla coat (*Cch*) or albino coat (*c*). Mice which are heterozygous for X^t/X^n may therefore carry the wild type alleles of these genes on the X^t chromosome and may carry the recessive alleles on an autosome. Variegation is produced when the recessive alleles are expressed in cells in which X^t is the inactivated X chromosome.

The extent of white coat colour in the translocation heterozygote is usually about 30%. Cattanach and Isaacson (1965) carried out a breeding programme over eight generations to select for

(1) mice with a high amount of white coat;
(2) mice with a low extent of variegation and little white coat.

The first selection pressure allowed the isolation of mice with upto 50% white coat colour; the second selection procedure retained mice with 30% white coat but did not lead to isolation of any lines with less variegation.

The recessive gene for white coat carried on the autosome should be expressed only when the autosomal $+^{Cch}$ gene on the X^t chromosome has been inactivated. The failure to derive mice which have more than 50% white coat suggests that inactivation of the X chromosomes is random; the maximum level of expression of the white coat gene is therefore set by the 50% maximum level of inactivation of the X^t chromosome. In the mice which have 50% white coat colour, the activity of the $+^{Cch}$ gene is always that of the X^t chromosome; if the X^t chromosome is active the cells have a wild type coat colour, but if it is inactivated they have white coat colour.

Selection in the other direction, for low extent of variegation, is in effect selection for an active wild type gene on the X^t chromosome. In mice which have 30% white coat colour, the wild type genes must be active in 70% of the cells. But the X^t chromosome is active in only 50% of the cells. This implies that in the mice with low white coat colour, 50% of the cells contain an active X^t chromosome and therefore an active $+^{Cch}$ gene, 20% contain an inactive X^t chromosome but nevertheless have an active $+^{Cch}$ gene, and the remaining 30% have an inactive X^t chromosome with an inactive $+^{Cch}$ gene.

One model to explain these results is to suppose that the inactivation of the X chromosome is random, but that the extent to which the inactivation spreads into the inserted autosomal region is under genetic control. In the high (50%) white line, the autosomal genes attached to the X chromosome are always inactivated when the X^t chromosome is inactivated. But in the low (30%) white line, the inserted autosomal genes are only inactivated in 60% of the cells in which X^t is inactivated—that is 60% of 50% gives the overall 30% white colour—and remain active, because the inactivation fails to spread into the autosomal genes, in the remaining 40% of the cells in which X^t has been inactivated. The two lines must therefore differ in some genetic factor(s) controlling the spread of inactivation.

By following the inheritance of high and low variegation when reciprocal crosses are made between the selection lines, Cattanach and Isaacson (1967) showed that the controlling factor must be located on the X chromosome itself; heterozygous progeny always have the same degree of variation as the parent which provided the X^t chromosome. When males of either selection line are crossed to normal females, the daughters must have the X^t chromosome characteristic of the parental selection line. Males from the high selection line always give daughters with a high white coat colour content. But the low white males have two types of daughter; some have the expected low white content but others have a high white content. Since the two types of coat variegation can be found within one family, the properties of X^t seem to be able to vary. The two sublines tend to breed true, but show a low rate of exchange between high and low coat colour types.

The difference between the translocation chromosomes of the two selection lines might lie either in the regions of the X chromosome itself or in the autosome segment which is transferred to it. It seems more likely that the controlling factor is located within the X region of X^t since any changes in the autosome region could take place only by mutation, whereas a controlling element in the X part of the chromosome could undergo recombination with a counterpart in the X^n chromosome. However, this model does not explain why the high line X^t is stable but the low line X^t is unstable.

By crossing a male with Cattanach's translocation carrying two wild type X chromosome genes $(+ +)$ with a female carrying two X^n chromosomes both of which are mutant at these loci (m1 m2), Cattanach, Pollard and Perez (1969) showed that the high and low control elements—termed the Xce factors—act on genes within the X chromosome as well as on genes within the autosomal segment. The high and low levels of variegation are found for the loci m1 and m2, showing that the wild type genes on X^t must be inactivated to the same degree as the autosomal elements. The Xce control element appears to act only on alleles located within the same X chromosome.

By screening the X chromosomes of several inbred strains for their effects on variegation of the genes Ta (tabby) and Vbr (viable brindled)—which are X linked genes subject to control by Xce on the X^t chromosome—Cattanach and Williams (1972) found that the substitution of one normal X chromosome for another influences the expression of these genes. These strains appear to contain three types of X^n chromosome with regard to variegation control. These results therefore suggest that the Xce locus first identified on the X^t chromosome is present on all X chromosomes and represents one of the controls of single X inactivation. The first step in inactivation must be the decision on which chromosome is inactivated. The Xce element of the inactivated chromosome must then control the spread of inactivation from the inactivating centre.

By introducing the high and low X^t chromosomes into common inbred backgrounds, Cattanach, Perez and Pollard (1970) have followed changes in its state. The X^n chromosomes of the inbred lines carry a low state (30% variegation) of the controlling element; no change of state is therefore seen when the X^t-low chromosome passes through a female in association with X^n-low. But in females which carry X^t-high and X^n-low, crossing over at meiosis can cause changes in the state of the Xce element of the X chromosomes inherited by the next generation. The frequency of change is about 3%; this implies that the map distance from the autosomal insertion in the X^t chromosome to the Xce site is about 3 map units.

The changes in state observed in these experiments can be explained by genetic recombination. However, the instability of the controlling element in the earlier experiments of Cattanach and Isaacson (1967) is greater than would be expected to result from recombination; changes of state seemed to be clustered in individual families, within which the frequency of changes is much greater than 3%. This observation led to the suggestion that the control of X-inactivation might be analogous to the changes of state observed in the control elements of maize chromosomes by McClintock (1965).

Models of this class rely upon the insertion into the X chromosome of new information (see Cooper, 1972). Excision and replacement events can then account for high frequencies of changes of state which cannot be accommodated by either genetic recombination or mutation. Brown and Chandra (1973) have proposed a model which reconciles the paternal inactivation of marsupials with the random inactivation of eutherians. Figure 4.26 shows that two classes of site control the inactivation of the X chromosome.

In eutherians, a donor site which is sensitive to parental origin is located on an autosome; this site is inactivated during meiosis of the male, but is unaffected in the female. The function of this site is to produce an informational entity which attaches to a receptor site on an X chromosome during early embryogenesis. In eutherians, the informational entity is produced by the active donor site of the maternal autosome set and becomes associated with one of the two X chromosomes at random. This X chromosome remains in an active state; for at some stage later in development all X chromosomes lacking the informational entity are inactivated. This explains why only one X chromosome is active no matter how many are present in the cell.

In marsupials, the donor and receptor sites have a closer relationship and lie adjacent on the X chromosome. The paternal donor site is inhibited during meiosis and therefore loses its ability to activate the adjacent receptor site. The maternal donor site, however, remains active and thus can activate its adjacent receptor site during early embryogenesis, so that the maternal X chromosome remains active. The random inactivation of eutherians may therefore have evolved from the paternal inactivation of marsupials (suggested also in the model of Cooper, 1972) by translocation of the donor site from the X chromosome to an autosome; in marsupials the donor site acts upon only the adjacent receptor site; in eutherians it produces only one copy of the informational entity, which is able to act upon the receptor site of an X chromosome chosen at random.

This theory predicts that the number of active X chromosomes should equal the number of maternal autosome sets in the cell; this is consistent with the results observed in polyploid cells. In mice, there appears to be one major site on the X chromosome responsible for inactivation. However, the

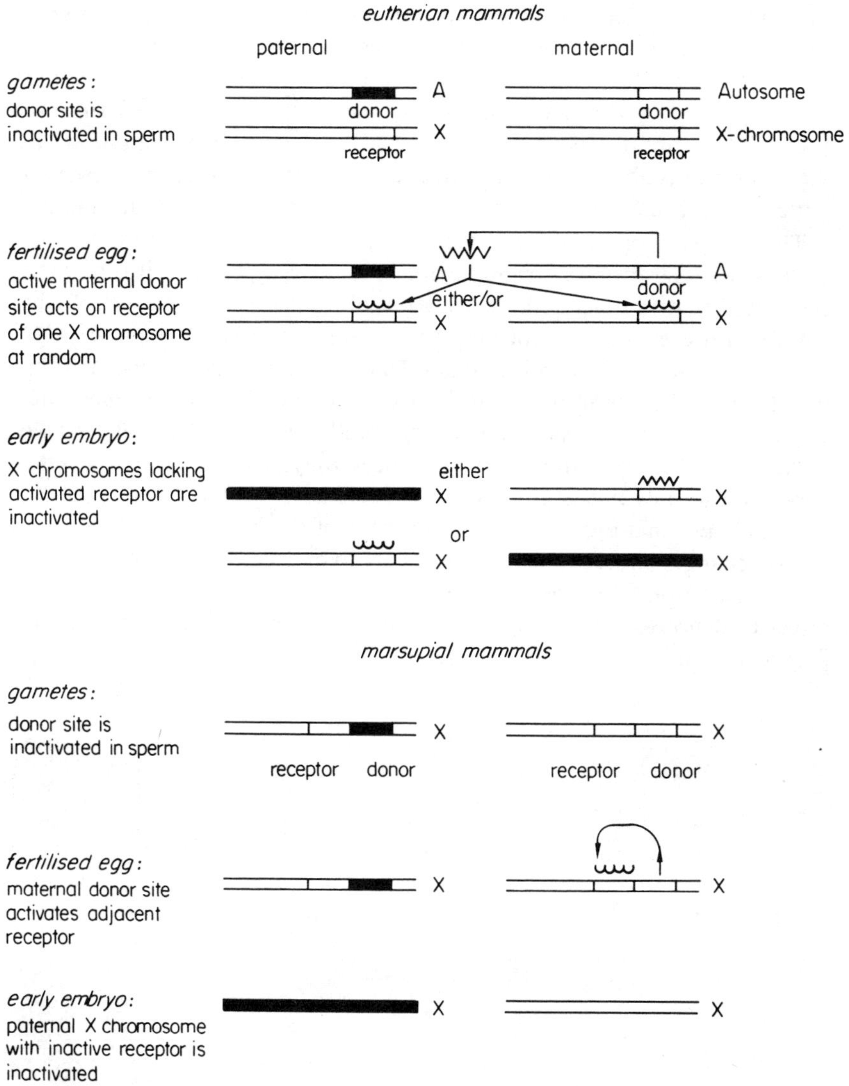

Figure 4.26: model for single X inactivation. Mammals possess a donor site which produces an informational entity which acts at a receptor site. In eutherians (upper) the donor site is located on an autosome (A) and the receptor site on the X-chromosome. The informational entity may act upon the receptor of either X chromosome; this chromosome remains active whereas the other is inactivated later development. In marsupials (lower) both donor and receptor sites are adjacent on the X chromosome. The active maternal donor site can act only upon the adjacent receptor site, so that the paternal X chromosome is always inactivated. After Brown and Chandra (1973).

Cattanach's translocation chromosome contains an autosomal insertion which in effect divides the X chromosome into two parts; both parts are inactivated. This suggests that in addition to the primary receptor site there may be secondary sites from which inactivation can spread. Deficiencies in both the long and the short arms of human X chromosomes are known, but the other arm is able to suffer inactivation; this suggests that the presence of more than one inactivating centre may be a common occurrence in mammals.

This model does not specify the molecular nature of the informational entity which is transferred from the donor to the receptor site. It is of course possible that it represents the association of an activator molecule of some nature. However, the concept that it may represent a DNA sequence which is inserted into the X chromosome offers two advantages in explaining X inactivation. The addition of new information to the chromosome could explain how its state is perpetuated so faithfully through many cell divisions; postulating that each donor site produces only one informational entity explains why only one X chromosome is active. A second feature of models of this class is that excision and replacement of the informational entity can explain changes in state which do not appear to be caused by either recombination or mutation. However, we should emphasize that there is at present no direct evidence to support models of this class, although they can account for the known facts of single X inactivation.

Expression of Genetic Information

CHAPTER 5

Transcription and Processing of RNA

Characterization of RNA Sequences

Selective Transcription of DNA

That all the somatic cells of an organism appear to possess the same complement of chromosomes implies that their genetic information is identical and that differences between them result from the differential expression of this information. The renaturation kinetics of DNA isolated from the different tissues of an organism support this conclusion; the same spectrum of sequences appears to be present in all cells. Direct measurement of gene frequencies is possible only when the messenger RNA of a gene has been isolated; in these instances, however, hybridization with RNA shows that the same number of copies of the gene is present in the DNA of all tissues, irrespective of the extent of expression of the gene in the tissue. This excludes differential amplification of genes as a mechansim for genetic differentiation (with the exception, of course, of the amplification of ribosomal RNA genes in oocytes). Each eucaryotic cell therefore carries information not only for the proteins which it synthesises, but also for all the proteins synthesised by the other cells of the organism.

Genetic differentiation must therefore be explained by the control of gene expression. The activity of a gene might be controlled at any stage from its transcription into RNA to its translation into an active protein. The very high informational complexity of the eucaryotic genome implies that some control at least must be exerted over the selection of DNA sequences for transcription; it is difficult to imagine that more than a small proportion of the genome is transcribed in any eucaryotic cell. However, this does not imply that all the sequences which are transcribed are also translated; in addition to transcriptional control, the utilization of the transcripts may also be controlled.

That different messenger RNAs are present in different cells is confirmed by hybridization experiments to compare the RNA sequences associated with the ribosomes. But eucaryotic messenger RNA passes through several stages before it reaches ribosomes in the cytoplasm. It is initially detected in the nucleoplasm in the form of a large precursor molecule; competition hybridization experiments suggest that the precursor may contain both the sequence of messenger RNA which is transported to the cytoplasm for translation and also an extensive sequence which is degraded within the nucleus. To show that gene expression is in the first place controlled at the level of transcription requires an assay of the precursor sequences present in the nuclei of different cells.

Maturation of a large precursor into a much smaller RNA is a feature of the synthesis of both ribosomal and messenger RNA in eucaryotes, although different mechanisms appear to be used in the nucleolus and nucleoplasm to distinguish the sequences of the precursor which are degraded from those which are transported to the cytoplasm. One possible control of gene expression therefore lies in the maturation process; specific control of degradation might ensure that different sequences are selected from the precursor population for transport to the cytoplasm. Control of gene expression at this level predicts that the relative stabilities of individual precursor sequences may be different under different circumstances; however, hybridization techniques are not sufficiently precise to test this prediction.

Specific control of the translation of messenger RNA present in the cytoplasm seems to occur in early development of a fertilised egg, which possesses many RNA sequences which are not expressed until the appropriate stage of embryogenesis. Some of the RNA of the cytoplasm of adult somatic cells is found in ribonucleoprotein particles and does not appear to be associated with the ribosomes; we do not know whether this represents messenger RNA whose translation is controlled or whether it represents another species of RNA with some other function.

Comparison in hybridization experiments of the sequences of RNA found in the cells of different eucaryotic tissues therefore represents the overall properties of many species of RNA. And the early experiments which showed that there are some similar sequences and some differences between the RNA populations of different cell types used nitrocellulose filters, which allow reaction only at the equivalent of low cot values. These experiments therefore showed that the DNA sequences of intermediate repetition are transcribed and that the populations of RNA sequences of different cells are in part the same and in part tissue specific. Subsequent experiments have shown that the populations of transcripts of unique sequence are also tissue specific but with some overlap. In any particular cell, there is therefore a specific restriction upon the sequences of the genome which are transcribed. However, this may not represent the sole control of gene expression, for the transcripts may be further controlled at any or all of the stages prior to association of messenger RNA with ribosomes in the cytoplasm.

Hybridization to Saturation with Excess RNA

The sequences of DNA which have been transcribed into RNA can be identified by hybridizing the RNA of the cell with the denatured strands of DNA. Hybridization between the DNA and RNA of eucaryotes suffers from the same restrictions as the renaturation of DNA itself. Because the complexity of the genome is high, any unique sequence of DNA is in effect present in only very low concentration. Hybridization with unique sequences therefore requires a high product of nucleic acid concentration and time whether reaction involves renaturation of single strands of DNA or hybridization between DNA and RNA. Preparations of immobilised DNA can therefore be used in hybridization reactions only to measure repeated sequences; reaction between unique sequences must be performed in solution.

Hybridization between RNA and DNA may be performed in excess of either component. The first technique developed to assay the sites on DNA which are transcribed into RNA was hybridization between a small amount of DNA and an excess of radioactively labelled RNA. These are "RNA-driven" reactions. As the concentration of RNA is increased, the radioactivity bound to DNA increases until a plateau of saturation is reached when all available sites on DNA are hybridized with RNA. A very large excess of RNA, of the order of 100 times the concentration of DNA, is needed to achieve saturation; and only a small proportion of the RNA becomes bound to the DNA. The proportion of the DNA which is bound by the radioactive RNA should correspond to the sites in the genome which are represented in the RNA population; cellular RNA usually saturates of the order of 5 % of the genome.

The proportion of the DNA which is hybridized is a meaningful value only when transcripts of unique sequences alone are implicated in reaction, since the transcripts of repetitive sequences may bind not only to the sites from which they are derived but also to the DNA sequences representing other members of their family. When repeated sequences of DNA are hybridized to saturation with excess RNA, the proportion of DNA which is hybridized therefore overestimates the number of sites in the genome which are transcribed, since it may include many members of a family only some of which are transcribed. Absolute values of saturation hybridization involving repeated sequences are therefore of doubtful meaning.

Competition experiments provide more information since they compare the similarities of sequence of two populations of RNA. When a preparation of radioactively labelled RNA is bound to DNA in the presence of increasing concentrations of a competing, unlabelled preparation, the extent of competition relies upon the similarities in the two RNA preparations (see page 158). The precision of this technique is also reduced by the repetition of sequences in the genome; for when two RNA populations are competing for sites within repeated sequences of DNA, two RNA molecules representing different members of one family may have sufficient homology to bind to the same sequences of DNA. This means that non-identical RNA molecules may compete with each other for hybridization to repetitive DNA sequences. Two RNA populations are therefore likely to appear more similar to each other in competitive hybridization than they really are.

That RNA is transcribed from repeated sequences of DNA is suggested by the extensive amounts of hybridization which can take place between RNA and DNA immobilised on nitrocellulose filters. Comparison of the initial rates of hybridization between RNA and DNA in bacterial and in eucaryotic systems suggests the same conclusion noted for DNA renaturation; Church and McCarthy (1967b, 1968) found that the rates of the reaction of RNA with DNA are similar in both B. subtilis and in mouse at 60°C. This shows that the first RNAs to hybridize to the mouse DNA must be derived from repetitive sequences and must be able to hybridize with all the members of their family in the genome.

The reaction rate decreases as the temperature of incubation is raised, because conditions for the recognition of homologous sequences become more demanding; the effect is much more pronounced with mouse than with B. subtilis, because increased temperature provides conditions of greater stringency which reduce the number of members of each family with which an RNA may hybridize. At low temperatures, an RNA transcript of a repetitive sequence may anneal to a large number of the other members of its family; at higher temperatures annealing is restricted to only the more closely related members. Taking the greater complexity of the mouse genome into account shows that the RNA–DNA hybridization reaction proceeds at a rate two orders of magnitude greater than would prevail for interaction with unique sequences, a situation analogous to the DNA renaturation reaction.

The amounts of RNA which bind to the DNA at saturation also vary with the temperature of incubation as shown in figure 5.1. At 60°C, Church and McCarthy found no real saturation in the range of nucleic acid concentrations which they used. At 67°C, a plateau of saturation is reached at which the level of RNA bound to mouse DNA is about half of that bound at the same concentrations at 60°C. And when the temperature of hybridization is raised to 75°C, the amount of RNA bound at saturation is reduced again to half of that saturating the DNA at 67°C. When E.coli RNA is hybridized to DNA, by contrast, fairly similar saturation values are obtained at 60°C and 70°C. These results are consistent with the interpretation that in bacteria only unique sequences can hybridize, whereas in eucaryotes the RNA transcribed from a repeated sequence of DNA can hybridize with many members of its family of sequences, although the reaction becomes restricted to more closely related sequences as the temperature is increased, when both the rate of reaction and saturation values are reduced.

The product of RNA–DNA hybridization in bacteria is stable; perfectly paired RNA–DNA duplex molecules have a Tm only a few degrees below that of the DNA–DNA renaturation product. But the stability of mammalian RNA–DNA hybrids shows a significant dependence upon their temperature of formation. The higher is the temperature of formation (the more stringent the conditions for complementary sequence recognition) the greater is the resistance to thermal denaturation. But even at 75°C the renatured mouse

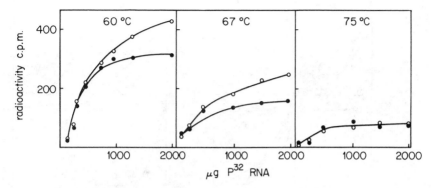

Mouse RNA - DNA hybridization

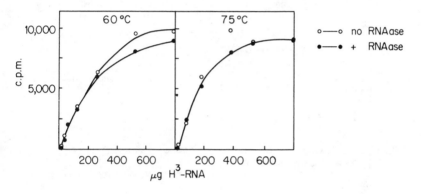

E.coli RNA - DNA hybridization

Figure 5.1: dependence of hybridization of RNA and DNA upon temperature. Conditions of hybridization were: 7 μg DNA incubated with excess amount of RNA shown in 0·2 ml of 4 × SSC for 18 hours. The curve with open symbols shows the radioactivity retained on a nitrocellulose filter; the curve with closed symbols shows the RNAase-resistant hybrid which remains when the filter is washed with 10 μg RNAase. The RNAase wash removes any non-paired sequences of RNA which have not been properly hybridized. Upper: the extent of mouse RNA bound to mouse DNA decreases as the temperature is increased to provide more stringent conditions of reaction. At 60°C and 67°C much of the hybrid is sensitive to RNAase, confirming that it does not represent accurate annealing; at 75°C, RNAase treatment has little effect, suggesting that reaction conditions are stringent enough to demand accurate pairing. Lower: hybridization of E.coli RNA with DNA is much less dependent upon temperature and is much less affected by RNAase. Bacterial hybridization can therefore take place only by accurate base pairing between unique sequences; that is, the repetitive sequences which account for the mammalian reaction are not present in the bacterial genome. Data of Church and McCarthy (1968b).

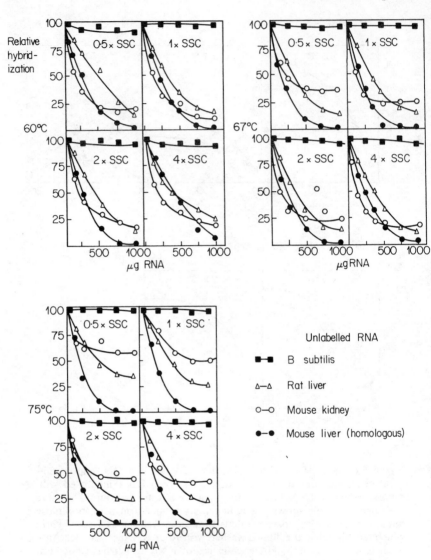

Figure 5.2: dependence of hybridization-competition reaction upon conditions of incubation. Temperature and concentration of SSC varied as shown; other conditions were: 5 μg of P^{32} labelled mouse liver RNA was incubated with 12 μg of mouse DNA in 0·5 ml for 18 hours in the presence of increasing amounts of unlabelled RNA from the sources shown. B. subtilis RNA never competes well; the homologous mouse liver RNA always competes well. Competition by rat liver and mouse kidney RNAs depends upon the reaction conditions. At each temperature, the greatest competition is displayed at high ionic strength when conditions are the least stringent, the extent of competition decreasing when the ionic strength is lowered.

hybrids have a stability below that of the B. subtilis hybrids. At low temperatures, an RNA molecule treats all members of the family from which it is derived as identical; hybridization thus forms duplex molecules of low stability and with appreciable mispairing. At high temperatures, recognition is restricted to more closely related members of the family, but there are still many sequences with which the RNA can hybridize but which are different from the sequence coding for it.

All criteria of hybrid duplex formation suggest that hybridization of transcripts of repetitive sequences to DNA almost entirely represents binding of an RNA to a sequence of DNA different from, although related to, its complement. The rate of initial hybridization is too great to be the result of recognition between complementary sequences; the extent of hybridization achieved by saturation depends greatly upon the stringency of the conditions of incubation; and the stability of the duplex hybrids which are formed depends in a similar manner upon the conditions of hybridization, but always suggests that there has been appreciable mispairing between the RNA and DNA.

Competition experiments exhibit a similar dependence upon the stringency of the reaction conditions. False competition increases as the conditions become less stringent, that is as the temperature of incubation is reduced or the salt concentration is raised. Figure 5.2 shows that mouse liver RNA always competes effectively with labelled liver RNA for mouse DNA; but the extent of competition shown by increasing concentrations of unlabelled mouse kidney or rat liver RNA depends critically upon the conditions; an increase in temperature from 60°C to 75°C reduces the competition of mouse kidney DNA from 75% to 50% in 2 × SSC. A decrease in salt concentration has a similar effect. It is therefore difficult to give absolute values for competition.

Another technique which can be used to measure the degree of similarity in two RNA populations is presaturation. Large amounts of unlabelled RNA are bound to DNA to saturate all the binding sites available to the population of RNA molecules; a labelled RNA preparation is then tested for its ability to bind to the unlabelled RNA–DNA hybrid. If it binds to the same extent as it does with free DNA, the two RNA populations must correspond to different sites on the DNA. But if the presaturation reduces the binding of the labelled preparation, the two RNA populations must have at least some binding sites

---

The apparent competition is greatly reduced when the temperature is increased, again raising the stringency of the annealing reaction. At conditions of low stringency (high ionic strength, low temperature) dissimilar RNAs may bind to the same sequences of DNA and therefore appear to compete; at conditions of high stringency (low ionic strength, high temperature) these RNAs must bind to different sites on DNA with which they show greater complementarity. Data of Church and McCarthy (1968b).

in common. The degree of overlap between the binding sites of the two RNA populations can be estimated from the extent by which binding of the second, labelled preparation is reduced by the presaturation.

Conditions of high stringency cannot in general be used with filter hybridizations because the reaction rate becomes miniscule. Under conditions which permit a reasonable rate of reaction, RNA/DNA hybridization therefore follows the reactions shown above and usually hybridizes with a DNA sequence related to, but not identical with, that from which it was transcribed. The results obtained in these experiments show that the mouse satellite DNA does not seem to be transcribed into RNA; but the 40% of the genome which consists of 100–100,000 repeated base sequences undergoes hybridization with RNA. Since these reaction conditions yield no information about the remaining 50% of unique sequences, the RNA of the intermediate fraction must to some extent at least be transcribed in a tissue specific manner.

Hybridization with DNA Fractions of Different Cots

One way to assay the transcription of different sequence components of the genome is to fractionate the DNA preparation according to its degree of repetition. Eucaryotic DNA is generally fractionated on hydroxyapatite columns into three components: a highly repetitive fraction which renatures at very low cot values and fails to hybridize with cellular RNA; an intermediate fraction which is obtained by renaturation to cot values of 10–100; and the unique sequences which can be recovered in duplex form only after renaturation at cot values greater than 1000. The accuracy of the fractionation can be followed by denaturing each isolated fraction and following its cot curve when it is subsequently renatured again. Such experiments show that it is possible to isolate reasonable pure fractions which are contaminated to only a small extent by more rapidly renaturing sequences. Both the intermediate and unique fractions of DNA hybridize with both nuclear and cytoplasmic RNAs in at least a partly tissue specific manner.

The reaction between the isolated DNA components and cellular RNA populations may be carried out by any of the usual methods for hybridization, that is in RNA excess, by saturation or competition, or in DNA excess. By performing the hybridization in solution it is possible to utilise more stringent conditions than can be achieved with immobilised DNA preparations; more accurate assays can therefore be made of the saturation levels of each fraction and of the competition between RNA molecules transcribed in different tissues. The accuracy of the hybridization reaction is usually assessed by measuring the Tm of the hybrid product; accurate hybridization generates duplex molecules with a Tm only one or two degrees below that of accurately renatured DNA (which in turn is a little below the Tm of native DNA).

By using hydroxyapatite columns, Hough and Davidson (1972) allowed sheared denatured Xenopus DNA to renature to a cot of 50 and thus isolated the intermediate fraction. Under their conditions of incubation, about 45% of the Xenopus DNA appears to be repeated and about 55% appears to be unique. (It is important to emphasize again at this point that the division of DNA into repeated and non-repeated sequences is not absolute and that the extent of repetition displayed by any DNA sequence of the intermediate fraction depends upon the stringency of the conditions used for renaturation—the more stringent the conditions, the smaller the repeated fraction.)

The repeated fraction of Xenopus DNA isolated on the hydroxyapatite column can be denatured into single strands; and then renatures completely by a cot of 10 to give a duplex which has a Tm 7°C below that of purified non-repetitive DNA. The hyperchromicity when this DNA melts is only 17·5% of the absorbance of the denatured DNA; the hyperchromicity of renatured unique DNA is 25% and a comparison of these figures suggests that about 30% of the DNA in the renatured repeated DNA duplex molecules is not paired. This confirms that the isolated intermediate fraction consists of families of related but non identical sequences. When the intermediate DNA is denatured and then hybridized with the RNA of Xenopus oocytes, the extent of hybridization increases with the RNA concentration and time of incubation; a plateau of binding is reached when about 3·5% of the Cot 50 DNA fragments have reacted with RNA.

The hybrids formed between RNA and repetitive DNA melt at about 70–71°C, whereas hybrids between RNA and non-repetitive DNA melt at about 80°C. This shows that there is appreciable mispairing in the repetitive hybrid. Hybridization between the RNA of oocytes and DNA bound to membrane filters gives slightly greater saturation values; filter experiments may overestimate the extent of hybridization because they assay retention of RNA, whereas experiments in solution allow the hybrid to be isolated and characterized under conditions when greater lengths of base pairing can be demanded.

The transcription of unique sequences has been measured by experiments in which Davidson and Hough (1971) incubated denatured Xenopus DNA to a cot of about 2500; the non-associated single strands remaining were taken as a sample of the unique component. About 0·9% of this DNA forms a hybrid when incubated with RNA in great excess (2500 μg RNA/μg DNA). An RNA/DNA titration curve taken to an RNA excess of about 70 fold shows saturation at about 0·6%, probably a more accurate estimate of the sites corresponding to the RNA. Hybrids between RNA and non-repeated DNA melt at only 1–2°C below the melting point of good DNA–DNA renatured duplex molecules and thus appear to represent an accurate hybridization in which each RNA has recognised its complementary site in DNA. The saturation value therefore provides a good estimate for the proportion of unique DNA sequences which are transcribed in the oocyte.

If transcription represents only a single strand of the genome—whereas both strands are present in the incubation mixture—then only half of the denatured DNA can potentially hybridize with RNA. This implies that about 7·0% of the repeated sequences and about 1·2% of the unique sequences of Xenopus DNA must be transcribed into RNA in the oocyte. Since the repeated sequences make up 45% of the genome and the unique sequences constitute the remaining 55%, this means that about 4% of the total DNA is represented in RNA, 3% from the repeated and 0·7% from the unique sequences of DNA.

Transcription of the unique sequences represents in total about 20×10^6 nucleotide pairs—about 20,000 genes of 1000 nucleotides. (This is a five fold greater information content than that of the E.coli genome, $4·5 \times 10^6$ base pairs.) The high ratio of RNA/DNA required to saturate the unique sequences of DNA implies that some of the transcripts may be present in only low amounts in the total population of RNA molecules.

Much lower ratios of RNA/DNA are needed to saturate the repetitive sequences of DNA with the corresponding RNAs; we cannot perform a similar calculation for the number of repetitive sequences transcribed since the accuracy of hybridization is much less, with the RNA molecules binding to DNA sequences which do not code for them. However, taking this mispairing into account, the number of repeated sequences which are transcribed may be of the same order as the number of unique sequences which act as templates.

Both the intermediate repetitive and the unique sequences of the genome appear to be transcribed in all eucaryotic cells. The unique fraction must certainly include genes coding for proteins; the functions of the transcripts of intermediate RNA are not known, although since some are found in polysomes they too may be concerned with translation. The proportion of the genome which hybridizes with cellular RNA at saturation estimates the maximum number of genes which may be expressed, since not all of the sequences of RNA which are transcribed need necessarily be translated into proteins. It is possible to make an accurate estimate of the extent of transcription only of the unique sequences; this estimate corresponds to a substantial number of genes not only in Xenopus oocytes, but also in sea urchin embryos and mammalian cells maintained in tissue culture.

The extent of transcription has been compared between embryonic tissues at different stages of development and in different tissues of the adult. Gelderman, Rake and Britten (1971) have examined the extent of transcription of unique sequences in new born mice. About two thirds of the genome behaves as though unique in their conditions of hybridization. When L cell DNA is denatured and incubated to a cot of 1500, about 56% binds to columns of hydroxyapatite and 44% is eluted as single strands which represent the unique fraction. If a labelled preparation of unique DNA is incubated with an excess of RNA, about 8% of the labelled DNA is retained on hydroxyapatite; since unique DNA represents two thirds of the genome, this represents hybridization of 5·6% of the total genome; allowing for transcription of only one strand of

DNA, this means that about 11 % of the genome of new born mice is represented in the form of transcripts of unique sequences.

Most tissues of the mouse show about 2 % hybridization when unique DNA is saturated with RNA, although brain is an exception which shows a much greater degree of hybridization. Brown and Church (1971) isolated unique mouse DNA by renaturing denatured single strands at a cot of 220 and isolating the remaining single strands from hydroxyapatite columns. Total RNA was isolated from brain, liver, kidney and spleen (without separation into nuclear and cytoplasmic fractions). Both hybridized and unhybridized DNA can be assayed by incubating the RNA preparations with denatured unique H^3 DNA and adsorbing the preparations on columns of hydroxyapatite. Single strand

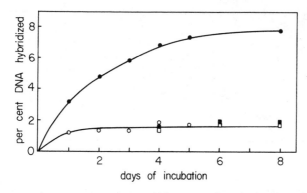

Figure 5.3: hybridization of unique sequences of mouse DNA with RNA extracted from brain (●), liver (○), kidney (□) or spleen (■). Samples contained 0·5 μg of H^3-DNA with 300 μg of RNA in 50 μl of 0·12 M phosphate buffer at 60°C. Data of Brown and Church (1971).

DNAs and RNAs eluted at low ionic strength constitute the unhybridized fraction; increasing the temperature melts the hybrid and the labelled DNA strands are eluted to assay the hybridized fraction.

Figure 5.3 shows that after 192 hours of incubation, 1·6 % of the unique DNA hybridizes with liver RNA, 1·8 % with kidney RNA, 1·9 % with spleen RNA, and 7·8 % with brain RNA. About the same proportion of the unique DNA is transcribed in liver, kidney and spleen; but about four times as much is transcribed in brain. The Tm of the hybrids is 79°C, compared with 84°C displayed by DNA–DNA duplexes renatured under the same conditions, so there appears to be little mismatching. Similar experiments have been performed by Hahn and Laird (1971), who found saturation values of 3 % with liver and with kidney and 9 % with brain RNA. Taking the transcription of only one strand of DNA into account, these results imply that in liver, kidney and spleen about 5 % of the unique sequences are transcribed; this corresponds to about 3 % of the whole genome, or about 75,000 genes of 1000 nucleotides each. In brain, the extent of transcription is even greater and could correspond to about 300,000 genes of 1000 nucleotides each.

These figures are very much greater than the number of genes which we can reasonably expect to be active in coding for proteins in any tissue. However, since these values are based upon extracts of total RNA, we do not know what proportion of the RNA represents sequences which are translated into proteins; much of the RNA may have some other function. It is reasonable to expect all cell types to transcribe about the same proportion of the genome, whatever the functions of the RNA which is transcribed; it is therefore difficult to account for the much greater extent of transcription in brain compared with other tissues.

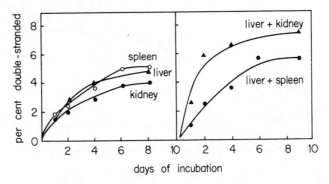

Figure 5.4: hybridization of unique sequences of mouse DNA with RNA. Left: 1·78 µg/ml of unique H³-DNA was incubated with 20 mg/ml of spleen (○), liver (▲) or kidney (●) RNA. Right: the same DNA concentration was incubated with a mixture containing 10 mg/ml liver + 10 mg/ml kidney RNA or 10 mg/ml of liver + 10 mg/ml of spleen RNA. Data of Grouse, Chilton and McCarthy (1972).

An important question is whether the same or different sequences are transcribed in different tissues. We should expect to find some similarities, since all tissues share certain functions—and this should be true whether the RNAs are translated or are related in some other way to gene expression—but there should also be differences related to the specialised cellular functions of each tissue. Grouse, Chilton and McCarthy (1972) and Brown and Church (1972) have performed additive experiments in which labelled DNA is hybridized with an excess of RNA derived from two tissues. If the RNAs are completely different, the total hybridization should be the sum of annealing shown by the separated RNAs of each tissue; if the RNAs are identical the mixture should show no increase in hybridization over the level of each separate RNA. Figure 5.4 shows that liver and kidney RNAs together hybridize to an extent greater than either separate RNA, but less than the sum of the two separate hybridizations; similar although less pronounced results are found for liver and spleen. The precise level of competition varies with the conditions of incubation; but in general more of the sequences appear to be in common than are different in these tissues.

That different sequences of RNA are transcribed during genetic development is suggested by the saturation levels shown by RNA extracted from mice of different ages. Synthesis of the RNAs corresponding to the large number of unique sequences which are transcribed in adult brain develops rapidly; Brown and Church (1972) found that newborn mouse brain shows $2 \cdot 5\%$ saturation, but this increases to a stable level of $8 \cdot 5\%$ within two weeks. In liver cells, however, newborn mice show $2 \cdot 5\%$ saturation, which increases to $3-4\%$ by two weeks of age, and then declines to 2% by 6 weeks of age. Similar results are found in kidney and spleen. The RNA of new born liver and adult liver cells together saturate at $3 \cdot 5\%$, more than their separate levels of saturation but less than their combined total of $4 \cdot 5\%$. This suggests that some RNAs are specific to each stage of development of the tissue, but others continue to be transcribed throughout the life of the mouse.

A similar conclusion is suggested for the transcription of repetitive sequences by the competition hybridization studies of Church and McCarthy (1967b) upon annealing to filter-bound DNA. The extent of hybridization declines as RNAs are extracted from liver tissues at stages from the 14-day embryo to the mature rat. The degree of competition of embryo liver RNA with mature liver RNA is at first low, showing that different sequences are present in 14-day embryo and mature livers, but increases with the time of extraction upto birth. The embryonic liver RNA population is thus gradually replaced by the population of RNA molecules characteristic of the adult tissue.

Transcription is therefore under specific control both in each tissue and during genetic development. No conclusions can be drawn from the levels of saturation when repetitive transcripts are hybridized with DNA (often of the order of 10%). But much larger proportions of the unique sequences also appear to be transcribed than might be expected to code for proteins (of the order of 2%). Although control of transcription must represent the first stage in control of gene expression, if we accept the saturation levels of hybridization of unique sequences as an indication of the extent of transcription, it becomes necessary to postulate that there are further controls of the utilization of the RNA transcripts.

Hybridization in Excess of DNA

The spectrum of DNA sequences represented in RNA can be assayed by hybridization in excess of DNA. A small amount of labelled RNA is incubated with an excess of single strands of denatured DNA in solution. In this situation the hybridization of RNA with DNA and the renaturation of DNA take place simultaneously; but the amount of DNA withdrawn into the RNA–DNA hybrid does not change the concentration of DNA sequences, which is instead governed by the predominating renaturation reaction between strands of DNA. Greenberg and Perry (1971) have pointed out that the reaction should therefore follow the kinetics of renaturation so that:

$$R/R_0 = \frac{1}{1 + k\,\text{Dot}}$$

where R_0 is the initial concentration of RNA, R is the concentration after time t, D_0 is the initial concentration of DNA and k is a rate constant.

The dot value at which any RNA preparation anneals can be characterized by its $\text{dot}_{1/2}$ and this term is strictly analogous to the $\text{cot}_{1/2}$ used to describe DNA renaturation. Indeed, the values of RNA hybridization in DNA excess have often been described by the cot applied to the DNA (rather than using the term dot). Melli and Bishop (1969) found that hybridization between E.coli DNA and RNA transcribed from it in vitro—that is a random representation of the genome—lags behind the renaturation curve, so that the rate constant for the RNA/DNA interaction is about half of that for DNA/DNA renaturation. Melli et al. (1971) have used the reaction in DNA excess to examine the spectrum of RNA sequences by comparing the reaction of eucaryotic RNA and DNA with a control of bacterial RNA and DNA; this calculation derives the reiteration frequency in the genome of a hybridizing RNA molecule in a manner exactly analogous to the calculation which compares renaturation of eucaryotic DNA with that of a control of bacterial DNA (see page 167). Thus:

$$\text{reiteration frequency} = \frac{\text{dot}_{1/2}\ \text{E.coli RNA}}{\text{dot}_{1/2}\ \text{eucaryotic RNA}} \times \frac{\text{complexity eucaryotic DNA}}{\text{complexity E.coli DNA}}$$

where the complexity of the DNA is measured in base pairs or molecular weight.

Hybridization in DNA excess has been used to assay the spectrum of sequences represented in cellular RNA. Gelderman, Rake and Britten (1971) found that rapidly labelled RNA—the precursor to messenger RNA—of embryonic mice contains the same proportion of intermediate and unique sequences as the genome. When RNA is incubated with DNA in the ratio of 1 mgm RNA/70 mgm DNA, the RNA/DNA hybridization curve closely follows the DNA renaturation curve as the cot of incubation is increased. This implies that the spectrum of DNA sequences from which RNA is transcribed does not differ greatly from the total spectrum of DNA sequences with respect to degree of repetition. Grouse, Chilton and McCarthy (1972) obtained similar results from the reaction of total tissue RNA of the adult mouse with DNA. This means that most of the RNA is transcribed from unique sequences and only the smaller part from the repeated sequences. The even representation of the intermediate and unique components in RNA implies that both may exercise functions at the level of transcription of RNA; the non repetitive sequences presumably code for proteins, but the repetitive sequences may have some other function.

Kinetics of Hybridization in Excess of RNA

Hybridization in excess of RNA was first used to measure saturation or competition values for the reaction. More recently, Birnsteil et al. (1972) have observed that the kinetics of reaction can be used to measure the complexity of the RNA population. The kinetics of hybridization depend upon the sequence complexity of the RNA; the greater the number of different RNA sequences present, the longer the reaction takes. Under these conditions the concentration of DNA is not a limiting parameter. A preparation of one messenger RNA therefore hybridizes more rapidly than a preparation containing several different messengers. Just as the $dot_{1/2}$ or $cot_{1/2}$ of reactions performed in excess of DNA can be used to measure the reiteration frequency in the DNA of the sequences which hybridize with RNA, so the $rot_{1/2}$ derived from the product of RNA concentration and time of incubation needed for half saturation provides a measure in RNA-driven reactions of the complexity of the RNA preparation.

The ratio of the $rot_{1/2}$ value of an experimental preparation to that of a standard of a single RNA molecule is therefore the ratio of their complexities; complexity is measured in nucleotide length or in molecular weight in the same way used for DNA renaturation. The equation used to calculate the kinetic complexity of an RNA is thus:

$$\frac{rot_{1/2} \text{ unknown RNA}}{\text{daltons RNA molecule}} = \frac{rot_{1/2} \text{ RNA standard}}{\text{daltons RNA standard}}$$

which may be rewritten as:

molecular weight unknown RNA =

$$\frac{rot_{1/2} \text{ unknown RNA}}{rot_{1/2} \text{ RNA standard}} \times \text{M.W. (daltons) standard RNA}$$

This method can be used to calculate the number of different RNA sequences present in a preparation of known molecular weight. A single RNA species of known molecular weight must be used as the control; the complexity calculated for the unknown preparation then gives the molecular weight—or the length in nucleotides—of its basic repeating unit. If the preparation consists of a single RNA molecule, then its molecular weight and calculated complexity should coincide. The ratio of the length of the repeating unit in hybridization to that of the physical length of the molecules therefore corresponds to the number of different types of molecule which are present.

Dividing the amount of DNA hybridized at saturation by the sequence redundancy of the RNA preparation gives the number of genes which code for each RNA. For example, one such experiment shows that Xenopus tRNA saturates the DNA at amounts corresponding to about 6500 cistrons. Comparing the complexity of Xenopus tRNA calculated from the kinetics of hybridization (1.05×10^{-6} daltons) with the average molecular weight of the tRNA (0.025×10^{-6} daltons) shows that it includes about 40 different sequences. To achieve a saturation value of 6500, each of these 40 sequences must hybridize with a family of about 160 related sequences in the genome.

Identification of Messengers in Polysomes

Messengers in Tissue Culture Cells

Messenger RNA has proved more difficult to detect in eucaryotic than in bacterial cells and has only recently been characterized in any detail. Isolation of the active messenger fraction from polysomes has been hindered by the association with this fraction of other RNAs which are not under translation by the ribosome; it has therefore been necessary to resolve mRNA from these contaminants. More recently, however, the discovery that messenger RNA contains a sequence of poly-adenylic acid has been used to isolate messenger fractions.

Isolation of individual species of messenger RNA cannot be achieved by hybridization with the specific sequences of DNA which code for it, as is possible with bacteria, and is therefore confined to cells which are so specialised that their protein synthesis is devoted largely to one protein. The messenger RNA coding for this protein is usually present in large amounts and may be resolved from the other messengers by physico-chemical methods. Although messengers can to some extent be characterized by their physical properties, the ultimate test for any purified mRNA is to translate it into protein and characterize the product.

A small proportion of the RNA in the cytoplasm of Hela cells was first identified as mRNA because it is associated with polysomes, has sedimentation properties corresponding to the right order of size to code for proteins, and has a base composition generally resembling that of DNA (Girard et al., 1965; Latham and Darnell, 1965). A technique for specifically labelling messenger RNA was developed by Perry and Kelley (1968), who found that a pulse label given to mouse L cells in culture reaches cytoplasmic mRNA within 5–10 minutes, appears in 18S rRNA after 25 minutes and in 28S rRNA only after 40–45 minutes. Short pulse doses can therefore be used to prepare cytoplasmic extracts in which only the mRNA should be labelled.

Using different methods to prepare the mRNA fraction showed that the polysomes from which it is derived are readily contaminated with labelled RNA from other sources. If the cells are extracted with detergent to yield a polysome preparation, all the label is found at the position of the polysomes on a sucrose gradient; and if the polysome fraction is recovered and then banded on a CsCl buoyant density gradient, it forms only one peak at the density of 1·54 g/ml characteristic of ribosomes (see figure 5.6). This fraction therefore represents messenger RNA bound to ribosomes.

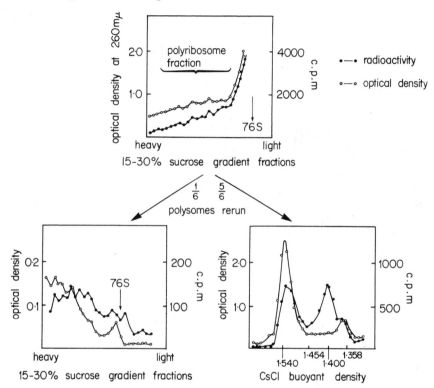

Figure 5.5: contamination of polysomes of L cells extracted by the hypotonic-mechanical method. The upper curve shows the isolation of polysomes on sucrose gradients. The fractions indicated were fixed with formaldehyde and then rerun to give the lower gradients. The left profile shows the polysomes in a second sucrose gradient; the right profile shows that the material which cannot be separated on sucrose divides into two bands on CsCl. The heavier band identifies the polysome structures; the lighter band identifies the contaminating peak. Data of Perry and Kelley (1968a).

But if the cells are instead disrupted mechanically, the single fraction obtained from the sucrose gradient—which separates molecules according to their rate of sedimentation, that is their size—separates into two bands when it is recovered and placed on a CsCl buoyant density gradient. Figure 5.5 shows an experiment in which the polysome fraction was recovered from a sucrose density gradient (upper) and then divided into two portions for centrifugation through further gradients (lower). The optical density of these fractions

Polysomes isolated, fixed and centrifuged

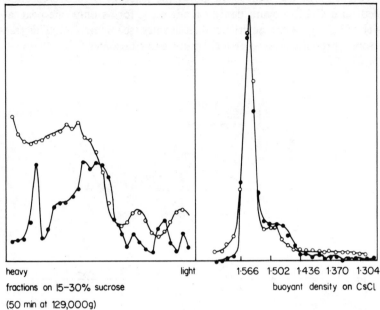

heavy   light   1·566   1·502 1·436 1·370 1·304

fractions on 15–30% sucrose   buoyant density on CsCl

(50 min at 129,000g)

polysomes treated with EDTA, fixed and centrifuged

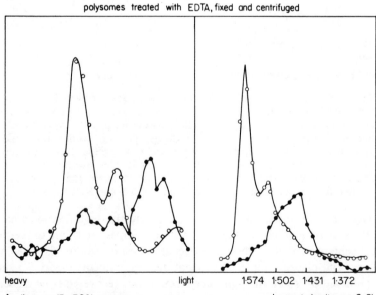

heavy   light   1·574   1·502 1·431 1·372

fractions on 15–30% sucrose   buoyant density on CsCl

(3hours at 129,000g)

identifies the bulk of the RNA, that is the ribosomal RNA, and the radioactivity identifies the labelled RNA which should include the messengers.

When rerun on a second sucrose gradient (left) a single polysome profile is generated. But when the polysomes are rerun on a CsCl gradient, the radioactive profile separates into two peaks with about the same radioactivity. One of these peaks bands with the optical density peak at the buoyant density of 1·54 g/ml characteristic of mRNA–ribosome complexes; this therefore identifies the messenger RNA which directs the ribosomes bound to it in the polysome structure. The second radioactive peak, however, displays a lower buoyant density of about 1·40 g/ml; this band must comprise a ribonucleoprotein fraction which happens by chance to sediment at the same rate as the polysomes on a sucrose gradient. Although the contaminating fraction cannot be separated from the polysomes on sucrose gradients, its different ratio of RNA to protein (compared with the polysomes) allows its separation by buoyant density on CsCl.

Because of this ready contamination, it is necessary to ensure that RNA extracted from the polysome fraction of sucrose gradients comprises only the messenger RNA species and does not include RNA derived from the contaminating ribonucleoprotein peak. Perry and Kelley found that when the polysome fraction isolated from the sucrose gradient is treated with 5 mM EDTA, the ribosomes dissociate into subunits and release messenger RNA. The contaminating ribonucleoprotein peak, however, is not affected by this protocol. Only the RNA released by EDTA is therefore considered to constitute the messenger fraction.

After treatment with EDTA, the released fractions can be fixed with formaldehyde—to prevent dissociation of RNA and protein—and then sedimented on another sucrose gradient to determine their sizes and banded to equilibrium on CsCl to determine their buoyant densities. Figure 5.6 shows the effect of EDTA upon pulse labelled Hela polysomes obtained by the detergent method which minimises contamination by additional ribonucleoproteins. (Open circles show absorbance, closed circles radioactivity.)

When polysomes are fixed and centrifuged without EDTA treatment to generate the control profiles shown in the upper gradients, most of the activity

---

Figure 5.6: release of messenger ribonucleoproteins from L cell polysomes by EDTA. Upper profiles show polysomes extracted by a detergent method in which they form the usual dispersed fractions on sucrose sedimentation (left) and only one band on CsCl (right). Lower profiles show the effect of treatment with EDTA before fixation and centrifugation. The left profile shows that ribosomes dissociate into 60S and 40S subunits, releasing a 16S peak of messenger ribonucleoprotein. The right profile shows the large subunits banding at 1·57 g/ml, the small subunits banding at 1·50 g/ml, and the messenger ribonucleoprotein separated from them at a band of 1·45 g/ml. Data of Perry and Kelley (1968a).

sediments at around 200S on sucrose gradients and forms only a single polysome peak on CsCl. Treatment with EDTA before fixation and centrifugation
generates the profiles shown in the lower gradients. The optical density traces
identify the now separated ribosome subunits. The radioactive RNA released
from the polysomes comprises an array of polydispersed ribonucleoprotein
complexes which sediment on sucrose gradients at values from 12S to 60S,
with the bulk displaying a peak at 16S. When centrifuged through CsCl, the
released messenger ribonucleoproteins band at about 1·45 g/ml, well separated
from the 1·50 g/ml and 1·57 g/ml bands of the small and large ribosome
subunits.

These values imply that the released messenger ribonucleoprotein particles
have a composition which is about 40% RNA and 60% protein. All of the
newly synthesised RNA of the polysomes appears to be released by EDTA in
this form. We shall come later to the problem of whether these complexes are
genuine or form as artefacts of extraction, but in general it seems likely that
they may represent the state of mRNA within the cell. The RNA of the
ribonucleoprotein messenger particles (mRNP) can be extracted in SDS-phenol
and generates a polydispersed profile with a peak at about 10S and extending
to 40S.

Similar conclusions were suggested by Penman, Vesco and Penman (1968)
from the results of experiments which used a different method for differentially
labelling messenger RNA. An important obstacle to resolving the kinetics of
messenger RNA synthesis is that of ribosomal RNA synthesis, which in any
lengthy labelling period becomes extensively labelled, both in the nucleus and
in the cytoplasm. The labelling of rRNA can be avoided, however, by using
low doses of actinomycin, which inhibits synthesis of rRNA at the nucleolus
but leaves synthesis of other RNAs from chromatin relatively uninhibited.
High doses of actinomycin inhibit all RNA synthesis. When Hela cells are
incubated with a labelled RNA precursor in the presence of a low dose of
actinomycin, a considerable amount of cytoplasmic RNA becomes labelled;
only a small proportion of the RNA becomes associated with polysomes, but
this can be identified as messenger RNA.

When cells are separated into nuclei and cytoplasmic fractions by mechanical
hypotonic disruption, the RNA which is extracted from the polysome fraction
isolated on sucrose gradients sediments from 8–30S, with a peak at about
18–20S and a tail up to about 70S. But this RNA extract includes a contaminant
from the ribonucleoprotein particles which sediment with the ribosomes on
sucrose gradients. When the polysomes are treated with EDTA to dissociate
their ribosomes into subunits and to release mRNA, they form two fractions
on sucrose gradients in addition to the ribosome subunits. A peak of polydispersed ribonucleoprotein sediments in the region of the gradient where the
polysomes would have sedimented if they were still intact. This represents the
contaminating material, unchanged by treatment with EDTA. The mRNA

fraction released from the polysomes, however, sediments in a range of values lower than that of the small ribosome subunit.

The extent of contamination of the polysomes can be assessed by comparison of the RNA extracted from the isolated polysome fraction (which includes both messengers and the contaminating molecules) and the RNA extracted from the region of a gradient in EDTA where the polysomes would sediment if present, although in fact they are dissociated into ribosome subunits (so that this region now includes only the contaminating RNA molecules). Figure 5.7 shows the sedimentation on SDS gradients of these two populations of RNAs;

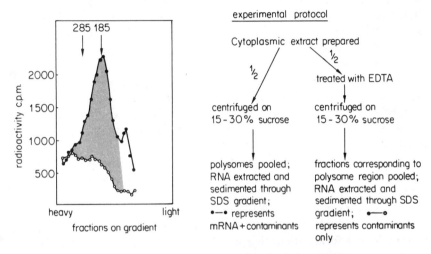

Figure 5.7: extent of contamination of Hela cell polysomes with non-messenger RNA. Comparison of the total RNA present in isolated polysomes with the contaminant remaining after the polysomes have been released from the heavy part of the gradient with EDTA identifies the shaded area as comprising the messengers. Data of Penman, Vesco and Penman (1968).

the shaded area between the two curves represents the amount of the control curve which corresponds to messenger RNA.

The messengers display a peak at about 18S, virtually all the RNAs displaying a size of less than 30S when the contribution to the control curve of the contaminating polydispersed RNAs is subtracted. Confirmation that the RNA fraction released by EDTA represents the messengers whereas that remaining in the heavy polysome region of the gradient is a contaminant is provided by the responses of the fractions to puromycin; the inhibitor causes the putative mRNA fraction to sediment in smaller polysomes (because some ribosomes have been released from the messenger) whereas the contaminating RNA retains the same position on the gradient.

Messenger RNAs in eucaryotes therefore appear to comprise fairly small molecules which are contained in polysomes and can be released from them with EDTA. The mRNA is released in the form of a particle which contains protein, but the RNA and protein components of the ribonucleoprotein may be separated by treatment with SDS and phenol. We do not know the function of the RNA contained in the contaminating particles which sediment with the polysomes on sucrose.

Messenger Ribonucleoprotein Particles

Both messenger RNA in the cytoplasm and its precursor in the nucleus have been extracted from many cell types of various species in the form of ribo-nucleoprotein particles. It is of course possible that such particles may form as artefacts during preparation of mRNA by a non-specific association between the negatively charged RNA and basic proteins. However, the binding between mRNA and proteins in the particle seems to be more stable than that of randomly formed complexes and in some instances specific proteins have been identified in the RNP particle. This suggests that mRNA is associated in vivo with particular proteins. One possible function for this association might be to transport mRNA from the cytoplasm to the nucleus; another might be to control its translation.

Early experiments which used sucrose gradient sedimentation alone to characterize the rapidly labelled RNA showed only that it sediments more rapidly than free RNA, without revealing the nature of the components associated with it. Joklik and Becker (1965) and McConkey and Hopkins (1965), for example, found that rapidly labelled RNA first appears in the cytoplasm in a form sedimenting at 45S; this led to suggestions that the com-plex might represent a 40S subunit of the ribosome bound to messenger RNA, perhaps as a means to transport the messenger from nucleus to cytoplasm. However, centrifugation to equilibrium on CsCl density gradients has since shown that these particles represent messenger RNA bound to proteins; they can be distinguished from ribosome subunits by their much lower buoyant densities.

That the mRNP particles are distinct from ribosomes was shown by Perry and Kelley (1968) in their analysis of the ribonucleoproteins released from ribosomes by EDTA. The lower part of figure 5.6 shows that when messenger RNA alone is labelled by a pulse dose of H^3-uridine, the mRNP particles sediment at 16S on sucrose gradients and band at 1·45 g/ml on CsCl. Figure 5.8 shows an experiment in which L cells were labelled with uridine for 30 minutes after which a chase for an equal time with unlabelled uridine was allowed; this means that both mRNA and rRNA become radioactively labelled. Buoyant density analysis shows three peaks of material; ribosome subunits which band at 1·57 g/ml and 1·50 g/ml are clearly separated from the mRNP peak which bands at 1·45 g/ml.

Similar mRNP particles have since been identified in many cell types. Parsons and McCarty (1968) injected rats with labelled RNA precursors and isolated RNP particles which sediment on sucrose within the rates shown by the ribosome subunits, but which display much lower buoyant densities on CsCl. Henshaw (1968) showed that mRNA is released from rat liver polysomes in the form of an RNP particle. Kumar and Lindberg (1972) found that cytoplasmic extracts of KB (human tumour) cells contain RNP particles which can

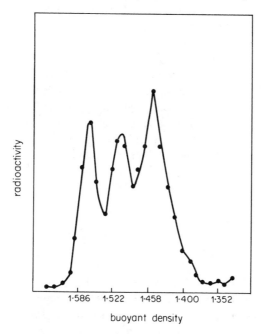

Figure 5.8: separation by buoyant density of ribosome subunits and messenger ribonucleoprotein particles. After a labelling and chase period long enough to label both rRNA and mRNA, EDTA separates polysomes into three constituents: large ribosome subunits of density 1·57 g/ml, small ribosome subunits of density 1·50 g/ml, and mRNP particles of density 1·45 g/ml. Data of Perry and Kelley (1968a).

be released from polysomes and which band between 1·35–1·47 g/ml. Although mRNP particles were first found to sediment within the range of velocities shown by the ribosome subunits, they have since been demonstrated over a much wider range of values. Henshaw and Loebenstein (1970) found that rat liver cytoplasm contains a range of RNP particles which sediment at values from 20S to 80S.

The question of whether these RNP particles represent the state of mRNA in vivo or whether they are artefacts of preparation has been resolved only recently. That particles of this general nature can be formed during extraction has been shown by Baltimore and Huang (1970), who found that cytoplasmic extracts of uninfected Hela cells or of poliovirus infected Hela cells cause poliovirus RNA to be retained by nitrocellulose filters; this implies that the RNA must have complexed with proteins. But this binding is inhibited by 0·5 M salt and Lee and Brawerman (1971) found that the mRNP particles released from rat liver polysomes by EDTA are stable in this condition. Perry and Kelley (1968) suggested that the mRNP particles released from L cell polysomes are not artefacts, because when labelled DNA-like RNA extracted from the nucleus was added to unlabelled polysomes, little of the label could be recovered from the mRNP fraction.

If the mRNP particles exist in vivo, we might expect that they contain specific proteins, whereas if they are formed during extraction of the mRNA their protein content should be non-specific. And if mRNA exists in the form of an mRNP particle in vivo, we may also ask where these proteins attach to the mRNA—presumably they cannot be bound to regions which must be translated since this might interfere with the action of the ribosomes. Another problem is whether all messengers of a cell bind to the same proteins or whether different messengers bind different proteins (in which case many proteins might be associated with mRNA even in a specific manner).

The mRNP fraction isolated from most eucaryotic cells must contain a large number of different messengers, so that attempts to characterize mRNP complexes have been made in specialised cell systems devoted to the synthesis of large amounts of one protein. Reticulocytes contain a predominant RNA fraction of messengers coding for globin protein. This fraction can be released from polysomes in the form of an mRNP particle. Jacobs-Lorena and Baglioni (1972) found that a 20S particle of reticulocyte lysates contains an RNA sedimenting at about 10S which directs synthesis of globin protein in vitro. Although the size of the mRNP particle has been reported to vary, the haemoglobin mRNA is always found in a form associated with protein.

Two reports have identified proteins associated with globin mRNA. Lebleu et al. (1971) isolated a 15S particle by EDTA dissociation of recticulocyte polysomes; the mRNP bands at about 1·46 g/ml, which corresponds to about 45 % RNA, 55 % protein. Acrylamide gel electrophoresis of the proteins of the particle identifies three species, of 135,000, 68,000 and 45,000 daltons each. The amount of the smallest protein varies and can be removed by sucrose centrifugation in association with the 5S rRNA; it is probably therefore a ribosomal protein which binds the 5S rRNA and is released by EDTA. The mRNP particle thus appears to contain only two proteins.

Globin mRNA isolated from the particle binds to 40S subunits less efficiently when free than when in the form of an mRNP complex. Free mRNA can be released when polysomes are washed with SDS before their dissociation with EDTA; free globin mRNA cannot bind to 40S subunits washed with SDS whereas the mRNP can do so. This suggests that at least one function of the proteins associated with the globin mRNA is to promote binding to the ribosome; these proteins must be removed from the mRNA—ribosome complex by washing with SDS. This leaves open the possibility that the proteins may be initiation factors which bind mRNA to the ribosome and then remain attached to it in the polysome; in contradiction to this concept, however, is the observation of Nudel et al. (1973) that under conditions in which initiation factors are removed from polysomes, mRNA remains attached to the proteins of the RNP particle.

Different proteins have been characterized as part of the globin mRNP complex by Blobel (1972). Rabbit reticulocyte ribosomes dissociate into subunits and release an mRNP particle upon treatment with puromycin in high concentrations of KCl. The particles contain only two proteins, one of which electrophoreses in SDS–acrylamide gels at a position corresponding to a molecular weight of 78,000 daltons, the other displaying a molecular weight of 52,000 daltons. That these complexes are not artefacts is suggested by their resistance to low salt concentrations which dissociate mRNA from proteins bound non-specifically during extraction.

That these proteins are specific components of the mRNP particle is suggested by the experiments of Blobel (1973) which have shown that the ribonucleoproteins released from rat liver and mouse L cell polysomes with puromycin-KCl treatment contain only two proteins, one of 78,000 daltons and one close to 52,000 daltons. This suggests that the same proteins may be associated with all mRNA molecules in these cells and that similar proteins may exercise the same function in different eucaryotic species.

But these results are also consistent with an alternative model that the proteins are translation factors or components of the ribosome. This is rendered less likely, however, by the finding that the 78,000 dalton protein appears to bind to a specific sequence, the length of poly-A, located at the 3' end of the message (see page 297). Blobel has suggested that the 52,000 dalton protein may be bound to a non-translated sequence located at the 5' end of the messenger. These results therefore show that the proteins found in the globin mRNP particle comprise specific components, although they do not completely exclude the possibility that their function is concerned with the action of the ribosome. If the function of these proteins is independent of the action of the ribosome, they might also be associated with the nuclear precursor to globin mRNA.

That the formation of mRNP particles is concerned with transport is suggested by experiments which show that the large nuclear RNAs which are the precursors to messenger RNA also exist as ribonucleoprotein particles. Lukanidin et al. (1972) found that treatment with RNAase converts these particles to a size of 30S, which they suggested might be the basic unit of association. When the nuclear RNA and proteins are dissociated in 2 M NaCl, the isolated proteins can aggregate in vitro to form complexes similar in size to the RNP particles. Each of the 30S units of rat liver seems to consist of 20–40 protein molecules, all of about 40,000 daltons in weight. That there may be a direct relationship between nuclear and cytoplasmic RNP particles has been suggested by Olsnes (1971) who has also found that removal of the proteins of nuclear RNP particles of rat liver yields only one principal molecular species. In his experiments, the proteins released by SDS from the mRNP particles isolated from polysomes comprise about six molecular species. One of these proteins, however, appears identical by the criterion of molecular weight to that found in the nuclear RNP particles.

In addition to the mRNP particles released from polysomes by EDTA, ribonucleoproteins which are not associated with the ribosomes have been detected. The heavy RNP particles which sediment with the polysomes on a sucrose gradient probably represent a nuclear contaminant, but other, lighter ribonucleoproteins have been found in the cytoplasm. The role of the free cytoplasmic RNP particles is not clear; it is difficult to exclude the possibility that they represent artefacts of the preparation procedures, nuclear contaminants etc. But in Hela and L cells and in certain embryonic systems a relationship has been demonstrated between the free RNP particles and the mRNP particles of polysomes.

In Hela cells, the kinetics of entry into mRNA of a radioactive label suggest that it may first enter the cytoplasm as a free RNP particle and may then associate with ribosomes for translation. Spohr et al. (1970) found that when Hela cells are grown in the presence of low doses of actinomycin, rapidly labelled mRNP may be released from the polysomes by EDTA, but some labelled RNP is also found in the supernatant of the preparation in which the polysomes are pelleted. About 50% of the cytoplasmic RNA does not seem to enter protein synthesis, but the label remains associated with the free RNP particles. The label which enters mRNP particles appears to pass through the pool of free RNP particles. These results would be consistent with a model in which mRNA enters the cytoplasm as an mRNP, which may either remain in this state or may associate with ribosomes for translation. However, kinetic studies are far from conclusive and it is possible that some or all of the free RNP particles play some other role. To define their function requires characterization of their RNA molecules and of the proteins associated with them.

That there may be a precursor–product relationship between the ribonucleoprotein particles and polysomes of L cells is suggested by the analysis of Schochetman and Perry (1972) of the effects of incubation at high temperature. Such treatment results in a rapid disaggregation of polysomes; subsequent reduction of the temperature to its normal level is accompanied by a reversal of the change in the polysomes when messenger RNA again associates with ribosomes. In cells incubated at 47°C, the mRNA fraction sediments as fast as or faster than the monosomes which result from polysome disaggregation. Extraction of particles from a sucrose gradient and rebanding in CsCl generates three peaks: polysome associated mRNA, a peak with a density of 1·47 g/ml and a peak with a density of 1·40 g/ml. Treatment of these fractions with EDTA converts the 1·47 g/ml density peak to a density of 1·40 g/ml and does not alter the behaviour of the 1·40 g/ml density peak. The 1·47 g/ml peak thus appears to constitute a complex of mRNP with ribosomes and the 1·40 g/ml peak represents the mRNP complex alone.

Both the 1·47 g/ml and the 1·40 g/ml peaks are found in normal as well as in heat shocked cells, although in lesser amounts. When the temperature of incubation is increased from the usual 37°C to 47°C, there is an immediate decrease in the proportion of polysomes, with an increase first in the peak at 1·40 g/ml, followed by an increase in the peak at 1·47 g/ml which restores the usual ratio between the 1·40 and 1·47 g/ml peaks. When the temperature is then decreased to normal, the 1·40 g/ml peak declines, the 1·47 g/ml peak increases concomitantly and then in turn declines as the peak of polysomes at 1·54 g/ml is restored. These sequences of events suggest that the peak at 1·40 g/ml may be the usual precursor to the peak at 1·47 g/ml. That is, when the temperature is raised ribosomes run off existing polysomes to yield mRNP complexes (buoyant density 1·40 g/ml) which then associate with free ribosomes to give the mRNP–ribosome complex of 1·47 g/ml. This complex is inactive in protein synthesis, achieving the translational inhibition caused by high temperature (and caused also when cells enter stationary phase, mitosis (see page 329) or are treated with sodium fluoride). When the temperature is later decreased, so that polysomes reform, the mRNP particles forming the 1·40 g/ml peak first associate with ribosomes to give the 1·47 g/ml particles and then form polysomes of 1·54 g/ml. A similar sequence of events may take place in the other situations in which cells suffer translational inhibition and reversal. These experiments suggest that mRNA not suffering translation in the cell is always maintained in the state of a ribonucleoprotein particle, so it displays a buoyant density of 1·40 g/ml rather than that characteristic of free mRNA.

Free RNP particles have been most clearly identified in certain embryonic tissues where it seems likely that they represent mRNAs which are not yet ready for translation. In these cells, many messenger RNAs are synthesised well before they are translated, and it is possible that translational control may be exercised by the proteins associated with them. Early work in demonstrating the presence of RNP particles which may be of this nature has been reviewed by Spirin (1969, 1972).

Cytoplasmic extracts of the loach Misgurnis fossilis taken from the blastula or gastrula stages of embryonic development contain newly synthesised RNA in the form of RNP particles which sediment from 20S to 110S on a sucrose gradient. Since the embryo utilises ribosomes inherited by the egg until the end of gastrulation, when it begins to synthesise its own ribosomes, these particles cannot represent ribosomes. After fixation with formaldehyde, they band in CsCl at a buoyant density of 1·35–1·50 g/ml with a major band at 1·39 g/ml. The particles are retained by nitrocellulose filters and digested by pronase and ribonuclease. The RNA extracted from them sediments in the range 4–35S; there is no DNA in the particles since all their nucleic acid content is destroyed by treatment with RNAase. All the proteins of the RNP preparation have about the same molecular size, sedimenting at around 9S. The heterogeneity of the particles is not continuous, for there are definite peaks within the distribution which may represent particular size classes of RNA.

By examining the part of a sucrose gradient lighter than the ribosomes, Infante and Nemer (1968) demonstrated that particles containing newly synthesised RNA in sea urchin embryos sediment from 15–60S. The RNA of these particles sediments in the same size range as the mRNA of polysomes, at about 10–40S, with an average of 20S. Spirin (1969) has proposed that the particles found in the sea urchin and loach embryos are "informosomes" which contain a masked form of mRNA. The embryos contain mRNAs which suffer delays before translation and although there is no proof that the RNP particles contain these messengers, it is tempting to suppose that they do so and to equate inert mRNA with the RNA of informosomes.

The kinetics of incorporation of a label into RNA are consistent with this interpretation, for most of the newly synthesised messenger-like RNA of loach embryos enters the informosomes rather than the polysomes. Only about 20–30% of a label enters the polysomes directly. Informosomes may therefore represent the state in which mRNA is transported from nucleus to cytoplasm. The proteins associated with the mRNA might have protection against degradation and nucleocytoplasmic transport as their sole function; or they might play some regulatory role. However, since informosome-like particles are found in the adult cells of eucaryotes, it is unlikely that their function is concerned solely with regulation of a developmental programme; it seems likely that mRNA is transported in this form from nucleus to cytoplasm and remains in it upon association with ribosomes.

Globin Messengers and the Genes for Globin

The first eucaryotic messenger RNA to be identified was the RNA fraction of reticulocytes which specifies synthesis of the α- and β-polypeptide chains which form the globin protein. Because the principal function of reticulocytes is the production of haemoglobin, much of their non-ribosomal RNA codes for globin protein and can therefore be isolated in bulk. Early work with this system has been reviewed by Chantrenne, Burny and Marbaix (1967).

When polysomes from reticulocytes are dissolved in a solution containing 0·5% SDS and centrifuged through a sucrose gradient, peaks of RNA are found at 23S, 16S and 4S (corresponding to the ribosomal and transfer RNAs) and at 9–11S. The precise sedimentation coefficient of the 9–11S peak depends upon the species from which the reticulocytes are derived and also to some extent upon the isolation procedure. Several types of experiment indicate that

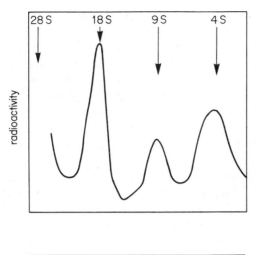

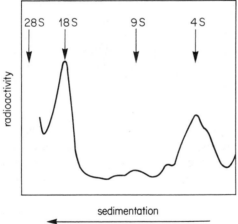

Figure 5.9: sedimentation of P³²-labelled RNA extracted with SDS from reticulocyte polysomes. Upper: RNA extracted from untreated polysomes. Lower: RNA extracted from polysomes treated before extraction with ribonuclease. Only the 9S RNA constituting the messenger fraction is degraded; both the 28S and 18S rRNAs and 4S tRNA are resistant to degradation. Data of Chantrenne, Burny and Marbaix (1967).

the 9–11S RNA fraction, now usually denoted 9S RNA, comprises the messenger fraction.

Treatment of polysomes with RNAase degrades messenger RNA alone, leaving the ribosomes intact; figure 5.9 shows that when reticulocyte polysomes are treated with RNAase before RNA is extracted from them, only the 23S, 16S and 4S RNA peaks remain—there is no 9S RNA. Dissociation of the polysomes with EDTA also allows the isolation of an RNA fraction of 9S in the form of a ribonucleoprotein in which mRNA is usually found. That 9S RNA corresponds to globin messengers is suggested also by the observation of Evans and Lingrel (1969a, b) that when 9S RNA is extracted from polysomes of different sizes, there is always one molecule of 9S RNA in each polysome. The 9S RNA is synthesised in erythroid cells of the mouse after the synthesis of rRNA has ceased, so it must comprise messengers which are translated by pre-existing ribosomes. The 9S RNA fraction provides about 2% of the total cellular RNA, a much larger fraction than can be obtained for any particular mRNA in a less narrowly specialised cell synthesising many different types of protein.

An RNA fraction of about 9S seems to be present in reticulocytes of all animals. A common feature amongst globin messengers is therefore that the RNA molecule is always longer than would be needed to code for a chain of globin. The α-globin polypeptide of the rabbit, for example, is 141 amino acids long; β-globin is 146 amino acids long. This implies that the corresponding messengers should be about 425 and 440 bases long. The similarity in the lengths of α and β-globins explains why the 9S RNA appears to comprise only one class of molecule; it would not be possible by present techniques to separate two RNA chains differing in only a very small number of nucleotides.

The 9S RNA is substantially longer than the 425–440 nucleotides required for globin (although not long enough to code for both globin chains). Labrie (1969) has shown that the 9S globin messenger of rabbit reticulocytes has a molecular weight of about 1·95 daltons, or 550 nucleotides—more than one hundred bases longer than needed to code for a globin chain. Gaskill and Kabat (1971) have analysed the length of 9S globin mRNA of the mouse on acrylamide gels; and suggested an even greater weight, of $2·2 \times 10^5$ daltons, corresponding to 650 bases. Whatever the exact length of the messenger, it is therefore clearly appreciably longer than the sequence which codes for the globin protein. Some at least of the excess represents a length of poly-adenylic acid at the 3' end of the messenger (see later); the extent of the extra sequence must be the same in both the α-globin and the β-globin messengers.

That the 9S fraction includes the two separate messengers which code for α-globin and β-globin is suggested by the results of Laycock and Hunt (1969) and Lockard and Lingrel (1969), who showed that it can be translated by cell free systems into globin protein containing both chains. The rabbit globin mRNA can be translated by an E.coli system to give a protein which cochromatographs with globin (in a system in which the two globin chains are not separated) and which generates the same tryptic peptides as the two globin chains. The mouse globin mRNA can be translated by a cell free system derived from the rabbit; and produces the β-globin of the mouse. The production of greater amounts of β-globin than α-globin has been found in many cell free systems.

That two different mRNAs are contained in the preparation is suggested by the results of Temple and Housman (1972), who found that they respond differently to inhibition with L-O-methyl-threonine (a competitive inhibitor of ile-tRNA synthetase; there are three isoleucine residues in α-chains and only one in the β-chain). The effect of treatment with this drug is to decrease the size of the α-polysomes from their usual 4–5 ribosomes to a smaller 2–3 ribosomes, but to increase the size of the β-polysomes from their usual 4–5 ribosomes to 7–10 ribosomes. Messenger RNA extracted from the larger polysomes directs the synthesis in vitro of β-globin by a cell free system of Krebs II ascites cells.

Translation of the Hb 9S RNA into globin protein by heterologous cell free systems both confirms that this RNA fraction comprises the messenger for the globin chains and also shows that its translation is neither tissue nor species specific. Sampson et al. (1972) have shown that rabbit or mouse globin mRNAs can be translated into globin proteins by crude cell free systems of mouse liver, rat liver, or mouse Landschutz ascites cells. The mRNP particle obtained by dissociating polysomes with EDTA is also translated into globin, so that the proteins attached to the mRNA do not seem to influence its activity. This suggests that their role is more probably concerned with transport than with control of translation. Matthews (1972) showed that cell free extracts of mouse Krebs II ascites cells can translate several purified mammalian messengers, including those for globin. By using a highly purified protein synthetic system, derived either from mouse liver or guinea pig brain, Schreier and Staehelin (1973) demonstrated that translation of rabbit globin depends upon the provision of initiation factors; however the factors of the protein synthetic systems are adequate. This implies that translation of this messenger at least does not depend upon tissue or species specific factors.

The rate of production of α and β-globin chains in lysates of rabbit reticulocytes is the same. But in accordance with the observation that β-globin is the predominant product of in vitro systems for translation, Lodish and Jacobson (1972) have shown that each β-globin messenger on average carries some one third more ribosomes than each α-globin messenger. The parameter responsible for this difference is the frequency of initiation. For some reason at present unknown, therefore, the β-globin mRNA must function more efficiently in binding of ribosomes and initiation of protein synthesis. This implies that if reticulocytes are to produce the same numbers of α and β-globin chains, they must possess about one third more α-globin messengers than β-globin messengers.

Another technique for demonstrating heterologous translation of a messenger has been developed by Lane et al. (1971, 1973) and Marbaix and Lane (1972), who injected 9S RNA of rabbit or duck reticulocytes into oocytes of Xenopus. After incubation for 6 hours with H^3-histidine, the radioactive proteins synthesized within the oocyte can be extracted and analysed. When the cofactor haemin is injected at the same time as the RNA, about 60% of the labelled protein elutes from a sephadex column at the position of haemoglobin; this material includes both α-chains and β-chains, although there is always a greater amount of the β-chain. Globin chains can also be synthesized in the absence of added haemin cofactor.

These experiments therefore confirm that the 9S RNA fraction includes messengers for both globin chains and, because the efficiency of translation is several hundred times greater than that of the cell free systems, demonstrate strikingly that tissue or species specific factors are not needed for translation of globin mRNA. They show also that messenger RNA is stable in the Xenopus oocyte. Marbaix and Gurdon (1972) have observed further that injection into the oocyte of additional initiation factors, derived either from homologous oocytes or from rabbit reticulocytes, does not alter the relative efficiencies of translation.

The purified globin messenger of reticulocyte RNA can in principle be used to estimate the number of genes for globin in the haploid genome, either by the proportion of the genome saturated by an excess of RNA, or by following the kinetics of reaction in excess DNA. An alternative to using the 9S globin mRNA itself is to use its complementary sequence of DNA; using globin mRNA as template, the reverse transcriptase enzyme of RNA tumour viruses can synthesise a single strand of DNA complementary to the messenger. This DNA can be used in hybridization in place of the RNA, with the advantage that the DNA–DNA reaction produces a more stable product; use of the "anti-messenger" DNA therefore allows a more accurate estimation of the number of globin genes.

Saturation experiments in RNA excess with nitrocellulose filters give 0·084–0·18% saturation of mouse DNA, depending upon the stringency of the conditions of reaction. Under these conditions, repeated sequences anneal to DNA and the hybrid products display low stability, so that this value appears to be a very considerable overestimate (Morrison, Paul and Williamson, 1972). The high saturation value and rapid rate of hybridization between globin mRNA and filter-bound DNA implies that the messenger must, in addition to its (unique) coding sequence, possess a region of repetitive sequence.

Hybridization in DNA excess has been performed between the genome and either 9S RNA itself or its complement of DNA. Harrison et al. (1972) found that about 10–20% of the DNA complement appears to anneal at low cot, probably in a non-specific manner. The remaining DNA anneals with a $cot_{1/2}$ of 1500, close to the $cot_{1/2}$ of the unique fraction of mouse DNA as a whole, 1400. This suggests that the globin mRNA of the mouse is transcribed from unique sequences of DNA.

Using the globin mRNA of the duck, Bishop, Pemberton and Baglioni (1972) annealed an excess of the DNA genome with the RNA in solution and found a $cot_{1/2}$ of 250 for the messenger. By comparison with the $cot_{1/2}$ of 8 shown by E.coli RNA, there appear to be about 10 copies of the globin genes in the duck genome. However, because the excess of DNA was less than the 100 fold required by this technique, this value is probably an overestimate and sets an upper limit to the number of globin genes which is probably closer to 2 than to 10.

Using the synthetic DNA complement to the duck globin mRNA—the reverse transcription copies a length of about 250 bases, so that it is less than the length of a complete globin gene—Packman et al. (1972) found a $cot_{1/2}$ of 730 in reaction with an excess of duck reticulocyte DNA and a $cot_{1/2}$ of 540 with excess DNA of duck liver. These values are not significantly different within the limitations of the hybridization technique. These $cot_{1/2}$ values correspond to about 2–3 copies of each globin gene in the genome. A similar estimate has been made by Bishop and Rosbash (1973).

The globin gene frequency can also be calculated from the amount of the H^3-labelled complementary globin DNA hybridized with the unlabelled DNA of the genome when the reaction is complete. The duck has four types of globin chain and, assuming that each type of chain is equally represented in the globin mRNA, the extent of hybridization corresponds to about 4 copies of each gene in each haploid genome. Harrison et al. (1972) found that in their conditions of hybridization about 20% of the globin complement failed to hybridize, even at very high cot values; by assuming that this DNA represents globin sequences which cannot find complements in the genome because all the globin sites have already hybridized, it is possible to calculate that there is one copy of each globin gene in the mouse genome.

Hybridization is not sufficiently precise to take these values as absolute figures for the numbers of globin genes in the mouse and duck. However, it is clear that the number is small, certainly less than five per genome, and perhaps as little as one copy. Taken together with the results of genetic analysis of globin mutations, these results imply that there may be only one copy of each globin gene in the haploid genome. That the same number of globin genes is found in the DNA of duck reticulocyte and duck liver shows that there is no differential amplification of the globin genes in the tissue in which they are expressed.

Messengers of Specialised Cells

That the properties of the mRNA fraction released from the polysomes of tissue culture cells are typical of messenger RNAs in general is confirmed by the isolation and characterization of individual messengers from several narrowly specialised cells in addition to the reticulocyte system. These include calf lens, which synthesises the four components of α-crystallin; the silk gland of the silk worm Bombyx mori, whose principal product is fibroin; and myeloma cells of the mouse, which synthesise large amounts of immunoglobulins.

The α-crystallins comprise more than one third of the total protein synthesised by calf lens and consist of four chains, αA_1, αA_2, αB_1 and αB_2, of which αA_2 is the major component. The αA chains have a molecular weight of 19,000 and overall are synthesised in about twice the amount of the αB chains whose molecular weight is 22,000 daltons. In addition to these proteins, the calf lens synthesises several polypeptides of β-crystallin, with molecular weights from 23,000 to 27,000 daltons. These various protein products can be separated on acrylamide gels.

Two major RNA components which act as messengers have been isolated from the calf lens by Matthews et al. (1972) and Berns, Strous and Bloemendal (1972). These have sedimentation coefficients of 10S and 14S; in a protein synthetic system from ascites cells, each mRNA seems to be translated once. The 14S messenger produces a protein product which bands on acrylamide gels at a position corresponding to the weight of αA crystallin, 19,000 daltons; the 10S fraction directs the synthesis of the αB chains of 22,000 daltons, although it also generates some material which runs together with the αA and the β-crystallin chains. With a protein synthetic system derived from rabbit reticulocytes, the 14S RNA directs synthesis of a protein whose N-terminal sequence is that common to all four α crystallin chains and which bands at the electrophoretic position of αA crystallin.

That 14S calf lens RNA codes for $\alpha A2$ crystalline has been shown also by its translation after injection into Xenopus oocytes; Berns et al. (1972) found that the αA component isolated on SDS gels by its molecular weight can be separated on basic urea gels from the other polypeptides and corresponds to $\alpha A2$ alone. The protein chain usually contains an acetylated N-terminal methionine and this modification appears to be made by the oocyte.

The 14S messenger RNA can therefore be translated into its protein product, $\alpha A2$ crystallin, by heterologous systems drawn from the mouse, rabbit and frog; this shows very strikingly that there is no tissue or species specificity in its translation. The length of the 14S mRNA appears to be about 1100 nucleotides; this could code for two polypeptide chains of 20,000 daltons. The messenger is therefore about twice the length needed to code for the only protein product whose synthesis it appears to direct. The excess length does not seem to be due to attachment of a poly-A sequence, for the message is poor in adenine. It is possible that much of the RNA is not translated; or it may be bicistronic, with a second gene representing either a duplication of $\alpha A2$ crystallin, or some other protein which is not produced in vitro.

The posterior silk gland of Bombyx mori devotes most of its protein synthesis to the production of silk fibroin during the terminal stages of the fifth larval instar. The protein has a highly repetitive structure; glycine is situated at every other position throughout most of the protein and provides about 45% of the total amino acid content; another 45% is provided by alanine, serine and tyrosine. The crystalline region which comprises about 60% of the protein consists of these four amino acids alone in a simple repeating structure; the amorphous region which constitutes the rest of the protein has a similar organization but includes other amino acids.

The codons for these amino acids predict that silk fibroin mRNA should be at least 57% G–C; since the genome of Bombyx mori is 39% G–C and rRNA is only 50% G–C, the mRNA should have a buoyant density very different from that of RNA molecules. Suzuki and Brown (1972) were therefore able to isolate the mRNA after injecting larvae with a P^{32} orthophosphate label. Following the decay of the label showed that this mRNA is stable, with a half life of hours or days. The sequence of the messenger can be predicted for the major repeating unit from the genetic code; and T1 ribonuclease digests produced oligonucleotides which agree with the predicted sequence (on a non-random use of the degenerate third base of each codon). These sequence studies therefore define this RNA as the fibroin messenger.

The fibroin genes can be separated from the rest of the genome by their high buoyant density in CsCl; they also bind fewer silver ions than other DNA sequences and can therefore also be separated in Ag^+–Cs_2SO_4 gradients by their lower buoyant density, a reverse of the usual relationship. By isolating the DNA fraction which contains the fibroin genes, Susuki, Gage and Brown (1972) were able to measure the hybridization of P^{32}-labelled fibroin mRNA with DNA with greater precision than would be possible with unfractionated bulk DNA. The internal repetition of sequence within fibroin means that very stringent conditions must be used to ensure proper hybrid formation. The same results are obtained when the RNA is hybridized with denatured DNA which has been extracted from the posterior silk gland (which synthesises fibroin), the middle silk gland, or larvae lacking the silk gland and gut (which are not able to synthesize fibroin).

Saturation is achieved at a level which corresponds to about 0·0022% of the genome. The precision of hybridization is not great enough to set an exact value on the number of fibroin genes per haploid genome, but this result shows that it must at least be no more than three. The silk gland is a polyploid tissue in which there have been many duplications of the normal genome; and the presence of the silk fibroin genes in the same concentration in this and in other tissues shows that differential gene amplification is not used as a mechanism for controlling gene expression, even in tissues in which the genome in any case suffers replication.

A general feature of mammalian messengers is therefore that they are of a small size. The globin messengers are monocistronic, presumably comprising two molecular species each of which codes for one of the globin chains. A similar situation appears to prevail in the synthesis of immunoglobulins; mouse myeloma cells, which synthesise large amounts of both light and heavy immunoglobulin chains, possess two prominent species of messenger. One class appears to code for the light immunoglobulin chain and the other codes for the heavy immunoglobulins (Kuechler and Rich, 1969; Stavnezer and

Huang, 1971; Swan, Aviv and Leder, 1972; Mach et al., 1973; Stevens and Williamson, 1973). The globin and immunoglobulin messengers, however, are longer than needed to code for their protein product; and in each case some at least of the extra length is accounted for by the presence of a sequence of poly-A at the 3' terminus. The presence of poly-A is common to all the messengers of tissue culture cells, with the exception of the histones (see later).

Another exception to the presence in mRNA of poly-A is provided by the messengers of the specialised tissues of calf lens and the silk gland of Bombyx mori. Each of these tissues contains one messenger which does not include poly-A and which is about twice the size needed to code for its protein product. Since only one protein has been identified as the product of each of these messengers, the extra length may represent either a duplicate copy of the gene or some sequence which is not translated.

In the systems which have been characterized at present, therefore, each messenger RNA appears to direct the synthesis of only one polypeptide chain. No messengers have been isolated which are comparable to those of bacteria and code for several proteins. When several related proteins must be synthesised by the cell at the same time, different messengers are transcribed and translated together. This is shown both by the two globin chains, which aggregate into one functional protein, and also by the five classes of histone, which are all synthesised during S phase and then associate with chromatin. In addition to any quantitative control of the translation of these messengers, the two globin genes must suffer a common control of transcription, as must the five classes of histone gene.

Isolation of Histone Messengers

Unlike the other individual messenger RNA species which have at present been characterized, the histone mRNAs are not the product of a specialised cell type devoted to the synthesis of a restricted number of proteins, but must be transcribed in all dividing cells. That histone messengers may be similar in many eucaryotic cells is suggested by the close evolutionary relationship of the histone proteins of different species and by the common feature that histones are synthesised only during S phase. Very similar polysomes appear to be responsible for histone synthesis in cells so distantly related as the Hela and the sea urchin embryo. Histone messengers differ from the other messengers of mammalian cells in two important respects. They do not contain a sequence of poly-A at their 3' termini (see page 297); and they form polysomes to direct the synthesis of histones only during the S phase of the cell cycle (whereas other messengers are synthesised continuously throughout the cycle).

Polysomes synthesising histones are distinguished by their small size. Moav and Nemer (1971) found that sea urchin embryos, in which synthesis of histones occupies a large part of the protein synthetic activity, contain a class of poly-somes which have only 3–7 ribosomes. Although these polysomes are at first present in only small amounts, they increase to occupy more than one third of the total ribosome population by the 10 hour, 200 cell blastula stage. When nascent proteins are isolated from the polysomes upon incubation in vitro, they include histone-like proteins, which comprise about 70% of the product. Formation of the polysomes depends upon RNA newly synthesised by the early embryo, not upon pre-existing RNA inherited by the egg. The RNA which can be isolated from the polysomes sediments at about 7–9S.

A similar fraction of polysomes has been found in Hela cells. Borun, Scharff and Robbins (1967) found that during S phase Hela cells contain a fraction of small polysomes which is absent from G1 cells and which incorporates an excess of lysine compared with tryptophan—a characteristic of histone proteins. These polysomes are the only class of the polysome population to disappear when the cells are incubated with cytosine arabinoside, a drug which inhibits both DNA and histone synthesis. The polysomes appear in the cell about 2 hours before DNA synthesis commences and contain a 7–9S RNA which has a half life of about 3 hours (much less than the other messengers of the Hela cell). Jacobs-Lorena et al. (1972) have translated the 7–9S RNA in a cell free protein synthetic system from the mouse ascites tumour. Gel electrophoresis shows that the products have the small sizes and electrophoretic mobilities typical of histones and also generate the same pattern of tryptic peptides.

The 7–9S RNA can be characterized by electrophoresis on acrylamide gels. Adesnik and Darnell (1972) found that on gels which do not separate the histone messengers from each other, the size appears to be about 1.4×10^5 daltons. Breindl and Gallwitz (1973) have separated the RNAs of the small polysomes synthesised by Hela cells released from a double thymidine block. Acrylamide gels resolve the histone messengers into three species:

$$1.55 \times 10^5 \text{ daltons (about 470 nucleotides)}$$
$$1.8 \ \times 10^5 \qquad\qquad 550$$
$$2.1 \ \times 10^5 \qquad\qquad 630$$

These are of the right order of size to code for the histones: f2a1 (IV) would require a messenger of 309 nucleotides; f2b (IIb2), f2a2 (IIb1) and f3 (III) correspond to messengers of from 378–408 nucleotides; and the f1 group demands messengers of 630–650 nucleotides. When these RNAs were translated in a cell free system from rabbit reticulocytes, the protein products cochromatographed with f3, f2b and f2a2 and to a lesser extent with f2a1. None of these messengers appears to code for f1 histone. It therefore seems very likely that each of the classes of the RNA separated on the gels includes mRNA for at least one histone. The apparent excess in the lengths of the mRNA is not accommodated by poly-A but its significance cannot be assessed until individual messengers have been equated with each histone protein.

Stability of Messenger RNA

No single value can be given for the stability of messenger RNA in eucaryotic cells, for messenger lifetimes may be specific for each cell type and even for classes of messenger within a cell. This contrasts with the situation in bacteria, where all messengers appear to suffer degradation very rapidly after their synthesis (see chapter nine of volume one). Specialized tissues of eucaryotes which are devoted to the synthesis principally of one or a very small number of proteins may possess rather stable messengers. The messenger RNA coded by the BR2 locus in Chironomus salivary glands, for example, displays a half life of the order of 35 hours (see page 342). Globin, lens and silk fibroin messengers may all possess half lives measured in days. The half lives of messengers of cells grown in tissue culture may be less than this, but still appear to be of the order of some hours.

Early experiments upon the stability of messenger RNA made use of protocols in which actinomycin was added to block synthesis of new messengers so that the decay of pre-existing messengers could be followed. Half lives in general of about 3 hours were inferred from the rate of decline of protein synthesis. A critical assumption upon which this technique relies is that actinomycin inhibits only the synthesis of messenger and not its translation by ribosomes. However, Singer and Penman (1972) showed that addition of actinomycin inhibits the attachment of ribosomes to pre-existing messengers, so that the ability of cells to synthesize proteins declines due to reduced translation of pre-existing messengers as well as because of their degradation.

If actinomycin inhibits only synthesis of new messengers, the size of the polysome population in Hela cells treated with actinomycin should decline after addition of the drug; but the size of the individual polysomes should remain the same. However, during the first five hours after its addition, the principal effect of actinomycin is to cause a decline in the size of the polysomes rather than to reduce their number. That the size reduction is caused by inhibition of initiation relative to the rate of ribosome movement upon the messengers is suggested by the observation that addition of cycloheximide—a drug which slows the rate of elongation of growing polypeptide chains— restores the usual size of the polysomes (presumably by slowing the rate of ribosome movement to a lower level corresponding to the lower rate of initiation).

The messenger population of Hela cells can be isolated by taking advantage of the ability of the sequences of poly-A within mRNA molecules to react with poly-dT immobilized on cellulose (see later). Using this method of isolation, Singer and Penman (1973) labelled cells briefly with H^3-uridine and then followed its presence in the messenger population over several days. The decline of the radioactivity in messenger RNA does not follow simple exponential kinetics but displays a more complex curve, which can be fitted to the exponential decline of two different components shown in figure 5.10, one possessing a half life of 6–7 hours and the other exhibiting greater stability at a half life of about 24 hours (close to the generation time of the cells). (When the decay of a label in mRNA is followed in the presence of actinomycin, both messenger populations turnover more rapidly, with half lives of 4·5 and 12 hours; this experiment confirms directly that measurements of messenger stability in the presence of actinomycin do not reflect the parameters prevailing in uninhibited cells.)

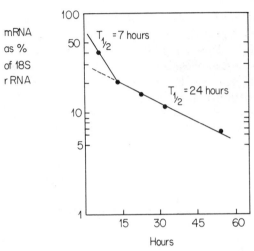

Figure 5.10: decay of radioactive label in messenger RNA of Hela cells. Hela cells were labelled with H³-uridine for 3 hours and the label present in the poly-A–containing fraction of the cytoplasm was assayed after increasing periods of incubation in unlabelled medium. The decay curve can be fitted to two populations of mRNA, one with a half life of about 7 hours, the other with a half life of about 24 hours. Data of Singer and Penman (1973).

When cells are incubated in the presence of a radioactive precursor for long enough to label all species of RNA to uniform specific activity, a steady state is reached. At this time—after six days of labelling—the amount of messenger RNA relative to ribosomal RNA can be measured; this gave a value of 5·2%, somewhat greater than earlier estimates of the proportion of mRNA. When the decay of a pulse label is followed, about 60% of the mRNA appears to comprise the more rapidly decaying component (half life 6–7 hours) and about 40% decays more slowly (half life 24 hours). When measured under steady state conditions, of course, the less stable component forms a smaller proportion of the messenger population; about 33% of the total cellular mRNA constitutes the rapidly decaying component and about 67% provides the more stable component.

Messenger RNA with a half life of some hours has also been observed in L cells growing in culture. By following the approach to steady state labelling of polysomal RNA containing poly-A, Greenberg (1972) showed that the mRNA behaves as a single population with a half life of about 10 hours. This means that the average lifetime of a messenger is about 15 hours, close to the generation time of the cells. Messenger RNA therefore turns over once in each generation. The reason for the difference in behaviour of Hela cell and L cell messengers is not clear; but in both cells, the messenger population displays a stability of several hours.

The presence within Hela cells of at least two populations of mRNA with different stabilities implies that some mechanism for selective degradation of messengers must exist. This may constitute some form of translational control. The stability of mRNA implies also that rapid cessation of the synthesis of any protein cannot be achieved by control of transcription but must require inhibition of translation of its stable messengers. We do not know whether the coincidence of the half life of the more stable messenger component with the generation time of the cell (24 hours) is fortuitous or has any significance for the systems which control messenger stability.

Processing of Heterogeneous Nuclear RNA to Messenger RNA

Fate of Rapidly Labelled Heterogeneous Nuclear RNA

Synthesis of RNA in the nucleus falls into two general categories: transcription of ribosomal RNA precursors and their maturation in the nucleolus; and transcription of heterogeneous nuclear RNA and its processing into messenger RNA in the nucleoplasm. Although taking place in morphologically distinct parts of the nuclear structure and catalysed by different enzymes, both types of transcription event share in common the production of a nuclear precursor much larger than the mature RNA of the cytoplasm. The precursor comprises regions which are degraded and regions which are preserved and transported to the cytoplasm. Different criteria are used in the nucleolus and nucleoplasm, however, to distinguish the sequences which are retained from those which are degraded: the sequences destined to comprise mature ribosomal RNA are methylated whereas the rest of the molecule is not modified; the 3' end of nucleoplasmic RNA suffers attachment of a sequence of poly-adenylic acid after which this region is cleaved from the rest of the molecule.

Much of a radioactive pulse label incorporated into nuclear RNA appears to turn over in the nucleus without entering the cytoplasm; and Harris (1963) first suggested that at least a large proportion of the nucleoplasmic RNA must be degraded in the nucleus and is never transported to the cytoplasm. If a radioactive precursor of RNA is given to Hela cells for a short time, most of it enters 45S RNA and then matures through a series of precursors to yield ribosomal RNA. But Penman (1966) found that some of the label enters a heterogeneous nuclear RNA species which shows a wide range of sedimentation values, including molecules much larger than are found in cytoplasmic messengers.

The label which enters the heterogeneous nuclear RNA (Hn RNA) must be distinguished from that in ribosomal RNA precursor molecules before the processing of messenger RNA can be followed. Warner (1966) found that the Hn RNA and 45S RNA molecules display different kinetics of labelling. If H^3-uridine is given to Hela cells for 5 minutes and the cells are then grown in the presence of unlabelled uridine, incorporation of the label into RNA continues for the next 30–60 minutes. Figure 5.11 shows that after 5 minutes the labelled RNA displays a heterodispersed distribution sedimenting from 20S to 75S with only a faint peak at 45S; this distribution represents the Hn RNA molecules. After 15 minutes, label appears in the 45S precursor peak and subsequently its maturation generates the rRNA precursor at 32S and the mature molecules at 28S and 18S.

The heterogeneous RNA visible at 5 minutes occupies a range of molecular sizes varying from much larger than the 45S rRNA to below that of the mature rRNAs. Once label has begun to appear in the 45S precursor, it becomes impossible to distinguish the Hn RNA which sediments below 45S from the ribosomal RNA molecules. Experiments in which the synthesis of ribosomal RNA is not inhibited are therefore usually restricted to following the behaviour of the Hn RNA molecules of sizes greater than 45S RNA. This fraction, which is taken as typical of the entire Hn RNA population, reaches a plateau of labelling within 15 minutes and declines in label between 45 and 60 minutes.

The kinetics of Hn RNA synthesis can also be followed by treating Hela cells with a low dose of actinomycin to block ribosomal RNA synthesis selectively whilst the cells continue to incorporate labelled precursors into Hn RNA. After a period of incubation, all RNA synthesis can be blocked by adding a high dose of actinomycin. Although this technique has the disadvantage that the drug may interfere with cellular metabolism, it offers the advantage that all size classes of Hn RNA can be followed. Penman, Vesco and Penman (1968) found that after incubation in low actinomycin, all the nuclear label is located in Hn RNA; after the addition of high actinomycin doses, the label decays rapidly during the next 60 minutes. Since RNA extracted from the cytoplasm does not increase in label during the incubation, the RNA lost from the nucleus cannot have been transported to the cytoplasm; it must turn over within the nucleus. The stability of the Hn RNA is very low and Soeiro et al. (1968) found that some 90% of it turns over in the nucleus with a half life which is measured in minutes.

Electron microscopic visualization of RNA shows that the Hela cell Hn RNA consists of linear molecules of considerable length and Granboulan and Scherrer (1969) showed that the range of sedimentation values of Hn RNA appears to represent genuine variations in length and not changes in conformation. The larger molecules of Hn RNA are therefore several times the length of messenger RNA.

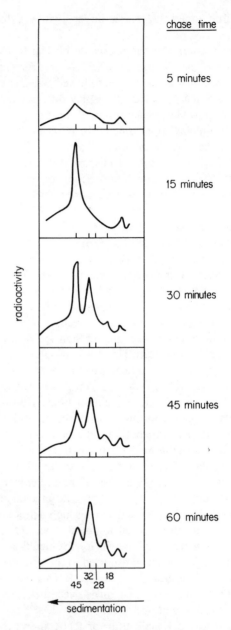

Figure 5.11: sedimentation profile of Hela cell RNA after pulse chase. A five minute period of labelling with radioactive RNA precursor is followed by incubation with unlabelled precursors for varying lengths of time before RNA is extracted and assayed by sedimentation. After 5 minutes of chase, much of the label is found in large structures which appear to be hetero-

Analyses of the base sequences and location of the Hn and 45S RNAs within the nucleus support the idea that Hn RNA is synthesised in the nucleoplasm whilst ribosomal precursors are transcribed within the nucleus. Penman, Smith and Holtzman (1966) developed a method for fractionating nuclear preparations into nucleolar and nucleoplasmic fractions by exposing the nuclei to DNAase in high ionic strength—centrifugation then locates 45S RNA in the pellet and a large fraction of the Hn RNA is found in the supernatant. Using this technique, Soeiro, Birnboim and Darnell (1966) showed that the Hn RNA of the nucleoplasm has a G–C content of about 48 %, which is close to the 44 % of Hela cell DNA. Ribosomal RNAs, by contrast, have a much greater proportion of G–C base pairs. The Hn RNA therefore fills the requirement for messenger RNA that its base composition is DNA-like.

That messenger RNA is derived from Hn RNA is suggested by their similar responses to actinomycin. Penman, Vesco and Penman (1968) found that low concentrations of actinomycin leave the number of polysomes almost unaffected even after 7·5 hours of incubation; this implies that the Hn RNA which is synthesised in the presence of low actinomycin doses provides the precursor to messenger RNA. High doses of actinomycin which halt Hn RNA synthesis cause the polysomes to disaggregate.

The response of the synthesis of different RNAs to inhibition with actinomycin has been measured in mouse L cells by Perry and Kelley (1970). Because actinomycin interacts with DNA to provide a physical block to transcription, the response of an RNA to inhibition should depend upon its length—the greater the length, the more sensitive its synthesis should prove to inhibition by the drug. The concentration of actinomycin needed to inhibit the synthesis of 45S RNA, 5S RNA and tRNA, expressed per unit length of RNA, falls into a small range; this is some 50–100 times less than the concentration per unit length needed to cause the same extent of inhibition of Hn RNA. This supports the idea that different polymerase systems undertake the synthesis of ribosomal and Hn RNA (see chapter 6). The sensitivity of messenger RNA is in the same range as the larger molecules of Hn RNA, which is consistent with the idea that many of the mRNAs are derived by cleavage of much larger precursors. The dose response curve for mRNA is complex, however, so that some mRNAs might be derived from small precursors.

Sequences of Nuclear and Cytoplasmic RNAs
No species corresponding to messenger RNA itself appears to be synthesized

---

geneous in size; this is Hn RNA. After 15 minutes, the 45S rRNA precursor is labelled; by 30 minutes maturation to 32S rRNA precursor takes place; and by 45 and 60 minutes 18S rRNA is present also. The presence of the rRNA precursors at these later times obscures the heterogeneous nuclear RNA profile. Data of Warner et al. (1966).

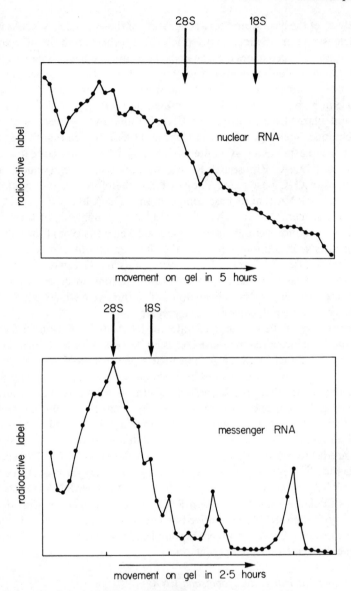

Figure 5.12: size distribution of nuclear and messenger RNAs of L cells analysed on acrylamide gels. The positions of 28S and 18S rRNA control are indicated by the arrows. Nuclear RNA, obtained from cells incubated in the presence of low levels of actinomycin to inhibit nucleolar synthesis of rRNA, consists largely of molecules greater in size than 28S ($1 \cdot 6 \times 10^6$ daltons); whereas messenger RNA, obtained from polysomes, consists of much smaller molecules. Data of Greenberg and Perry (1971).

in the nucleus. Greenberg and Perry (1971) reported that messengers isolated from polysomes in general display molecular sizes of upto about 10^6 daltons (of the order of 3000 nucleotides); heterogeneous nuclear RNA consists of much larger molecules, ranging from $1-15 \times 10^6$ daltons (upto about 45,000 nucleotides). Figure 5.12 shows an analysis on acrylamide gels of the nuclear and messenger RNAs synthesized in L cells in which nucleolar synthesis of rRNA was inhibited by low doses of actinomycin.

Most of the Hn RNA fraction decays rapidly within the nucleus; the unstable proportion depends upon the conditions of labelling but is usually from 70–90% of the total Hn RNA. Scherrer et al. (1970) found that Hn RNA molecules larger than 35S appear to decay initially with a half life of about 20 minutes; those Hn RNA molecules smaller than 35S decay with a half life of 80 minutes. These kinetics are consistent with a model for decay in which the larger Hn RNA molecules are first cleaved to yield molecules of less than 35S and are then further degraded to give small acid soluble fragments.

Early experiments with Hela cells failed to demonstrate a product–precursor relationship between Hn RNA and mRNA; a radioactive label incorporated into Hn RNA appeared for the most part to decay rapidly but could not be chased into messenger RNA. That Hn RNA provides the precursor to cytoplasmic mRNA, however, is suggested by three classes of experiment. Synthesis of Hn RNA and of mRNA displays the same response to actinomycin, which is ineffective at low concentrations and inhibitory at high concentrations. More recent kinetic studies of the addition of poly-A to Hn RNA and of its appearance in messenger RNA suggest that a poly-A sequence is added to the 3' end of an Hn RNA molecule, which is then cleaved to release a stable 3' fragment containing the poly-A; this is transported to the cytoplasm. And a comparison by nucleic acid hybridization of the sequences present in Hn RNA and mRNA show that all messenger sequences are found in the nuclear fraction.

Several models can be constructed for the derivation of messenger RNA from Hn RNA. The discrepancy in size between Hn RNA and mRNA suggests that the precursor molecule must consist of two regions: a sequence which is unstable and is degraded rapidly within the nucleus; and a comparatively short sequence at the 3' end of the molecule which may be cleaved from it in a stable form and transported to the cytoplasm. The very large proportion of Hn RNA which is unstable implies that some Hn RNA molecules may be degraded in their entirety and may not donate a sequence which matures into messenger RNA. One possible model for the fate of Hn RNA is therefore to suppose that the large transcription product is cleaved into smaller segments, most of which are rapidly degraded within the nucleus but one of which may instead of suffering degradation give rise to a stable messenger RNA.

If Hn RNA is the precursor of mRNA, all the sequences which are found in cytoplasmic mRNA should also be present in the nuclear RNA. Of course, not all the sequences of Hn RNA may be found in mRNA, for some may be degraded within the nucleus and so fail to appear in mRNA. Hybridization experiments in which Hn RNA and mRNA are used as competitors can define the proportion of the Hn RNA sequences which may be used as messengers. Hybridization assays can also be directed to two further questions about the roles of these molecules. What is the difference in the sequences present in the mRNA populations of different tissues? And how much difference is there between the regions of Hn RNA molecules which are degraded in different cell types—are they tissue specific and what role, if any, do they play in gene expression?

Using the nitrocellulose filter technique to assay repeated sequences, Shearer and McCarthy (1967, 1970) compared the hybridization in RNA excess of the nuclear and cytoplasmic populations of mouse L cells. Nuclear RNA saturates the DNA at about 4% of its sites; but cytoplasmic sequences reach saturation at about 1%. Scherrer et al. (1970) reported saturation values of 4·3% for the nuclear RNA of Hela cells and only 0·5% for the mRNA fraction of polysomes. This implies that many of the sequences of DNA transcribed into nuclear RNA never leave the nucleus.

Using competition hybridization assays between Hn RNA and mRNA such as those shown in figure 5.13, Shearer and McCarthy demonstrated that all the sequences found in cytoplasmic RNA are competed by nuclear RNA; but only about two thirds of the sequences of nuclear RNA can be competed by cyto-plasmic RNA. This suggests that all the messenger sequences of the cytoplasm are derived from the nucleus; but some of the nuclear sequences are not transported to the cytoplasm. The transcripts in Hn RNA of the repetitive fraction of the genome thus show the behaviour implied by kinetic studies; some are restricted to the nucleus and some are transported to the cytoplasm. When RNA is labelled for 100 minutes and the decrease in the hybridization of labelled RNA with DNA is followed over a period of time, the sequences found only in nuclear RNA—that is those which are not competed by cyto-plasmic RNA—decline with a half life of less than 30 minutes. All the sequences restricted to the nucleus are therefore unstable and turn over rapidly.

Instead of using simultaneous competition between two RNA populations for DNA, Soeiro and Darnell (1969, 1970) used a presaturation technique. They found that when unlabelled and labelled populations of RNA are present together in solution, an apparent competition which does not reflect binding to the same sites of DNA may be produced by an interference between the RNAs. They therefore first saturated DNA with an unlabelled preparation of RNA and then followed the ability of a second, labelled preparation to bind to the saturated DNA. Under their conditions, both Hn RNA and messenger RNA separately saturate about 5% of the DNA sites. Presaturation with Hn RNA is equally effective in preventing either homologous Hn RNA or messenger RNA from binding to DNA. This suggests that all the sequences of mRNA are present in the Hn RNA.

A considerable spectrum of reiterated sequences is found in Hn RNA. By comparing the rate of hybridization between fractions of Hn RNA and excess DNA with that shown between rRNA and excess DNA, Darnell et al. (1970) and Pagoulatos and Darnell (1970b) showed that the most rapidly hybridizing Hn RNA sequences react at a rate some 30 times that of rRNA. When Hn

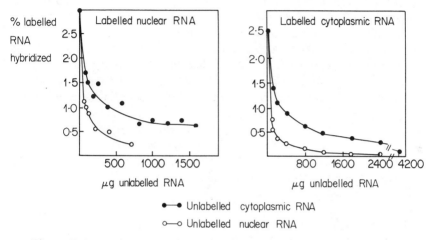

Figure 5.13: competition hybridization between repetitive sequences of nuclear and cytoplasmic RNAs of L-cells. Incubation mixtures contained 4 μg of the labelled RNA indicated and 12 μg of mouse DNA (left) or 8 μg of mouse DNA (right) in the presence of excess amounts of unlabelled RNA. Left: nuclear RNA is effectively competed by a second sample of nuclear RNA but cannot be completely competed by cytoplasmic RNA. About 23% of the labelled nuclear sequences continue to bind to DNA at the plateau of competition; other experiments by Shearer and McCarthy (1970) have yielded similar results, with a plateau of 36%. Right: labelled cytoplasmic RNA is effectively competed by either unlabelled cytoplasmic or unlabelled nuclear RNA. All the sequences of cytoplasmic RNA must therefore be present in the preparation of nuclear RNA. Data of Shearer and McCarthy (1967).

RNA is hybridized with excess DNA, the most reiterated fraction hybridizes during the first 6 hours of incubation in the conditions used; and the initial rate of reaction—expressed as the fraction of input RNA which hybridizes—depends only on the RNA concentration. The RNA which does not hybridize during this incubation can be recovered and then incubated with a fresh preparation of DNA; its initial rate of hybridization measures its degree of repetition in the genome.

In an alternating series of hybridization and isolation of the non-hybridized RNA for re-testing with DNA, the rate of reaction of the RNA declines in each successive exposure to DNA. Each successive cycle of hybridization thus measures a less repetitive fraction of transcripts. In the first incubation, RNA was hybridized for 5 hours to isolate the most rapidly reacting sequences. In the second cycle, the non-hybridized sequences were recovered and hybridized with a fresh preparation of excess DNA for 17 hours; the RNA which hybridizes constitutes the 22 hour fraction (total time of exposure to DNA is $5 + 17 = 22$ hours). In a third cycle, the non-hybridized RNA was again recovered and annealed with fresh DNA, this time for 40 hours to give the 62 hour fraction. About 10% of the total RNA is included in the three fractions.

The initial rate of reaction of the 5 hour RNA fraction is some 30 times faster than that of 28S rRNA; the initial rate of reaction of the 62 hour fraction is about one half that of the 28S rRNA. Since rRNA has about 400 sites in the genome, it is possible to make a crude estimate that the most repeated fractions of the Hn RNA are represented in about 10,000 copies in the genome; the lowest degree of repetition measured in these experiments must correspond to about 200 fold repetition. Of course, these estimates are only approximate, since the rRNA hybridizes to identical sites whereas the repeated sequences of Hn RNA hybridize to related sites. Most of the RNA sequences do not hybridize at all under these conditions and must therefore be considerably less reiterated than rRNA—they may comprise sequences transcribed from unique DNA.

That both non-repeated and repeated sequences of the genome are transcribed into Hn RNA and that both are represented in the mature messenger RNA molecules has been shown by hybridization experiments in DNA excess performed by Greenberg and Perry (1971). In L cells incubated with low doses of actinomycin to inhibit nucleolar RNA synthesis, both the Hn RNA and the mRNA contain two classes of RNA transcript. Rapidly annealing sequences have a $cot_{1/2}$ of about 1·0, corresponding to about 10,000 fold repetition in the genome; and more slowly annealing sequences have a $cot_{1/2}$ of about 1000, probably corresponding to unique sequences. The population of Hn RNA molecules consists of 32% repetitive and 68% unique transcripts; the mRNA molecules include 20% repeated sequences and 80% unique transcripts. There is thus some tendency for transcripts of unique sequences to be conserved when Hn RNA matures; but not all the unique sequences are conserved—much less than 68% of the Hn RNA is represented in mRNA—so that the nucleus-restricted sequences must include both repeated and unique transcripts.

The distribution of repeated and non-repeated sequences in the messenger population has not been defined. We do not know, therefore, whether some messengers comprise solely transcripts of unique sequences whereas others represent repeated sequences, or whether some or all messengers contain transcripts of both repeated and non-repeated sequences. Nor can we say what proportion of the length of each messenger is used to code for protein.

When labelled mRNA is extracted from polysomes of Xenopus embryos dissociated at the neurula stage in medium lacking divalent cations—a treatment which is thought not to alter the pattern of RNA synthesis and which is necessary to achieve incorporation of labelled precursors—a spectrum of molecules is obtained, sedimenting from 6–12S and displaying some 15–20 peaks on gels. This preparation therefore comprises a population of many messengers.

A discrepancy was noted by Dina et al. (1973) between two assays for following the hybridization reaction between this RNA and excess DNA. When hybridized with DNA in excess in solution and then treated with pancreatic ribonuclease to digest any unreacted RNA, about 13% of the RNA becomes RNAase-resistant by a cot of 100 and 61% hybridization is achieved by the cot range 5×10^3–2×10^4. But when the hybrid formed in solution is isolated by filtration through membrane filters, the mRNA behaves as two components each comprising 50% of the population, the first retained with a $cot_{1/2}$ of 5×10^{-2} and the second with a $cot_{1/2}$ value of 20–30. Reaction is complete by a cot of 100.

The discrepancy between the two methods of following the reaction suggests that each messenger molecule comprises in part a transcript of a unique sequence and in part a transcript of a repeated sequence. In solution, only the repeated sequences react at low cot values, leaving the unique transcripts in the form of single stranded RNA; digestion with ribonuclease before counting removes all sequences of RNA except the molecules which have hybridized with the repeated regions of DNA. But when the hybrid is isolated by its retention on membrane filters, the complete RNA molecule becomes bound to the filter—and therefore appears to be hybridized—once its repeated part has formed a duplex with DNA. Because essentially all the RNA is retained at low cot values, this implies that almost every messenger molecule must possess a repeated sequence.

But different results have been obtained in an analysis of the sequences of the messenger RNA fraction of sea urchin embryos. Goldberg et al. (1973) found that the mRNA released from polysomes by puromycin possesses few or no sequences corresponding to the repetitive fraction of the genome, but hybridizes at the rate expected of transcripts of the non-repeated fraction of DNA. This suggests that most of the genes which are expressed at the gastrula stage of embryonic development are present in only one copy per haploid genome. Although repetitive sequences lie adjacent to non-repetitive sequences in the sea urchin genome (see page 203), they must therefore either fail to be transcribed or must be removed from messengers before entry into the polysome fraction.

Further support for this concept has been provided by the experiments of Galau, Britten and Davidson (1974), who analysed the sequence complexity of the sea urchin gastrula polysomes. Hybridization analysis showed that the messenger fraction, derived entirely from non repetitive DNA sequences, displays a sequence complexity corresponding to about 14,000 genes. Since the gastrula embryo comprises only about 600 cells, this is a substantial number and, presumably, an adult organism would display expression of many more genes. The frequency with which each messenger sequence is represented in the polysome population varies appreciably, and some sequences are present in small amounts only. The sequence complexity of nuclear RNA is some ten fold greater than that of messenger RNA, and includes transcripts representing repetitive sequences of the genome.

The sea urchin system therefore displays the same general characteristics exhibited by mammalian cells in culture: nuclear RNA contains many sequences that are not transported to the cytoplasm to be translated, and these include sequences derived from the repetitive DNA component. An apparent difference between mammalian and sea urchin cells is that some repetitive sequences appear to be conserved in the mammal and are represented in mRNA, whereas none is found in the sea urchin mRNA. However, before deciding the significance of this observation it will be necessary to characterize the mammalian messengers more stringently, in particular to analyse the nature of the repetitive sequences present in the cytoplasmic RNA.

It is important to note that artefacts introduced during preparation of messenger fractions might lead to the apparent presence of repetitive sequences; for example, only a small degree of contamination by nuclear RNA rich in repetitive sequences would suggest the presence of repetitive sequences in the messenger fraction. The differences observed between the sea urchin and Xenopus embryonic systems may therefore represent a genuine difference in the characteristics of mRNA, or alternatively may have arisen because of some difference introduced during extraction of the messenger RNA.

No general model can yet be constructed for the sequence organization of messenger RNA, therefore, although it is clear that the messenger fraction is always rich in transcripts of non-repeated sequences which presumably represent structural genes. However, in their review of the organization and transcription of the eucaryotic genome, Davidson and Britten (1973) noted that there is increasing evidence to support the idea that most mammalian messengers carry a sequence coding for only one protein (that is they represent only a single gene); this single coding sequence is probably almost always derived from a non-repeated structural gene and Davidson and Britten suggested that any additional, non-translated sequences (such as transcripts of repetitive sequences) may be rather short.

The histone messengers provide the only exception so far discovered to the general restriction of coding sequences to the unique fraction. By hybridizing the histone mRNA of sea urchin embryos with an excess of DNA extracted from sperm, Kedes and Birnsteil (1971) showed that the histone messengers appear to be present in many copies, with a repetition frequency of the order of 400. (The frequency of repetition may be somewhat lower in mammals, but still appears to be greater than one.) The histone mRNA-DNA hybrids display a high melting temperature, which suggests that they are formed by accurate pairing, so that there must be little heterogeneity among the repeated copies of each histone gene; since the degeneracy of the genetic code would allow appreciable variation to occur between the sequences of repeated genes coding for the same protein, this implies that some mechanism may exist to suppress variation between them. The distinct buoyant density of the DNA molecules complementary to histone mRNAs suggests that the genes coding for them may be organized in clusters on the chromosome; this situation may therefore perhaps be analogous to the suppression of sequence variation in the clusters of genes coding for ribosomal RNAs (see pages 153 and 316).

That the rapidly labelled Hn RNA molecules of the nucleus contain all the sequences present in the messengers of the cytoplasm and in addition contain sequences restricted to the nucleus suggests a precursor–product relationship for Hn RNA and mRNA. However, these experiments involve populations of RNA; confirmation of this model requires the characterization of particular Hn RNA molecules which contain the sequences of individual messengers. The reticulocyte system is particularly suitable for such experiments because its rapidly labelled Hn RNA molecules can be isolated in quantity and because globin messenger has been purified.

The presence of globin sequences in Hn RNA molecules has been directly tested by Imaizumi, Diggelman and Scherrer (1973), who have used as a probe the "anti-messenger" DNA complement which can be synthesised from the globin message by the reverse transcriptase enzyme of RNA tumour viruses (see page 260). This single stranded DNA hybridizes specifically with the sequences of globin mRNA. The reaction between anti-messenger DNA and an excess of the RNA population to be tested can be used to measure the $cot_{1/2}$ of reaction. By comparing the $cot_{1/2}$ of reaction with Hn RNA and the $cot_{1/2}$ shown with globin mRNA itself, it is possible to calculate the number of equivalents of the globin sequence which is present in the Hn RNA.

Globin messenger RNA and anti-messenger DNA have a $cot_{1/2}$ of reaction under these conditions of 7.5×10^{-4}; the largest fraction of Hn RNA has a $cot_{1/2}$ of reaction of 2.5×10^{-1}. The Hn RNA therefore hybridizes more slowly with anti-messenger because its effective concentration of globin sequences is lower than that of the message. If the globin mRNA fraction contains only the sequences for globin and provided that the anti-messenger DNA pairs only with these sequences—whether its partner is mRNA or Hn RNA—the ratio of the two $cot_{1/2}$ values estimates the dilution of the globin sequences in the Hn RNA. Thus only 0.3% of the Hn RNA corresponds to globin mRNA.

Globin mRNA sediments at a rate which corresponds to a size of 2×10^5 daltons; since it must include the sequences for both α-chains and β-chains, this corresponds to a hybridizing length of 4×10^5 daltons. If only 0.3% of the

Hn RNA corresponds to globin messenger, then the length of Hn RNA which includes both globin sequences must be $4 \times 10^5/0.003 = 1.3 \times 10^8$ daltons; the length of Hn RNA containing one globin sequence, of either type, must therefore be 6.5×10^7 daltons. The average weight by sedimentation of this Hn RNA preparation was 3.5×10^7 daltons. If each Hn RNA can contain only one globin sequence—any other model implies that multiple globin genes lie adjacent to each other in the genome, which is unlikely from genetic studies—then $3.5/6.5 = 52\%$ of the Hn RNA molecules contain a globin sequence.

The Hn RNA population of the duck reticulocytes was divided into three classes by centrifugation through a sucrose gradient; the two smaller classes had molecular weights by sedimentation of an average of 7.5×10^6 and 4.0×10^6; their $cot_{1/2}$ values of 4.2×10^{-1} and 5.0×10^{-2} respectively correspond to 0.18% and 1.5% possession of globin sequences; this means that 7% of the molecules of intermediate size and 30% of the molecules of smallest size contain globin sequences. Considering the entire population of Hn RNA, from $7–52\%$ of the molecules therefore possess a globin sequence, depending upon the size class.

The relationship between the Hn RNA molecules of different sizes which contain globin RNA sequences is not clear; one factor complicating the analysis is that two types of globin sequence are present. However, it seems likely that the immediate transcription product containing either type of chain should comprise a single species; in this case there should be no more than two initial types of precursor. This suggests that the smaller molecules may have been derived from the larger precursors by cleavage events.

That sequences for globin chains are present in the Hn RNA molecules of the nucleus has also been shown by Williamson et al. (1973), who injected Xenopus oocytes with the large Hn RNAs extracted from the livers of 14 day embryonic mice (this tissue synthesizes large amounts of haemoglobin). Globin chains characteristic of the mouse are then produced within the oocyte. As a control, Hn RNA of brain does not cause the production of globin when injected into oocytes. This confirms that synthesis of Hn RNA is tissue specific and includes sequences destined to become messengers.

That other species of RNA molecules containing poly-A include messenger sequences has been shown by Stevens and Williamson (1972) in experiments in which RNAs isolated by their content of poly-A were injected in Xenopus oocytes. When myeloma cells grown in culture are labelled with radioactive precursors of RNA in the presence of low levels of actinomycin, 40% of the cytoplasmic RNA binds to millipore filters in the conditions which allow poly-A retention and 30% of the label in H^3-adenosine is resistant to RNAase. Some 20% of the nuclear RNA binds to the filters, about 10% of an H^3-adenosine label displaying resistance to RNAase. When cytoplasmic and nuclear RNA molecules possessing poly-A sequences were isolated by elution from a column of poly-U and injected into oocytes, both preparations caused the production of the myeloma immunoglobulins. This confirms that the cytoplasmic RNAs possessing poly-A sequences include the messengers; and demonstrates also that the nuclear RNAs containing poly-A include messenger sequences and may therefore be their precursors.

The simplest model for the production of globin and other messenger RNAs is therefore to suppose that they are cleaved from large Hn RNA precursors. The remainder of the Hn RNA molecule appears to be degraded rapidly within the nucleus although it comprises a sequence much longer than that of the messenger region itself. The structure and function of this sequence is not known; but the ability to isolate specific precursor molecules by their hybridization with the anti-messenger implies that it may become possible to investigate the sequences comprising the remainder of the Hn RNA molecule.

By hybridizing labelled mouse L cell Hn RNA with membrane bound DNA in the presence of competing (unlabelled) cytoplasmic mRNA, Shearer and McCarthy (1970) followed the behaviour of the nucleus-restricted sequences. If the nucleus-restricted sequences are related to those which are transported to the cytoplasm, a reduction in the stringency of hybridization should allow greater competition. However, the RNA restricted to the nucleus of the L cell appears to comprise a constant 40% of all nuclear RNA under a wide range of hybridization conditions. This implies that quite different sets of families of repeated sequences are restricted to the nucleus and transported to the cytoplasm from amongst the Hn RNA population—the families involved are sufficiently different not to be mistaken for each other even under conditions of only low stringency.

Diversity within gene families can be followed by measuring the thermal stabilities of the RNA/DNA hybrids formed under conditions of low stringency; the greater the diversity within a family, the lower the thermal stability. Under such conditions the nucleus-restricted sequences have a Tm about 2°C higher than that of the cytoplasmic sequences. This implies that the gene families from which the nucleus-restricted sequences are transcribed are less diverse than those which are represented in messenger RNA.

That both nucleus-restricted and cytoplasmic sequences are controlled in gene expression is shown by the changes in both classes of sequence which can be detected in regenerating mouse liver. If only some of the liver tissue is excised, the remaining cells are stimulated into cell division to restore the lost mass. When Church and McCarthy (1967a) compared the RNA populations of normal and regenerating liver, they found that new RNA molecules are produced by the regenerating cells. The extent of DNA which is saturated by RNA increases from 11% in normal liver to 23% soon after regeneration starts; the saturation level then declines over the next few days to its previous value of 11%. These experiments concern repeated sequences and although the absolute values are therefore much greater than the number of sites which is transcribed, they show that there is a substantial increase in transcription of RNA when liver cells regenerate.

The sequences present in normal liver all appear to be present in regenerating liver; but regenerating liver also contains new sequences which are not found in normal liver. The new sequences appear within 1–3 hours after tissue excision, and many of them disappear by 6 hours; a smaller number of new molecules is synthesised during later periods. Church and McCarthy (1967c) have shown that some of the new RNAs comprise messengers found in the cytoplasm; but others correspond to fractions of Hn RNA which are not transported to the cytoplasm. Changes in gene expression therefore involve both nucleus-restricted and cytoplasmic sequences of RNA.

All messenger RNAs except the histones appear to be derived from large nuclear precursors by a process of cleavage. It seems likely—although there is at present no proof—that the only region of the Hn RNA precursor which is preserved is located at the 3' end and that the remainder of the molecule is degraded rapidly within the nucleus. At least some of the nucleus-restricted sequences must be provided by the unstable regions of these precursors. This suggests a model in which the unit of transcription is very much larger than the mature messenger molecule, which may correspond to only one protein chain; all of the transcript except the (comparatively) short messenger region serves some function other than coding for protein and is degraded within the nucleus.

One unresolved problem is whether all Hn RNA molecules are treated alike in that the 5' sequence is degraded and the 3' sequence is preserved; or whether only some molecules donate messenger sequences and others are degraded entirely. (It is also possible that some precursors carry many coding sequences, all of which are stable and are transported to the cytoplasm, although studies of the maturation process in which poly-A is added make this less likely.) It is possible that a control of gene expression may be exerted in the nucleus if a selection is made amongst the Hn RNA molecules so that some are utilised to donate messengers whereas others, although potentially coding for proteins, are in fact chosen for degradation. In this case, the nucleus restricted fraction might include sequences which in changed circumstances or in other cell types are utilised as messengers in the cytoplasm. We do not at present know whether maturation of Hn RNA is purely mechanical, or whether it may act as a parameter controlling gene expression.

Location of Poly-A in Messengers

Most eucaryotic messenger RNAs possess a sequence of poly-adenylic acid at the 3' end. This part of the messenger appears to be synthesised separately from the Hn precursor and is added to it after transcription but before cleavage of the mRNA. One function of poly-A may therefore be concerned with the selection of messenger sequences and their transport from nucleus to cytoplasm. But because poly-A sequences are also found in viruses which reproduce within the cytoplasm and in mitochondrial messengers which are translated in situ, at least part of their function must also be concerned with stages of expression subsequent to any transport. Since messengers are translated from 5' to 3', translation must be terminated before ribosomes reach the poly-A region so that its role is presumably unrelated to the action of the ribosome.

One of the first observations of the existence of poly-A was made by Edmonds and Caramela (1969) when they isolated the RNA of Ehrlich ascites cells grown in mice injected with P^{32}. After extraction with phenol, some of the RNA is retained by poly-dT cellulose; this is due to the presence of poly-A sequences which can hybridize with the poly-dT. About 1 % of the total RNA labelled over a period of 12 hours contains poly-A. By measuring the poly-A content of RNA fractions isolated from Hela nucleus and cytoplasm, Edmonds, Vaughan and Nakazoto (1971) showed that poly-A is absent from the nucleolar (ribosomal RNA) fractions but is present in the RNAs of both nucleoplasm and cytoplasm.

The presence of poly-A sequences in these RNA fractions has since been confirmed by several methods of isolation. Lee, Mendecki and Brawerman (1971) found that RNA containing poly-A is adsorbed by membrane filters at high ionic strength. That the RNA is retained because it possesses a sequence of poly-A is confirmed by the effect of pancreatic ribonuclease, which cannot degrade regions of poly-A. When the messenger fraction of mouse sarcoma cells is labelled with adenosine and then treated with pancreatic ribonuclease, the label is retained on the filter. But if the RNA is instead labelled with uridine before treatment with the RNAase, the label is not retained. It is therefore the region of poly-A which adsorbs to the filter. All the fractions of mRNA from a sedimentation gradient are retained by a membrane filter. Figure 5.14 shows the similar experiments of Darnell et al. (1971) in which the presence of poly-A was followed by comparing the total radioactivity with that remaining acid-precipitable after digestion with pancreatic ribonuclease.

By using fibreglass filters impregnated with poly-U, Sheldon, Jurale and Kates (1972) have shown that Hela cell RNA purified from nucleoplasm or cytoplasm contains poly-A. Shutz, Beato and Feigelson (1972) found that at sufficiently high ionic strength, cellulose itself can retain poly-A and RNA molecules containing regions of poly-A. Fractions of RNA extracted from rabbit reticulocyte or chick oviduct polysomes are retained on such columns and can be eluted as a sharp peak by applying water. That the RNAs act as messengers is shown by their stimulation of globin and ovalbumin synthesis respectively in cell free systems.

The location of the poly-A region in mRNA has been identified by experiments with a purified exoribonuclease enzyme of ascites cells which is specific for 3'–OH termini. If poly-A is located at the 3' terminus of mRNA, it should be degraded first. Molloy et al. (1972) found that when mRNA of Hela and L cells is degraded with this enzyme, poly-A is degraded exceptionally rapidly relative to the other sequences. The regions of poly-A can be isolated by degradation of the messengers with pancreatic ribonuclease; the poly-A sequences remaining intact after this treatment should be susceptible to the exoribonuclease only if they contain free 3′–OH ends, for if they have been released from the internal parts of the messenger by pancreatic RNAase they should have 3′ phosphate termini conferring resistance to the enzyme. About 90% of the isolated poly-A is susceptible to exoribonuclease; this implies that a maximum of 10% can be internal, although a more probable interpretation

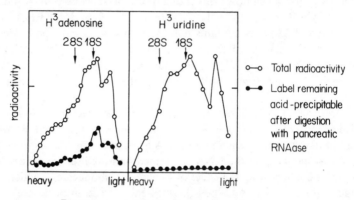

Fractions on sucrose gradients

Figure 5.14: presence of poly-A in messenger RNA. Hela cells were labelled with H³ adenosine (left) or H³ uridine (right). The mRNA fraction extracted from polysomes was sedimented through a sucrose gradient. The total radioactivity in each fraction was determined after acid precipitation of RNA; the label remaining acid precipitable (that is in polynucleotide form) was determined after digestion with RNAase. The distribution of total messenger remains the same with either labelled adenosine or uridine. But polynucleotides resistant to pancreatic RNAase remain labelled only with the H³-adenosine precursor. This identifies the poly-A. Data of Darnell, Wall and Tushinski (1971).

is that these fragments have lost their free 3′–OH termini because of internal cleavage during extraction.

That globin mRNA contains a poly-A sequence is suggested by its ready isolation on oligo-dT cellulose; Aviv and Leder (1972) used this technique as a preparative method for separating globin mRNA from rRNA in crude extracts. More than 90% of the RNA of polysomes is eluted directly from the column; the 3% which is retained represents globin mRNA and can be eluted by a reduction in ionic strength. This fraction sediments at 9S and produces rabbit globin in an in vitro system. Swan, Aviv and Leder (1972) have also used this method to isolate the messenger for a light chain immunoglobulin from myeloma cells. The molecular weight of the mRNA is about 850 nucleotides, about 200 greater than needed to code for the protein which Stavnezer and Huang (1971) identified as its in vitro product. The extra length presumably at least in part comprises poly-A.

By incubating globin mRNA with pancreatic ribonuclease, Lim and Canellakis (1970) showed that the resistant fragment possesses an AMP content of more than about 80% and displays a mobility on gels comparable to that of tRNA. Pemberton and Baglioni (1972) have isolated a resistant sequence from duck globin mRNA which by this criterion appears to be about 150–200 nucleotides long. By labelling the 3′ terminal nucleotide of rabbit globin mRNA with periodate oxidation and tritiated borohydride, Burr and Lingrel (1971) isolated the 3′ fragment released by digestion with T1 and pancreatic ribonucleases. The fragment is a nine base length of oligo-A; this shows that there is no guanine residue in the last nine bases and suggests that the poly-A is probably located at the 3′ end of the message.

That the poly-A sequence is 3′-terminal is implied also by experiments in which Ross et al. (1972), Verma et al. (1972) and Kacian et al. (1972) have utilised the reverse transcriptase enzyme of RNA tumour viruses to produce globin anti-messenger DNA from mRNA. Reverse transcriptase requires a primer from which it can extend a DNA chain on an RNA template, growing in the direction 5′ to 3′. When oligo-dT is bound to globin mRNA, it can be used as a primer for this chain extension; the product is a length of DNA complementary to the globin message and with a sequence of oligo-dT at its 5′ end (that is corresponding to the 3′ end of the message).

RNA viruses also contain sequences of poly-A. Yogo and Wimmer (1972) found that RNAase treatment of polio virus yields resistant fragments of about 90 nucleotides long which are at least 90% adenine. These sequences are located at the 3′ ends of the molecules; there do not appear to be any regions of poly-A within the virus genome. The RNA genomes of the tumour viruses also contain about 2% of poly-A according to Green and Cartas (1972). That this sequence is concerned with gene expression is suggested by the observation of Johnston and Bose (1972) and Gillespie, Marshall and Gallo (1972) that RNA viruses in which the mature strand is translated (positive strand viruses) contain poly-A; but viruses where the transcript of the mature strand of the particle is translated (negative strand viruses) do not appear to contain poly-A.

Poly-A sequences are also found in viral coded RNAs which are transcribed within the cytoplasm of an infected cell. These sequences must therefore be concerned with some cytoplasmic function of mRNA and not with nucleocyto-plasmic transport. Vaccinia is a DNA virus which replicates in the cytoplasm or in enucleated cells. Kates and Beeson (1970) found that transcription of the viral DNA with E.coli RNA polymerase produces a product which contains large amounts of poly-A. When labelled poly-A is hybridized to denatured strands of vaccinia DNA, 0·7% of the DNA hybridizes with poly-A; 1·3% of the DNA hybridizes with labelled poly-U (this reaction should detect dA clusters on the strand complementary to the dT clusters coding for poly-A). This shows that the poly-A sequences found in the viral RNA are coded by the virus genome.

When RNA is extracted from Hela cells infected with vaccinia virus, some of the high molecular weight RNA contains fragments resistant to pancreatic RNAase; this shows that the poly-A sequences are transcribed as part of larger functional units. The poly-A recovered from the viral RNA sediments at about 4S, in the same size range as the poly-A of cellular messengers. Sheldon et al. showed that all of the poly-A in the RNA coded by vaccinia virus is located at the 3' end of the messenger and at least 98 % of the Hela cell messenger poly-A is also found in this position: polynucleotide phosphorylase, which degrades RNA from 3' to 5', causes loss of A residues much more rapidly than loss of U residues; and the ascites cell exoribonuclease also attacks the poly-A, showing that it must have a free 3'–OH terminus.

Poly-A found in these viral messengers is therefore coded by the genome and its transcription completes synthesis of a messenger. Mitochondrial protein synthesis appears quite different from that of the cytoplasm and Perlman, Abelson and Penman (1973) isolated a fraction of Hela cell mitochondria which includes mRNA; it is synthesised during incubation with camptothecin, which inhibits all nuclear RNA synthesis. The 12S and 21S mitochondrial ribosomal RNAs labelled under these conditions do not bind to poly-U glass fibre filters; but a heterogeneous fraction sedimenting from 15–30S is retained. The length of poly-A in this RNA is about 50–80 nucleotides, under half the length of the poly-A sequence of cytoplasmic messengers. Since this fraction comprises messenger RNAs which are translated within the mitochondria, no transport is involved and the poly-A must therefore be concerned with some other function; the same conclusion is true of the poly-A sequences found in RNAs coded by viruses which enter only the cytoplasm during infection.

The messenger RNAs released from Hela cell cytoplasmic polysomes by EDTA contain some 2·5–5·0% poly-A. Edmonds, Vaughan and Nakazoto (1971) and Darnell, Wall and Tushinski (1971) observed an inverse relationship between the proportion of poly-A and the size of the messenger; this is consistent with the idea that all messengers, irrespective of their overall size, contain the same length of poly-A. When mRNAs of different sizes are isolated from a gradient and digested with pancreatic ribonuclease, the product always sediments at a rate about 10% faster than the tRNA of E.coli. This corresponds to a size of less than 200 nucleotides and suggests that each messenger contains only one sequence of poly-A, located at its 3' terminus.

Few if any bases other than adenine are present in the sequence of poly-A. By degrading the poly-A with alkali, Lee, Mendecki and Brawerman (1971) detected one 3'-OH terminus for every 200 adenine residues. By using gel electrophoresis to analyse the poly-A recovered from newly synthesized Hela cell messengers, Sheiness and Darnell (1973) demonstrated that the most

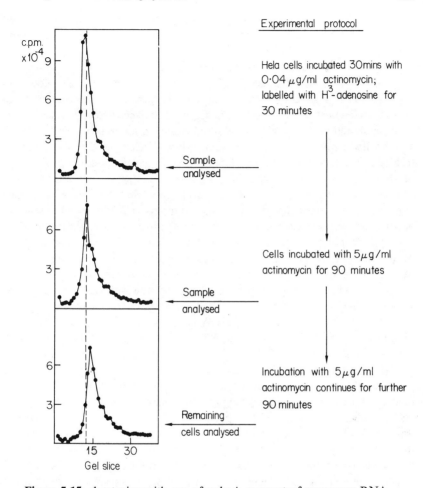

Hela cells incubated 30mins with
0·04 μg/ml actinomycin;
labelled with H^3-adenosine for
30 minutes

Sample
analysed

Cells incubated with 5μg/ml
actinomycin for 90 minutes

Sample
analysed

Incubation with 5μg/ml
actinomycin continues for further
90 minutes

Remaining
cells analysed

Figure 5.15: shortening with age of poly-A segment of messenger RNA. The polysomal RNA in each cell sample was extracted and treated with pancreatic and T1 RNAase. The resistant RNA—that is the poly-A fraction—was electrophoresed on the gels shown. Upper: immediately after labelling period, poly-A fragment migrates farthest, that is it is longest. Centre: after 90 minutes during which no further RNA synthesis can take place, the length of poly-A is reduced, as shown by its more rapid migration. Lower: after a further 90 minute chase period, the length of poly-A is shorter still. Data of Sheiness and Darnell (1973).

prominent peak migrates at a position corresponding to 190 ± 20 nucleotides. When the poly-A released from Hela cell nuclear or messenger RNA by T1 ribonuclease was freed from contaminating oligonucleotides, Molloy and Darnell (1973) identified the presence of 2 UMP and 1–2 CMP residues for every 195 residues of AMP. Since T1 ribonuclease cleaves RNA adjacent to guanine residues, this means that a sequence containing the U and C residues may lie between the first G residue and a sequence containing only A residues at the ends of both nuclear and mRNAs.

The poly-A liberated by ribonuclease digestion of cytoplasmic mRNA migrates in a heterogeneous band upon gel electrophoresis. Sheiness and Darnell (1973) showed by actinomycin chase experiments that the length of poly-A attached to a messenger decreases with its age. After Hela cells have been incubated with a low dose of actinomycin to block synthesis of rRNA and labelled within H^3-adenosine, a large dose of actinomycin may be added to block all RNA synthesis. The labelled messengers isolated from the cytoplasm during incubation with high actinomycin therefore represent those synthesized before its addition. Figure 5.15 shows that the poly-A recovered from messengers at the time when the high actinomycin is added—this identifies newly synthesized messengers—is the longest; after 90 and 180 minute chase periods, the poly-A decreases successively in length. The shortening of the poly-A does not seem to be related to the utilization of the messenger during protein synthesis, for it continues at the same rate in cells in which protein synthesis is inhibited. It is tempting to speculate, however, that the shortening of poly-A may be related to the lifetime of the messenger.

Although the function of the poly-A is unknown, it seems to comprise a unique part of the messenger not only for its sequence but also for its association with protein. Blobel (1973), who has characterized two proteins found in mRNP preparations of globin messenger, showed that the larger of these proteins (78,000 daltons) retains a radioactive label when the cells have been incubated with H^3-adenine; the protected material includes about 80% adenine. This suggests that the poly-A sequence of globin messengers is associated with a specific protein in the cell. Kwan and Brawerman (1972) found that poly-A can be released from polysomes of mouse ascites cells by treatment with pancreatic RNAase and EDTA in the form of a 12–15S component; this is caused by attachment of a protein(s). It is therefore possible that poly-A may play some role in the formation of the mRNP particle.

Role of Poly-A in Processing of Hn RNA

Sequences of poly-A are found in nucleoplasmic Hn RNAs as well as in messengers; Edmonds, Vaughan and Nakazoto (1971) found that the distribution in density gradients of Hela cell nuclear RNAs containing poly-A—isolated by adsorption to poly-dT–cellulose—parallels that obtained by rapidly labelling the Hn RNA fraction. Since the poly-A cannot be separated from the Hn RNA by denaturation or other techniques which disrupt noncovalent bonds, it must be an integral part of the Hn RNA molecule.

The proportion of poly-A in Hn RNA decreases as the length of the RNA molecule increases; since the sequences of poly-A recovered from RNAs in all parts of a gradient show a uniform size of about 150–200 nucleotides, this suggests that each Hn RNA molecule may have only one length of poly-A irrespective of the length of the rest of the molecule. The overall content of poly-A in Hn RNA is some ten fold less than in mRNA, which implies that the sequence of poly-A may be conserved together with the messenger sequence when the other regions of the Hn RNA are lost during processing. Since the poly-A is located at the 3' of the Hn RNA as well as of the mRNA, this suggests that the sequence of the Hn RNA is:

5' non-conserved regions—messenger sequence—poly-A 3'

According to this model, the addition of poly-A to the 3' end of the molecule may play an essential role in the conservation and transport to the cytoplasm of the messenger sequence when the Hn RNA matures.

Estimates of the length of poly-A and its frequency in Hn RNA vary in different experiments and artefacts reducing the apparent content of RNA containing poly-A sequences may occur rather readily. If each Hn RNA or mRNA contains only one sequence of poly-A, whose presence is needed for the assay used to isolate the RNA, one break in the RNA chain is sufficient to release one of the two fragments of the RNA molecule from its attachment to poly-A. These assays are therefore very sensitive to endonucleolytic cleavages in the RNA and the occurrence of such events may explain variations in the parameters reported for poly-A content.

In general, the proportion of poly-A in Hn RNA is within the range $0.5-1.0\%$ and its length upon sedimentation appears about 200 nucleotides. By electro-phoresing poly-A isolated from Hela Hn RNA on acrylamide gels, Sheiness and Darnell (1973) showed that its length is fairly homogeneous at about 210 ± 20 nucleotides. This is a little longer than the poly-A of mRNA, which decreases from about 195 nucleotides when newly synthesized to about 100 nucleotides some six hours later. A short length of poly-A may therefore be removed when Hn RNA matures to mRNA if isolated poly-A remains the same length as that in Hn RNA and mRNA in vivo.

Several observations suggest that Hn RNA and mRNA have a precursor-product relationship in which the addition of poly-A to the Hn RNA plays an essential role. Mendecki, Lee and Brawerman (1972) noted that the first RNAs to appear in the cytoplasm after addition of labelled adenosine to mouse sarcoma ascites cells have labelled poly-A sequences. This is consistent with the idea that poly-A is either transcribed last or is added after transcription to the 3' end of the molecule. Darnell et al. (1971) found that labelled adenosine is first incorporated into the poly-A of Hela cell Hn RNA and is later found in polysomal RNA.

That poly-A is not transcribed as part of the Hn RNA molecule but is added later is suggested by experiments utilising mouse sarcoma cells infected with adenovirus. RNA coded by the virus genome can be assayed by hybridisation with denatured adenovirus DNA; both nuclear and polysomal RNAs hybridize when extracted from infected cells but there is no reaction with the RNA of uninfected cells. Philipson et al. (1971) found that when the adenovirus RNA–DNA hybrids are digested with both DNAase and RNAase, a resistant fragment remains; this is a poly-A sequence of some 150–200 nucleotides which contains 80–95% adenine residues and is completely sensitive to T2 RNAase (which preferentially attacks ApA linkages).

The RNA transcribed from adenovirus DNA therefore contains a poly-A sequence similar to that of cellular RNAs. But the poly-A does not hybridize with denatured adenovirus DNA. This suggests that the poly-A is derived from a cellular source and is added to the viral coded RNA after its transcription has been completed. About 65–80% of adenovirus specific RNA is retained on millipore filters under conditions in which poly-A binds—this is the same proportion bound in uninfected cells. Bearing in mind the susceptibility of this assay to artefacts introduced by cleavage, this probably means that poly-A is attached to all nuclear RNAs before they are processed to yield messenger RNA. Polysomal adenovirus RNA contains about 6–8% poly-A whereas nuclear adenovirus RNA has about 2%; these proportions are similar to those of uninfected cells. Experiments in which Wall, Philipson and Darnell (1972) labelled adenovirus infected cells with H^3-A at different times after infection showed that poly-A is implicated in nuclear processing at all times. Nuclear RNAs assayed by hybridization show both restricted sequences and parts which are transported to the cytoplasm; poly-A may therefore play the same role in attaching to messenger sequences of Hn RNA which it appears to play in uninfected cells.

It is not possible to perform similar hybridization experiments to follow the transcription and maturation of RNA molecules in the uninfected Hela cell; the advantages of the adenovirus system is that sequences coded by the virus can be identified, but the greater complexity of cellular DNA does not allow individual cellular messengers to be assayed. However, several indirect lines of evidence suggest that cellular Hn RNA is also synthesised as a molecule to which a length of poly-A is added after transcription. Two models can be postulated for this addition. A length of poly-A may be synthesised separately—either by transcription of a dT sequence in the genome or in some other way—and then added to the completed Hn RNA; or an enzyme may add residues of adenine to the Hn RNA one at a time until a length of about 200 is achieved.

The synthesis of poly-A appears to have characteristics different from those of Hn RNA synthesis and may therefore be catalysed by a different enzyme; it seems likely that the adenine residues are added gradually to the completed Hn RNA since no pool of free poly-A is present in the nucleus. A poly-A polymerase which may undertake this action has been isolated by Winters and Edmonds (1973a, b) from calf thymus nuclei; this enzyme is activated by Mg^{++} ions, has a specific demand for ATP as substrate, and depends upon added polynucleotides to which it can add 100–200 AMP residues from the 3'-OH terminus.

A common observation is that concentrations of actinomycin which inhibit synthesis of Hn RNA do not efficiently inhibit production of poly-A. Darnell et al. (1971) labelled Hn RNA with C^{14}-uridine by incubating cells in the presence of low doses of actinomycin which inhibit ribosomal RNA synthesis. The level of actinomycin was then increased to inhibit Hn RNA synthesis and H^3-A was added. In these conditions the synthesis of Hn RNA is suppressed by 96% and that of mRNA by 93%; but the incorporation of poly-A into Hn RNA shows only 60% inhibition and its incorporation into mRNA is inhibited by 75%. The incorporation of poly-A can be stopped completely by longer periods of incubation and higher doses of actinomcyin. This suggests that poly-A is not transcribed as part of the Hn RNA molecule—in which case its synthesis would be inhibited together with that of the Hn RNA—but that it is separately added to Hn RNA after its transcription.

The nucleotide analogue precursors 3'-deoxyadenosine (also known as cordycepin) and 3'-deoxycytidine inhibit RNA synthesis, presumably by incorporation on the end of the growing chain so that no further nucleotides can be added. Cordycepin acts at the level of transcription; Penman, Rosbash and Penman (1970) found that it inhibits synthesis of nucleolar RNA by causing production of molecules of RNA shorter than the usual 45S precursor. Its effect upon nucleoplasmic RNA synthesis can be followed in cells in which nucleolar synthesis has been blocked with low doses of actinomycin; 3'-deoxyadenosine then proves to have different effects upon the synthesis of Hn RNA and mRNA. Figure 5.16 shows that the drug does not inhibit entry of a labelled precursor into Hn RNA, but does inhibit the appearance of the label in cytoplasmic messenger RNA.

That the differential effect of the 3'-deoxyadenosine is exerted on poly-A dependent processing of Hn RNA is suggested by the failure of the analogue 3'-deoxycytidine to have the same effect. Abelson and Penman (1972) found that 3'-deoxycytidine has the same effect on nucleolar synthesis as does 3'-deoxyadenosine; it inhibits production of the 45S precursor. The 3'-deoxycytidine shows the same failure as 3'-deoxyadenosine to inhibit Hn RNA synthesis. But as figure 5.16 shows, 3'-deoxycytidine differs from 3'-deoxyadenosine in that it fails to inhibit appearance of a label in messenger RNA.

This suggests that both analogues are accepted by the nucleolar RNA polymerase, when they are incorporated into rRNA and thus cause premature termination of the growing polynucleotide chain. The nucleoplasmic polymerase which transcribes Hn RNA is presumably able to discriminate against the deoxynucleotides so that its activity is not inhibited. However, 3'-deoxyadenosine must be able to inhibit mRNA production by interfering with the addition of poly-A to Hn RNA; whereas 3'-deoxycytidine has no influence on this process. The failure of a label in Hn RNA to enter messenger RNA when

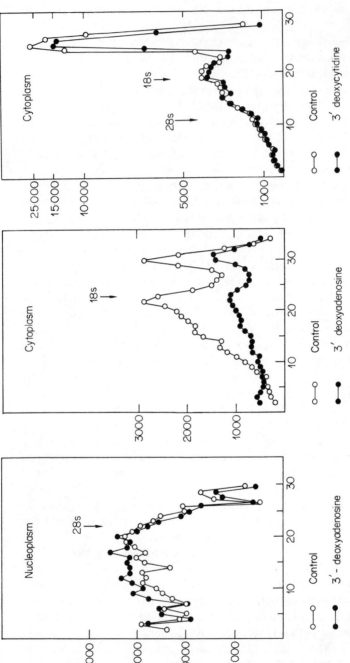

Figure 5.16: influence of 3′ deoxyadenosine and 3′ deoxycytidine upon processing of Hn RNA to messenger RNA. In the left and centre panels one half of the culture of Hela cells served as control and the other was incubated with 25 μg/ml 3′ deoxyadenosine for 30 minutes; in the right panel, half the cells were treated with 3′ deoxycytidine at 25 μg/ml for 10 minutes. All cultures were then labelled with H³-uridine for 30 minutes. Left: 3′ deoxyadenosine does not inhibit synthesis of Hn RNA in the nucleoplasm. (3′ deoxycytidine also fails to inhibit nucleoplasmic RNA synthesis; both drugs also have the same effect upon nucleolar synthesis of rRNA, which they inhibit effectively—data not shown.) Centre: 3′ deoxyadenosine inhibits the appearance of messenger RNA in the cytoplasm. Right: 3′ deoxycytidine fails to inhibit production of mRNA. Data of Penman, Rosbash and Penman (1970) and

poly-A addition is prevented suggests that this step is essential for the maturation of the messenger sequence.

After several hours of labelling, Hn RNA contains poly-A in all size classes, although more of the larger molecules include a sequence; Jelinek et al. (1973) found that at least 30–40% of the largest Hn RNAs labelled for 3–4 hours contain poly-A but as few as 10–20% of the smaller molecules do so. Labelling for only 45 seconds with H^3-A—a time shorter than is needed to synthesise an Hn RNA—also includes molecules of all size classes; poly-A must therefore be added directly to the 3′ termini of available Hn RNA molecules of all sizes. Total RNA synthesis is reduced by 70–80% when cells are treated with high concentrations of actinomycin for 1–2 minutes; but in a subsequent exposure to H^3-A for 1·5 minutes, incorporation of poly-A into Hn RNA is reduced by only 20%. This shows that the synthesis of poly-A is independent from the transcription of Hn RNA and that the Hn RNA molecule remains intact and able to accept poly-A for at least 2 minutes after its synthesis is complete. After incubation with actinomycin for 5–10 minutes, however, there is a large decrease in the incorporation of poly-A, perhaps because the Hn RNA has been degraded or possibly because poly-A has been added to all available Hn RNA molecules.

When nuclear poly-A is labelled with H^3-A and further synthesis is prevented with 3′-deoxyadenosine, the cellular redistribution of the label can be followed. During the next 30 minutes, the label of poly-A in Hn RNA declines and the label in the cytoplasmic RNA fraction increases; although complete equality cannot be demonstrated, it seems very likely that poly-A added in the nucleus is conserved and is transported to the cytoplasm. About 70% of the cytoplasmic poly-A is present in polysomes and about 30% in RNA which is not associated with the ribosomes.

By labelling cells with H^3-U for 20 minutes and then incubating with high doses of actinomycin to inhibit all RNA synthesis, Penman, Rosbash and Penman (1970) found that the addition of 3′-deoxyadenosine together with the actinomycin does not inhibit the appearance of label in cytoplasmic messengers. This is in contrast to the effect of adding the drug before addition of the H^3-U label, in which case no label enters the cytoplasm. This suggests that the Hn RNAs which are labelled during a 20 minute incubation can be processed for transport of messengers to the cytoplasm before the 3′-deoxyadenosine is added; in other words, the stages of maturation in which poly-A is involved occupy less than 20 minutes.

By shortening the labelling time in such experiments to 7·5 minutes, Adesnik et al. (1972) found that the subsequent addition of actinomycin together with 3′-deoxyadenosine causes a 70% reduction in appearance of labelled messenger RNA. The essential step in which poly-A is implicated must therefore be unable to take place within 7·5 minutes of the completion of transcription of an Hn RNA molecule, but can take place within 20 minutes. The labelled molecules which succeed in reaching the cytoplasm during inhibition with 3′-deoxy-adenosine contain some poly-A, although the lengths of the fragments recovered after RNAase digestion are reduced by about half. This shows that the label which enters polysomal messengers in the presence of 3′-deoxyadeno-sine does not correspond to a class of mRNAs lacking poly-A, but represents escape from the inhibition; it is possible for molecules with less than their usual length of poly-A to suffer maturation. These experiments therefore suggest that all Hela cell mRNA ends in a sequence of poly-A whose addition is essential for its cleavage from Hn RNA and transport to the cytoplasm. (Histone mRNAs appear to be the only exception; see below.)

By hybridizing RNA with poly-U in solution, Greenberg and Perry (1972) isolated the molecules containing poly-A by their adsorption to hydroxyapatite columns. The messenger RNA which is retained corresponds to about 90% of the cellular content; the RNAs of the messenger fraction which do not contain poly-A include histone mRNAs and contaminants of ribosomal RNA and the polydispersed RNA which cosediments with the ribosomes. This again suggests that all functional messengers except the histones must contain a 3′ terminal length of poly-A.

Heterogeneous nuclear RNA, however, has a much lower content of mole-cules containing poly-A; more than 80% of the Hn RNA fails to react with poly-U. If the addition of poly-A is essential for conservation of the adjacent messenger sequence, this implies that only 20% of the Hn RNA molecules act as precursors to messenger RNA; the rest are presumably degraded in toto. We do not know how the addition of poly-A to Hn RNA ensures that the adjacent sequence of messenger fails to be degraded and is subsequently transported to the cytoplasm. However, the role of the poly-A may be analogous to the methylation of ribosomal RNA which distinguishes sequences to be conserved from the unmethylated sequences which are degraded; the presence of poly-A may in some way act as a signal to the degradation system, possibly by its association with specific proteins, which protects the adjacent messenger region.

Molecules of Hn RNA appear to have some secondary structure and this might of course also play some role in its maturation. Jelinek and Darnell (1972) found that after ribonuclease treatment the undegraded fragments of Hn RNA include some duplex sequences as well as poly-A. The duplex frag-ments can be isolated by adsorption to hydroxyapatite at low ionic strength and elution at higher ionic strength. Fractions of Hn RNA sedimenting at 50S and 100S both contain about 3% duplex regions; and the sequences recovered from different size fractions of Hn RNA are about the same length. This suggests that each Hn RNA molecule contains a number of short duplex regions in proportion to its overall length.

When the RNA is denatured by boiling in low salt and is then chilled in high salt, the duplex regions reform; this implies that they are intramolecular and represent pairing between complementary regions within a molecule of Hn RNA. (If the duplex regions were formed by pairing between different Hn RNA molecules, they would be unable to reform after this treatment). When the isolated duplex regions are hybridized to DNA, at least half show the characteristics of repetitive sequences. However, we do not know the location of these duplex fragments within the Hn RNA molecule; nor is their formation in vitro proof that they exist in the cell. It is therefore impossible at present to say whether they are implicated in the processing of Hn RNA.

Production of cytoplasmic messenger RNA therefore passes through the four stages shown in figure 5.17 and may occupy about one hour. Transcription of the large precursor Hn RNA appears to take less than five minutes. When transcription has been completed, a sequence of poly-A is added to the 3' end, probably one base at a time by an enzyme distinct from the RNA polymerases; this occupies some 10–15 minutes. Processing of the Hn RNA is the least well characterized part of messenger production, but is probably undertaken by a combination of endonucleases and exonucleases; if poly-A has been added to a molecule of Hn RNA, the 3' region including the messenger is in some way protected against degradation. Processing may take of the order of 10–60 minutes. The cleaved messenger sequence is then transported to the cytoplasm in the form of a ribonucleoprotein where it may associate with the ribosomes and be translated.

That addition of poly-A is essential is shown by the failure of messengers to be transported to the cytoplasm when poly-A synthesis is blocked. But one particularly important problem which has not yet been resolved is whether the role of poly-A is mechanistic or selective. Because only 20% or so of the Hn RNA molecules suffer addition of poly-A, we can define two classes of Hn RNA in the nucleus: those which gain poly-A and donate a messenger sequence for nucleocytoplasmic transport; and those which fail to gain poly-A and are degraded in their entirety. If the addition of poly-A to a molecule of Hn RNA is a matter of chance, the same sequences of Hn RNA must be included in both classes and any individual molecule of Hn RNA will have a certain probability of entering either class. In this case the nucleus restricted RNA should consist only of the 5' regions of the Hn RNA molecules.

An alternative model is to suppose that poly-A is selectively added only to certain Hn RNA sequences; this would mean that different molecules of Hn RNA should enter the totally degraded and partially conserved classes. Nucleus restricted RNA would then consist of the 5' degraded regions of all Hn RNA molecules and in addition the 3' regions of those molecules not selected by the poly-A addition system. If gene expression is controlled at this level, changes in the phenotypic state of a cell which require changes in gene expression might be achieved at the level of selection of messenger sequences for nucleocytoplasmic transport in addition to changes in transcription; in this case some formerly nucleus restricted sequences would enter the cytoplasm.

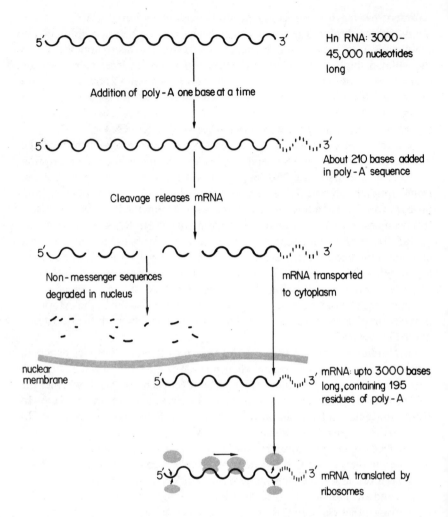

Figure 5.17: maturation of messenger RNA from Hn RNA. The Hn RNA precursor is synthesized from chromatin by the nucleoplasmic RNA polymerase. A sequence consisting only of adenosine residues is added to it, probably by a separate enzyme. The Hn RNA molecule may be degraded in more than one location exonucleolytically, one of these breaks releasing messenger RNA. The fragments of the Hn RNA derived from its 5′ regions are degraded to acid soluble nucleotides very rapidly within the nucleus; the mRNA–poly-A molecule is transported to the cytoplasm. The mRNA and probably also the Hn RNA is in the form of a ribonucleoprotein complex during this time.

Poly-A must play more than one role in the function of messenger RNA; although clearly essential for nucleocytoplasmic transport, its presence in viral messengers found only in the cytoplasm and in mitochondrial messenger translated in situ implies that it also exercises some function in the cytoplasm. It is possible that either or both functions are mediated by the attachment of specific protein(s) to the poly-A sequence. Although poly-A is indispensable for the production of almost all messengers of Hela cells, an alternative pathway for messenger production must exist for histone mRNAs which provide the sole exception. Adesnik and Darnell (1972) found that the 9S histone mRNA fraction of Hela cells leaves little resistant material upon treatment with RNAase—the longest tracts recovered are less than 10 nucleotides long.

If cells are labelled with H^3-uridine for 10 minutes and a high dose of actinomycin is added to inhibit all RNA synthesis, the small polysomes which synthesise the histones can be separated from the other polysomes so that the label which enters histone and other messengers can be compared. Label appears in histone messengers within 5 minutes, when it is essentially complete; the label enters other messengers only after some delay. This suggests that poly-A processing takes some 10–30 minutes and that histones appear in the cytoplasm more rapidly because they omit this step. Either histones must be transcribed as small messengers, or if cleaved from a larger precursor some mechanism other than poly-A addition must be used. Histone messengers are synthesised only during S phase, unlike other messengers which are synthesised continuously through interphase, and it is tempting to speculate that the absence of poly-A is in some way linked to this difference.

Two important qualifications which must be introduced in the model for production of messenger RNA from Hn RNA are that not all Hn RNA molecules may serve as precursors to mRNA and that not all messengers possess poly-A. Indeed, Davidson and Britten (1973) have criticized the concept that Hn RNA serves as precursor to mRNA and have suggested that further experiments are necessary before any firm model for messenger production can be established. The data on which the model is based have been reviewed by Darnell, Jelinek and Molloy (1973).

Certainly, the kinetic relationship between the total Hn RNA population and the total messenger population is not close enough to support the idea that all Hn RNA molecules suffer cleavage to generate an mRNA sequence. However, it is possible that a better kinetic relationship may apply to subpopulations of each of the nuclear and messenger fractions.

The model presented in figure 5.17, however, remains our best working concept for messenger production, although it is likely that at least some of its features may in future be modified and it may not apply to all Hn RNA and mRNA molecules. One alternative model is to suppose that an Hn RNA molecule might carry more than one messenger sequence; following internal cleavage(s), poly-A might be added to the newly generated 3' end(s) as well as to the original 3' end. This would predict that poly-A is added to some older as well as newly synthesized RNA sequences and implies also that some smaller Hn RNA molecules may be intermediates in the processing of larger molecules that represent primary transcripts.

Organization and Transcription of Ribosomal RNA Genes

Synthesis and Processing of 45S Precursor RNA

The genes which code for ribosomal RNA in eucaryotic cells are clustered together on the homologues associated with the nucleolus and, indeed, there has been evidence for some time to suggest that synthesis of ribosomal RNA and assembly of the ribosome takes place within the nucleolus (reviewed by Perry, 1967, 1969). Indirect evidence is that rapid protein synthesis—which demands the production of many ribosomes—is correlated with an increase in the size of the nucleolus; the composition of nucleolar RNA resembles that of ribosomal RNA and the nucleolus contains particles which resemble ribosomes when viewed in the electron microscope. According to Birnsteil et al. (1966) and Vaughan et al. (1967), these particles have the sedimentation properties characteristic of ribosomes and contain RNA very like that of, if not identical to, ribosomes.

Kinetic studies suggest that these particles are precursors to the ribosomes of the cytoplasm and their proteins appear on gel electrophoresis to be similar to, although not identical with, those of ribosomes. Cytological studies suggest that ribosomal RNA is first transcribed at the fibrillar core of the nucleolus, for a newly formed nucleolus is composed largely of fibrillar components and the granular cortex accumulates as the synthesis of ribosomal RNA continues. Das et al. (1971) have suggested that the precursor to ribosomal RNA may be synthesised and processed in the core region, after which it moves to form the granular cortex together with ribosomal and other nucleolar proteins. The precursors to ribosomal RNA are found in the nucleolus in the state of ribonucleoprotein particles; their maturation to give the ribosome-like particles takes place within the nucleolus and it seems likely that the ribosome subunits released into the cytoplasm are identical with mature ribosomes apart from their content of a few proteins which must be replaced by those of the ribosome.

The immediate product of transcription of the ribosomal RNA genes is a large precursor molecule which sediments at about 45S in mammalian cells. Both the major molecules of RNA in eucaryotic ribosomes are produced from this one precursor RNA, which is about 14,000 nucleotides long and is therefore about twice the size of the two mature species together; 28S rRNA consists of about 5000 nucleotides and 18S rRNA of about 2000 residues. In other species of organism, as table 5.1 shows, the precursor is somewhat smaller. The process by which ribosomal RNAs are produced from this precursor has been followed by observing the fate of a radioactive pulse label during its conversion through the various stages of maturation.

Mature ribosomal RNA is methylated and the pattern of methylation on the 2'-O position of ribose—which comprises some 80% of the total methylation—shows that the base methylation frequencies do not correspond to the base composition frequencies. This suggests that methylation is a specific rather than random process. By pulse labelling with C^{14} methionine, Greenberg and Penman (1966) demonstrated that the methylation of Hela cell rRNA takes place on the nucleolar 45S precursor. This accounts for all the methylation of ribosomal RNA, apart from a secondary methylation which Zimmerman (1968) showed to take place later to generate dimethyl-adenine in 18S rRNA.

The early stage at which this modification takes place allows the maturation of ribosomal RNA precursors to be followed by the use of radioactive methyl groups; Greenberg and Penman (1966) and Zimmerman and Holler (1967) used labelled methionine, which enters 6-methyl-adenine and 2'-O-methylcytidine. At 10 minutes after incorporation, only the 45S RNA in the nucleolus is labelled. Its specific activity declines rapidly after 40 minutes, when a 32S nucleolar RNA reaches its maximum labelled level. This in turn then declines and the label in 28S rRNA reaches a maximum after 80 minutes; 18S rRNA seems to reach the cytoplasm somewhat more rapidly.

This suggests two possible schemes for the synthesis of ribosomal RNA. The 45S precursor might contain the sequence of both 28S and 18S rRNAs, as well as a large amount of surplus material which must be degraded when the mature molecules are cleaved from the precursor. An alternative model—which proved difficult to exclude conclusively—is to suppose that there might be two types of RNA precursor, which by chance both happen to sediment at 45S. But hybridization experiments and direct sequence analysis suggest that the 45S RNA molecules are homogeneous, each molecule containing the sequences for both 28S and 18S rRNA. When Amaldi and Attardi (1968) and Willems et al. (1968) compared the base compositions of these molecules, they found that the 45S and 32S precursors have a rather higher content of G and C than the mature ribosomal RNA species. This implies that the fraction of RNA which is discarded during maturation of the precursor must be extremely rich in G–C content—of the order of 75–77%.

By separating nucleolar RNAs by gel electrophoresis, Weinberg and Penman (1970) were able to show that the 45S and 32S precursors which are readily found are not the only intermediates implicated in maturation. When Hela cells are infected with polio virus, processing of precursor RNAs in the nucleolus is inhibited so that they accumulate. Figure 5.18 shows that radioactive labels given under these conditions enter other species also. As the 45S precursor is cleaved to successively smaller intermediates, the ratio of the H^3 label in methyl groups to a P^{32} label in the nucleotides of the RNAs

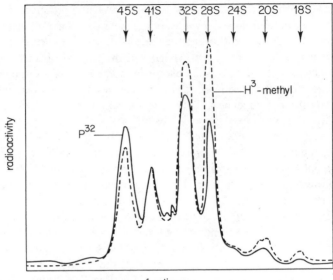

fractions

RNA	45S*	41S*	32S*	28S†	20S*	18S†	units
weight	4·1	3·1	2·1	1·65	0·95	0·65	× 10⁶ daltons
length	14,000	10,500	7000	5500	3000	2000	nucleotides
GC content	70		70	67		58	per cent
methylation	100	94	68	67	36	39	relative number of groups

Source: * nucleolus; † cytoplasm.

Figure 5.18: precursors to ribosomal RNAs in Hela cells. The upper curve shows an electrophoretic separation of precursor and ribosomal RNAs, identifying the molecules by a P^{32} label in their bases and an H^3-methyl label in modified groups. The proportion of methyl groups/unit length (the $H^3:P^{32}$ ratio) increases as the precursors are cleaved. The lower table shows the base compositions and methyl group content (relative to 45S RNA) of the precursors and mature rRNAs. The sequences which are lost during maturation appear to be rich in guanine and cytosine; the methyl groups appear to be conserved since the total present in 28S and 18S rRNAs is the same as that of the 45S RNA. Data of Weinberg and Penman (1970) and Jeanteur and Attardi (1969).

increases; this implies that regions of the molecule which do not contain methyl groups are lost, thereby increasing the relative proportion of the methyl groups which were present on the 45S RNA.

Calculating the relative number of methyl groups on each intermediate, taking the 45S RNA as 100, shows that the 28S and 18S rRNAs together have the same number; this suggests that it is the methylated regions of the 45S precursor which are destined to become the mature ribosomal RNAs, and the

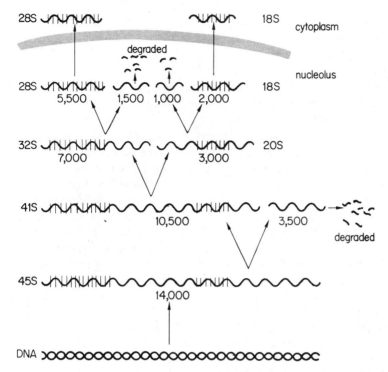

Figure 5.19: Maturation of ribosomal RNA precursors in Hela cells. A series of cleavages leads to the degradation of about half of the original precursor molecule—the figures on each chain represent its approximate length in nucleotides. The methylated regions—indicated by the short cross lines—are conserved from attack. The three fragments which are presumably released from precursor molecules during maturation must be degraded rapidly because they cannot be isolated. The order of the ribosomal and degraded parts of the precursor molecule is arbitrary, except that 28S is at the 5′ terminus of 45S RNA of Novikoff Hepatoma cells and is therefore probably located in the same position in the Hela cell precursor. False cleavages sometimes take place to generate 36S and 24S RNAs which do not seem to lie on the usual pathway of maturation (not shown). The precursor and ribosomal RNAs are not free in the cell as depicted here but are associated with proteins during maturation. Based on data of Weinberg and Penman (1970).

non methylated regions which are discarded. Methylation may therefore serve to indicate to the enzymes which cleave the larger precursors into ribosomal RNAs which sequences are to be preserved and which are to be lost.

The equalities of the numbers of methyl groups found on the other precursors suggest the maturation scheme illustrated in figure 5.19. The 41S precursor appears to have the same number of methyl groups as 45S RNA; and the model suggests that it represents cleavage of a sequence from the 3′ end of the first precursor (see below). The 32S RNA and the 28S rRNA each contain the same number of methyl groups; likewise the 20S RNA and 18S rRNA each contain the same number. This suggests that the 41S RNA is split into a 32S precursor to the large ribosomal RNA and a 20S precursor to the small ribosomal RNA. Each cleavage involves loss of non-methylated sequences. The 32S and 20S precursors are finally cleaved into ribosomal RNAs which are then exported to the cytoplasm.

None of the non-ribosomal cleavage products has been isolated, which suggests that they may be rapidly degraded after their loss from the maturation pathway. Cleavage appears to be accurate on the whole, but two species of RNA present in the nucleolus, sedimenting at 36S and 24S, do not seem to have appropriate sizes or methyl group content to lie on the maturation pathway; they may represent mistaken cleavages at incorrect sites in the precursors. That errors in processing may result from mistakes in methylation is suggested by the observation of Caston and Jones (1972) that aberrant processing of rRNA precursors in Rana pipiens occurs in conditions of under methylation.

The processing scheme has been confirmed by sequencing studies to characterize the oligonucleotides produced by ribonuclease digestion of the different precursors and ribosomal RNAs. Using a radioactive methyl group label to identify fragments by autoradiography, Maden, Salim and Summers (1972) found that the fragments produced by digesting 45S RNA with T1 ribonuclease are virtually identical to the sum of the fragment patterns of 28S and 18S rRNAs, as shown in part *a* of figure 5.20. This indicates that the 45S precursor carries the sequences of both mature ribosomal RNAs and implies that virtually none of its non-ribosomal regions are methylated.

Part *a* of figure 5.20 shows that the fragment pattern of 41S RNA is virtually identical with that of 45S RNA; this confirms that 41S RNA is a precursor to both ribosomal RNAs and must be derived by loss of one end of the molecule which is not methylated. As part *b* of the figure shows, the 32S RNA has the same pattern as that of 28S rRNA; and 20S RNA, although slightly contaminated with 28S rRNA, displays the same fingerprint pattern as that of 18S rRNA.

A total of about 110 methyl groups are introduced into 45S RNA and Salim and Maden (1973) reported that all except 5–6 represent methylations of ribose. Most of these modifications are made to different nucleotide sequences.

In addition to the methylations of the 45S RNA, about 6 base methylations are made to 18S rRNA after its cleavage from the precursor. Part *c* of figure 5.21 shows the result of an experiment in which only labelling at a later stage of maturation was permitted, by the use of actinomycin to block synthesis of new 45S RNA molecules. A notable difference between 18S rRNA and its precursors is that one ribonuclease fragment spot is labelled very strongly in mature 18S rRNA but is much fainter in 20S, 41S and 45S rRNAs; this is the prominent spot of the late methylation pattern of part *c* of the figure. This represents a sequence in which four methyl groups are introduced to generate two adjacent dimethyl-adenines.

When fragments of 45S RNA are identified by means other than methyl group content, the extra sequences which represent the non-ribosomal RNA regions are revealed. Jeanteur, Amaldi and Attardi (1968) found that many additional fragments are present in 45S RNA digested with ribonuclease; and Birnboim and Coakley (1971) found that some of the tetra-, penta- and hexanucleotides which survive this treatment are characteristic of 28S rRNA whereas others identify sequences of 18S rRNA. Although the 45S precursor RNA contains both sets, the 32S RNA contains only those characteristic of 28S rRNA, which is consistent with the cleavage scheme of figure 5.19. Choi and Busch (1970) and Seeber and Busch (1971) reported that the many nucleotide fragments common to 45S, 32S and 28S RNAs include the 5' terminal sequence. This implies that the sequence of the 28S rRNA must be located at the 5' end of the 45S molecule.

Competition hybridization experiments also suggest that 45S RNA contains one sequence of 28S rRNA and one sequence of 18S rRNA. When Jeanteur and Attardi (1969) tested the competition for Hela cell DNA between 45S RNA and 28S or 18S rRNA, they found that both sequences compete; the sequences of 28S rRNA seem to occupy some 35% of the 45S precursor and those of the 18S rRNA correspond to another 13% of the molecule. The remainder is not ribosomal. There is no competition between 28S and 18S rRNAs, which must therefore comprise quite different sequences. Although 32S RNA and 18S rRNA do not compete with each other, 32S RNA competes with 28S rRNA to an extent which suggests that some 70% of the 32S RNA consists of sequences of 28S rRNA.

This maturation scheme appears to be similar for all mammalian cells; and a similar model has been proposed for the Novikoff hepatoma system by Quagliarotti et al. (1970), with the difference that the RNA molecules are slightly smaller and have lower sedimentation values. The initial 45S precursor is about 11,800 nucleotides long and is cleaved to a 35S precursor of 7400 nucleotides and an 18S rRNA of 1600. The 35S precursor in turn matures to a nuclear 28S RNA of 4900 nucleotides, which is the direct precursor of the cytoplasmic 28S rRNA of 3900 bases. It is not certain whether the sequence of cleavages is precisely the same in all cells, but similar steps ensure the loss of about half of all mammalian precursors. In other eucaryotic cells, the maturation process is more economical for as table 5.1 shows the initial precursor is smaller so that less of it is degraded. The identification of sequences which are to be conserved by methylation is not unique to the processing of ribosomal RNA, for DNA sequences of bacteria are identified in a similar manner (see chapter 11 of volume 1).

Maturation of the 28S and 18S rRNA molecules from the 45S precursor must implicate both endonuclease and exonuclease activities. Since the 28S

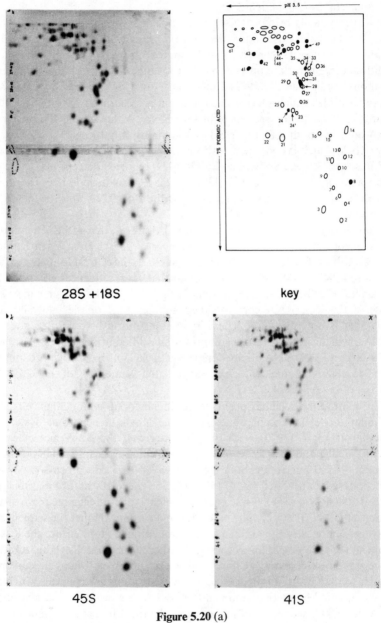

28S + 18S

key

45S

41S

Figure 5.20 (a)

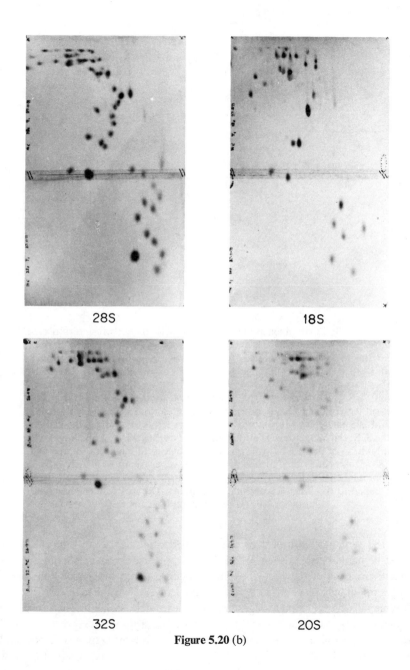

28S

18S

32S

20S

Figure 5.20 (b)

Figure 5.20 (c)

Figure 5.20: T1 ribonuclease digests of nucleolar precursors and ribosomal RNAs. The molecules were labelled with C^{14} methyl groups and the nuclease fragments identified by autoradiography. (a) A mixture of 28S and 18S rRNAs has the fingerprint pattern shown above; the key marks the spots unique to 18S rRNA in black. The lower row shows that the 45S and 41S precursors have the same pattern as the 28S and 18S rRNAs together. (b) Comparison of the isolated 28S and 18S rRNAs with the 32S and 20S nucleolar precursors. The 28S fingerprint pattern is the same as the 32S; the 18S pattern is almost the same as the 20S. (c) Labelling during inhibition of 45S RNA synthesis with actinomycin identifies fragments of 18S rRNA which are labelled after its cleavage. The most heavily labelled spot is labelled extensively in the 18S fingerprint of part *b*, but is only lightly labelled in the 20S, 41S and 45S precursors. Data of Maden, Salim and Summers (1972) and Salim and Maden (1973).

and 18S sequences are part of one polynucleotide chain, an internal break must be made between them when the 32S precursor separates from the 20S precursor. Loss of the non-ribosomal sequences from the various precursor stages might take place either by an endonucleolytic cleavage to release discrete fragments which are then degraded by an exonuclease (as implied in figure 5.19); or an exonuclease might attach to the appropriate precursor and remove only the unmethylated end from it. Three different non-ribosomal sequences of the 45S precursor must be removed in toto but we do not know which mechanism applies to each of these events.

No intact non-ribosomal fragments have yet been isolated, but Inagaki and

Table 5.1: molecular weights of ribosomal and precursor RNAs. The size of the ribosomal RNAs themselves remains fairly constant in eucaryotic cells, although the 28S rRNA is largest in mammals and birds, intermediate in size in reptiles, amphibians and fishes, and smallest in insects and plants. The large 45S precursor is found only in mammals and birds in which almost half of it is degraded during maturation of the ribosomal RNAs. Other organisms have a much smaller precursor, much less of which is degraded. Data of Perry et al. (1970) and Loening, Jones and Birnsteil (1969).

species	nucleolar precursor RNAs		ribosomal RNAs		% conserved of 45S RNA	organism
	first (45S)	second (32S)	28S	18S		
mammal	4.10×10^6	2.1×10^6	1.65×10^6	0.65×10^6	56	Hela cell (human)
"	4·19	2·16	1·70	0·65	56	mouse
bird	3·92	1·98	1·61	0·63	57	chicken
reptile	2·74	1·58	1·51	0·62	78	iguana
amphibian	2·76	1·65	1·58	0·61	79	frog
fish	2·70	1·60	1·55	0·65	81	trout
insect	2·85	1·60	1·40	0·65	72	*Drosophila*
plant	2·76	1·50	1·29	0·66	71	tobacco

Busch (1972) have identified some of the sequences of 45S and 32S precursors which are absent from mature rRNA; if maturation involves a series of discrete cleavage events, it should prove possible to identify some of the sequences in small unstable RNA molecules of the nucleolus. Enzymes of the required specificity have not yet been identified, although Prestayko et al. (1972) found an endoribonuclease activity in 78S ribonucleoprotein particles which contain rRNA precursors; this is consistent with the idea that cleavage takes place in the state of a precursor RNP. Perry and Kelley (1972) identified an exoribonuclease of mouse nuclei which degrades the 3′ ends of 45S and 32S precursors very rapidly, but then functions more slowly when it reaches a methylated region. The enzyme is much less active with 28S and 18S mature rRNAs. This suggests that both 45S and 32S precursors have a 3′ sequence which is unmethylated; it is some 600 nucleotides long in the 45S molecule. It is possible that this enzyme removes these sequences specifically, but since the enzyme appears to be located in the nucleoplasm rather than the nucleolus it is unlikely that it functions as an activity specific for ribosomal RNA maturation.

Although different criteria may be used to distinguish conserved and degraded sequences of precursor RNAs in the nucleolus and nucleoplasm, maturation in each part of the nucleus involves degradation of extensive regions of the precursor. About half of the 45S ribosomal precursor is lost; the proportion of Hn RNA which is lost during messenger production is more variable but is if anything larger than this. Maturation of eucaryotic RNA is therefore a very wasteful process and we do not know what purpose it serves in cellular metabolism; its occurrence is peculiar to the eucaryotic cell, for in bacteria messenger RNA is synthesised directly and ribosomal and transfer RNAs require only a rather limited amount of processing (see chapters 4 and 5 of volume 1). Development of the apparently uneconomical processing of the nucleus must therefore lie in the evolution of eucaryotic cells.

Assembly of Ribosome Subunits

The protein components of eucaryotic ribosomes are not well enough defined to allow the assembly of the ribosome to be followed in the detail possible in bacteria, but precursor particles which contain various precursor and ribosomal RNAs can be identified in the nucleolus. The proteins of the ribosome must presumably be synthesised on polysomes in the cytoplasm, after which they must enter the nucleolus to associate with ribosomal RNA, Hela cells appear to contain a pool of ribosomal proteins, for Warner et al. (1966) found that stopping protein synthesis by the addition of cycloheximide does not immediately prevent ribosome assembly; some ribosomal RNA can still be synthesised and finds its way into mature ribosomes in the cytoplasm. But the overall rate of production of ribosomes in Hela cells is linked to protein synthesis, for Pederson and Kumar (1971) found that the addition of cycloheximide decreases the rate of appearance of 45S RNA and lengthens the time taken for it to mature to the 32S precursor; this slows the formation of ribosomes but maintains the usual balance of 45S : 32S RNA in the nucleolus.

When Hela cells are incubated in a hypertonic medium, protein synthesis declines slowly over the 1·5 hours after the transfer, rather than stopping abruptly as it does when inhibited by cycloheximide. This allows the fate of the precursors to be followed more readily. During the first 60 minutes after transfer to hypertonic medium, transcription of 45S RNA continues at the usual rate, but its half life is extended from the usual 6·5 minutes to 16·0 minutes. Ribosomal RNA which is pulse labelled during this incubation appears only in the form of 45S precursor molecules in nucleolar particles which sediment at 80S. In normal medium, the label enters 80S particles and then shifts into particles sedimenting at 55S. Hypertonicity causes a substantial reduction in pulse labelled protein in both 80S and 55S particles; buoyant density measurements show that these precursors have less protein content than in normal cells, so any pool of ribosomal proteins must be used up rather rapidly, or its rate of use must be linked to the entry of new proteins.

The processing of ribosomal RNA seems to take place entirely in these nucleoprotein particles and the precursors are never found in free form. According to Yoshikawa-Fukada (1967) the 70–90S particle which contains the 45S RNA of mammalian FL cells has some of the proteins which are characteristic of ribosomes. Kumar and Warner (1972) found that the 80S particles in Hela cell nucleoli contain 45S RNA, 5S RNA and proteins similar to those found on cytoplasmic ribosomes. The 55S particles contain 32S RNA, 5S RNA and most of the proteins found in the 60S subunit of the ribosome. Other nucleolar proteins are also found in this particle; these proteins seem to be restricted to the nucleolus and they may be maturation proteins which help the assembly of successive ribosomes.

The proteins of the 80S and 55S precursors have been compared with those of the ribosome subunits by Shepherd and Maden (1972), who labelled Hela cells with S^{35} methionine and then digested the proteins of precursor or ribosome particles with trypsin and characterized the radioactive tryptic peptides. Although the pattern of the 55S precursor is very similar to that of the 60S subunit, the 80S particle contains both 60S proteins and also some characteristic of the 40S subunit, although those of the small subunit show only about one third of the labelling of those of the large subunit. This implies that some of the proteins of each subunit associate with the appropriate part of the 45S precursor before it is split into precursors to each individual ribosomal RNA. The particle which precedes the 40S subunit has not been isolated, but the 55S particle seems to be a direct predecessor of the 60S subunit. Modification and cleavage of the precursor RNAs must therefore take place whilst they are associated with proteins; and the effect of inhibiting protein synthesis on maturation of the 45S precursor implies that this association may play some part in regulation of the maturation process.

Arrangement of Genes for Ribosomal RNA

The amount of DNA in the genome which codes for ribosomal RNA is usually redundant, for there are about six genes for each major ribosomal RNA in E.coli and up to several hundred in the cells of higher organisms as table 5.2 shows. So far as we know, all the genes coding for each of the RNA species are identical or at least very closely related so that there is little or no heterogeneity amongst the RNA molecules in the ribosome subunits. In Hela cells, as McConkey and Hopkins (1964) found, the DNA complementary to ribosomal RNA is concentrated in the chromatin associated with the nucleolus. When Huberman and Attardi (1967) developed a method for fractionating isolated metaphase chromosomes of Hela cells by sedimentation at low velocities through glycerol-sucrose density gradients, they were able to purify DNA from the different chromosome fractions. Hybridization with ribosomal RNA revealed that only the smaller chromosomes—which are associated with the nucleolus—contain sequences complementary to 18S and 28S RNA. By contrast, the remaining RNA—including the presumptive messengers—hybridizes more or less uniformly with chromosomes of all classes.

The genes coding for ribosomal RNA are usually located in one large block. The *anucleolate* mutant of the frog Xenopus laevis is a recessive lethal which lacks a nucleolus altogether; heterozygotes have only one nucleolus compared with the two found in normal diploid animals. Brown and Gurdon (1964) discovered that the mutation does not allow any synthesis of 18S and 28S RNA, although other species of RNA are synthesised as normal. Wallace and Birnsteil (1966) showed that DNA from homozygous mutant animals anneals very poorly to ribosomal RNA, whilst the DNA from heterozygous animals anneals at a level between that of the homozygous mutant and wild type. This implies that the anucleolate deletion of the nucleolar region removes all of the genes of X. laevis which code for both 18S and 28S rRNA. Knowland and Miller (1970) have been able to isolate mutants which have partial deletions of this region and as a result form smaller nucleoli.

A similar situation exists in Drosophila melanogaster, in which *bobbed* mutants, representing deletions in one restricted region of the X chromosome, lack some or all of the genes which code for both the major ribosomal RNAs. By using a range of mutant flies derived from a mutant containing an inversion of the nucleolar region, Ritossa and Spiegelman (1965) showed that the amount of DNA present which hybridizes with ribosomal RNA is proportional to the number of doses of the nucleolar region. It is interesting that Hallberg and Brown (1969) found that the deletion of the nucleolar region in Xenopus also results in the failure of cells to synthesise some, if not all, of the ribosomal proteins. Because the deletion appears to cover only the ribosomal RNA genes, this implies that some mechanism may exist to link the synthesis of ribosomal proteins to that of rRNA.

Table 5.2: Number of genes coding for transfer and ribosomal tRNAs in different organisms. Most, if not all, of the 18S, 28S and 5S genes appear to be identical because there is little heterogeneity in their RNA products. Data of Hatlen and Attardi (1971), Tartof and Perry (1970), Spadari and Ritossa (1970), Brown and Weber (1968a), Brown, Wensink and Jordan (1971).

species	number of genes coding for			size of genome	organism
	tRNAs	18S/28S rRNA	5S rRNA		
bacterium	50	6	7–14	$2 \cdot 8 \times 10^9$ daltons	E.coli
"	42	9–10	4–5	$3 \cdot 9 \times 10^9$	B.subtilis
yeast	320–400	140	—	$1 \cdot 25 \times 10^{10}$	S. cerevisiae
insect	860	180–190	195–230	$1 \cdot 2 \times 10^{11}$	D. melanogaster
frog	1150	450	24,500	$1 \cdot 8 \times 10^{12}$	X. laevis
mammal	1310	280	2000	$3 \cdot 1 \times 10^{12}$	Hela (human)

Because the ribosomal RNA of Xenopus laevis has a greater content of G and C than the average of the DNA, the regions of the genome which code for rRNA have a buoyant density which differs from that of the bulk of the DNA. After fragmenting the chromosomal DNA, these segments can be separated by centrifugation in CsCl density gradients. The different fractions separated on the gradient can be recovered and then denatured to give single stranded DNA which can be tested for its ability to hybridize with rRNA. Brown and Weber (1968a) were able to show that the DNA which hybridizes with ribosomal RNA is located in a separate band of buoyant density, corresponding to a G–C content of about 67%, whereas the bulk of the DNA bands at a density corresponding to only 40% G and C. The ribosomal RNA band is absent from the DNA of the anucleolate mutant.

By fragmenting the DNA preparations to a specified range of molecular weights, it is possible to test whether regions homologous to 18S and 28S rRNA are present on the same molecule of DNA. If genes representing the two types of RNA lie in two contiguous blocks, one for 28S rRNA genes and one for 18S rRNA genes, there should be molecules of fragmented DNA which bind to both ribosomal RNAs; for under the conditions used, blocks of 400 genes should be broken into many fragments. But if the two types of gene alternate, then any stretch of DNA longer than a 28S rRNA gene should hybridize to both types of rRNA. These are extreme models, of course, and intermediate structures can be pictured; it is also possible that regions of DNA non-homologous to either ribosomal RNA are interspersed in the region which codes for rRNAs.

The ability of these fragments of DNA to hybridize with both 18S and 28S rRNA has been tested by Brown and Weber (1968b) and by Birnsteil et al. (1968). The 18S and 28S rRNA molecules have different GC/AT ratios of 57% and 65%—distinct from the average of the genome, 40%—so that the cistrons of DNA which code for each ribosomal RNA should have different buoyant densities, both greater than that of the non-ribosomal DNA. As figure 5.21 shows, the buoyant density of the DNA hybridizing with ribosomal RNAs depends upon the size of the fragments to which it is degraded. When the broken pieces of DNA are large (6–7 × 10^6 daltons), both 18S and 28S rRNA hybridize with DNA characterized by the same position on the density gradient. The DNA fragments corresponding to the two ribosomal RNAs show different densities only when the fragments are smaller (530,000 daltons), on average, than the genes themselves.

Another assay is to look at the buoyant densities of the DNA–RNA hybrids made by annealing either of the ribosomal RNAs to denatured DNA. The results obtained are analogous to those which measure the density of the DNA itself; by either criterion, at high molecular weight the fragments are always homogeneous and the same DNA molecules respond to both 18S and 28S rRNA. This means that there are no blocks of genes for either 18S or 28S

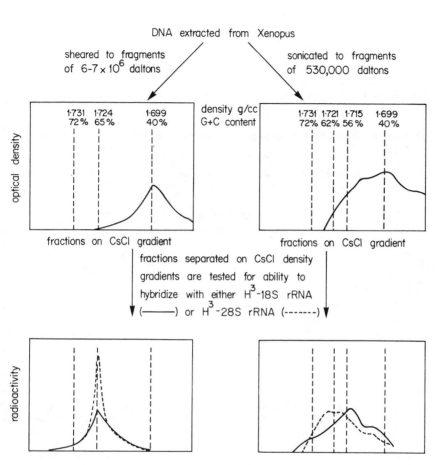

Figure 5.21: ability of fragmented Xenopus DNA to hybridize with ribosomal RNAs. Upper: after fragmentation to different sizes—6–7 × 10⁶ daltons on left, 530,000 daltons on right—regions of different buoyant density can be isolated by centrifugation through CsCl. Fragments of greater buoyant density are released from their associated sequences of lower buoyant density more effectively as the fragment size is reduced. Lower: large fragments (left) only of 1·724 g/ml can hybridize with rRNAs. The genes for rRNA are therefore located in a part of the genome with the atypically high G–C content of 65%. Fragmentation to a small size (right) separates the peaks binding 28S and 18S rRNAs into buoyant densities of 1·721 and 1·715 g/ml. The decrease in buoyant density of small fragments binding rRNA from that of the complementary large fragments shows that the larger fragments contain non-ribosomal RNA regions rich in G–C.
Data of Brown and Weber (1968b).

DNA extracted from Xenopus

↓

fragmented to size shown below and denatured

pre-hybridized with 18S rRNA   pre-hybridized with 28S rRNA

↓   ↓

separated by density on CsCl gradient   separated by density on CsCl gradient

↓   ↓

fractions hybridized with H³-18S rRNA   fractions hybridized with H³-18S rRNA
(——) or H³-28S rRNA (·····)   (——) or H³-28S rRNA (·····)

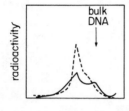

fragment size of
denatured DNA

3×10^6 daltons

length of 2 28S genes
     or 4 18S genes

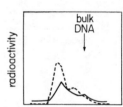

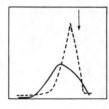

$1·5 \times 10^6$ daltons
length of 1 28S gene
     or 2 18S genes

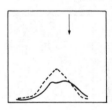

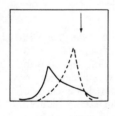

$0·6 \times 10^6$ daltons
length of ½ 28S gene
     or 1 18S gene

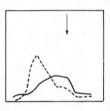

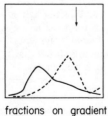

$0·2 \times 10^6$ daltons
length of ¼ 28S gene
     or ½ 18S gene

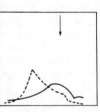

fractions on gradient   fractions on gradient

rRNA. The rDNA must always be fractionated to about the size of the cistron itself to release segments of DNA which will bind only 28S or 18S rRNA and whose buoyant densities—measured either as duplex DNA or as DNA–RNA hybrids—are different.

Linkage experiments, which measure the effect of hybridization with one of the RNAs on the buoyant densities of the DNA molecules which can bind rRNAs, also show that the 18S and 28S rRNA genes are intimately linked. Ribosomal DNA can be separated from the main peak of denatured DNA by hybridization with rRNA; RNA–DNA hybrids have a greater buoyant density than the unhybridized DNA fragments and so should form a separate peak. After degrading the DNA to a specified size, it can be denatured and hybridized with either 18S or 28S rRNA. The preparation is then centrifuged through a CsCl gradient and the different fractions are tested for their ability to hybridize with 28S and 18S rRNA. Figure 5.22 shows that when the fractions are small—$0 \cdot 2$–$0 \cdot 6 \times 10^6$ daltons—prehybridization with 18S (or 28S) rRNA moves the complementary DNA to a heavier position on the gradient but does not change the position of the DNA fractions hybridizing with the other rRNA. But when the DNA fragments are larger ($1 \cdot 5$–3×10^6 daltons), the fractions which hybridize with one class of rRNA also hybridize with the other type. In other words, when the DNA is small, prehybridization with either type of rRNA separates the complementary DNA from the rest of the DNA, that is to say, it isolates one class of rRNA genes. When the DNA is larger, this technique cannot separate the two types of rRNA.

These results show that the two types of rRNA gene must be intimately linked, for DNA fragments must be broken to the order of the length of the genes themselves before either type of gene can be separated from the other. The most likely arrangement of rRNA genes to fit with these conclusions is for sequences of 18S and 28S rRNA genes to alternate. Because the buoyant density of the large fragments containing rDNA ($1 \cdot 724$ g/ml) shown in figure 5.21 is greater than that expected of cistrons coding for rRNA alone ($1 \cdot 721$

Figure 5.22: linkage between genes for 18S and 28S rRNAs. Left: at large fragment sizes the sequences which are complementary to 18S rRNA are also complementary to 28S rRNA. Separation of the two classes of sequence begins at a fragment size of $1 \cdot 5 \times 10^6$ daltons and is efficient at a size of $0 \cdot 6 \times 10^6$ daltons. Any sequence of DNA carrying an 18S rRNA gene therefore also carries a 28S rRNA gene unless the DNA is fragmented below the size of the 18S gene. Right: at fragment sizes of 3 or $1 \cdot 5 \times 10^6$ daltons the sequences of DNA carrying a 28S rRNA gene also carry an 18S rRNA gene. Fragments of $0 \cdot 6$ or $0 \cdot 2 \times 10^6$ daltons show separation of the two classes of gene. It is therefore necessary to degrade DNA below the size of a 28S rRNA gene to release the 18S sequences. Data of Brown and Weber (1968b).

and 1·715 g/ml), it is likely that this region includes DNA sequences of a higher G–C content. Each pair of rRNA genes may therefore be separated from the next by a region of about the same length which does not code for rRNA.

The model which this immediately suggests is that genes coding for the precursor RNA—which sediments at about 40S in Xenopus and is smaller than, but equivalent to, the 45S RNA of mammals—are contained in one large block on the chromosome. This arrangement would provide non-ribosomal regions of high GC content between pairs of rRNA genes; but it is also possible, of course, that spacer regions which are not transcribed exist between the precursor genes themselves. The rapidity of the annealing reaction between rRNA and rDNA suggests that all the precursor genes are identical, at least in their rRNA coding regions, so that any 18S (or 28S) rRNA molecule can hybridize with any 18S (or 28S) coding segment.

Visualisation of Active Ribosomal RNA Genes

The arrangement of ribosomal RNA genes on the chromosome has been directly examined by another technique which discriminates between regions of DNA according to their content of G and C. When DNA is partly denatured, either by heating or by alkali treatment, the first regions of the duplex to separate into single strands are those low in GC. The separated strands are prevented from re-annealing by the presence of formaldehyde. Using this technique, Wensink and Brown (1971) have constructed a denaturation map of Xenopus rDNA by electron microscopy. They found repeating structures along each denatured rDNA, the average distance between repeats being $5·4 \pm 0·5$ μ. This gives a repeat length for the DNA corresponding to about $8·7 \times 10^6$ daltons, which seems to fall into two parts, one coding for the precursor RNA and one a "spacer" region between precursor RNA genes.

Electron micrographs of DNA in the throes of transcription suggest that the ribosomal DNA region consists of sequences which code for the precursor RNA alternating with inactive spacer sequences which separate these genes from one another. Miller and Beatty (1969a, b, c) have made use of the extra-chromosomal DNA in frog eggs which is engaged in a very intensive transcription of ribosomal RNA alone. Amphibian oocytes pass through a lengthy growth period which in Xenopus laevis may occupy some six months and in Triturus viridescens may take upto two years. During this period, the nucleolar organiser region of the genome which is associated with the nucleolus and codes for ribosomal RNA is multiplied to produce some 1000 nucleoli within each nucleus. These nucleoli comprise a compact fibrous core (containing DNA, RNA and protein) surrounded by the granular cortex which contains only RNA and protein. By using a solution of very low ionic strength, the core and cortex can be separated and the core DNA dispersed for electron microscopy.

When the cores are completely unwound, each consists of a thin axial fibre

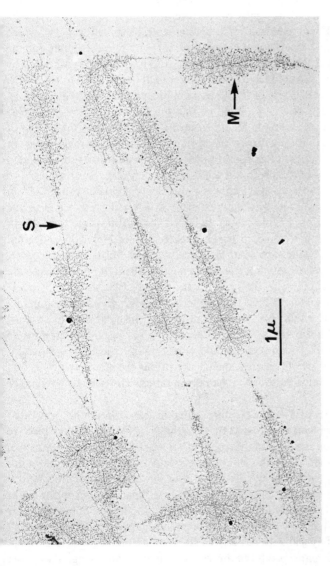

Figure 5.23: Nucleolar genes of *Triturus viridescens* oocytes engaged in transcription of ribosomal precursor RNA. Each longitudinal core axis comprises DNA in deoxyribonucleoprotein form. The axis is periodically coated with a matrix (M) which consists of ribonucleoprotein fibrils 50–100Å in diameter and up to 0·5 μ long, attached to the axis by spherical granules of about 125Å in diameter. Each matrix is some 2–3 μ long and the length of its attached fibrils increases steadily from one end to the other. The successive matrixes along the axis show the same polarity and are separated by genetically inactive segments of DNA whose function is unknown. Each matrix represents a gene engaged in the simultaneous transcription of about a hundred RNA molecules; the shortest of its fibrils is attached to a polymerase molecule that has just initiated transcription and the longest to one that has just completed synthesis of its RNA. Since the maximum length of the fibril is much less than that expected of an RNA corresponding to a gene 3 μ long, the RNA must be coiled within the ribonucleoprotein structure. Data of Miller and Beatty (1969a, b, c); photograph kindly provided by Dr O. L. Miller jr.

some 100–300Å in diameter periodically coated with a matrix as revealed by figure 5.23. The axial fibre of each core forms a circle; treatment with DNAase breaks the core axis and treatment with trypsin reduces its diameter to about 30Å. This suggests that the axis is a single duplex of DNA—the diameter of a DNA double helix is 20Å—coated with protein.

The length of the core axis covered by one matrix segment is 2–3 μ, which agrees well with estimates for the length of DNA required to code for the expected ribosomal RNA precursor molecule; the RNA chain synthesised by a gene of about 2·5 μ should have a molecular weight of about $2·5 \times 10^6$, which is close to the estimated size of the amphibian precursor RNA (see table 5.1). Each matrix consists of about a hundred thin fibrils connected by one end to the core axis, and increasing in length from the "thin" to the "thick" end of the unit. After incubation with a tritiated ribonucleoside, the label appears in the matrix, with kinetics of incorporation which agree well with those found in biochemical studies of the precursor RNA molecule itself.

Each fibril terminates upon the core in a spherical granule of about 125Å in diameter; this must almost certainly represent an RNA polymerase molecule engaged in transcription. This means that about one third of the total length of the gene must be covered in enzyme molecules—a high concentration. Treatment with ribonuclease or trypsin removes the fibrils from the core, implying that they are ribonucleoprotein in composition. Protein specific staining shows that as the RNA molecules are synthesised, newly made portions are immediately coated with protein. An RNA chain corresponding to a gene 3 μ long would be about 6 μ in length; since the maximum fibril length observed (at the "thick" end of the matrix) is 0·5 μ, the chain must be coiled so as to give a roughly twelve fold reduction in its length. The pattern of a hundred fibrils steadily increasing in length along the matrix indicates that successive RNA chains must be sequentially initiated before the completion of earlier chains.

The matrix units of 2–3 μ which are active in transcription are separated from one another by inactive matrix free lengths of DNA. Most of the inactive segments are about one third of the length of the adjacent matrix covered units, but may be upto ten times the length of the active segment. There does not appear to be any pattern to the distribution of the larger matrix-free regions along the axis. The average length of the inter-gene "spacer" segments is about two thirds of the length of the gene, indicating that about 60% of the nucleolar DNA codes for precursor RNA, with the remainder inactive. This agrees well with the results of hybridizations between rRNA and isolated rDNA, although the denaturation maps produced by electron microscopy suggest that the spacer regions are fixed in size between genes and do not vary as they appear to when viewed in this way.

Although the major ribosomal RNAs are represented by genes which are closely linked in one region of the genome, no such pattern is found for the genes which code 5S rRNA. Aloni, Hatlen and Attardi (1971) found that 5S rRNA hybridizes with virtually all of the fractionated metaphase chromosomes of Hela cells and Tartof and Perry (1970) have shown that the genes for 5S rRNA in Drosophila are not linked to the sex chromosomes as are the 18S and 28S rRNA sequences. The large number of genes for 5S rRNA in many species (see table 5.2) is a mystery, for there is no reason apparent why it should be represented many more times in the genome than the major rRNA species. Nor does its synthesis seem to be coordinated with that of the other ribosomal RNAs. The 5S rRNA of Xenopus should be represented by DNA which has a lower buoyant density in CsCl than the bulk of Xenopus DNA, but the difference is not great enough to rely upon to achieve a separation of sequences for 5S rRNA in DNA. Both the major rDNAs and 5S rDNA bind less silver than bulk DNA, however, and 5S rDNA binds much less actinomycin, which Brown, Wensink and Jordan (1971) have used as an isolation method for 5S rRNA genes. They have been able to isolate molecules of DNA which appear to be able to code for many 5S rRNA molecules, so there may be some clustering of these genes in the genome, although they are unrelated to the other sequences of ribosomal RNA.

CHAPTER 6

Control of Transcription

Organization of the Transcription Apparatus

Activities of RNA Polymerases

Transcription of DNA in eucaryotic cells falls into two major categories (considering only the nucleus and excluding transcription in cytoplasmic organelles). The two synthetic activities correspond to the division of the nucleus into the nucleolus and nucleoplasm. The genes which code for ribosomal RNA are associated with the nucleolus, which is devoted to the synthesis and maturation of the rRNA precursor molecules. One RNA polymerase appears to be associated with the nucleolus and engaged in transcription of the precursor rRNA. Another RNA polymerase activity is associated with the chromatin and can therefore be characterized by its presence in nucleoplasmic fractions after the nucleolus has been removed by fractionation. In addition to these two polymerases, a third activity—sometimes characterized as extranucleolar because it appears to be located in the region of the nucleolus although not within it—may synthesise some of the smaller RNAs of the cell such as transfer RNA.

Two types of experiment indicate the independence of the enzyme systems which synthesize ribosomal RNA at the nucleolus and nuclear RNA from chromatin. Drugs such as actinomycin D may discriminate between them by preferentially inhibiting the synthesis of only one type of RNA. And the enzymes which can be isolated from the two fractions of the nucleus have different ionic requirements for activity.

The first identification of eucaryotic RNA polymerases detected enzyme activities firmly bound to chromatin which incorporate the four nucleoside triphosphates into RNA. In this form, however, RNA polymerase is not amenable to isolation and characterization. In attempting to obtain a soluble form of the enzyme, Widnell and Tata (1966) incubated rat liver nuclei in the presence of ammonium sulfate, which stimulates RNA synthesis. (One possible reason for the effect of the salt is that it removes some of the histones from chromatin.) In the presence of ammonium sulfate, RNA synthesis is most active when Mn^{2+} is provided as the divalent cation required by the reaction. But in the absence of the salt, the reaction works best with Mg^{2+}.

One possible explanation for these results is that two different enzyme reactions are responsible for RNA synthesis, one catalysed by an enzyme which is most active in the presence of ammonium sulfate and Mn^{2+} ions, and the other catalysed by an enzyme which is most active in the absence of the salt but requires Mg^{2+} ions. In support of this idea, the RNA synthesis performed under the two different conditions shows different sensitivities to actinomycin; the ammonium sulfate–Mn^{2+} reaction is little inhibited, and the Mg^{2+} stimulated reaction is more inhibited. And the RNA synthesised in the presence of salt and Mn^{2+} has a more DNA-like base composition, whereas the RNA synthesised in the presence of Mg^{2+} appears generally similar in base composition to ribosomal RNA.

That a salt–Mn^{2+} stimulated enzyme is responsible for synthesising RNA from chromatin whilst an Mg^{2+} stimulated enzyme transcribes ribosomal RNA at the nucleolus is suggested also by the autoradiographic observations of Maul and Hamilton (1967). When nuclei are incubated with RNA precursors such as H^3-uridine in the presence of ammonium sulfate and Mn^{2+} ions, most of the grains become incorporated over the chromatin; when Mg^{2+} ions are provided instead, most of the incorporated grains are located over the nucleolus. The simplest explanation of these experiments is that the nucleus contains two enzymes, a salt–Mn^{2+} stimulated activity which is located in chromatin and an Mg^{2+} stimulated activity which is sited in the nucleolus. However, since the experiments are indirect they leave open the possibility that only one enzyme system is involved, but that it has different ionic requirements for activity which depend upon whether it is transcribing RNA from chromatin or in the nucleolus.

The activity in RNA synthesis of extracts of nuclei suggests that the two activities correspond to two different enzymes. Roeder and Rutter (1969) prepared extracts from the nuclei of sea urchin embryos or rat liver, solubilized the RNA polymerase activities by sonication at high ionic strength, and then reduced the ionic strength and centrifuged the chromatin out of the preparation. Fractionation of such preparations with ammonium sulfate reveals three peaks on DEAE-sephadex chromatography of the sea urchin extract (termed I, II and III in order of elution) and two peaks from rat liver nuclei (corresponding to I and II of the sea urchin).

The fractions display different optimal levels of divalent cations for maximum activity. The peak I enzyme activity functions well at low ionic strength (below 0·05) with either Mg^{2+} or Mn^{2+} ions; peak II functions efficiently at higher ionic strength (about 0·1) with Mn^{2+}; and peak III shows a slight preference for Mn^{2+} compared with Mg^{2+}. This means that assay with Mg^{2+} in conditions of low ionic strength measures the activity largely of peak I, but assay with Mn^{2+} at higher ionic strength measures the activity largely of peak II. Ammonium sulfate stimulates the peak II enzyme by increasing the ionic strength, which implies that at least some of its effect on isolated nuclei must be to provide a suitable milieu for the peak II enzyme, even if it also assists its activity by influencing the structure of the chromatin template.

By using these assays to measure directly the relative proportions of the two enzyme activities in different fractions of rat liver nuclei, Roeder and Rutter (1970a) found that peak I is concentrated in the nucleolus and peak II in the nucleoplasm. When peak III is present it is found in the nucleoplasmic fraction. Multiple RNA polymerase activities have now been detected in the nuclei of many eucaryotic cells. Two peaks, corresponding to I and II, are always found; the appearance of other peaks varies—although a third peak has now been detected in rat liver, for example, it is somewhat variable in content. Blatti et al. (1970) and Chambon et al. (1970) have characterized two peaks from calf thymus nuclei; Jacob et al. (1970) have identified the two principal peaks of rat liver. A more recent terminology for these activities is enzyme A (peak I) to describe the nucleolar polymerase and enzyme B (peak II) to describe nucleoplasmic RNA polymerase of chromatin.

None of the eucaryotic RNA polymerase enzyme activities is inhibited by rifampicin, which is a potent inhibitor of the bacterial enzyme (see chapter 6 of volume 1). That there is little similarity between the eucaryotic and bacterial enzymes is suggested also by the effect of α-amanitin, a bicyclic octapeptide isolated from a poisonous mushroom, which inhibits one of the eucaryotic polymerases but does not inhibit the bacterial polymerase. The effect of α-amanitin is specific for the enzyme B of chromatin; it does not inhibit the activity of the nucleolar enzyme A in vitro (nor that of the peak III activity when it is present). This distinction between enzyme A (resistant to amanitin) and enzyme B (sensitive to amanitin) has proved applicable to the in vitro activities of the RNA polymerases extracted from all eucaryotes at present (see also Lindell et al., 1970).

In isolated rat liver nuclei, the RNA which is synthesised in the low salt–Mg^{2+} conditions of assay for enzyme A, in the presence of amanitin, is 62% G + C in base composition; this suggests that it is ribosomal. The activity of peak III can be assayed in high salt–Mn^{2+} conditions if amanitin is present to inhibit enzyme B; the RNA which it synthesises has about 49% G + C. The activity of enzyme B can be deduced by subtracting the activity of enzyme III from the total activity displayed in high salt–Mn^{2+} when no amanitin is present; this RNA has a G + C content of 45%. These results support the concept that enzyme A is nucleolar and transcribes rRNA whereas enzymes B and III are nucleoplasmic and synthesise DNA-like RNA.

Hybridization experiments have confirmed this conclusion by demonstrating that the RNA synthesized in the absence of amanitin (that is in conditions of assay for the enzyme B nucleoplasmic activity) saturates more sites on DNA than the RNA synthesized in the presence of the inhibitor (that is in conditions of assay for nucleolar enzyme A). The RNA synthesized by uninhibited nuclei is competed by RNA extracted from the nucleoplasm but not by rRNA; the RNA synthesized in nuclei inhibited with amanitin is competed well by rRNA.

Amanitin must act on the polymerase enzyme, rather than on the DNA template, because it is equally effective whether added before the start of transcription or ten minutes later; this implies that it must act at some stage of the enzyme action subsequent to initiation. That this is its site of action is suggested directly by the observation of Chan, Whitmore and Siminovitch (1973) that mutant Chinese hamster ovary cells resistant to amanitin may possess an altered RNA polymerase activity which is resistant to the drug in vitro.

The localization of the enzyme activities resistant to actinomycin and amanitin has been confirmed by the autoradiographic analysis performed by Moore and Ringertz (1973). When human fibroblasts are incubated with H^3-uridine, grains are found over both the nucleoplasmic and nucleolar regions. The addition of 0.5 $\mu g/ml$ actinomycin inhibits nucleolar labelling; addition of 5 $\mu g/ml$ of amanitin inhibits incorporation of uridine in the nucleoplasm.

Results have differed on the demand of the enzymes for template DNAs. Some preparations are more active with denatured DNA templates, some show a preference for native DNA. There is a general tendency for enzyme A (nucleolar) to prefer native DNA and for enzyme B (nucleoplasmic) to prefer denatured DNA. Of course, we do not know what condition prevails in nuclear DNA when it is transcribed, but it is usually assumed that the action of the polymerase must involve unwinding the DNA as it does in bacterial systems (see chapter 6 of volume 1). Even in those preparations of enzyme which show greater activity with duplex DNA, however, the specificity of enzyme action lacks the degree of preference for duplex DNA which might be expected.

One possible explanation for the increased activity of eucaryotic polymerases, especially the nucleoplasmic enzyme, with denatured DNA is that some component necessary for transcription of duplex DNA may be lost during preparation. For example, one subunit of the enzyme might be largely responsible for unwinding the duplex; enzyme molecules lacking this component might be able only to transcribe single stranded DNA. Some hints that this may be the reason for the preference of the polymerases for denatured DNA are provided by the isolation from fractions discarded during preparation of factors which increase the activity of the enzyme with duplex DNA. Lentfer and Lezius (1972) found that polymerase B of a mouse myeloma is 2–5 times more active with denatured than with native DNA; but the addition of a factor lost during preparation increases its ability to transcribe duplex DNA. Stein and Hausen (1970) observed that crude preparations of enzyme B of calf thymus prefer native DNA, but a more purified enzyme preparation prefers denatured DNA. Addition of one of the fractions discarded during preparation of the purified enzyme stimulates its activity with native DNA some three fold. The factor seems to bind to the enzyme only at low ionic strength so that in its presence the optimum ionic strength of the enzyme is changed from 0·2 M KCl to 0·02 M KCl. Sugden and Keller (1973) have similarly reported that two factors isolated from material discarded during preparation of RNA polymerase B of Hela and KB cells allow the enzyme to function with duplex DNA as template, an activity it is otherwise virtually unable to undertake.

It is not clear whether the specificity of the A and B enzymes for nucleolar and nucleoplasmic transcription relies largely upon their localization in the nucleus or whether the enzymes themselves recognise specific sequences of DNA. Both parameters probably play some role in controlling transcription. Butterworth, Cox and Chesterton (1971) found that the nucleolar enzyme prepared from rat liver does not transcribe rat liver chromatin to any measurable extent. However, if purified nucleoplasmic enzyme B is added to the chromatin, it supplements the endogenous enzyme activity, which shows that sites are available for transcription by an appropriate enzyme. Bacterial RNA polymerase can also transcribe RNA from the chromatin, although it appears to use different sites from those transcribed by the homogeneous enzyme. The failure of the nucleolar enzyme in transcription therefore implies that some mechanism prevents its recognition of the available sequences of DNA in chromatin. That each enzyme may act on specific recognition sites in DNA is suggested also by the observation of Smuckler and Tata (1972) that the nucleolar enzyme of rat liver transcribes a product from rat liver DNA which is more ribosomal-like than the product transcribed from the same template by the nucleoplasmic enzyme. However, experiments in other systems have failed to demonstrate such specificity. It is likely in any case that the conditions of incubation must approximate the physiological before recognition becomes specific, and of course DNA is packaged in chromatin within the nucleus; transcription in vitro may therefore differ appreciably from that achieved under the conditions prevailing in vivo.

More precise definition of the catalytic activities undertaken by each enzyme requires a better characterization of the protein molecules and identification of any additional factors which may be needed for enzyme activity in vivo. And the nucleolar (A) and nucleoplasmic (B) enzyme activities may each comprise more than a single enzyme entity. Both fractions of calf thymus and rat liver consist of proteins which sediment at about 13S and appear to be some 500,000 daltons in weight. Chesterton and Butterworth (1971) have fractionated polymerase A of rat liver into two fractions of the same molecular weight and catalytic activities, but the relationships of these fractions to each other and their subunit compositions are not known. The A fraction seems to include two predominant subunits of about 200,000 and 135,000 daltons.

Enzyme B of calf thymus has been separated into two fractions, B1 and B2, by gel electrophoresis. Kedinger et al. (1971) reported that the first fraction contains two subunits of molecular weights 215,000 and 150,000 daltons;

the second fraction contains two subunits of 185,000 and 150,000 daltons. The relationships of these subunits are not known. Sugden and Keller (1973) reported a different structure for human RNA polymerase B extracted from Hela or KB cells, which appears to possess three predominant subunits, of 220,000, 140,000 and 35,000 and two further subunits which appear in variable amounts at 170,000 and 25,000 daltons.

Correlations between enzymes found in different eucaryotic species have not in general proved very productive, for there are very few common parameters apart from the general division into A (nucleolar) and B (nucleoplasmic) activities; the same relative responses to ionic levels, actinomycin and amanitin are found in all cells but the physical forms of the enzyme seem to vary widely. We do not know whether this represents a genuine variation in the molecular structure of the enzyme or whether it reflects differences due to the effects of the protocols for enzyme preparation on different cell types. It is possible that the different fractions which have been reported may represent aggregates of one or a very small number of active enzyme molecules, or they may represent enzyme molecules which have lost some of their components. It is therefore impossible to define the molecular organization of the two principal polymerases until the in vivo form of each enzyme has been identified and characterized in different cells.

In addition to the enzyme B activity which appears to transcribe chromatin, the nucleoplasm may also contain another polymerase which is responsible for the synthesis of small RNAs. Zylber and Penman (1971a) found that the effect of amanitin depends upon ionic strength; at high ionic strength the synthesis of nucleoplasmic RNA is 95% inhibited but at low ionic strength there is only 50% inhibition. Zylber and Penman suggested that there are two enzyme activities in the nucleoplasm which respond differently to ionic strength and amanitin. In the presence of amanitin, an increase in ionic strength has little effect on the nucleoplasm because the enzyme activity B which is stimulated by high ionic strength is inhibited by the drug. But the enzyme activity which is not inhibited by amanitin does not respond to ionic strength—it shows much the same level of enzyme activity at either low or high ionic strength. This enzyme might correspond to peak III.

That small RNAs are synthesised by an enzyme distinct from the B polymerase is suggested by the continuation of their synthesis during mitosis, when the principal nucleoplasmic and nucleolar enzyme activities are inactive (Zylber and Penman, 1971b). Synthesis of small RNA also appears to be distinct during infection of Hela cells with adenovirus. The principal product of transcription of the virus is a large nuclear RNA molecule which is processed to give messenger RNAs; its synthesis is sensitive to amanitin and is presumably undertaken by enzyme B. Late during the lytic cycle, however, a small RNA is transcribed; it sediments at 5·5S, is 151 nucleotides long, and is synthesised in large amounts in its mature form (it is not cleaved from a larger precursor). Price and Penman (1972) found that its transcription is not inhibited by amanitin; the transcription of 5S and 4S RNA is also resistant to the toxin. This is consistent with the idea that a second nucleoplasmic polymerase transcribes small RNA molecules.

Transcription of chromatin must be under a specific control which determines that the appropriate sequences are transcribed in each differentiated cell. We do not know whether this control is exercised by the introduction of changes in the activity of the nucleoplasmic polymerase molecule or whether the sequences available for transcription are controlled in a manner such that the only role of the enzyme is the catalysis of RNA synthesis. The nucleolar enzyme, however, presumably has an identical function in all cells since its role appears to be confined to the transcription of ribosomal RNA precursors.

But the relative activities of the two enzymes may be established at a level appropriate for the cell type. By following the pattern of RNA synthesis and enzyme activity during early development of the sea urchin, Roeder and Rutter (1970b) showed that the level of RNA synthesis declines from early blastula to post gastrula stages and that the total polymerase activity declines in parallel. The decline results largely from a decrease in enzyme B and peak III, the nucleolar A enzyme remaining constant in activity. The increase in enzyme A relative to B during this period correlates with the relative increase in rRNA synthesis. Production of the two enzymes may therefore itself be controlled.

The activity of the nucleoplasmic enzyme is independent of that of the nucleolar enzyme, for transcription from chromatin continues when the transcription of ribosomal RNA is prevented by the addition of actinomycin. However, the activity of the nucleolar enzyme may depend upon continuing nucleoplasmic transcription. The effect of amanitin upon transcription appears to be more drastic in vivo than in vitro. With isolated nuclei or enzyme fractions, amanitin inhibits the nucleoplasmic enzyme but has no effect on the nucleolar activity. But Jacob et al. (1970) found that when rats are injected with amanitin, both nucleoplasmic and nucleolar enzymes of the liver are inhibited.

The discrepancy between the in vivo and in vitro effects of amanitin depends upon the interval after addition of the inhibitor before its influence is assessed. Tata, Hamilton and Shields (1972) injected amanitin into rats and assayed the RNA polymerase activities of isolated nuclei. Maximum inhibition of nucleoplasmic RNA synthesis (about 70%) is reached within 15–20 minutes after addition of the toxin; there is less than 30% inhibition of the nucleolar enzyme. The inhibition is reversed rapidly so that the polymerase activities of extracts return to normal after about an hour.

However, a long period of inhibition with amanitin in vivo has a different effect when the activity of the polymerases is assessed in vivo. In these experiments, rats were injected with amanitin and an H^3-labelled precursor of RNA added 10 minutes before the animal was killed; incorporation of the label depends upon the activity of the enzyme in vivo. By this criterion, amanitin continues to inhibit RNA synthesis from chromatin for much longer than when enzyme activities are assayed in vitro; inhibition increases during the first two hours after addition of the inhibitor and is still considerable even after 4 hours. The toxin at first inhibits nucleoplasmic synthesis, but the synthesis of rRNA becomes impaired after a lag of 1–2 hours of uninhibited synthesis. The apparent action of amanitin therefore depends upon whether it is assessed in vivo or in vitro and upon the delay after addition of the inhibitor. Its effect on RNA polymerase activity in vitro is rapid and is confined to the nucleoplasmic enzyme; its inhibition of synthesis continues for longer when assessed in vivo and eventually applies to rRNA synthesis also.

These experiments exclude the possibility that amanitin inhibits one enzyme in vitro but both in vivo because the milieu of the nucleolar enzyme renders it susceptible in vivo but resistant in vitro; nucleolar RNA synthesis is only inhibited after an appreciable lag and is probably therefore an indirect effect of the toxin. This also makes it unlikely that amanitin is converted by the animal into some further metabolite which inhibits nucleolar RNA synthesis. An explanation for the later inhibition of nucleolar RNA synthesis may be that inhibition of nucleoplasmic RNA synthesis for some time in turn causes an inhibition of nucleolar RNA synthesis, perhaps because the nucleolus depends for its activity upon the products of nucleoplasmic transcription.

Results consistent with this interpretation have been obtained by Yu and Fiegelson (1972), who found that inhibition of protein synthesis with cycloheximide has no effect upon the activity of nucleoplasmic RNA polymerase but inhibits the activity of the nucleolar fraction; the nucleolar enzyme behaves as though itself unstable and appears to decay with a half life of about 1·3 hours. If sufficient actinomycin is given in vivo to inhibit all RNA synthesis, the nucleolar enzyme continues to decline with a short half life. This suggests that the mRNA for the enzyme and the enzyme itself may be unstable, so that inhibition of nucleoplasmic transcription or translation of the messenger prevents nucleolar transcription.

RNA Synthesis During Interphase

Synthesis of RNA takes place during all three parts of interphase but halts at mitosis. When Zetterberg (1966b) followed the accumulation of RNA in the nucleus, he found that its level stays about constant; the RNA content of the cytoplasm, however, doubles during the cell cycle. This means that RNA synthesised in the nucleus must be transported to the cytoplasm before it can accumulate; there is no retention of stable RNA within the nucleus. Cell division restores the status quo of the G1 cell when the RNA of the parent cell is divided equally between its progeny; protein, RNA and DNA all therefore double during a cell cycle and are apportioned at division (see also page 72).

With the exception of genes whose functions are linked to the cell cycle itself—of which those coding for the histones are at present the only well characterized example—the sequences transcribed in any cell do not vary with the cell cycle. Using a hybridization assay, Pagoulatos and Darnell (1970a) showed that Hela cells synthesise the same sequences of Hn RNA throughout interphase. Although there appear to be no qualitative changes in gene expression as a cell passes through G1, S and G2 phases, the rate of RNA synthesis increases. Synthesis of DNA during S phase doubles the genetic content of the cell and might therefore stimulate an increase in transcription by the doubling of active gene loci. Another possible effect of S phase is to restrict transcription from those sequences which are undergoing replication; however, there is no evidence for any decline in either individual or overall RNA synthesis during the replication of DNA.

RNA synthesis appears to take place at a fairly constant rate during G1 phase, increases in S phase, and continues throughout G2 at a higher level than that of G1. The relationship between the increase in transcription and the replication of DNA in S phase is not clear. Klevecz and Stubblefield (1967) found that when Chinese hamster cells are released from a block in mitosis, the rate of incorporation of H^3-uridine doubles about 4 hours later, when DNA synthesis commences. Pfeiffer and Tolmach (1968) found that Hela cells increase the rate of RNA synthesis during the first half of S phase. Addition of hydroxyurea to block the start of DNA synthesis reduces the increase to one third of its usual value; treatment with the inhibitor during the first half of S phase blocks a proportionately lesser amount of the increase. That this effect is a general one and does not depend upon changes in the activity of one RNA polymerase is suggested by the observation of Pfeiffer (1968) that both ribosomal and messenger RNA synthesis display the same pattern of increase.

When DNA is replicated, the number of copies of each gene must double and it is possible that both old and newly synthesised copies are then transcribed; this may explain some of the increase in RNA synthesis but cannot account for all of it since a strict dependence of transcription rate on DNA content would demand a simple proportionality between increase in RNA synthesis and progress into S phase. (That is unless we assume that active gene sequences are replicated at a different time from inactive gene sequences.)

Although gene dosage effects may in part be responsible for an increase in transcription, cells grown in culture may have in any case deviated from their original diploid complement of chromosomes so that quantitative changes in the rate of transcription may be atypical of those prevailing in diploid cells. Normal somatic cells should possess two copies both of each structural gene and of whatever elements control its expression (with the exception of sex linked loci); this may not be true in established tissue culture lines, which may therefore differ in the quantitative control of gene expression from cells restricted to a diploid complement.

Acceptable evidence for a gene dosage effect would be the demonstration in diploid cells that the synthesis of some particular gene product doubles at some point during S phase. But such experiments have at present been performed only with bacteria, in which the doubling of gene expression can be used to follow the progress of replication around the chromosome (see chapter 10 of volume 1). Mitchison (1969) has pointed out that kinetics of increase of enzyme activity during the cell cycle may fall into any one of several classes, depending upon whether the enzyme is stable or unstable and whether it is synthesised continuously or is subject to a stepwise increase. However, no doublings in synthesis with gene duplication have been documented in mammalian cells.

Because the genes coding for ribosomal RNA are clustered on the chromosome, if gene duplication increases transcription at the nucleolus there should be a sudden doubling in rRNA production at one point in S phase. (Nucleoplasmic RNA synthesis, by contrast, takes place at many widely separated loci so that no precise correlation between overall synthesis and gene duplication can be made; the instability of Hn RNA also makes these measurements more difficult.) But ribosomal RNA represents a direct and stable gene product which can readily be assayed. Scharff and Robbins (1965) found that the rate of rRNA synthesis in Hela cells shows no discontinuous changes during S phase. Enger and Tobey (1969) followed the rate of methylation of 18S rRNA and the incorporation of H^3-uridine into total RNA in Chinese hamster cells and found only a continuous increase which doubles the rate of transcription during the first 10 hours of interphase.

Some experiments therefore show an increase in RNA synthesis which is at least loosely correlated with DNA replication in S phase; whereas others show only a continuous increase in RNA synthesis during interphase as a whole. It is possible that metabolic changes take place in some cells during S phase which lead to discontinuous changes in the rate of transcription; there is at present no evidence to prove that this relates directly to gene duplication. Assessment of possible gene dosage effects during the cycle requires assay of individual gene products in diploid cells; no such measurements have been made on nucleoplasmic RNA synthesis.

Inhibition of RNA and Protein Synthesis During Mitosis

The transcriptional apparatus of the nucleoplasm and nucleolus exists in an intact state only during interphase for it is disrupted by the reorganization of cellular structure which takes place at mitosis. The nucleolus disappears during prophase and the disintegration of the nuclear membrane marks the start of metaphase, so that by this time the cell is no longer divided into separate compartments. The chromosomes are coiled into a condensed state in which they are inert. Entry into mitosis is accompanied not only by a cessation of transcription, but translation of messengers synthesised during the preceding interphase is also halted. Mitotic cells are therefore largely inactive in macromolecular synthesis. It is only the transcription and translation of nuclear genes which is inhibited at division, for Fan and Penman (1970a) found that mitochondrial synthesis continues in dividing Hela cells.

By following RNA synthesis with autoradiography of cells incubated with a radioactive precursor, Prescott and Bender (1962) and Terasima and Tolmach (1963) showed that a five minute exposure to H^3-uridine labels only the nucleus of Hela or Chinese hamster fibroblasts in interphase. But incorporation of the label declines in early prophase cells and all RNA synthesis ceases by mid prophase. Transcription ceases before the disintegration of the nucleolus and the nuclear membrane and resumes late in telophase before the nucleolus has reformed. The timing of the inhibition of transcription therefore implies that it is under a specific control and is not simply a response to condensation of the chromosomes. Amino acid incorporation also falls, to about 25% of the interphase level, during mitosis.

Autoradiographic analysis also suggests that any segregation of RNA between nucleus and cytoplasm is lost during division and regained in the daughter cells. Prescott (1963) pointed out that the RNA which is present in the nucleus when the nuclear membrane disintegrates is released into the cytoplasm; but the label is found within the nucleus again after division has been completed. Davidson (1964) found that RNA synthesis in Vicia faba ceases during prophase as the chromosomes condense. In late prophase cells, the RNA synthesised in the earlier stages of the cycle is rapidly lost to the cytoplasm.

That inhibition of transcription is specific is implied by the cessation shown in figure 6.1 of synthesis of nucleoplasmic and nucleolar RNA synthesis alone. Zylber and Penman (1971b) found that metaphase Hela cells continue to synthesise 4S tRNA and 5S rRNA; the synthesis of transfer molecules proceeds at about one third of the interphase rate and synthesis of 5S rRNA is very little inhibited. It must therefore be possible to transcribe these chromosomal loci even from the condensed mitotic chromosomes. This is of course consistent with the concept that a third RNA polymerase synthesises these small RNA molecules and that only the activities of enzymes A and B are inhibited.

The fate of the components of the nucleolus between its disappearance during prophase and reappearance during telophase has been followed by Fan and Penman (1971) by labelling the RNA of Hela cells with H^3-uridine before mitosis. Extraction of RNA from the chromatin fraction of cells arrested in metaphase by colcemid shows that the chromosomes are associated with the 45S and 32S precursors to ribosomal RNA. Some 75% of the interphase content of 45S RNA and 50% of the usual interphase content of 32S RNA are present in chromatin. Fractionation of metaphase cells shows that 80% of the cellular content of 45S and 32S RNA is associated with the

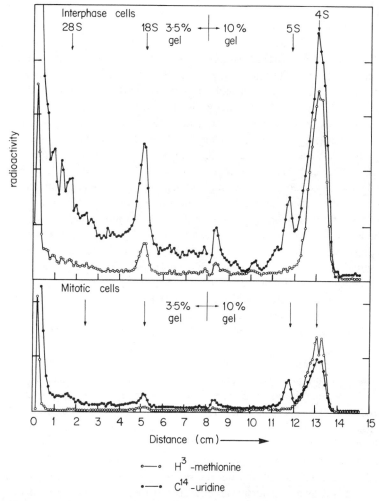

Figure 6.1: comparison of RNA synthesis in interphase and metaphase-arrested Hela cells. Cells were labelled with H³-methionine (enters methyl groups of RNA) or C¹⁴-uridine and the total cellular RNA analysed on two phase acrylamide gels. In interphase cells all RNAs are synthesized and the methylation of ribosomal and transfer RNAs is prominent. In mitotic cells only 5S and 4S RNA are synthesised and the tRNAs modified. Data of Zylber and Penman (1971b).

chromosomes, together with 11 % of the mature ribosomal RNA. The rRNA may be present as the result of cytoplasmic contamination, for it is readily removed by washing with detergent; but the presence of the precursors may reflect a more specific attachment in vivo for they are not removed by this treatment.

The precursors can be extracted from the chromatin of cells arrested in metaphase for 5–6 hours; since no nucleolar RNA is synthesised during this time the precursors must comprise molecules synthesised before the start of mitosis which remain stable. This contrasts with their instability in interphase cells in which their lifetimes average about 15 and 45 minutes respectively. Association of the ribosomal RNA precursors with the chromosomes must therefore halt their usual processing for the duration of mitosis.

The interphase nucleolus is fairly stable and can be isolated from the rest of the nucleus. It can be fragmented by the use of dithiothreitol to yield particles of 80S and 55S; in the presence of magnesium ions particles of 110S are also produced. When the chromatin fraction of mitotic cells is digested with DNAase and layered onto high ionic strength sucrose gradients, all the 45S and some of the 32S RNA molecules are found in particles sedimenting from 85–200S; these may correspond to the usual interphase precursor particles. It is therefore the particles containing precursor RNA which halt processing and associate with the metaphase chromosomes in an inert state.

If cells are labelled with H^3-uridine before metaphase and then released from metaphase arrest in the presence of actinomycin, no new transcription can occur. Under these conditions the precursors become labile when the cells are released from metaphase; their conversion to mature ribosomal RNA in ribosomes can take place in the presence of cycloheximide and must therefore be independent of protein synthesis. The components from which the nucleolus is regenerated after mitosis are thus preformed and not newly synthesised. Phillips (1972) found that when Chinese hamster fibroblasts are incubated with actinomycin for three hours prior to mitosis, no nucleolus can be reformed after division. This implies that the components of the nucleolus of a daughter cell are synthesised in the late part of interphase preceding division. The proteins which reconstitute the nucleolus appear to be drawn from the chromosomes to which they were dispersed at prophase.

When mammalian cells divide, the rate of protein synthesis drops to some 20–40% of the level of interphase cells, according to the decrease in incorporation of labelled amino acids detected by autoradiography. Hodge, Robbins and Scharff (1969) found that when Hela cells are synchronised by selective detachment, the addition of actinomycin does not inhibit cell division and protein synthesis resumes at the beginning of the next G1, although at a lower level than that displayed in untreated cells. Resistance to inhibition by actinomycin and the short duration (30 minutes) of mitosis compared with the stability of mRNA (lifetime of some hours) suggest that at least some of the premitotic RNA population persists through mitosis and is utilised again in the next division cycle.

Messenger RNA must therefore remain in the cell, although it is not translated during mitosis. Its failure to be translated appears to result from a loss of the ribosomes associated with it. Scharff and Robbins (1966) noted that polysomes disaggregate in mitotic cells to generate a pool of 80S ribosomes; Steward, Schaeffer and Humphrey (1968) reported that Chinese hamster cells in interphase have most of their ribosomes as polysomes engaged in protein synthesis, but metaphase cells largely contain 80S ribosomes, with a 75% reduction in the incorporation of labelled amino acids into the polysome part of the gradient. Figure 6.2 shows that as a cell population passes through mitosis, polysomes reform and a greater proportion of the label is found in their region until the normal state of the cell is restored some two hours after mitosis.

Addition of actinomycin to metaphase cells does not inhibit the reformation of polysomes, which suggests that this utilises messenger RNA synthesised during the previous cell cycle and present in an inert state through mitosis. Hodge, Robbins and Scharff (1969) found that when cells are pulse labelled with H^3-uridine in G2, some 40% of the label incorporated into the polysomes is again found in this fraction after a division. Messenger RNA must therefore lose its ribosomes at the onset of mitosis, is preserved in an inert but stable form during division, and is available for use as a template in the daughter G1 cells.

That the disaggregation and reformation of polysomes is responsible for the level of protein synthesis in mitotic cells is suggested by the measurements of Fan and Penman (1970b). By plotting the rate of amino acid incorporation against time after Chinese hamster cells have been released from a double thymidine block, they found that protein synthesis declines in parallel with the increase in the mitotic index. Under these conditions, protein synthesis in mitosis is about 28% of that in interphase. The drop in the number of polysomes in this culture parallels the drop in protein synthesis; in interphase cells about 75% of the ribosomes are in the form of polysomes, but mitotic cells contain about 25%.

Cycloheximide inhibits protein synthesis by preventing movement of ribosomes along the messenger; low levels of the drug can therefore be used to slow the rate of translation. The rate of attachment of ribosomes to mRNA is less inhibited than translation itself, so the addition of cycloheximide causes the accumulation of ribosomes on a messenger. Incubation with cycloheximide thus causes 87% of the ribosomes of interphase cells and 57% of the ribosomes of mitotic cells to accumulate as polysomes. These polysomes are active in protein synthesis—as shown by the incorporation of labelled leucine—although of course they function more slowly. The ribosomes present in mitotic cells are therefore active and can synthesise protein when they are associated with mRNA.

When interphase cells are labelled with C^{14}-uridine and then incubated with low doses of actinomycin to inhibit nucleolar RNA synthesis in the presence of H^3-uridine, both labels enter the polysomes. This implies that the newly synthesised messengers (identified by the H^3-label) can associate with the

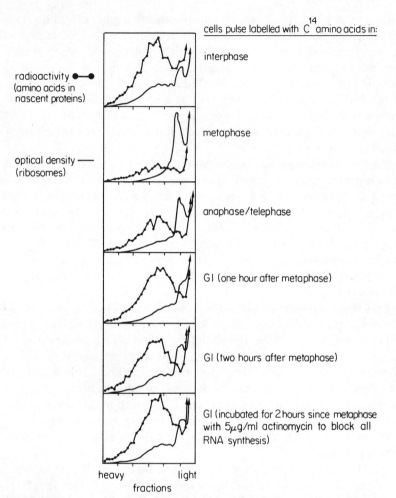

cells pulse labelled with C^{14} amino acids in:

interphase

radioactivity ●—●
(amino acids in
nascent proteins)

metaphase

optical density ——
(ribosomes)

anaphase/telophase

G1 (one hour after metaphase)

G1 (two hours after metaphase)

G1 (incubated for 2 hours since metaphase
with 5μg/ml actinomycin to block all
RNA synthesis)

heavy   light
          fractions

Figure 6.2: protein synthesis in interphase and mitotic Chinese hamster cells. The C^{14}-amino acid pulse label identifies nascent proteins on the polysomes; the optical density trace reveals the condition of the ribosomes. Interphase cells possess a large peak of polysomes which incorporate amino acids into proteins. Metaphase cells suffer disaggregation of polysomes and possess ribosomes in monomer state; incorporation of amino acids is less than 25% of that in interphase cells. Reformation of polysomes begins in anaphase/telophase together with increased use of amino acids. The interphase state is restored by one hour after metaphase, that is early in G1. Since the same results are obtained in cells incubated for 2 hours after metaphase with high actinomycin, polysome reformation and protein synthesis does not depend upon synthesis of RNA but must rely upon RNA molecules synthesized in the previous cell cycle. Data of Steward, Shaeffer and Humphrey (1968).

previously synthesised ribosomes (identified by the C^{14}-label). The same situation is found in the presence of cycloheximide, when extra ribosomes enter the polysomes, although in interphase cells no additional mRNA is utilised. But in mitotic cells addition of the drug doubles the amount of mRNA bound in the polysome fraction. This shows that shortage of mRNA is not responsible for the disaggregation of polysomes.

These experiments show that the amount of messenger RNA and the number of ribosomes are not reduced in mitotic cells; and when cycloheximide is used to cause ribosomes to accumulate on messengers, they function normally in protein synthesis. This suggests that the reduction in protein synthesis at mitosis may be caused by a specific inhibition of the ability of ribosomes to associate with messenger RNA; only the initiation of protein synthesis is inhibited.

Gene Expression in Polytene Chromosomes

Function of Puffs in Salivary Gland Chromosomes

Polytene chromosomes are found in more than one tissue of insect larvae but have been studied largely in the salivary gland. Each polytene chromosome displays a characteristic set of puffs, the largest of which are known as Balbiani rings. Beerman (1965) noted that puffing is in general a function of single bands, more occasionally of pairs of bands (although some bands which appear double under the light microscope have since been shown to comprise single units under the electron microscope; see page 45).

That the puffs represent sites of RNA synthesis was suggested by the autoradiographic studies of Pelling (1964), who showed that puffs stain with reagents specific for RNA whereas unpuffed bands fail to do so. When auto-radiography is performed on salivary gland cells previously exposed to labelled RNA precursors, the pattern of labelling corresponds with the puffing pattern and the extent to which each puff is labelled is related to its size. Beerman (1964) found that actinomycin inhibits puff formation and causes them to regress. Each puff therefore represents a site of synthesis of RNA, the size of the puff providing some measure of the activity of the band from which it derives.

That the pattern of puffs depends upon both the tissue and its stage of development was first shown by Beerman (1952). Four categories of puff have been distinguished by Berendes (1965b). Some puffs are restricted to particular stages of development; in particular, the major periods of puff activity precede larval moults. By comparing the puffing patterns of salivary glands, Malpighian tubules and the midgut of D. hydei, Berendes (1965a, 1966) showed that some puffs are tissue specific. A third class of puff is neither tissue specific nor characteristic of particular stages of development; these puffs presumably represent the expression of genes coding for functions which are needed in all the polytene tissues at all times. The environmental treatments such as temperature shock (Berendes, 1968) or injection of certain sugars (Beerman, 1973) also cause changes in the pattern of puffing; however, the relationship of this class of puffs to those usually observed in larval development is not clear. In reviewing the appearance of puffs in salivary glands, Ashburner (1970) noted that most (more than 80%) of the puffs in Drosophila are specific and appear or regress at particular stages of development; but in C. tentans the situation is reversed and only a minority of the puffs shows developmental specificity.

The pattern of puffing in salivary glands of D. melanogaster during the last ten hours of the third larval instar and the subsequent twelve hour prepupal period has been defined by Ashburner (1967, 1969a). In agreement with the earlier studies of Becker (1959), he found that puffs develop only at specific sites and that formation and regression of most puffs occurs in a well defined sequence. Some 108 loci form puffs on the autosomes during one or both of these developmental periods; the timing and size of 83 puffs depends upon the stage of development. The X chromosome has a further 21 loci which are active during these periods. Some 109 bands therefore represent the genetic loci which are active during these stages of larval development.

The pattern of puffing is also strain specific. When Ashburner (1969b) compared the oregon and vg-6 strains of D. melanogaster, only 64% of the puffs proved to be the same during the third instar and prepupal stages. Some 12% differ in size between the two strains and 19% in the time of activity; and 5% differ in both size and timing. However, the loci implicated in puffing remain the same and it is only the degree and timing of puff development which differs. The sole exception is one puff which is active only in vg-6. In genetic crosses this puff segregates in the same way as any other Mendelian locus. In heterozygotes of oregon and vg-6, the vg-6 specific puff develops in only the vg-6 homologue when the two homologues are asynapsed; but develops in both when they are synapsed. The nature of the factors which control development of this puff is not known, but this implies that puffing at this locus is controlled by interactions at the level of the strands of the polytene chromosome.

There is a striking increase in puff activity before larval moults and prior to metamorphosis, when sudden and multiple changes take place in the puffing pattern (see Berendes, 1967). Resetting the pattern of puffs coincides with the time when the moulting hormone ecdysone is released from the prothoracic gland. Two types of experiment show that ecdysone controls puff development.

Larvae can be ligatured so that the salivary gland is divided into two regions, the anterior part lying in the same half of the larva as the prothoracic gland and the posterior region separated from it. Becker (1962a) found that when larvae are ligatured before the last larval moult, the anterior region alone develops the puffing pattern characteristic of the puparium whereas the posterior region remains at the intermoult stage. Tying the ligature at different times shows that the hormone is released about 3–5 hours before the moult. That the puffing pattern of the salivary gland chromosomes is established by the developmental stage is also suggested by the transplantation experiments of Becker (1962b), which showed that a transplanted gland acquires the puffing pattern characteristic of the stage of development of its host.

Injection of ecdysone into larvae, first performed by Clever and Karlson (1960), confirms the role of the hormone. When ecdysone is injected into fourth instar intermoult larvae, the two puffs which appear are located at the sites which are usually the first to puff before moulting. Clever (1966) noted that these two puffs have different threshold levels for responding to hormone, corresponding to about 10 and 100 molecules of ecdysone per haploid genome. This dose must be increased very considerably—by more than two orders of magnitude—for maximum reaction. In addition to the two puffs which respond immediately to the injection of ecdysone, other puffs respond later. Berendes (1967) showed that injection of ecdysone into third instar larvae of D. hydei causes the same overall pattern of changes as that observed in normal development before pupation.

The puffs which respond rapidly—within 15–20 minutes—to injection of the hormone are presumably activated by the interaction with the cell of the ecdysone molecule itself. The puffs which respond only after a period of 4–6 hours may represent activation by some of the gene products of the puffs which respond immediately. Ashburner (1972) has shown that this biphasic puffing pattern can be reproduced when salivary glands are incubated with ecdysone in culture. The appearance of a series of puffs in proper developmental sequence may therefore depend upon an initial activation of a small number of genes by ecdysone, followed by the activation of other genes by the first gene products.

That puff formation is required for the transcription of RNA is shown by the incorporation of labelled precursors only at sites of puffing. Beerman and Bahr (1954) showed that puffing represents an uncoiling of DNA in the band and this may be necessary to allow access of RNA polymerase molecules. Holt (1971) observed by interferometric measurements that the formation of puffs is accompanied by an increase of mass due to protein as well as RNA. That histones are implicated in puff formation has been excluded by the observation of Gorovsky and Woodard (1967) that quantitative staining with alkaline fast green—a reaction specific for histones—shows no difference between puffs and bands. But Holt (1970) found a local increase in the capacity of puffs to bind naphthol yellow S, which relies upon reaction with the indole group of tryptophan. Since tryptophan is largely absent from histones, this suggests that non histone proteins accumulate at the site of a puff.

The order of events in development of a puff suggested by Berendes (1968) is: accumulation of non histone proteins; uncoiling of DNA; transcription of RNA. If the structure of the puff depends upon RNA synthesis, inhibitors of transcription should prevent formation of new puffs and cause those already existing to regress. Induction of new puffs has been clearly prevented by several reagents; regression of existing puffs is not always complete, but this may be caused by impermeability of cells to drugs. It is clear that puffs rely upon synthesis of RNA by an amanitin-sensitive enzyme. Beerman (1964) and Ellgaard and Clever (1971) observed that actinomycin inhibits puff activity; Ashburner (1972) showed that actinomycin, 3'-deoxyadenosine and amanitin all block induction of puffs. Beerman (1971) found that amanitin inhibits incorporation of H^3-uridine at puffs of Chironomus salivary gland nuclei but does not inhibit nucleolar synthesis of RNA—a situation precisely analogous to that prevailing in mammalian cells.

Although actinomycin inhibits RNA synthesis in the salivary glands of C. tentans, Clever, Storbeck and Romball (1969) and Clever (1969) found that protein synthesis is not inhibited at late larval stages. Doyle and Lauffer (1969) also found that RNA synthesis in fourth instar larvae can be inhibited for 12–24 hours with no effect on either protein synthesis or secretion of salivary proteins. This suggests that messenger RNA is stable in these cells. However, since the further development of larvae is inhibited by actinomycin, progression to the next stage must require synthesis of new RNA molecules.

Although the puff pattern remains constant during the last larval instar, the stability of messenger RNA in salivary gland cells appears to change. Clever and Storbech (1970) found that actinomycin causes polysomes to disaggregate in larvae at the beginning of the last instar but has no effect on polysomes in larvae at the end of this period. Rubinstein and Clever (1972) therefore suggested that messenger RNA may be unstable at the beginning of the last instar but stable at the end. Differences in the stability of total RNA are found between the two larval stages, but it is not yet clear how they relate to any changes which may take place in messenger RNA. However, one possible model is to suppose that gene expression is first controlled by selection of the bands which are to develop puffs and transcribe RNA; after which a selective degradation allows only some molecules to act as messengers. Changes in the transcript sequences which are degraded and in the stability of messenger RNAs may therefore impose a control on gene expression subsequent to transcription itself.

Puffs represent sites at which gene expression can be visualized; transcription seems to depend upon an accumulation of non histone proteins and uncoiling of the DNA. The tissue and temporal specificity of puffing patterns shows directly that genetic development of the larvae depends upon specific selection of the genes which are transcribed. That sequential changes in the puffing pattern—in particular activation of new loci—are set in train by the release of ecdysone suggests that the effect of the hormone is to control the pattern of gene expression. However, post-transcriptional controls may modify the relative activities of genes which are established by the puffing pattern activated at the previous moult.

Individual bands appear to be the units of gene expression since puffs usually derive from one band; but because the structure of the larger puffs obscures the underlying band pattern it is difficult to show conclusively that puffs are always equated with single bands. Although there is some evidence to suggest that bands represent individual genetic loci in at least many instances, the average DNA content of the band (per haploid genome) is much larger than needed to code for any one protein (23×10^6 daltons in D. melanogaster; 60×10^6 daltons in C. tentans—see page 46). We do not know how much of the DNA of each band is transcribed into RNA; definition of the sequences of DNA present in a band requires measurements of the activity of individual loci rather than the population as a whole.

Transcription of Balbiani Ring 2

That bands represent units of genetic organization is indirectly suggested by the development of puffs at single bands and by equation of the genetic and cytological maps of Drosophila (see page 45). Direct evidence to show that a puff represents a unit of transcription has been provided by the isolation and characterization of the RNA synthesised from one of the Balbiani rings, BR2, of the polytene chromosomes of the salivary gland of C. tentans.

By using microdissection to separate the chromosomes, nucleoli, nuclear sap and cytoplasm of Chironomus salivary gland cells, Edstrom and Daneholt (1967) were able to characterize the RNA of each fraction. The RNA synthesised by the chromosomes falls into the polydispersed distribution characteristic of mammalian Hn RNA. This comprises a separate peak of 4–5S RNA and a distribution continuously dispersed from 10–90S, with a peak at about 35–45S. Nucleoli produce the discrete classes of ribosomal RNA precursors and their processing products.

In a more detailed analysis of the transcripts of the polytene chromosomes, Daneholt et al. (1969a) showed that the large RNA molecules sediment rapidly through sucrose and electrophorese slowly on agarose. RNA treated by heat denaturation, chemical denaturation in dimethyl sulfoxide, or low ionic strength—all conditions which prevent the aggregation of RNA molecules by base pairing of complementary sequences—produce the same results. The rapidly sedimenting molecules must therefore represent large RNAs and not aggregates of smaller molecules.

After incubating salivary glands with a tritium labelled RNA precursor, Daneholt et al. (1969b) divided the RNA product into two fractions: that deriving from chromosomes I, II and III and that synthesised by chromosome IV alone. The heterogeneous distribution of the first fraction has a peak at 35–40S and the label in the RNA increases during the first 45 minutes of incubation. This sets an upper limit for its turnover time on the chromosome of 45 minutes. The RNA synthesised by chromosome IV falls into a similar distribution, although the main peak sediments more rapidly and maximum labelling is reached within 30 minutes.

When the salivary glands are labelled for 45 minutes and then incubated with actinomycin for 10 or 20 minutes, the distribution of label in the Hn RNA product remains the same, although the amount of radioactivity decreases with the time of the chase. This suggests that the transcription products are removed from the chromosomes but remain their original size. RNA molecules of the same size can be recovered from the nuclear sap and, after longer periods of incubation, from the cytoplasm. This suggests that there is no processing either on the chromosomes or in the nuclear sap; the Hn RNA transcripts are transported from the chromosomes, through the nuclear sap, to the cytoplasm with no change apparent in their size distribution. This contrasts with the selective degradation of part of each Hn RNA molecule observed in mammalian cells.

Transcription of chromosome IV comprises about 25% of the total chromosomal RNA synthesis of the salivary gland cell. Chromosome IV displays the three tissue specific Balbiani rings shown in figure 6.3; the average activities of these puffs in RNA synthesis is: BR1—25%; BR2—60%; BR3—15%. These proportions are the same as the relative sizes of the puffs. Daneholt et al. (1969c) isolated the largest ring, BR2, by microdissection and characterized the RNA product. In these early experiments, BR2 RNA showed a wide distribution upon electrophoresis, with a principal peak corresponding to 40–60S but with the overall distribution extending to the usual limits of 10S and 90S. That BR2 RNA comprises only one transcript has since been shown by Daneholt (1972); its electrophoretic profile is simpler than that of chromosomes I–III and under appropriate conditions displays one sharp peak of about 75S. The size of this RNA can be estimated as 35×10^6 daltons from its mobility on gels and as 15×10^6 daltons from its rate of sedimentation through sucrose.

A peak of 75S RNA is prominent in the distribution of the nuclear sap RNA and corresponds to the transcript of BR2. The synthesis and subsequent transport of this RNA can therefore be followed in salivary glands incubated with labelled precursors to RNA. Daneholt and Hosick (1973) reported that BR2 RNA is synthesised at its locus on chromosome IV, can then be recovered from the nuclear sap, and after a delay appears in the cytoplasm. The behaviour of this individual RNA molecule therefore conforms to that displayed by the distribution of chromosomal RNA as a whole; it enters the cytoplasm in the form in which it is transcribed with no significant reduction apparent in its size. The 75S RNA is a prominent component of salivary gland cells in which it corresponds to about 1·5% of the total RNA (this is comparable to the concentration of individual messengers in other narrowly specialised

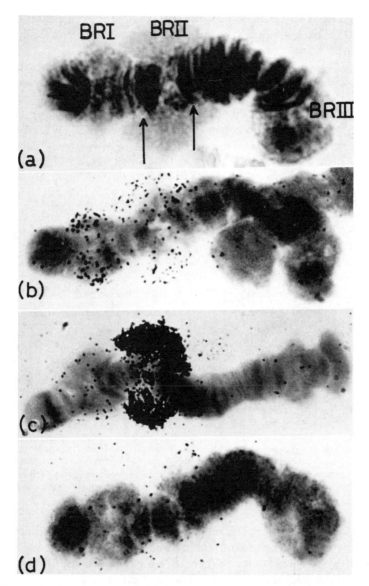

Figure 6.3: chromosome IV of C. tentans salivary glands, × 1200. (a) Only three loci form puffs, of which BR2 is the largest. (b) Labelled RNA extracted from BR1 hybridizes mostly with the BR1 locus under the conditions of cytological hybridization. (c) Labelled RNA extracted from BR2 hybridizes well with the BR2 puff and must therefore contain repeated sequences. (d) Labelled RNA extracted from BR3 does not hybridize under the conditions of cytological hybridization. Data of Lambert (1972).

cells; 9S RNA of reticulocytes comprises about 2% of the total content). The 75S molecule appears to be fairly stable, with a half life in the cytoplasm of about 35 hours.

The RNA coded by BR2 and by other chromosome regions has been examined both by cytological hybridization and by a microtechnique for RNA–DNA hybridization developed by Lambert et al. (1973a)—essentially a scaled down version of the filter method. Both these techniques therefore measure the reaction of repeated sequences. Lambert et al. (1972) found that the RNA synthesised by chromosome I labels all chromosomes when used in

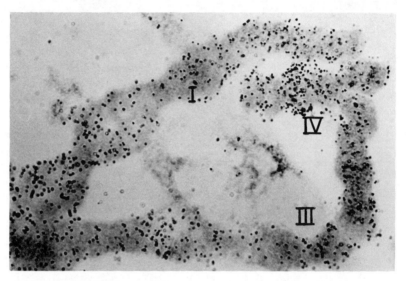

Figure 6.4: cytological hybridization between polytene chromosomes of C. tentans salivary glands and the RNA transcribed from chromosome I. All chromosomes are labelled, showing that the corresponding sequences in the genome are repeated at many loci. Magnification × 600. Data of Lambert et al. (1972).

cytological hybridization; this result is shown in figure 6.4. This implies that it is derived from members of families whose sequences are repeated in all regions of the chromosome. But the RNA synthesised by BR2 hybridizes only with the BR2 locus itself as shown in figure 6.5; this implies that the family of repeated sequences from which it derives is located entirely within the BR2 region. Nuclear sap RNA also hybridizes heavily with the BR2 locus, which supports the idea that nuclear sap is rich in BR2 RNA. Lambert (1972) found that BR1 RNA shows a comparable specificity in annealing largely only to the BR1 locus; but BR3 RNA anneals very poorly with DNA under these conditions and may therefore comprise transcripts of unique rather than repeated sequences.

Under conditions when only repeated sequences anneal, 0·17% of the DNA of C. tentans hybridizes with BR2 RNA. Lambert et al. (1973b) found that under the same conditions nuclear sap RNA saturates 0·25–0·30% of the DNA. Comparing these figures suggests that BR2 RNA may provide more than half of the repeated sequences of nuclear sap RNA. Since synthesis of BR2 RNA accounts for only 15–20% of total chromosome RNA synthesis, this implies that some of the sequences synthesised at other loci are not

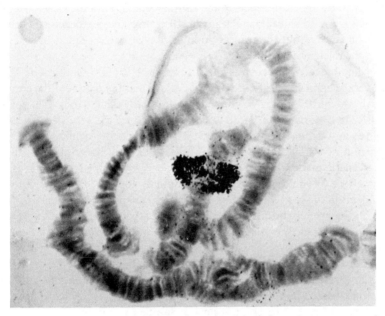

Figure 6.5: cytological hybridization between polytene chromosomes of C. tentans salivary glands and the RNA transcribed from locus BR2 of chromosome IV. Only the BR2 locus is labelled, showing that the repeated sequences which interact in this reaction are concentrated within the DNA around the BR2 site. Magnification ×900. Data of Lambert (1972).

represented in the nuclear sap RNA. That many chromosomal RNA sequences are selectively degraded before they can enter the nuclear sap is suggested by the hybridization level of chromosomal RNA; RNA extracted from the chromosomes does not saturate under the conditions used for hybridization, but an extrapolation of the hybridization curve suggests a saturation value of about 2·2%—an order of magnitude greater than the saturation level of nuclear sap RNA. Also consistent with this conclusion is the observation of Daneholt and Svedhem (1971) that the base ratio of nuclear sap RNA is similar to that of BR2 RNA but is different from that of chromosome RNA. These experiments therefore suggest that much of the RNA transcribed on the polytene chromosomes is immediately degraded but that BR2 RNA is one of the sequences which enters the nuclear sap, and because of its high rate of synthesis and immunity to degradation forms a much larger proportion of nuclear sap sequences than of chromosome RNA sequences.

By following the cot curve of renaturation of C. tentans DNA, Sachs and Clever (1972) showed that 95·5% of the DNA comprises single copy sequences and only 4·5% corresponds to repeated sequences, with an average degree of repetition of 120 fold. These values imply that all of the repeated sequences of the genome are transcribed in the salivary gland, although less than 10% of them survive to enter the nuclear sap. Of course, these experiments provide no information about the transcription of non-repeated sequences and their fate in the nucleus.

The haploid genome corresponds to about $1·2 \times 10^{11}$ daltons of DNA. The average band size is about 60×10^6 daltons and the region of the chromosome in which BR2 is located contains no unusually thick bands. It is possible that BR2 may derive from more than one band since the very size of the puff obscures any definition of the bands in the immediate vicinity of the puff. Saturation of 0·17% of the genome corresponds to transcription of 0·34% of the genetic material since only one strand is transcribed; this means that about 400×10^6 daltons of DNA corresponds to BR2 RNA. Cytological hybridization suggests that all this DNA is located in the BR2 locus.

Using the microhybridization technique, Lambert (1972) found that when nucleolar 38S precursor rRNA (the equivalent of the mammalian 45S precursor) is hybridized in excess of DNA its rate of reaction is 5% per hour. There are about 100 identical genes for this RNA in the haploid genome. BR2 RNA reacts at a rate of 10% per hour; comparison of these values suggests that there must be 200 binding sites per genome when BR2 RNA hybridizes with the DNA. The rate of reaction of nuclear sap RNA is about the same as that of nucleolar RNA; but the rate of reaction of chromosomal RNA is some one third less. This implies that nuclear sap RNA may be enriched in repeated sequences relative to chromosome RNA. However, definition of the non-repeated sequences of these fractions requires hybridization at high cot values; the initial rate of reaction in DNA excess describes only the properties of the repeated transcripts.

Since BR2 RNA is about $15–35 \times 10^6$ daltons, its length must be about 5–10 times less than the total length of DNA with which it hybridizes. It is therefore possible that there is a repetition of several identical sequences at the BR2 locus (although there is no proof that all of the repeated sequences are in fact transcribed; only some of them might comprise the DNA active in the Balbiani ring). However, in order to achieve the 200 fold degree of repetition implied by the rate of hybridization in DNA excess, each of these sequences must be internally repetitive. If all of the sequence of the BR2 RNA is translated into protein, this would mean that there must be repetition in the sequence of amino acids (perhaps analogous to that of silk fibroin).

The proteins coded by the salivary gland puffs have not yet been characterized. However, Grossbach (1969) reported a correlation between the Balbiani rings of chromosome IV and the synthesis of salivary polypeptides (which are secreted by the salivary gland). These comprise only a small number of proteins; Wobus, Panitz and Serling (1970) resolved nine fractions of secretory proteins of the salivary glands of some Chironomus species. The largest salivary polypeptide of C. tentans has a molecular weight of some 500,000 daltons; although exceptionally large for a protein, this would still require an RNA of only 5×10^6 daltons. This illustrates the large discrepancy between the length of BR2 RNA and that demanded of a messenger; BR2 RNA is clearly very much longer than needed to code for any one protein. We do not know whether the molecule carries information to code for several proteins—presumably either identical or with closely related sequences—or whether much of its length serves some other function. Nor can we yet define the organization of DNA sequences at the BR2 chromosome locus.

Models for Control of Gene Expression

Molecular Organization of Chromatin

The first step in controlling gene expression lies at the transcription of RNA from its DNA template. But in view of our lack of knowledge at the molecular level of the structural and functional organization of the eucaryotic chromosome, it is difficult to construct models to show how sequences are selected for transcription. Apart from the general conclusion that DNA is supercoiled into fibres in which it is largely enclosed by protein, we have no model for the molecular arrangement of DNA, histones and non histone proteins. Without defining the molecular architecture which contains the template, it is difficult to envisage the interactions by which regulator molecules may control transcription.

A general model for the state of eucaryotic DNA is to suppose that its intimate association with chromosomal proteins maintains it in a largely repressed condition. This contention is supported by the observation that chromatin is much less active as a template for RNA polymerase than DNA itself. And addition of histones to DNA in vitro inhibits its transcription by RNA polymerase; the enzyme is therefore unable to utilise DNA as template when it is bound by basic proteins. The small number of histones and their evolutionary conservation of sequence implies that their role in the chromosome is structural rather than to specifically control gene activity; and their inhibition of transcription in vitro seems to apply equally to all sequences of DNA. Genes in eucaryotic chromosomes may therefore be non-specifically repressed; so that gene expression must be controlled by the specific de-repression of sequences which are to be transcribed.

Models of this nature imply that sequences of DNA which are under transcription may have a structure different from that of the repressed sequences comprising most of the genetic material. Two changes in chromosome structure may be necessary to allow transcription. The supercoiled state of DNA may be at least partially unwound; puffing in polytene chromosomes provides a visual example of what may be a general prerequisite for RNA synthesis in diploid cells also. Relief of supercoiling may be essential both to allow the polymerase access to the DNA and then for its subsequent movement along the template when the two strands must presumably unwind locally.

And histones may be removed from DNA at active loci or, as perhaps seems more likely, may associate with the template in a somewhat looser configuration which does not prevent transcription by RNA polymerase. Progress of the enzyme along DNA seems to be inhibited when it encounters firmly bound proteins. Derepression of genes may therefore demand the binding of regulator molecules to change the local structure of the chromosome so that DNA is available for transcription. The accumulation of non histone proteins at puff sites in polytene chromosomes may achieve this purpose.

Models which visualise the general state of eucaryotic DNA as one of repression thus imply that euchromatin may comprise two classes of structure; most of its sequences are maintained in a compact inactive state in which they are closely associated with chromosome proteins, but the derepressed sequences may have a less coiled structure in which the proteins are more loosely bound to DNA. That the molecular architecture of the chromosome depends upon its function is suggested also by the more drastic features of structure which distinguish euchromatin as a whole from the permanently inactive and highly condensed regions of euchromatin.

In bacteria, by contrast, sequences suffering transcription are probably distinguished from the rest of the genome only by the presence of locally unwound sequences of DNA associated with RNA polymerase (see chapter 6 of volume 1). Sequences which are repressed are maintained in this state by the interaction with regulator proteins of a short sequence at the end of the operon where RNA polymerase initiates transcription; the sequences which code for protein do not participate in this reaction. The short control sequences are essentially available to RNA polymerase and are transcribed whenever the specific repressor protein is removed to allow the enzyme to proceed into the structural genes (see chapter 7 of volume 1). But we may speculate that in eucaryotic chromosomes genes are unavailable to RNA polymerase unless they are specifically derepressed and suffer a change in structure of the nucleoprotein fibre.

Following the example of bacterial genomes, however, we may suppose that specific sequences of DNA act as control elements for adjacent sequences which code for proteins. Many genes located at different chromosome loci must be coordinately controlled since there is no evidence to suggest any clustering of genes coding for functionally related proteins. One model is therefore to suppose that recognition sequences are repeated at separate loci which are under common control; this implies that control elements may form part of the repetitive component of the genome.

If eucaryotic gene expression requires derepression of sequences which are to be transcribed, we may suppose that regulator molecules are bound to the recognition elements in DNA at active loci, presumably causing a change in the local structure of the chromosome. In this sense the regulator molecules must be part of the structure of chromatin rather than, as in bacteria, molecules which are transiently associated with a control sequence of DNA. Of course, if regulator molecules control the local structure of the chromosome they must interact not only with DNA but also with other components; their role is probably therefore more complex than that of bacterial regulator proteins whose action is limited to recognising a sequence of DNA and/or interacting with RNA polymerase.

Whatever systems control the expression of eucaryotic DNA must be amazingly precise in operation. Starting with a single fertilised egg, a complete organism of many different cell phenotypes is achieved through a particular set of intermediate stages. The number of divisions and cell phenotypes developed during embryogenesis cannot be estimated with any accuracy but must certainly be very large. At each cell division, the genetic material must be replicated without error and its state of gene expression maintained or reset. Once an adult phenotype has been developed in any cell line, its state of expression is maintained without change through any subsequent generations.

That the systems which control gene expression are stable is shown by their perpetuation through cell division. The topological problems which confront transcription are no less important in replication since DNA polymerase (and any other proteins included in the replication complex) presumably have requirements for the state of the template which are more stringent than those of RNA polymerase. If DNA must progressively unwind to allow synthesis of new strands by complementary base pairing, replication must involve relaxation of the supercoiled state of DNA and its release from the proteins with which it is so tightly associated. The structure of each region of chromatin must therefore be disrupted as it passes through S phase.

If the state of expression of chromatin depends upon its molecular structure, the arrangement of chromosome components during G1 must be reformed exactly when their replication has been completed and regulator molecules must take up their appropriate sites in both the G2 daughter chromosomes. This implies that non histone proteins and/or specific regulator molecules as well as histones must be added to the chromosomes during S phase, a point upon which experimental evidence is not yet complete. Of course, when daughter cells are to enter a developmental state which differs from that of their parent, replication offers an opportunity to reset the state of expression of the chromatin and it has sometimes been suggested that a cycle of replication may be needed to help change the differentiated condition of a cell. In at least some cells, however, changes in transcription may certainly be made prior to any entry into S phase.

Because eucaryotic cells can neither be subjected to the genetic manipulations possible with bacteria nor utilised for biochemical isolation of gene systems and the regulator molecules which bind to them, it is not possible to follow changes in the control and expression of individual gene loci. It is therefore necessary to follow more general changes in gene expression. One approach has been to compare chromatins of different tissues or of different developmental stages of an organism to identify any differences which may be correlated with differences in transcription in these cells. Another has been to follow the specific changes in gene expression induced during early embryonic development or by hormones in their target tissues.

Experiments with isolated chromatin in vitro have been taken to support the concept that its state of expression is established by regulator molecules which are part of its structure. Two components of chromatin have been visualised in the capacity of regulator molecules: the non histone proteins and RNA associated with the chromosomes. Although chromatins of different tissues should, according to this model, possess different arrays of regulator molecules, no significant differences in any component have yet been displayed. Of course, in quantitative terms any differences may depend on only a small number of regulator proteins which derepress active loci, so that they are too slight to detect by comparing chromatin preparations as a whole.

That changes in the state of expression of chromatin implicate the non histone proteins has been suggested by experiments which take as a starting point a molecule which is known to influence gene expression; following the entry into their target tissues of steroid hormones suggests that their association with the chromosomes depends upon non histone proteins apparently present only in the chromatin of the target tissue. The chromatin of each eucaryotic cell may therefore include a characteristic array of regulator molecules which establishes its state of derepression; and the activities of some of these regulator molecules—and thus the pattern of gene expression—may depend on specific signals encountered by the cell. (Just as bacterial induction and repression are achieved by the interaction with a repressor protein of an inducer or co-repressor, with the difference that cells of different eucaryotic tissues may possess only those regulator molecules apposite for the genes which they expect to transcribe.) This means that each cell possesses a characteristic state of differentiation in which its chromatin is programmed to respond only to appropriate signals.

Defining the control of transcription cannot in itself provide a full explanation of how gene expression is controlled until the processing of messenger

RNA is better characterized. Until we know how messenger sequences are conserved from amongst the many more sequences transcribed, and in particular whether this process is mechanistic or selective, it is difficult to know how to view the immediate products of transcription. Since only a small part of the product comprises messenger sequences, characterization of the sequences which are degraded is also essential before control of transcription can be seen in perspective. It is therefore important to note that experiments in which chromatin is transcribed in vitro can at best expect to assay the immediate transcription products—that is Hn RNA if RNA synthesis in vitro reflects transcription in vivo—and not messenger sequences. Any comparison between sequences transcribed in vitro and in vivo must therefore concern nuclear RNA.

Template Activity of Chromatin

That the sequences of euchromatin active in transcription may differ in structure from those which remain repressed is suggested by the thermal denaturation profiles of the chick genome. McConaughy and McCarthy (1972) found that under conditions in which free chick DNA melts between 70°C–80°C, most of the DNA of chromatin from chick erythrocytes or liver is stabilized so that it melts in a higher temperature range (and see page 134). But in addition to the bulk of the stabilized sequences, each chromatin possesses a melting component whose Tm lies in the range of free DNA; this amounts to 2·5% of erythrocyte chromatin and 15% of liver chromatin.

By subjecting erythrocyte chromatin to thermal fractionation on hydroxyapatite columns, McConaughy and McCarthy separated the low melting from the bulk DNA components. When these fractions of DNA were hybridized with an excess of RNA from the erythrocyte, some 30% of the low melting DNA was saturated with RNA whereas the high melting fraction reacted only very poorly. This suggests that the fractionation may separate active from inactive sequences of chromatin. That the reaction is specific is suggested by the low value of 4% saturation achieved when the low melting erythrocyte DNA component hybridizes with RNA derived from chick liver.

Since the increased Tm of DNA in chromatin is a consequence of the stabilization of the duplex structure conferred by the proteins bound to DNA, these results accord with the concept that derepressed sequences of DNA may have a looser structure in which they are less tightly associated with chromosome proteins. The inclusion of 2·5% of erythrocyte chromatin but 15% of liver chromatin in the low melting range is consistent with the relative activities of these tissues in transcription. If the equation of the low melting and active DNA sequences is confirmed by characterization of the components of other tissues, differences in the state of DNA in repressed and derepressed regions of chromatin may prove to be appreciable, presumably extending over the entire sequence which is transcribed rather than restricted simply to any control loci.

Isolated chromatin usually contains an RNA polymerase activity which can be dissociated from DNA by high salt concentrations. The activity in chromatin of the endogenous enzyme is rather low; but chromatin can also be transcribed in vitro when RNA polymerase—either of the same species or from a heterologous source—is added to it. Chromatin is less active as a template for exogenous RNA polymerase than purified DNA (reviewed by Bonner et al., 1968b). Template activity displays some dependence on the activity in RNA synthesis of the tissue from which chromatin is derived and different chromatins may have activities ranging from 6–30% of that of DNA.

That chromatin is less active than free DNA as a template is clear from the relative extents to which they direct RNA synthesis in vitro. But the interpretation of experiments to characterize the sequences of the RNA products is more controversial. Two general problems have raised doubts as to whether the RNA synthesized in vitro can be taken to represent the molecules transcribed in vivo. Because of the absence of large amounts of purified mammalian RNA polymerase, chromatin derived from mammalian cells has usually been transcribed by RNA polymerase derived from bacteria; the specificity of transcription by the bacterial enzyme is probably very different from that of the endogenous polymerase. And the hybridization techniques which have been used to characterize the in vitro products provide conditions which allow annealing only to repeated sequences of DNA. We therefore have no information on the transcription in vitro of the non-repeated sequences which comprise an appreciable proportion of the products of transcription in vivo.

By incubating labelled precursors and bacterial RNA polymerase with a fine suspension of chromatin derived from different tissues of the rabbit, Paul and Gilmour (1968) obtained a labelled RNA product. The sequences present in the transcript can be characterized by hybridization between excess RNA and DNA. Figure 6.6 shows that the RNA transcribed from purified rabbit DNA saturates 47% of the genome; but RNA transcribed from bone marrow chromatin saturates only 6·8% and the transcripts of thymus chromatin saturate only 4·7%. Because repetitive sequences comprise only part of the genome, half saturation of DNA with its RNA transcript may mean that all sequences have suffered transcription. Although the absolute values of saturation of repetitive sequences have little significance, these results therefore show that most of the sequences of chromatin are inaccessible to RNA polymerase, whereas all are available in purified DNA.

Chromatin of each tissue appears to possess a characteristic activity as a template for transcription in vitro. Two important questions about the properties of chromatin have yet to be answered. Which components of chromatin restrict the availability of DNA for transcription? And how specific is the restriction of transcription in vitro—how is it related to the repression of genes within the nucleus?

That histones inhibit transcription is shown by the increase in the rate of

RNA synthesis and the number of sequences transcribed when histones are extracted from chromatin. Spelsberg and Hnilica (1971a, b) followed the transcription of repeated sequences in preparations of liver chromatin by measuring the extent to which an excess of the product competes with the labelled RNA transcribed from a template of purified rat DNA; the greater the competition, the more sequences must be transcribed in chromatin.

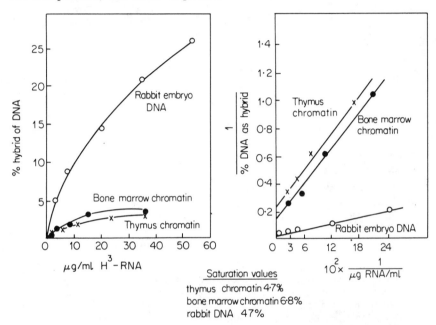

Figure 6.6: kinetics of RNA–DNA hybridization of RNA synthesized in vitro by bacterial polymerase from rabbit DNA or thymus or bone marrow chromatins. Left: filters bearing 1 μg of DNA were incubated with RNA at the concentrations shown. This reaction measures repeated sequences. Right: double reciprocal plot of the data suggests the saturation values shown above. Rabbit DNA is considerably more accessible for transcription than chromatin of either tissue. Data of Paul and Gilmour (1968).

Figure 6.7 shows that the RNA transcribed from isolated chromatin shows only limited competition for the RNA transcribed from purified DNA. When all the histones and a small proportion of the non histone proteins (upto about 20%) are removed with 2 M NaCl, the transcripts of the remaining preparation compete more effectively with the RNA synthesized from purified DNA.

Some restriction on transcription remains, however, for maximum competition is displayed only by the control experiment in which excess unlabelled RNA transcribed from purified DNA competes with labelled RNA transcribed

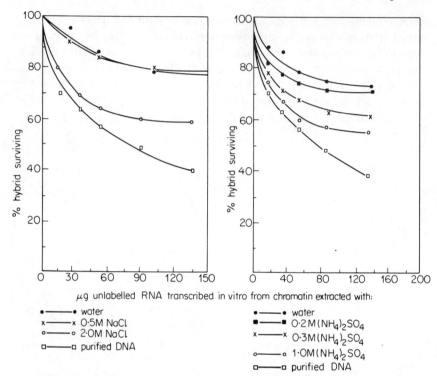

μg unlabelled RNA transcribed in vitro from chromatin extracted with:

•———• water
x———x 0·5M NaCl
o———o 2·0M NaCl
□———□ purified DNA

•———• water
■———■ 0·2M(NH₄)₂SO₄
x———x 0·3M(NH₄)₂SO₄
o———o 1·0M(NH₄)₂SO₄
□———□ purified DNA

Figure 6.7: reduction of template restriction achieved by extraction of histones from rat thymus chromatin. In each incubation 50 μg of labelled RNA transcribed in vitro from rat DNA was hybridized with 2 μg of filter bound DNA in the presence of the amounts shown of unlabelled RNA transcribed from partially extracted chromatin templates or (as a control) also from purified DNA. Competition becomes increasingly effective as the RNA is transcribed from chromatin preparations which have lost more of their histones. This suggests that accessibility to bacterial RNA polymerase increases with extraction of histones. These experiments assay repeated sequences only. Data of Spelsberg and Hnilica (1971b).

from the same template. Similar results are obtained when histone fractions are sequentially extracted by increasing concentrations of ammonium sulfate; extracted chromatin becomes increasingly active as a template for RNA polymerase, although not as active as purified DNA. Histones thus appear to be responsible for much of the repression of the template activity of chromatin, although non histone proteins also restrict the transcription of DNA when the histones have been removed.

Reconstitution of chromatin from its components also shows that histones repress gene expression. Using a procedure in which calf thymus nucleohistone is reconstituted by mixing purified DNA and histones in high salt and then dialysing against successively lower concentrations of NaCl, Huang, Bonner and Murray (1964) found that the ability of DNA to act as template for transcription by bacterial RNA polymerase declines as histones are added. In a review of early work upon the restrictive effect of histones in vivo, Allfrey and Mirsky (1963) noted that digestion of histones by addition of trypsin to suspended calf thymus nuclei stimulates RNA synthesis and addition of histones to the nuclei prevents transcription. But the effects of added trypsin or histones are not restricted to chromatin alone and the crude nature of this system is shown by the observation that the addition of DNAase can degrade upto 70% of the DNA before transcription is inhibited.

Are all histones equally active in repressing transcription? Early experiments produced conflicting results, some suggesting that arginine rich histones are the most effective, others implying that lysine rich histones exert a greater inhibition. Some of these discrepancies have been resolved by Johns and Forrester (1969), who found that the addition of lysine rich f1 (I) histones precipitates DNA from solution but that the addition of further histones renders the complex soluble again. At a given ratio of histone f1 to DNA, reversible changes in the solubility of the complex can result from small changes in the ionic environment. Johns and Hoare (1970) found that histone f3 (III) does not precipitate DNA until higher protein concentrations are reached (in the range in which histone f1 again increases solubility).

There has been a long debate about the solubility of preparations of chromatin and of complexes of histones with DNA (see page 130); and Johns and Hoare suggested that much of the repression of RNA synthesis caused by histones is due to the precipitation of DNA from its soluble state so that it is no longer physically available to RNA polymerase. In the range of histone concentrations and ionic strengths in which chromatin or DNA is precipitated, different histones may appear the most effective in different conditions because of changes induced in their relative effects on the solubility of the preparation. Although the physical state of chromatin preparations is not yet properly defined, it is clearly important to attempt to measure transcription by RNA polymerase in conditions in which the template is physically available to the enzyme.

Another possible cause of non-specific inhibition of transcription has been suggested by the observation of Spelsberg and Hnilica (1969a) that arginine rich histones interact with bacterial RNA polymerase and inactivate its catalytic activity. A similar but less pronounced reaction takes place with rat liver RNA polymerase. This means that the activity of nucleoprotein complexes must be measured in conditions in which free histones are absent.

In measurements of the solubility of chromatin after extraction of histones with increasing concentrations of deoxycholate or salt—which remove histones in a different order (see page 138)—Smart and Bonner (1971c) found that removal of the lysine rich histones is the most effective in reducing the precipitability of chromatin. But the rates of RNA synthesis by the partially extracted chromatin preparations increase in a manner which is nearly linear with the amount of histone removed by either reagent. In contrast with this observation, Spelsberg and Hnilica (1971a, b) found that extraction of lysine rich histones from chromatin with 0·5 M NaCl or 0·2 M ammonium sulfate does not increase the number of sequences transcribed.

Although in some experimental conditions the inhibition of RNA synthesis caused by histones may be due to changes induced in the solubility of chromatin preparations, in appropriate conditions their effect appears at least in part to be due to association with sequences of DNA which would otherwise be free to interact with RNA polymerase. There appears to be no specificity in this reaction. All sequences of DNA associate equally well with histones and there seems at present no reason to suppose that any particular class of histones binds more effectively to DNA; however, the lysine rich histones are a possible exception, consistent with the observation that their role in maintaining the structure of chromatin may be different in nature from that of the other histones (see page 133).

Non histone proteins appear to play two roles in controlling the template activity of chromatin. When histones are extracted from chromatin, the resulting preparation includes DNA and non histone proteins and is more active in RNA synthesis than untreated chromatin but less active than purified DNA. Paul and Gilmour (1968) found that when histones are added to such preparations of calf thymus chromatin from which they have previously been extracted, template activity is reduced to the level characteristic of native chromatin. But when histones are added to purified calf DNA, the resulting gelatinous preparation of nucleohistone is virtually inactive as a template for transcription. This implies that in chromatin the non histone proteins alleviate the repression caused by histones. Gilmour and Paul (1969) reported that the RNA transcribed from these preparations saturates the repetitive sequences of DNA at hybridization levels of:

free DNA	24%
DNA + histones	0
DNA + non histone proteins	14
DNA + histones + non histone proteins	7
chromatin	7

Although when associated with DNA alone the non histone protein fraction causes some inhibition of transcription, as part of the structure of chromatin it reduces the repression caused by the histones.

Specificity of Transcription In Vitro

Transcription of mammalian DNA by bacterial RNA polymerase is presumably non-specific in the sense that the enzyme does not recognise proper initiation signals; but we do not know if it simply transcribes all available sequences at random or whether it indulges in some preferential selection of sequences (for example, by recognising sequences related to the binding sites of bacterial DNA). If transcription by the bacterial enzyme is random, we may view its use as a probe to determine which sequences are accessible to it and which are physically protected from transcription by the presence of chromosome proteins. In this case, both strands of the template may be transcribed in these experiments, whereas the endogenous enzyme should select one strand at any locus. Comparing the RNA synthesized in vitro with that transcribed in vivo should in principle define whether the sequences exposed in chromatin and therefore available to the bacterial enzyme are the same as those expressed in the cell; because of the predicted failure of the bacterial enzyme to exercise strand selection, at most half of the sequences transcribed in vitro should represent sequences found in the cell.

That different sequences are transcribed in vitro from chromatins derived from different tissues was suggested by Tan and Miyagi (1970). By using hybridization experiments to characterize the product of the in vitro reaction, they found that the RNA transcribed from rat DNA by bacterial polymerase saturates 22% of its template; but RNA transcribed from liver or kidney chromatin saturates only 6% or 3% of rat DNA respectively. When DNA is saturated with the RNA transcribed from liver chromatin, addition of the RNA transcribed from kidney chromatin causes a further 2% hybridization. In the reverse experiment, in which DNA is first saturated with RNA transcribed in vitro from kidney chromatin and then tested for its ability to bind the RNA transcribed from liver chromatin, an additional 5% hybridization results. Since the RNA products show only limited competition, these experiments provide a crude demonstration that many of the repeated sequences available for transcription in vitro might be different in the two chromatins.

Competition experiments between the RNA transcribed in vitro and that extracted from the tissue used as the source of the chromatin have been taken to imply that similar sequences are transcribed in vitro and in vivo (reviewed by Paul, 1970). Tan and Miyagi (1970) found that when DNA is saturated with RNA transcribed in vitro from rat kidney or liver chromatin, addition of RNA extracted from the same tissue as the chromatin causes no extra hybridization. The same sites of DNA therefore seem to correspond to RNA synthesised in vitro and in vivo.

A lesser degree of competition is usually suggested by experiments using simultaneous competition. Figure 6.8 shows the competition of excess unlabelled RNA extracted from rabbit bone marrow or thymus with the labelled RNAs synthesised in vitro from the chromatins of these tissues. Paul and Gilmour (1968) found that bone marrow RNA competes for DNA about twice as effectively as thymus RNA against labelled RNA transcribed from bone marrow chromatin; in the reverse experiment, transcripts of thymus chromatin are competed more effectively by thymus RNA than by bone marrow RNA, although the difference in competition is much less.

Comparable results have been found in many other experiments using chromatins of a variety of tissues from different mammals. Smith, Church and McCarthy (1969) reported that the RNA synthesised in vitro from the chromatins of mouse embryo liver, kidney or brain behaves in competition experiments as though similar to RNA extracted from its homologous tissue but different from the RNA molecules of other tissues. In each case, labelled RNA transcribed in vitro is competed effectively by an excess of unlabelled

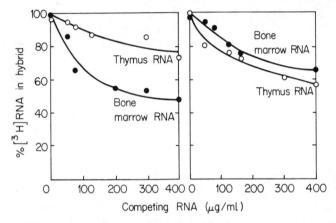

Figure 6.8: competition between labelled RNA transcribed in vitro from isolated chromatin and unlabelled RNA extracted from rabbit tissues. Left: RNA transcribed from bone marrow chromatin is better competed by RNA of bone marrow than that of thymus. Right: RNA transcribed in vitro from thymus chromatin is a little more efficiently competed by thymus RNA than bone marrow RNA. These experiments were performed with membrane bound DNA and therefore assay only repeated sequences. Data of Paul and Gilmour (1968).

RNA extracted from the same tissue as the chromatin, but is competed much less well by RNA derived from any other source. The RNA transcribed in vitro from liver chromatin is much better competed by RNA extracted from the nuclear fraction of liver than by RNA from the cytoplasm; this is consistent with the concept that cytoplasmic RNA represents only some of the sequences included in nuclear RNA.

How specific is the restriction of transcription in chromatin reconstituted from its dissociated components? When chromatin is dissolved in 2 M NaCl–5 M urea, all but a very small proportion of the chromosome proteins are dissociated from DNA. In reviewing early work on the transcription of chromatin, Bonner et al. (1968b) reported that an active preparation can be reconstituted by mixing chromosome proteins and DNA under these conditions and removing by gradual dialysis first the NaCl and then the urea. They suggested that the preparation reconstituted in this way transcribes the same sequences of RNA as those active in native chromatin.

The conditions of reconstitution are important. Using the NaCl-urea technique, Spelsberg and Hnilica (1969b) found that transcription is restricted to the level characteristic of chromatin when non histone proteins are mixed with histones to yield a preparation of chromosome proteins which are then added to DNA, or when non histone proteins are mixed with DNA before the addition of histones. But when histones are added to DNA to form a nucleohistone preparation to which non histone proteins are added, transcription remains at the almost completely inhibited level typical of nucleohistone. Binding of histones to DNA under these conditions is therefore non-reversible; the non histone proteins must alleviate the repression of transcription mediated by histones by preventing the binding of histones to DNA rather than by removing from DNA histones previously bound.

By reconstituting active preparations from components derived from different rat tissues, Spelsberg, Hnilica and Ansevin (1971) found that only the source of the non histone proteins influences the selection of sequences for transcription. Figure 6.9 shows that the labelled RNA synthesized in vitro from liver chromatin is competed better by the RNA of liver than by that of thymus. Preparations reconstituted from the components of liver chromatin appear to show the same specificity of transcription. When non histone proteins of liver and histones of thymus are provided for reconstitution, the specificity appears to remain that typical of liver chromatin; but when the non histone proteins are derived from thymus and the histones from liver, competition with liver RNA decreases to the level typical of that displayed by RNA transcribed from thymus chromatin. This suggests that the repression mediated by the histones is non-specific and that any elements which control the specificity of transcription in vitro may reside in the fraction of non histone proteins.

That RNA molecules associated with non histone proteins control the specificity of transcription has been suggested by Bekhor, Kung and Bonner (1969). In contrast with the results of other experiments upon the transcription of chromatin extracted with 2 M NaCl, they found that the product of RNA synthesis in vitro saturates 50% of the sequences of DNA, the same value achieved by RNA transcribed from purified DNA. The RNA transcribed from native pea bud chromatin hybridizes to a saturation level of 8%. They therefore suggested that the non histone proteins which remain associated with the DNA after salt extraction do not restrict transcription; specificity should reside in the fraction containing all the histones and RNA and less than 30% of the non histone proteins which is extracted.

When histones are added to the salt-extracted preparation in 2 M NaCl, gradual dialysis to low salt reconstitutes a preparation whose activity in RNA

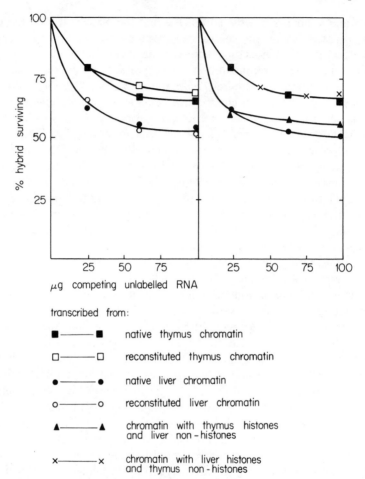

Figure 6.9: hybridization assays of RNAs transcribed from rat chromatins dissociated and reconstituted in vitro. Filters bearing 4 μg of DNA were incubated with 70 μg H^3-RNA transcribed in vitro from liver chromatin by bacterial RNA polymerase; incubations also included the amounts and classes of unlabelled RNAs shown. These reactions concern only repeated sequences. Left: the labelled RNA transcribed from liver chromatin is competed to a greater extent by unlabelled RNA transcribed from liver rather than from thymus chromatin. Native and reconstituted chromatin preparations appear to transcribe similar RNA sequences in vitro. Right: RNA transcribed from chromatins reconstituted with proteins derived from both liver and thymus preparations appear to direct synthesis of RNAs in the same way as the chromatins from which their non histone proteins are derived. Chromatin reconstituted with liver non histone proteins synthesizes RNA which is a better competitor than that transcribed from chromatin reconstituted with thymus non histone proteins. Data of Spelsberg, Hnilica and Ansevin (1971).

synthesis is that typical of chromatin. But the RNA products saturate DNA at 50 % and are not competed by the RNA transcribed from native chromatin; this implies that histones have bound at random sites, leaving different sequences uncovered on different molecules of DNA. When the preparation is instead reconstituted by dialysis out of 2 M NaCl–5 M urea, however, the sequences transcribed appear to be the same as those active when native chromatin is used as template. Consistent with the idea that the presence of urea is needed to allow the proper structure of chromatin to reform is the observation that salt-reconstituted chromatin has a gradual melting profile whereas that reconstituted in the presence of urea has the discontinuous profile characteristic of native chromatin.

An alternative to the idea that urea allows gradual reformation of hydrophobic forces needed to reestablish the proper structure of chromatin has been proposed by Bekhor, Kung and Bonner; they suggested that a fraction of chromosomal RNA (cRNA) may be responsible for the specificity of transcription and that the presence of urea allows these molecules to recognise their binding sites on DNA by base pairing. In support of this contention, they noted that treatment of the salt extracted protein fraction with ribonuclease (when cRNA is susceptible to attack) abolishes the specificity of reaction; but treatment of chromatin with ribonuclease (when cRNA is resistant to degradation) does not.

Interpretation of these experiments depends upon the stringency of the conditions used to characterize the RNA products. McConaughy, Laird and McCarthy (1969) noted that in the 30 % formamide used, the Tm of duplex hybrid molecules should be greatly above the incubation temperature used of 24°C. This implies that there may have been little specificity in the hybridization reaction. However, using a more stringent hybridization assay to characterize the products, Huang and Huang (1969) also observed that chick embryo chromatin reconstituted from NaCl–urea can be transcribed into RNA which competes with the nuclear RNA found in vivo. Degradation of the cRNA with zinc nitrate before reconstitution causes the transcript to lose the ability to compete. This implies that cRNA is responsible for selecting the sequences which are transcribed in vivo and in vitro.

That histones repress RNA synthesis in a non-specific manner which has no effect upon the selection of sequences for transcription is implied by all reconstitution experiments. Whether any specificity resides in the non histone proteins and/or RNA associated with them is more controversial. The presence of 5 M urea at 0°C used for reconstitution does not enable DNA to unwind for specific base pairing with RNA since its Tm in these conditions is about 50°C. Bekhor, Bonner and Dahmus (1969) observed that cRNA, but not other RNAs, associate with DNA in 5 M urea in a state resistant to ribonuclease; the form of this association and whether it is sequence specific remains a matter for speculation. And in contrast to experiments showing that degradation of RNA reduces specificity, Teng, Teng and Allfrey (1971) found that non histone proteins of rat liver contain no RNA and that the specificity of transcription of reconstituted chromatin is not influenced by treatment with ribonuclease.

The derivation of chromosomal RNA has been controversial. Chromatin contains upto 10% RNA by mass, although the value depends upon the tissue, species and method of preparation. At least some of this RNA represents nascent molecules whose synthesis has not yet been completed. Huang and Bonner (1965) reported that chromosomal RNA of pea bud can be distinguished from nascent RNAs by its association with the proteins released from chromatin by 4 M CsCl and by its low molecular weight, corresponding to a length of 40–60 bases.

That chromosomal RNA may be an artefact has been suggested by Artman and Roth (1971), who proposed that it is derived during preparation of chromatin by degradation of ribosomal or Hn RNAs, and by Heyden and Zachau (1971) who suggested that it is a contaminant derived from transfer RNA. Both these suggestions have been refuted by Holmes et al. (1972), who pointed out that cRNA is distinguished both by its characteristic level of hybridization with DNA and by its unusually high content of dihydrouridine (or dihydro-thymidine in some species), of the order of 10% of the bases. Similar cRNA fractions have been isolated from Novikoff ascites tumour cells of the rat by Dahmus and McConnell (1969). To exclude the presence of artefacts, Jacobson and Bonner (1971) prepared cRNA from several rat tissues by methods other than extraction with 4 M CsCl. They reported that it is present both in the chromosome proteins extracted by SDS and in association, apparently by covalent linkage, with a non histone protein extracted together with the histones by 2 M NaCl.

By hybridizing cRNA with DNA under conditions when repeated sequences anneal, Bekhor, Bonner and Dahmus (1969) found saturation levels of 3–5% of the genome. By hybridization in RNA excess, Sivolap and Bonner (1971) showed that pea bud cRNA corresponds to transcripts of repeated sequences of DNA. Mayfield and Bonner (1971, 1972) suggested that cRNA may be tissue and species specific and that increases in cellular content are correlated with gene activation in regenerating rat liver.

Whether cRNA influences the specificity of transcription in vitro, however, cannot be decided until experiments have been designed to show that cRNA molecules provide the only source of specificity when chromatin reconstituted from the components of different cell types is used as a template for RNA synthesis. This demands a better fractionation of chromatin into components which can be isolated for use in reconstitution experiments. And of course these experiments remain subject to the general limitations on transcription of chromatin in vitro. To show that cRNA controls gene expression in vivo therefore requires a much closer correlation between the cellular population of cRNA molecules and specific events in gene expression, in particular a demonstration that an individual species of cRNA is implicated in transcription of genes which can be identified.

One general interpretation of experiments which show some specificity in the restriction of transcription of chromatin in vitro is that gene expression is controlled by the organization of chromatin structure; active sequences of DNA may be exposed so that they are available for transcription by RNA polymerase whereas inactive sequences are covered by protein and are therefore inaccessible for transcription. The corresponding implication of reconstitution experiments is that all the information needed to specify the structure of chromatin is contained in its components, which under appropriate conditions may be dissociated and reconstituted without loss of structure (analogous to the dissociation and reconstitution of bacterial ribosomes). Since all these experiments use bacterial RNA polymerase as a probe to identify available sequences by non-specific transcription, these results imply that exposure of DNA may be sufficient as well as necessary for establishing gene expression.

Two features of experiments upon transcription of chromatin in vitro suggest that these conclusions may be premature: the characteristics of the transcription process differ considerably from those which presumably prevail within the nucleus; and the products of the reaction have not been sufficiently well characterized. The rate of transcription of chromatin in vitro remains linear for only a short time and the RNA products may be smaller than the molecules synthesized in the nucleus. Bacterial RNA polymerase probably synthesizes only rather short molecules of RNA from a chromatin template; its failure to initiate and terminate at the proper sites may therefore mean that even if transcription is confined to particular sequences of chromatin, the population of RNA products does not represent all available sequences equally.

Binding of bacterial RNA polymerase and the endogenous enzyme to chromatin appear to take place by different mechanisms. Keshgegian and Furth (1972) found that bacterial RNA polymerase displays the same K_m when it binds to either chromatin or DNA of calf thymus—so the same amount of either template is needed to bind the enzyme. The V_{max} of reaction is lower with chromatin than with DNA, which suggests that many enzyme molecules bind to sites which they do not transcribe. Similarly, Shih and Bonner (1970) showed that transcription of DNA by bacterial enzyme is inhibited by either poly-lysine or poly-arginine but that the concentration of template needed to saturate the enzyme is the same whether free DNA or DNA–polypeptide complexes are provided. These experiments therefore suggest that bacterial polymerase binds equally well to exposed DNA and to sequences protected by proteins; but the enzyme transcribes only the exposed sequences. To this extent the enzyme thus acts as a probe to detect exposed sequences.

Using calf thymus chromatin, Keshgegian and Furth found that the mammalian enzyme has a much lower K_m for binding to chromatin than to free DNA; some 2·5 times more chromatin than DNA is needed to achieve half saturation of the enzyme. But the V_{max} is similar with either chromatin or DNA as template. This suggests that the endogenous enzyme is restricted in the sites to which it binds in chromatin but that it transcribes RNA from all the binding sites which it recognises. The interaction with chromatin of the bacterial and mammalian RNA polymerases is therefore different in nature.

A critical issue about the use of bacterial polymerase is whether it transcribes the same sequences of DNA which would be transcribed within the nucleus by the endogenous enzyme. By comparing the activities of increasing amounts of bacterial and nucleoplasmic polymerases with limiting amounts of chromatin, Butterworth, Cox and Chesterton (1971) showed that the two enzymes appear to be independent in their selection of binding sites; the polymerases must transcribe different sequences of DNA since their activities are additive rather than competitive. It must therefore at present remain doubtful whether transcription by the bacterial enzyme can be taken to identify the sequences of chromatin which are expressed in the cell.

Another serious difficulty in interpreting experiments upon transcription of chromatin in vitro is that the conditions of hybridization used to characterise the transcripts permit annealing only with repeated sequences of DNA. Both the levels of saturation and the extent of competition between different preparations are probably overestimated appreciably in this reaction. And the accuracy of hybridization may be further reduced by the incomplete nature of the molecules transcribed by bacterial enzymes; size is especially important when repeated sequences are concerned. Until molecules of the same length found in the cell are synthesized in vitro by the endogenous enzyme and are characterized by hybridization with both repeated and non-repeated sequences of DNA, we cannot assess the specificity of transcription in vitro.

And in particular, comparisons of the populations of RNA transcribed from chromatins of different tissues provide an insufficiently specific assay for gene expression; only the synthesis in vitro of specific gene products can be accepted as evidence that preparations of chromatin retain their native control of gene expression. Reticulocyte chromatin, for example, should direct the synthesis of precursor molecules to globin mRNA, whereas chromatin of other tissues should fail to do so. The same criticisms apply to attempts to identify the molecules which regulate gene expression; particular non histone proteins (or cRNA molecules) must be identified with the activation of individual genes whose products can be characterized.

Although chromatin may perhaps retain some of its native characteristics in vitro, it would therefore at present be premature to conclude that the same sequences are transcribed in vitro as in vivo and that the original structure is reformed after its dissociation and reconstitution. It is thus not possible to attribute specificity of the restriction of gene transcription to any particular component of chromatin. As the typical data of figures 6.7–6.9 emphasize, the extent to which populations of RNA transcribed from chromatin can be characterized by hybridization assays is limited; and competition experiments allow only a very crude estimation of the tissue specificity of RNA sequences and thus of the fidelity of transcription in vitro.

It seems very likely that the active sequences of euchromatin possess a molecular architecture different from that of the inactive euchromatic sequences, but we cannot say whether maintenance of this structure although necessary is in itself adequate to ensure proper gene transcription. We may speculate also that if the transient modification of histones and non histone proteins is required in vivo to achieve their proper binding in chromatin (see page 129), the probability of achieving reconstitution of the native structure in vitro from non modified components may be quite low.

Induction of Transcription by Steroid Hormones

Steroid hormones commonly cause changes in gene expression in their target tissues and therefore offer an opportunity to follow the activation of a restricted number of genes in cells of established phenotype. A system widely used to follow the uptake of hormone by its target cells has been the interaction of 17β-estradiol with the uterus. Hen oviduct has proved a particularly good system in which to study gene activation since estrogen causes the synthesis of two specific proteins, ovalbumin and lysozyme, and progesterone induces synthesis of the protein avidin. In this tissue it may therefore prove possible to follow the activation of individual genes.

That estrogens cause changes in gene expression at the level of transcription has been demonstrated in many target tissues; actinomycin, puromycin and cycloheximide all usually block the response to the hormone (reviewed by Mueller et al., 1971). Diethylstilbestrol (DES) acts as an estrogen to induce the morphological changes in hen oviduct cells which are followed by synthesis of ovalbumin; and O'Malley et al. (1969) showed that there is an increase in the number of RNA sequences hybridizing with repeated DNA after treatment of chicks with DES. By isolating the RNA released from oviduct polysomes with SDS, Means et al. (1972) showed that hormone treatment results in the appearance of messengers which can be translated to give ovalbumin in vitro.

That steroid hormones influence gene action directly by entering the nucleus has been known for some time; Maurer and Chalkley (1967) first observed that labelled 17β-estradiol becomes bound to chromatin in the nuclei of its target tissue, calf endometrium. Gorski et al. (1968) and Jensen et al. (1968) first suggested that estrogen has a two step interaction with uterus. The steroid enters the cell to bind to a cytoplasmic receptor protein which sediments at about 9S but reversibly dissociates to a size of 4–5S on sucrose gradients in low ionic strength; the hormone then passes from the cytoplasm into the nucleus, from which it can be recovered in association with a protein sedimenting at about 4S. Shyamala and Gorski (1969) found that when estradiol binds to uterine cells, the 9S binding component of the cytoplasm disappears as a binding activity appears in the nucleus. The form of the receptor depends upon the ionic conditions, for Yamamoto and Alberts (1972) showed that in low ionic strength estradiol is bound by a cytoplasmic protein which sediments at 4S; after entry to the nucleus it is found in a form sedimenting at 5S. The state in which the receptor protein exists in the cell is not defined, but it seems likely that its subunit structure or conformation (or both) changes with transport from cytoplasm to nucleus.

The two step action of steroid hormones suggests two questions about the mechanism through which they influence gene transcription. Are the cytoplasmic receptor proteins present only in target tissues so that other types of cell are unable to respond to the hormone? And do the receptor proteins alone mediate the actions of the steroids, or are the nuclei of target tissue differentiated to possess the ability to respond, perhaps in the properties of their chromatin? That hormone action is mediated only through the interaction with cytoplasmic receptor is suggested by the observation that the affinities of uterine receptor for analogues of estradiol corresponds well with their overall biological activities.

When progesterone is injected into chicks or added to an aqueous extract of chick oviduct, it is bound by a receptor protein. O'Malley, Sherman and Toft (1970) showed that the receptor is present only in oviduct but is absent from non-target tissues such as lung or spleen. Sherman, Corvol and O'Malley (1970) reported that the hormone–receptor complex sediments at about 4S; a label in the hormone appears first to enter the cytoplasm but within eight minutes reaches a peak in the nucleus, in both fractions continuing to sediment at 4S. O'Malley, Toft and Sherman (1971) showed that oviduct segments which have been exposed to progesterone have receptor protein in both cytoplasm and nucleus; but tissue which has not interacted with hormone possesses receptor protein only in cytoplasmic extracts.

This suggests that receptor proteins are present in the cytoplasm of oviduct cells and move into the nucleus only after binding to the hormone; consistent with this concept, the cytoplasmic content of receptor declines as the nuclear content increases. Models of this nature in effect visualize the progesterone molecule as a way to move the receptor protein into the nucleus; the role of the hormone may be to activate a programme for differentiation which has previously been determined in the cell.

The progesterone–receptor protein complex binds to chromatin, displaying some specificity. Steggles, Spelsberg and O'Malley (1971) demonstrated that binding is most effective with chromatin of oviduct; it is four times less effective with that of erythrocytes and seven times less effective with spleen chromatin. Progesterone alone, or progesterone which has been incubated with cytoplasmic extracts of cells other than its target tissue, does not bind to chromatin. Similar results have been obtained by Steggles et al. (1971) with receptor bound complexes of dihydrotestosterone and estradiol; receptor proteins exist only in target tissues and when complexed with hormone bind preferentially to homologous chromatin. This suggests that target tissues may be distinguished from non-target tissues first by the presence of a cytoplasmic receptor protein which carries the hormone into the nucleus and second by an organization of chromatin which specifically binds the receptor–hormone complex.

By using 0·3 M KCl to extract labelled progesterone previously bound to chromatin, Spelsberg, Steggles and O'Malley (1971) showed that the label continues to display the properties typical of the hormone–receptor complex; this implies that it is the complex as a whole which binds to chromatin and reinforces the view that the role of the hormone may be to activate a programme for gene expression rather than itself to interact with the genome. In experiments in which chromatins of oviduct and spleen were dissociated and reconstituted from their components, each retained its characteristic ability to bind the progesterone–receptor complex. This suggests that it is some particular component of oviduct chromatin which is responsible for binding the complex.

In reconstitution experiments using heterologous components, the level of binding of the receptor–hormone complex appears to be established by the source of the non histone proteins. And omission of this fraction greatly reduces the level of the binding of the complex. Spelsberg et al. (1972) confirmed that it is the protein components which have binding activity by demonstrating that treatment of dissociated chromatin with ribonuclease does not inhibit binding. By sequentially extracting four fractions of non histone proteins from chromatin, they showed that it is the loss of one fraction alone (AP3) which is correlated with loss of ability to bind the progesterone–receptor complex. The implication of these results is that some non histone protein(s) present in oviduct chromatin but absent from other chromatins may specifically bind the hormone–receptor complex; this binding presumably results in changes in gene expression.

The cytoplasmic receptor protein for progesterone can be separated on DEAE-cellulose into two components, each of which sediments at 4S. Schrader, Toft and O'Malley (1972) showed that both components bind the steroid in the cytoplasm, enter the nucleus and become bound to chromatin. The complex of component A with progesterone binds to DNA when mixed with it on a sucrose gradient, whereas component B continues to sediment at its usual position and does not interact with added nucleic acid. But if chromatin is mixed with preparations of either component bound to progesterone, only component B binds and component A is inactive.

Whether the two components of the progesterone receptor protein interact with each other and how they bind to chromatin to change gene transcription is not yet defined. But it seems likely that their action is at least in part mediated by a non histone protein present in oviduct chromatin and probably absent from the chromatins of other tissues. Other receptor proteins also display the ability to bind to DNA; Baxter et al. (1972) found that the glucocorticoid receptor of cultured HTC cells binds to a limited number of sites on DNA and Yamamoto and Alberts (1972) showed that when the estradiol receptor binds to DNA its sedimentation rate changes from 4S to 5S, implying an alteration of conformation. However, we do not know whether this binding itself is sufficient for hormone action in vivo; it seems more likely that the structural organization of chromatin may control the sites of DNA which are available for binding. It is of course possible to speculate that the progesterone receptor proteins function in a manner in which the interaction with chromatin of one component helps make available DNA sites for binding of the second component.

Association with progesterone is needed for the receptor protein(s) to enter the nucleus. But we do not know what role the hormone molecule itself may play in the interaction with chromatin; because the assays for binding of the protein components to DNA or chromatin rely upon following a radioactive label in the progesterone molecule, we do not know whether the isolated proteins have the same abilities or whether they require the presence of the hormone moiety for their actions.

Models for the Unit of Transcription

An essential question which we cannot yet answer is the identity of the unit of transcription. Defining the units in which eucaryotic genes may be expressed falls into two parts: deducing the formal network of control circuits; and demonstrating the molecular interactions by which they are executed. Neither is understood in eucaryotic cells. But several models for control of the eucaryotic genome have been proposed, in general based upon a rethinking of the concepts of the Jacob-Monod model for the bacterial operon to suit the characteristics of eucaryotic cells. Although at present we lack the techniques to test the predictions of these models, they provide a set of concepts which are useful for considering the nature of the systems which may control eucaryotic genes.

All models for eucaryotic gene control postulate the existence of the two types of sequence identified in bacterial operons: structural genes which are transcribed into RNA (and subsequently translated into proteins, with the exception of the cistrons for tRNA and rRNA); and sequences which are not transcribed but whose function is to be recognised by regulator molecules or by enzymes of nucleic acid synthesis. It is usual to assume a similar organization to that of bacterial DNA in which recognition elements lie adjacent to structural genes which they control.

Genetic analysis of bacterial mutants has traced control networks by identifying and mapping the structural genes which code for proteins, the adjacent promotor and operator elements which are recognised by RNA polymerase and repressor proteins, and the regulator genes which code for the repressor proteins (discussed in chapter 7 of volume 1). Genetic loci have of course been mapped in some eucaryotic species, notably in Drosophila, by their mutant phenotypes; but structural genes have been identified with their protein products in only a small number of instances. And in mammals where mutant proteins have been identified by their effects on metabolism, gene mapping is at best rather crude. None of the classes of bacterial regulator sequences has been identified in eucaryotic cells; although mutations in some loci of eucaryotic chromosomes cause pleiotropic effects which imply they may have a regulator function, none has been characterized. There are therefore no molecular maps of DNA which relate regulator elements to structural genes.

One difficulty in analysing gene control in eucaryotes is their multi-cellular organization; any particular gene is usually active in only a limited number of cells of the organism. Another problem is the lack of gene systems which are dispensable under appropriate conditions, as for example are the inducible and repressible operons of E.coli; this means that regulator mutants may often be lethal to the organism. A further problem would lie in distinguishing any putative regulator mutations from adjacent structural genes, for their low frequencies of recombination would mean that the small number of progeny of eucaryotes would be inadequate to separate them.

From the mutations which have been identified and mapped, a general feature of the eucaryotic genome appears to be the lack of any functional clustering of genes analogous to that of bacterial DNA. Functional clustering appears to have been lost during the evolution of eucaryotes (see chapter 8 of volume 1). Any model for control must therefore provide a means to regulate together genes located at different sites in the genome. This implies that the unit of transcription may include only a single structural gene; we may therefore visualise each unit as comprising recognition elements adjacent to a structural gene. However, the large size of the immediate product of transcription (Hn RNA) compared with the messenger shows that the unit of transcription includes more information than that of the structural gene itself.

Control of individual bacterial operons appears in general to be exercised by repressor proteins which respond to small molecules of the environment. These operons therefore lie under a negative control in which the system is active unless specifically repressed; failure of the control system therefore leaves the operon in a state of continuous expression (discussed in chapter 8 of volume 1). In addition to this negative control, positive control seems to be utilised as a general coordinating system (activated by cyclic AMP); a feature of positive control systems is that the operon cannot function unless it is specifically activated, so that failure of the system leaves the structural genes in a permanently repressed state. One of the critical questions about eucaryotic genomes is whether they are under negative control or positive control.

Turning from control circuits to molecular mechanisms, we must define how transcription relates to the structure of chromatin—are regulator molecules part of the architecture of the chromosomes or are they superimposed on it? In the absence of regulator mutants, it is not possible to deduce by genetic analysis whether control is positive or negative; but if chromatin in general possesses sequences of DNA non-specifically repressed by histones, the eucaryotic genome must be subject to a positive control which specifically activates sequences to be transcribed. We must therefore seek to identify activator molecules; both non histone proteins and RNA have been proposed for this role (see page 357). Models for formal networks of control make few implications about the molecular nature of the interactions which must take place with DNA; in view of the structure of chromatin it is likely that these interactions will prove to be more complex than in bacteria. It is possible to speculate that activator molecules change the local structure of the chromosome, so that sequences of DNA become exposed for transcription; or we may suppose that they must interact specifically with both sequences of DNA and with RNA polymerase itself.

The model proposed by Britten and Davidson (1969) and reviewed by Davidson and Britten (1973) postulates four classes of sequence of eucaryotic DNA whose interactions constitute a system of positive control. *Producer genes* comprise sequences of DNA which code for proteins; these are directly analogous to the *structural genes* of bacterial operons, the term we shall use here. Adjacent to each structural gene is at least one *receptor site*; a structural gene can be transcribed only when an activator molecule recognises an adjacent receptor site. (We shall describe sequences which are transcribed into RNA as *genes* and sequences which serve only as recognition elements as *sites*; an alternative nomenclature is to describe all functional units of DNA as genes, whether or not they are transcribed.)

The regulator molecules which control gene activity are postulated to be *activator RNAs*; however, the model remains formally the same if activator proteins are substituted. The loci which code for the activator RNAs are the *integrator genes*; if proteins control transcription, it is necessary only to postulate that the transcripts of the integrator genes are translated instead of themselves possessing activator activity. The integrator genes are analogous to the *regulator genes* of bacterial operons.

To provide for the specific control of gene expression, it is necessary for either the synthesis or the activity of regulator molecules to respond to the milieu of the cell. The activities of the repressor proteins of bacterial operons are controlled by their interactions with small molecules of the environment. Britten and Davidson proposed that the activities of the integrator genes of eucaryotic DNA are controlled by adjacent *sensor sites*; an integrator gene can be transcribed only when its controlling sensor site is activated. Sensor sites are recognised by agents which change the pattern of gene expression;

hormone–protein complexes, for example, may bind to sensor sites to cause transcription of the adjacent integrator genes. The basic control circuit of the model shown in figure 6.10 therefore comprises a sensor site–integrator gene complex whose synthesis of activator RNA controls the activity of a receptor site–structural gene complex.

To account for the concerted control of many genes in each state of differentiation of a cell, Britten and Davidson proposed that the receptor sites and integrator genes may be repeated; there may be many copies of any one receptor or integrator sequence in the genome. These two types of repetition allow a small number of control elements to activate the large number of genes which may be active in any one cell.

Because structural genes whose protein products serve related functions may be located at separate chromosome sites, their common control requires

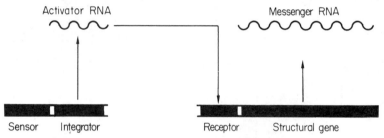

Figure 6.10: Britten-Davidson model for eucaryotic gene control. Stimulation of a sensor site causes the adjacent integrator gene to synthesize an activator RNA. The activator RNA is recognised by a receptor site and this interaction causes the production of messenger RNA from the adjacent structural gene.

each to possess a copy of the same receptor sequence. All these genes may then be transcribed in response to a single species of activator RNA. Repetition of receptor sites might be utilised to ensure that all the enzymes of some metabolic pathway are coordinately synthesized; to control their activities independently would be wasteful if the functions of each gene product are effective only as part of the intact pathway. (This is analogous to the genes of the arginine system of E.coli, which although dispersed on the chromosome all respond to a single repressor protein; see chapter 8 of volume 1.)

Transcription of any gene, or set of genes, may be demanded in more than one differentiated state of the cell. It must therefore be possible to activate structural genes in differing circumstances. This may be achieved by redundancy of either the receptor sites or the integrator genes. The model postulates that each structural gene may possess several adjacent receptor sites, any one of which may activate transcription by responding to its appropriate activator RNA. Figure 6.11 shows that redundancy in the receptor sites allows overlapping combinations of structural genes to be controlled by activator

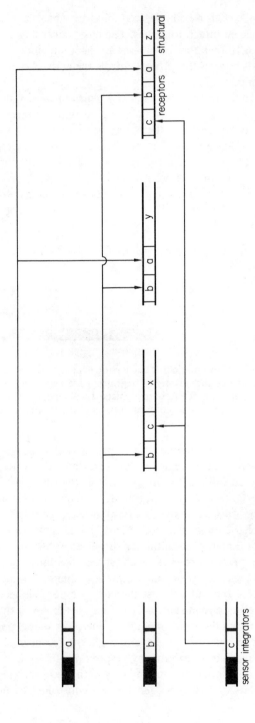

Figure 6.11: redundancy of receptor sites in the Britten-Davidson control model. Each structural gene may be activated in response to the binding of activator RNA at any one of several adjacent receptor sites. Activator a thus controls structural genes y and z; activator b controls structural genes x, y, and z; and activator c controls structural genes x and z. Each structural gene therefore belongs to more than one set.

RNAs. Production of any particular activator RNA causes transcription of all the structural genes which possess a copy of the receptor site which it recognises; any structural gene may therefore be included in the *set* recognised by any activator RNA by possession of the appropriate receptor sequence. Two activator RNAs thus control overlapping sets of structural genes when some of the genes possessing the receptor sequence for one activator RNA also possess the sequence for the other. We may imagine that the genes of each set code for proteins whose function is related; the product of a gene may be needed in more than one pathway and may therefore be included in more than one set.

When extensive changes are to be introduced in the pattern of gene expression, it may be necessary to activate many sets of genes as, for example, during early embryonic development or in response to hormones. Integrator genes may therefore be organised in clusters, each cluster falling under the control of a single sensor site. Activation of a sensor site causes transcription of all the integrator genes under its control; the corresponding activator RNAs then switch on the several sets of structural genes which they may recognise. The sets of structural genes controlled by any one sensor site have been termed a *battery*. Any particular state of cellular differentiation may in principle be defined by the batteries of structural genes which are active.

Any particular set of genes may be required in response to more than one stimulus. This can be achieved by repetition of the appropriate integrator genes. Figure 6.12 shows that when an integrator sequence is repeated under the control of more than one sensor, its activator RNA may be synthesised in response to more than one set of cellular conditions. Repetition of integrator genes under the control of sensor sites achieves a similar result to the repetition of receptor sites at several structural gene loci; just as it is possible to include one structural gene in more than one set, so it is possible to include one set in more than one battery.

Since all the genes of any one set must always be expressed coordinately, we may suppose that each set comprises a comparatively small number of genes whose products serve quite closely related functions. And the number of sets in which any one structural gene may be included is probably limited; since it is possible that receptor sites must lie physically close to the structural genes which they control, there might be a small limit on the number of receptor sites which can activate a structural gene. The number of sets controlled by a sensor gene is potentially great, since many integrator genes might be transcribed into RNA under control of a single sensor. The battery controlled by a sensor site may therefore specify the more widely disparate sets of gene products necessary to achieve some particular state of differentiation.

Embryonic development demands the sequential operation of different sets of genes; and a programme for ordered expression of the genome can readily be imagined by supposing that the product of a structural gene in an

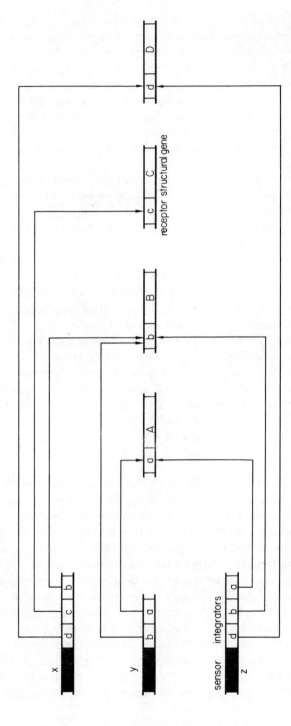

Figure 6.12: redundancy of integrator genes in the Britten-Davidson control model. Each sensor site causes synthesis of activator RNA at each of the integrator genes adjacent to it. By including the same integrator gene sequence under control of more than one sensor, the set of genes responding to its activator RNA may be activated in response to more than one cellular signal. Thus sensor *x* controls structural genes *B*, *C* and *D*, sensor *y* controls structural genes *A* and *B*, and sensor *z* controls structural genes *A*, *B* and *D*.

early battery acts upon a sensor site to cause transcription of the next battery to be expressed. Davidson and Britten (1971) suggested that an inequality in the cytoplasm of the egg of the factors which activate the first batteries might be sufficient to ensure that different sensors and hence different developmental programmes are followed to differentiate individual tissues. Some such compartmentalization must be invoked in company with models for gene control to explain the initial division of two cell types from one.

The repetition of receptor sites and integrator genes implies that these control elements must form part of the repeated sequences of the genome; this model implies that most, perhaps all, of the intermediate sequence component is concerned with coding for or being recognised by control elements. Although the model makes no prediction about the location in the genome of the groups of integrator genes controlled by each sensor site, it implies that each structural gene—presumably consisting of a non-repeated sequence of DNA—must be adjacent to the repeated sequences of the receptor sites. This is consistent with the alternation of repeated and non-repeated sequences observed in the Xenopus and sea urchin genomes (see page 203); since we may suppose that recognition sequences need not be long, the model is also in accord with the observation that repeated sequences appear on average to be rather short.

One advantage to the cell of using RNA rather than protein as the regulator molecule is that the entire control network might be contained in the nucleus; if activator proteins control transcription the RNA products of the integrator genes would need to be transported to the cytoplasm for translation into the regulator proteins. Whatever the molecular character of the regulator, the model predicts that there should be two classes of units of transcription. The predominant class should comprise the non-repeated DNA of a structural gene adjacent to the repeated DNA of the receptor sites. A different class consists of the group of integrator genes under control of a sensor site.

With RNA as the activator, this must correspond to a repeated sequence which is transcribed but not translated. Although the model does not imply any particular molecular interaction for the role of activator RNA, one interesting prediction is that if activator RNAs were to function by base pairing with unwound DNA, the nucleotide sequences of the integrator genes and the corresponding receptor sites might be identical. With protein as the activator, each integrator gene must presumably be longer; and the group as a whole must be coordinately translated as well as transcribed so that it should be represented in the messenger fraction. In this case, the multiple copies of each receptor might comprise related but not identical sequences, providing a good fit with the composition of the repetitive component.

In addition to any control loci which interact with activator RNA or with any other regulator molecule, each unit of transcription must contain a sequence to which RNA polymerase binds to initiate RNA synthesis and a sequence which subsequently terminates synthesis. The initiation site might be located either within the unit of transcription or may define its start. Any model to account for the control of eucaryotic genes must explain the very large size of the Hn precursor RNA; since messenger RNA is derived from its 3′ end, transcription must start at a point distant from the structural gene itself and terminate at its end. Of course, models which rely upon repeated sequences as control elements remain formally the same whether applied to control of transcription or translation. It is thus possible that some repeated sequences may be recognised in Hn RNA (rather than in DNA) as elements which control its processing to messenger RNA.

A model to reconcile the discrepancy in the sizes of Hn RNA and mRNA has been proposed by Georgiev (1969) and Georgiev et al. (1972), who suggested a negative control system in which binding of repressor molecules at any one of a number of receptor loci lying between the initiation site and the structural gene may prevent RNA polymerase from progressing (analogous to the action of the lactose repressor in E.coli; see chapter 7 of volume 1). This predicts that all the receptor sites adjacent to a structural gene should be transcribed; in this case the 5′ ends of Hn RNA should correspond to the repeated sequences of receptor sites whereas the 3′ end should contain the non-repeated sequence of the message. In contradiction to this model is the general concept that control must be positive to lift the repression caused by the histones; but of course it is also possible that an activation event might demand transcription of non-coding sequences preceding the structural gene.

A general problem in control is to construct a model for the interaction of activator molecules with DNA. A topological difficulty in recognising specific sequences of DNA resides in its intimate association with protein; at present we can only propose the general solution that the molecular structure of each chromatin is such as to allow access by activators to those sequences of DNA which may need to be transcribed. This organization must develop in each tissue during embryogenesis. A problem peculiar to models which rely upon activator RNA is to explain recognition of specific DNA sequences; this must presumably be achieved by base pairing with one of the single strands produced by unwinding DNA. Once unwound, interaction with RNA might, for example, maintain the DNA in an unwound state especially receptive to RNA polymerase; but it is necessary first to explain how DNA becomes unwound so that it is available for base pairing with RNA.

In a model proposed by Crick (1971), eucaryotic DNA is considered to fall into two general classes: "globular" DNA representing a compact structural organization of control sequences; and adjacent "fibrous" DNA representing uncoiled sequences coding for proteins. Crick proposed that the sequences recognised by activator molecules might be maintained in an unwound single stranded state by the topology of the globular structure. Chromosome proteins might coil the DNA in such a way that loops become forced into a single stranded conformation; these loops would provide the receptor sites recognised by activator molecules. Implicit in this model is the concept that an appreciable proportion of DNA may be maintained in a single stranded state in a sequence specific reaction dependent upon chromosome proteins, an idea inconsistent with our present view of the chromosome (see page 213); and it is not clear how recognition of the single stranded sequence—whether by activator RNA or by activator protein—may control transcription of the appropriate coding sequences.

A model also relying upon the division of DNA into compact and looser regions has been proposed by Paul (1972), who suggested that the less compact regions active in transcription might contain an additional chromosome component which prevents tight supercoiling—this might be a non histone protein(s). *Address sites* adjacent to structural genes would therefore have the function of serving as sequences recognised by this protein(s). One reason for the large size of the transcription unit might be a topological requirement for a minimum length of DNA which can be activated, perhaps because the process of transcription itself participates in unwinding the DNA. The evolution of such units might involve the acquisition of mutations to render non-coding extra lengths of DNA derived from gene duplications; this model therefore predicts that much of the non-message sequences of Hn RNA comprises "nonsense" which is removed in order to prevent its interference with translation.

CHAPTER 7

Interactions between Nucleus and Cytoplasm

Transplantation of Nuclei

Reciprocal Nucleocytoplasmic Interactions

A question commonly asked in the early days of developmental biology was whether nucleus or cytoplasm controls development. Such a question now seems too direct, of course, for we understand development to constitute a series of cyclic interactions between nucleus and cytoplasm. It is the cytoplasm of the egg which contains the information necessary to elicit from the nucleus the response appropriate for the start of embryonic development; and the products of the genes expressed at this time in turn modify the cytoplasm so that new patterns of gene expression are elicited from the nucleus during subsequent development.

The nature of the information in nucleus and cytoplasm can be studied in experiments which place a nucleus in a cytoplasm different from that in which it is usually located. This has been achieved both by micromanipulation and by somatic cell fusion. Micromanipulation enables the nucleus of a differentiated somatic cell to be placed in the cytoplasm of an egg of the same species; hybridization between two somatic cells yields a hybrid product containing both nuclei in a common cytoplasm.

The nuclei of all eucaryotic cells are restricted so that they express only certain genes; the nature of this restriction is one of the critical questions of developmental biology. The nucleus of the fertilized egg is *totipotential*; although itself expressing only certain genes, when its descendents subsequently generate the differentiated functions particular to each cell type they express all the genes utilised by the organism. As embryonic development proceeds, however, the range of gene functions which may be expressed in the descendents of any nucleus is progressively reduced as it enters pathways of increasing cell specialization.

Whereas the pattern of gene expression in embryonic cells is transient in the sense that their descendents may express different sets of genes, a fully differentiated cell (and its descendents, if any) may maintain a fixed state of gene expression. Transplantation of nuclei from differentiated cells to embryonic cytoplasm has been used to determine whether irreversible restrictions on gene expression accompany differentiation or whether nuclei which do not usually express certain genes nonetheless retain the potential to do so. These experiments show that changes in nuclear gene expression in differentiation are reversible.

Information in the cytoplasm must be molecular in the sense that it consists of regulator molecules which may enter the nucleus to influence its pattern of gene activation. The nature of this information can be studied by following the molecules which enter a nucleus after its transplantation. The specificity of these signals can be deduced from the properties of the hybrid cells produced by fusing two different somatic cells, in which genes in an inert nucleus of one species may respond to the cytoplasm of another species.

Positional information must also be present in the cytoplasm of the egg and early embryo. Development requires the differentiation of nuclei with various patterns of gene expression; this could not take place if all descendents of the first nucleus continue to respond in the same way to the cytoplasm. One model to explain the development of nuclei expressing different genes is to suppose that regulator molecules are not dispersed uniformly through the cytoplasm but are concentrated unequally in such a way that different nuclei gain different regulator molecules (or different quantities of them) from the cytoplasm and therefore start upon different developmental pathways. We know very little about how positional information may be established and maintained.

The reciprocal interplay of feedback signals between nucleus and cytoplasm is well illustrated by experiments with the unicellular organism Acetabularia. The zygote of Acetabularia germinates to form a stalk which grows to a total length of some 3–5 cms in about three months. During the next month, a *cap* of characteristic morphology develops at the end of the stalk, with a diameter of about 1 cm compared with the stalk diameter of 0·3–0·4 cms. At the lower end of the stalk, a *rhizoid* forms containing a single nucleus; this nucleus increases considerably in size during development, from some 4×10^{-9} mm³ in the zygote to 1×10^{-3} mm³ when the cap is fully developed.

Information about the state of the cap must be transmitted to the nucleus, for in cells with mature caps the primary giant nucleus begins to disintegrate, extruding a small diploid secondary nucleus which divides mitotically many times (reviewed by Hammerling, 1963). The secondary nuclei produced by these divisions are then transported by cytoplasmic streaming to the cap, where additional mitoses are followed by meiosis; the final number of secondary nuclei is of the order of 7000–15,000. Formation of secondary nuclei depends upon signals from the cap, for nuclear division can be completely inhibited by removing the full grown cap before the nucleus begins its reproductive cycle. This process can be extended indefinitely by removing the cap each time it regenerates.

The development of the cap in turn depends upon the nucleus. Experiments with Acetabularia provided a direct demonstration of the role of the nucleus in heredity and of the need for a messenger molecule to transport its information to the cytoplasm. When the nucleus is removed from a cell, the anucleate parts may survive for several months, with varying morphogenetic capacities (reviewed by Hammerling, 1953). The capacity of a stalk is proportional to its length—there is a concentration gradient of capacity from cap to rhizoid. This implies that messengers are produced by the nucleus and transported to the cap where they accumulate in stable form.

Each of the several varieties of Acetabularia forms a cap of characteristic appearance. If the nucleus of one plant is removed, that of another type may be inserted in its place. When nuclei are exchanged between A. mediterranea and A. crenulata, each plant develops a cap of the type characteristic of its new nucleus. Intermediate caps, showing characteristics of both the old and new nucleus, form in some instances because of mixing between the messengers synthesized by the new nucleus and those remaining from the old nucleus; but when they are removed the cap which regenerates displays the appearance coded by the new nucleus. When more than one nucleus is present in the cell— because further nuclei are added to instead of replacing the original nucleus— the type of cap formed depends on the ratio of the different classes of nuclei. These experiments thus demonstrate that each nucleus carries its own specific hereditary information, which can be expressed in the cytoplasm of another variety of the plant. The responses elicited from the nucleus by the state of development of the cytoplasm are species specific and determine the phenotype.

Developmental Potential of Differentiated Nuclei

Whether irreversible changes take place in the nuclei of somatic cells during differentiation is a question which has occupied developmental biologists since the last century. In most organisms, no change is apparent in the genetic material; all somatic cells appear to possess the same diploid chromosome content. So far as we can judge from nucleic acid hybridization studies, all differentiated cells contain the same sequences of DNA; there seems to be no addition of sequences representing particular genes in the chromosomes of the specialised cells in which they are expressed (see chapter 5), nor any loss of functions which are not utilised. (The amplification of ribosomal RNA genes which takes place at oogenesis in some organisms and their magnification in Drosophila to compensate for mutations reducing their number appear to be the only exceptions to the inflexibility of genetic content; see Gall, 1968, 1969; Brown and Dawid, 1968; Brown and Blackler, 1972; Henderson and Ritossa, 1970; Ritossa et al., 1971; Tartof, 1973.) Of course, the precision of the hybridization technique is too low to prove that all genes are invariant in representation, but it seems unlikely that any changes in genetic content occur during differentiation. Selective expression of genes alone must therefore be responsible for the development of differentiated phenotypes.

Maintenance of genetic content, however, does not preclude the possibility that other changes in the genome itself may take place, involving not the loss of genetic material but the permanent inactivation in each cell line of the genes which are not expressed. All differentiated chromatins may differ in their content of regulator molecules and thus in the pattern of gene expression. Alternative models invoking reversible or irreversible changes in gene expression during development make different predictions about these changes in chromatin structure. If gene expression is reversible, all chromatins may be characterized by a comparable organization, although differing in the molecular arrangement of their components. If irreversible restrictions on expression are imposed during differentiation, the structure of differentiated chromatin should differ in kind from that of totipotential embryonic nuclei; for example, nucleotide sequences might be modified by enzymes or proteins might be irreversibly associated with DNA.

To decide whether a differentiated nucleus is totipotential or restricted in its developmental capacity requires placing it in a situation in which it has the opportunity to express functions which have been turned off in its differentiated state. Such a test may be made in either of two ways. Cells of different types may be fused together by the mediation of the Sendai virus to give hybrid cells; the ability of an inactive nucleus to regain repressed activities in this situation shows that at least some functions may be activated by the new cytoplasm (see later). A test of the ability of a differentiated nucleus to give rise to other cell types may be achieved by micromanipulation experiments in which the nucleus of an unfertilised egg is removed or inactivated and the diploid nucleus of a somatic cell is inserted in its place. The ability of the transplanted egg to undertake normal embryogenesis is a measure of the extent to which differentiation of the transplanted nucleus may be reversed in egg cytoplasm.

Unfertilised eggs of amphibia have been used for most transplantation experiments since they are large enough to manipulate directly. In the frog Rana pipiens, the resident egg nucleus is usually removed with a glass needle and replaced by a nucleus taken from another cell. The egg must be stimulated into development by pricking with a needle. With Xenopus laevis, the eggs do not need activating and the resident nucleus is usually inactivated with an ultraviolet microbeam, after which a new nucleus may be implanted and utilised for subsequent development (reviewed by Gurdon, 1964).

Injection of a donor nucleus is performed by placing donor tissue in a saline solution lacking calcium and magnesium ions; the donor cell is sucked into a micropipette of a size which breaks the cell but retains the nucleus. The nucleus is surrounded by a small amount of cytoplasm whose presence is essential if it is to survive transplantation—exposure to the medium is invariably lethal.

It is more difficult to transplant nuclei from cells more advanced in differentiation because they have little surrounding cytoplasm compared with embryonic cells and are therefore more readily damaged.

In experiments with Rana pipiens, King and Briggs (1956) found that large numbers of eggs which are injected with nuclei extracted from the blastula stage of embryogenesis grow to yield normal tadpoles. Successful transplantations can also be achieved with nuclei taken from different parts of the gastrula embryo when cells have begun to develop specialized functions. Briggs and King (1960) later reared some of these tadpoles through metamorphosis.

Transplantations with Xenopus laevis follow a generally similar course, although greater numbers of the transplants survive to at least the tadpole stage. Gurdon (1962a) used as donors endoderm nuclei taken from stages between the late blastula and swimming tadpole. Although the cells from which they are derived have developed differentiated functions, upon transplantation the specialization of their nuclei is reversed and many of the eggs into which they are placed develop into normal frogs. These adults contain all the usual differentiated cell types of the frog; the endoderm nuclei are therefore totipotential, possessing the ability to respond to egg cytoplasm by generating descendents which exhibit all the usual differentiated functions of the frog, including those which would never be expressed in the endoderm cell.

A high proportion of transplanted blastula nuclei in both Rana and Xenopus give normal development; usually more than half pass the blastula stage and more than one third continue to develop into young frogs. Nuclei extracted from differentiating embryos and tadpoles have a lesser capacity to support normal development after transplantation; frequencies of growth to the adult may be less than 10% (reviewed by Gurdon, 1964).

Nuclei have also been extracted from fully differentiated adult cells which have—in the normal course of development—reached their final state of specialization. Gurdon (1962c) found that when nuclei are transplanted into eggs from the intestinal epithelium cells of feeding tadpoles, some of the transplanted nuclei support development of normal tadpoles, although the great majority shows abnormalities varying from lack of cleavage to nearly normal tadpoles. However, if the abortive cleavages—which may result when nuclei are taken from the donor at an unsuitable stage of mitotic cycle—are excluded from the results, it is possible to calculate that at least 24% of intestinal epithelium cells, which represent the final state of differentiation for many endoderm cells, have a full developmental potentiality. In these cells, differentiation is reversible.

Does the decline in successful development of transplants as donor nuclei are extracted from increasingly differentiated cells reflect some restriction on potentiality or is it a consequence of the transplantation technique itself? Figure 7.1 shows that although a decline is seen with both Rana and Xenopus nuclei, it is more pronounced with Rana and there is a difference in both its onset and rate of increase with differentiation. The incidence of incapacity is therefore not related to the stage of development in the same way in the two genera.

Each specialised tissue may consist of a variety of cell types; one explanation of the decline is therefore that some of their nuclei are totipotential but that

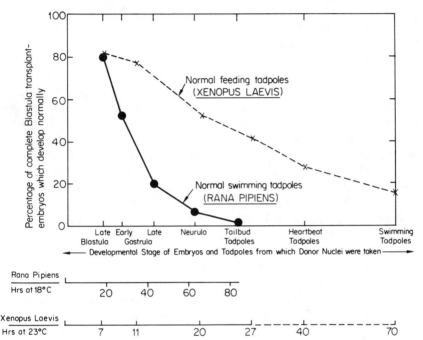

Figure 7.1: decline in successful transplantations with increasing age of donor nuclei for transplantation. Data of Gurdon (1963).

others bear irreversible changes which restrict developmental potential, the restricted proportion increasing with specialization. An alternative is to suppose that all nuclei are totipotential, but that with specialization cells become increasingly liable to extrinsic factors which inhibit successful transplantation; for example, more advanced nuclei appear to be more susceptible to damage during preparation and injection. A critical difference between these two postulates is that any genuine restriction of potential should be specific for the cell in which it occurs; cells of different specializations should possess restrictions of different genes and should therefore display different and characteristic inabilities to prosecute successful development after transplantation. Irreversible damage induced during transplantation itself, however, should be random.

Since no tissue or cell-specific restrictions have been found, but the same classes of abnormality may result in transplants derived from all types of nucleus, it seems likely that the decline in successful transplantation with development of donor nuclei represents a limitation of the technique. Some of the abnormalities which develop during transplantation have been discussed by Gurdon (1960). (Of course, it is possible that cell specific as well as technical restrictions of potential accompany development, but that the cell-specific restrictions are obscured by the random restrictions induced by the technique itself; however, the most likely explanation of these results is that nuclei retain their full developmental potential during specialization.)

That changes induced in an unsuccessfully transplanted nucleus may be irreversible is shown by serial transfer experiments. In this technique, first used by King and Briggs (1956), a nucleus is transplanted into an egg and allowed to grow to the blastula stage. Nuclei are then extracted from this *first transfer* blastula and in turn transplanted into many more eggs; all of these *second transfer* transplants represent descendents of the one somatic nucleus used to provide the first transfer blastula. Gurdon (1963) reviewed experiments of this nature with Xenopus, which show that nuclei taken from the first transfer blastulae, developed from transplants receiving embryonic nuclei, fare no better through several generations of serial transfer than the original nuclei. This implies that the original nucleus has suffered irreversible damage during transplantation which is perpetuated in all its descendents.

Irreversible damage may be induced in donor nuclei in several ways and its perpetuation through serial transfer implies that it is located in the genetic material. The likelihood of physical damage to the integrity of the donor nucleus during transplantation certainly increases with cell specialization. Another cause of restriction may be the replication of DNA which transplanted nuclei are compelled to undertake very shortly after entering the egg; some nuclei, especially those which have just completed a mitosis, may be unable to complete division in the time required and may therefore suffer an abnormal cleavage in which chromosomes are lost or other damage is suffered by the daughters. A more subtle form of damage was induced by Gurdon (1962b) by transplanting a Xenopus laevis nucleus into the cytoplasm of Xenopus tropicalis (the two species cannot interbreed). Such transplants develop only to the late gastrula stage; when nuclei are recovered from the hybrid blastula and retransplanted to eggs of Xenopus laevis, they are unable to support development past the neurula stage. The X. laevis nucleus must therefore suffer irreversible damage as the result of its passage through X. tropicalis cytoplasm.

Even cultured cells may retain full developmental potential. When Gurdon and Laskey (1970) transplanted nuclei from a monolayer culture of Xenopus epithelial cells, they found that about 75% of the transplants fail to cleave altogether or show abortive cleavage. Most of the remaining eggs suffer partial cleavages in which part of the egg appears to consist of normal blastomeres and the rest is uncleaved or abortively cleaved. Less than 5% of the transplanted nuclei cleave regularly to form blastulae and most of these die in the blastula stage. In all, fewer than 0·1% of the transplanted nuclei develop into normal or nearly normal tadpoles. Nuclei from cultured cells therefore seem more restricted in potential than nuclei of the cell type from which the culture was originally derived, emphasizing that cultured cells cannot be taken as typical of their ancestral population in a living tissue.

With donor nuclei of the cultured cells, serial transplants from a first-transfer blastula show much better development than that of the original donors. Gurdon (1962c) found that serial transfer of intestinal epithelial nuclei also increases the number of tadpoles which can be reared. In these cases, the first-transfer blastulae probably contain both normal and abnormal nuclei derived from the original transplanted nucleus and the serial transplantation allows some of the normal nuclei to develop unhampered by the damaged nuclei. This implies that the original donor nucleus is not itself damaged by transplantation but that irreversible mistakes may be made when it divides so that some but not all of its descendents are damaged.

Transplantation experiments suggest two general conclusions about the nucleus of the somatic cell. The ability of nuclei derived from specialised cells to support development of transplanted eggs into adult frogs confirms that all the genetic information of the nucleus is retained during differentiation. The proper development of each tissue of the adult shows further that differentiation is reversible; the chromatin of each specialized nucleus can be reset to the pattern of gene expression characteristic of the embryo and its descendents then generate the usual states of gene expression characteristic of each cell type.

Molecular Changes in Transplanted Nuclei

That the nucleus of a differentiated Xenopus cell can give rise to a normal adult organism when transplanted into egg cytoplasm in itself implies that all the usual molecular controls of gene expression are exercised at the appropriate stages of development. It is difficult to follow directly the synthesis of specific protein gene products, but measurements of the synthesis of RNA and DNA confirm the idea that all activities of the nucleus are very precisely controlled by the cytoplasm in which it is placed.

Distinct changes take place in the pattern of RNA synthesis during development of Xenopus. Synthesis of DNA-like RNA (representing the messenger fraction) and tRNA starts at late cleavage and continues at a rapid rate during gastrulation and neurulation. Synthesis of rRNA is quiescent at first and starts at the beginning of gastrulation, increasing in rate during later development. Neural endoderm nuclei therefore engage in extensive synthesis of rRNA. Gurdon and Brown (1965) found that when such nuclei are transplanted into eggs, the resulting blastulae do not synthesise rRNA (the usual situation at this stage). Synthesis of rRNA is therefore switched off when the neurula nucleus is transplanted into egg cytoplasm. Transcription of RNA is switched on again at the appropriate time in development, for if the transplants are allowed to continue to the neurula stage they produce rRNA in the same way as developing embryos of fertilised eggs.

The genes coding for rRNA can therefore be switched off and on again in response to signals from the cytoplasm. A similar control is exerted over replication of chromosomal DNA. By using autoradiography to follow the ability of transplanted nuclei to synthesize DNA, Graham, Arms and Gurdon (1966) showed that label is incorporated within 20–40 minutes of transplantation; by 70 minutes nearly all the transplanted nuclei have completed their first mitotic division and continue to incorporate label in the manner characteristic of the normally developing egg.

Nuclei of adult liver, brain and blood cells—all of which have virtually ceased replication of DNA and cell division—can recommence DNA synthesis under the stimulation of egg cytoplasm. Some 10% of mouse liver nuclei can be persuaded to start DNA synthesis by injection into Xenopus eggs, which argues that the signals controlling replication may be of low specificity. Nearly all Xenopus nuclei commence DNA synthesis after transplantation so that failure to do so cannot in general be responsible for unsuccessful development.

In reviewing transplantation experiments, Gurdon and Woodland (1970) observed that replication must be completed within one hour of transplantation, whereas S phase in the somatic cell nucleus usually occupies about six hours. Synthesis of rRNA in neurula nuclei is also suppressed within one hour of transplantation. Major changes in both RNA and DNA synthesis must therefore take place in the activity of chromatin within a very short time after transplantation; failure to complete replication in time for the first division may be responsible for many of the transplanted eggs which fail to show normal development.

Nuclear activities of growing and maturing oocytes differ from those of the egg. Growing oocytes engage in RNA synthesis but do not synthesize DNA; maturing oocytes complete meiotic division and therefore possess condensed chromosomes inactive in both RNA and DNA synthesis. Comparing the responses of nuclei injected into growing or maturing oocytes therefore tests their abilities to respond to two further sets of cytoplasmic conditions. If the activities of nuclei are established by their surrounding cytoplasm, those placed in growing oocytes should synthesize RNA but not DNA; whereas those placed in maturing oocytes should lose all synthetic activities, but should divide. Nuclei transplanted into eggs, of course, immediately synthesize DNA and divide.

The resident nucleus of an oocyte has a very large volume, diffuse chromatin and multiple (upto 1500) large nucleoli; nuclei from later embryonic or adult tissues have a much smaller volume and lack multiple nucleoli. Gurdon (1968) found that when nuclei from mid-blastula, late gastrula or adult brain, identified by a previous label with H^3-thymidine, are injected into oocytes, they suffer a pronounced enlargement. Swelling is experienced by all injected nuclei, but its rate depends upon their previous stage of differentiation; more differentiated nuclei swell more slowly. Blastula nuclei may enlarge more than 250 times within three days after transplantation so that they increase from 440 μ^3 to 110,000 μ^3; brain nuclei may swell from a starting size of 100 μ^3 to one of 4000 μ^3. The chromatin of injected nuclei becomes more dispersed in parallel with the increase in nuclear volume.

That the nuclei which have enlarged after injection into growing oocytes synthesize RNA is revealed by autoradiography of transplants given labelled precursors. Transcription can therefore be induced in mid-blastula nuclei, which usually synthesize little RNA, but are active in DNA synthesis. Replication continues in the blastula nuclei for about 30 minutes after injection into growing oocytes and then ceases. Growing oocyte cytoplasm therefore switches on RNA synthesis but switches off DNA synthesis and prevents the blastula nuclei from undergoing the divisions which they usually experience.

Brain nuclei divide very rarely—less than 0·1 % of the nuclei in adult frog brain are in division at any time. Gurdon found that when a suspension of brain nuclei is injected into maturing oocytes—each oocyte receiving some 20–40 nuclei—they form asters, spindles and very condensed chromosomes (although division often appears abnormal, as might be expected in an oocyte containing so many nuclei). It is not clear whether the injected nuclei enter mitosis or meiosis; but they are induced to divide by the cytoplasm of the maturing oocyte.

At each stage of development of the oocyte and egg, the cytoplasm must therefore contain specific information which elicits the appropriate response from any nucleus placed in it, even when that response is very different from the functions displayed by the nucleus before its transplantation. DNA or RNA synthesis may be either suppressed or induced according to the type of cytoplasm; and division may be induced. Nuclei therefore establish activities characteristic of their surrounding cytoplasm; nuclei of all cells appear to respond to the cytoplasmic signals in the same way, although at different rates (presumably because it takes longer to reverse changes in chromatin in more highly specialized cells.)

The cytoplasms of Xenopus oocytes and eggs can themselves support appropriate synthetic activities as well as dictating the state of gene expression of injected nuclei. This includes the translation of heterologous messengers (see page 259) and the replication of DNA injected into the cell. Fertilization of the egg induces DNA synthesis in both maternal and paternal genomes— this represents a considerable change in function since neither oocyte nor sperm synthesizes DNA during their maturation. Gurdon, Birnsteil and

Speight (1969) found that when H³-thymidine is injected together with DNA into egg cytoplasm, the label is incorporated into DNA; the injected DNA can replicate. Denatured DNA is two or three times more as effective as template than native DNA; and the demand for DNA is not species specific. In a subsequent experiment, Gurdon and Speight (1969) demonstrated that purified DNA is not replicated when it is injected into oocytes. The enzyme activities required to replicate DNA are therefore produced in egg cytoplasm only at the appropriate time and must include all the components necessary for replication of DNA; nuclear factors (apart from the DNA template itself) are not necessary.

Nuclear enlargement is essential for derepression of the genomes of differentiated cells after either nuclear transplantation or somatic cell fusions (see later). In nuclei injected into growing oocytes, RNA synthesis parallels the increase in nuclear volume and nucleolar-like bodies develop in nuclei previously lacking them (Gurdon, 1968). Nuclei injected into eggs also show a correlation between enlargement and replication. Active intestinal epithelial nuclei may increase in size from 160 μ^3 to 4500 μ^3 within 40 minutes and any nucleoli present disappear (Graham et al., 1966). Nuclei which fail to enlarge do not incorporate label into DNA and cannot support development.

Enlargement of a transplanted nucleus depends upon the specific uptake of proteins from egg cytoplasm. Arms (1968) followed the enlargement of nuclei of Xenopus liver in cleaving eggs (which contain 2–12 cells at the time of injection) and Merriam (1969) observed the enlargement of brain nuclei in earlier eggs. The injected nuclei can be distinguished by their smaller size and more intense staining. By labelling the recipient eggs with radioactive amino acids before injection of a donor nucleus, it is possible to follow the accumulation of proteins in the injected nuclei as they enlarge. Nuclear uptake of proteins from the cytoplasm is not changed by the injection of puromycin together with donor nuclei, which implies that it depends upon pre-existing proteins present in the cytoplasm before injection.

Proteins begin to enter injected nuclei even before there are any signs of enlargement, which suggests that the uptake may be needed for activation; certainly it does not appear to be a consequence of the enlargement. The increase in size of injected brain nuclei and the uptake of cytoplasmic proteins is accompanied by morphological changes in chromatin, which loses its heterochromatic clumps and acquires a more dispersed structure. This suggests that the uptake of cytoplasmic proteins may represent changes in the organization of chromatin necessary to reprogramme its state of gene expression. That the uptake is specific for the cytoplasm in which the nucleus is placed is suggested by observations of blastula nuclei injected into oocytes, when entry of the label takes place at a much slower rate.

The proteins entering the nucleus add to rather than replace those previously present. Gurdon (1970) found that when embryos are labelled with radioactive amino acids from late blastula to neurula, nuclei extracted for transplantation retain the label during their enlargement in egg cytoplasm. There is no loss during development of proteins labelled with arginine and alanine (largely histones) or of those labelled with tryptophan and phenylalanine (non histone nuclear proteins). This implies that it may not be necessary to remove proteins from chromatin in order to change its state of expression.

It is difficult to interpret experiments in which proteins of the egg cytoplasm are labelled before injection of a donor nucleus, because many different classes of protein are labelled and those synthesized prior to the labelling period may remain unlabelled. Gurdon therefore iodinated calf thymus histones and bovine serum albumin with I^{125} in vitro and injected the modified proteins into eggs containing a transplanted nucleus. Histones appear to be taken up by both the resident and transplanted nuclei, but the bovine serum albumin enters neither nucleus. Although these proteins are derived from another species, these results suggest that only selected proteins may enter the nucleus from the cytoplasm and that histones may be prominent amongst them.

Somatic Cell Hybridization

Growth and Division of Hybrid Cells

Cells of different specialized types and from different species may be fused together by the Sendai virus to generate *hybrid cells*. These hybrids contain genetic information of both the parental cell types and therefore offer the opportunity to follow its expression in circumstances which differ considerably from those of the cell in its native state. Hybrid cells formed by fusion of two parents derived from different species have been termed *heterocaryons*, in contrast to the *homocaryons* produced by fusion of like cells.

Sendai virus—inactivated by ultraviolet so that it cannot promote infection— adsorbs to the surfaces of cells, causing them to adhere to each other. This leads to the establishment of cytoplasmic bridges, after which an extension of the region of contact sees a dissolution of the cell membranes over their common area. The two cytoplasms ultimately coalesce to yield a hybrid cell which contains the individual nuclei of its parents in the mixed cytoplasm of both. Fusion is not limited to two cells, for several may fuse together simultaneously or sequentially to yield multinucleate cells containing varying numbers of nuclei. Studies of the ability of genes to function in a nucleus located in an alien cytoplasm have revealed the lack of species specificity of the mechanisms of gene expression and suggest that control signals of cytoplasm of one species may be recognised by the nucleus of another.

Chromosome replication and nuclear mitosis may take place in hybrid cells. In cells containing many nuclei, nuclear mitosis is usually not accompanied by formation of a spindle and division of the cell; and in such hybrids only some of the nuclei in the cell enter mitosis together, the others remaining in interphase. Those nuclei which enter mitosis together, however, are usually reconstituted as a single unit; when all the nuclei of a cell enter mitosis together —either immediately after fusion or after intervening mitoses and reconstitutions—post mitotic reconstitution leads to the formation of a single nucleus containing chromosomes of both parental species.

In cells containing only a small number of nuclei, synchronous nuclear mitoses may be accompanied by formation of a spindle and division of the cell. This happens most commonly in binucleate cells but is sometimes observed in cells containing more than two nuclei. Binucleate cells which undertake joint mitosis and division may form only one spindle so that the daughter cells each contain a single nucleus which is larger than usual and possesses the chromosome complements of both original nuclei. The propensity of such hybrid cells for losing the chromosomes of one parent in subsequent growth and division has been used to locate genes on chromosomes; loss of an enzyme marker coded by one parent only can be correlated with the loss of the chromosome which carries its gene.

Many types of cell may be fused to give viable progeny in which both parental genomes are active. By using autoradiography, Harris et al. (1966) showed that after fusion between Hela cells and Ehrlich ascites tumour cells of the mouse—in which the two types of nucleus can be distinguished by their appearance—radioactive precursors are incorporated into RNA in both human and mouse nuclei. Labelled leucine is also taken up and the hybrid cytoplasm synthesizes proteins. At least those genes which code for essential cellular functions must therefore be transcribed into RNA, transported to the hybrid cytoplasm, and translated into proteins.

Hybrid cells in which all the nuclei have fused into a single nucleus (*syncaryons*) synthesize DNA, RNA and proteins and can undertake mitosis. Cell division is usually more successful when the hybrid is derived from a small number of cells. Ehrlich ascites cells have a chromosome mode of 73–80 and most of the Hela cells possess 54–59 chromosomes. (Tissue culture cells suffer variation in chromosome content and rarely maintain a constant diploid complement.) Harris et al. found that the hybrid cells undergoing mitosis 72 hours after fusion possess from 111 to 275 chromosomes, corresponding to the fusion together of varying numbers of nuclei. Most of the multinucleate cells fail to give rise to viable progeny—often because mitoses fail to be completed—and the population of such cultures remains stationary for some days and then declines gradually. Only hybrids produced by fusion of a small number of nuclei can propagate successfully and most of the recent experiments with hybrid cells have made use of fusions between only two cells.

Differentiated somatic cells as well as tissue culture "undifferentiated" lines may be used for cell fusion. In one series of experiments, Harris et al. used rabbit macrophages, rat lymphocytes and hen erythrocytes, which display different characteristic activities in protein and nucleic acid synthesis. Rabbit macrophages synthesize RNA but have ceased division and can no longer synthesize DNA. Rat lymphocytes show a similar restriction of synthetic capabilities, although they can be induced to divide in vivo by suitable antigenic stimuli. The nucleated erythrocytes of birds represent completely inert nuclei— in many animals the red blood cells lack nuclei—which are rather small, contain condensed chromatin and synthesize neither DNA nor RNA.

Fusion between Hela cells and rabbit macrophages or rat lymphocytes displays the same characteristics as that with the Ehrlich ascites cell. Varying numbers of cells may fuse together and the products may undergo both synchronous and asynchronous mitoses. But the differentiated features of the macrophage or lymphocyte are lost and the hybrids behave as "undifferentiated" tissue culture cells.

Products of fusion between Hela cells and hen erythrocytes, however, do not contain the cytoplasm of the erythrocyte; because the Sendai virus haemolyses the erythrocyte to give a red blood cell ghost during fusion, its cytoplasm is lost. Fusion therefore in effect places the erythrocyte nucleus in the cytoplasm of the Hela cell, in contrast to the other fusions in which the two nuclei reside in a mixed cytoplasm rather than that of one of them. The state of a dicaryon possessing one Hela nucleus and one erythrocyte nucleus is shown in figure 7.2; the erythrocyte nucleus is compact, with deeply staining nuclear bodies. Figure 7.3 shows that the erythrocyte nucleus then begins to enlarge and figure 7.4 shows a further state of enlargement at which the nuclear bodies are no longer visible. This enlargement accompanies the acquisition of synthetic activities by the formerly inert nucleus.

When either the rabbit macrophage or rat lymphocyte is fused with a Hela cell the differentiated nucleus continues to synthesize RNA; and it commences the synthesis of DNA. The erythrocyte nucleus commences synthesis of both RNA and DNA under the stimulation of the Hela cytoplasm. The response of these inert nuclei to fusion with a Hela cell is therefore to switch on synthetic activities which had been repressed in their parent tissue.

That this activation is not a random response to the cell fusion has been shown by the fusion together of differentiated cells. When rabbit macrophages are fused with rat lymphocytes, both nuclei in the hybrid continue to synthesize RNA but neither is stimulated to synthesize DNA. The characteristic changes which take place in the lymphocyte nucleus after fusion with a Hela cell—enlargement of the nucleus, dispersion of chromatin, appearance of structures resembling nucleoli—do not take place after fusion with the macrophage. When rabbit macrophages are fused with hen erythrocytes, both nuclei in the hybrid synthesize RNA but neither replicates its DNA. This argues that activation is specific and that each cytoplasm contains signals which stimulate only those activities usually expressed by its nucleus and which may be induced in any other nucleus introduced into the cell. The Hela

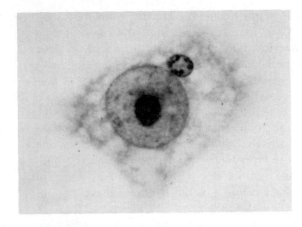

Figure 7.2: dicaryon containing one Hela nucleus and one hen erythrocyte nucleus in which the erythrocyte nucleus remains compact and displays deeply staining nuclear bodies. Photograph kindly provided by Professor Henry Harris.

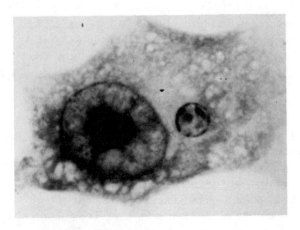

Figure 7.3: dicaryon containing one Hela nucleus and a hen erythrocyte nucleus which has begun to enlarge. Nuclear bodies of the erythrocyte stain less deeply and are more diffuse. Photograph kindly provided by Professor Henry Harris.

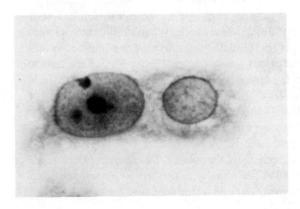

Figure 7.4: dicaryon containing Hela nucleus and erythrocyte nucleus at stage of enlargement in which it is much less compact and has lost its nuclear bodies. Photograph kindly provided by Professor Henry Harris.

cytoplasm contains signals which induce both transcription of RNA and replication of DNA.

Reactivation of Chick Erythrocyte Nuclei

The active partner of a cell fusion induces synthetic activities in its dormant partner; the inactive cell does not appear to inhibit the active one. The ability of fused cells to maintain their integrity and to continue growth when the nuclei fuse together implies that the genetic mechanisms and structural organization of cells of different species are compatible; DNA may be replicated, genes expressed and the organization of the cell maintained. The cytoplasmic signals which activate the nucleus therefore appear to lack species specificity. Two classes of model may account for this result (reviewed by Harris, 1970). If the cytoplasmic control signals recognise specific sequences of chromatin, the same sequences must be present in the nuclei of different species and must have compatible (although not necessarily identical) meanings. Alternatively, the signals themselves may be less specific—comprising, for example, changes in ionic or metabolite levels—but may elicit from each chromatin a similar response. More recent research has shown that proteins of one species may enter the nucleus of another during its reconstruction.

The incorporation of precursors into nucleic acids and proteins shows that a dormant nucleus can reactivate previously inert functions under the stimulus of the cytoplasm into which it is transferred by the cell fusion. Of course, this does not prove that specific genes—rather than a random selection of sequences—have been activated. By following the reactivation of hen erythrocyte nuclei in Hela or mouse A9 L cell cytoplasm, Harris et al. (1969) identified some of the hen specific protein products; the incorporation detected by autoradiography thus represents transcription and translation of meaningful sequences.

When Hela cells are fused with hen erythrocytes, for the first 12–16 hours after fusion all the hybrids—and also Hela cells in the population which failed to gain an erythrocyte nucleus because of failure of fusion—possess hen specific antigens on their surface. Isolated populations of Hela cells show no reaction, which suggests that the antigens have been introduced onto the surface of the hybrid during the cell fusion; they appear initially to be located in a limited area but are then redistributed to cover the entire cell surface.

The hen specific antigens begin to disappear from the surface of the heterocaryon within 24 hours after the fusion, although by this time the erythrocyte nuclei have begun to enlarge and synthesize RNA. By three days after fusion, 90% of the hybrid cells appear to have lost their antigens. Since the same rate of loss is displayed by Hela cells which initially gained these antigens but failed to complete cell fusion, antigens introduced at the fusion are lost from the surface irrespective of the presence of a hen nucleus. Similar results are found when chick embryo erythrocytes are used for fusion, although their antigens are lost more slowly, perhaps because they are initially introduced in greater amounts. That removal is an active temperature dependent process is implied by the observation that its rate is reduced by a decrease in the temperature of incubation of the hybrid cells. Loss of hen antigens therefore presumably results from their gradual displacement by Hela surface antigens because the erythrocyte nucleus in the heterocaryon does not direct synthesis of hen surface antigens.

The fate of the antigens on these cells can be followed for only four days, for by this time the Hela nuclei in virtually all the heterocaryons enter mitosis. Either the mitosis is irregular, causing cell death; or the erythrocyte nucleus also divides and thus fuses with the Hela nucleus. Both events see the disappearance of cells which contain a discrete erythrocyte nucleus so that it becomes impossible to follow its further reactivation. Mouse fibroblasts irradiated with ultraviolet light, however, can grow for upto three weeks without dividing. When erythrocyte nuclei are transferred to this cytoplasm by cell fusion, hen specific antigens are lost from the hybrids more slowly than from the erythrocyte–Hela fusions, but the loss is virtually complete within 6 days.

Virtually all of the reactivating erythrocyte nuclei produce nucleoli, which begin to appear on the third day and then become progressively larger. By day 11, more than 80% of the hen nuclei contain nucleoli. After 8 days, hen specific antigens begin to reappear on the surfaces of the hybrid cells and their distribution over the surface becomes progressively more widespread. Figure 7.5 shows that the reappearance of the antigens follows the production of nucleoli. Similar results are obtained by using chick embryo erythrocytes, when reactivation is more rapid; nucleoli begin to appear on the second day after fusion and new chick specific antigens begin to be synthesized before those introduced in the fusion have been completely lost. The time at which new antigens appear therefore depends upon the rate of reactivation of the erythrocyte nucleus, in particular upon the time at which it forms its nucleolus.

Other gene functions as well as the surface antigens are expressed by the hen nucleus in hybrid cells. The A9 mouse fibroblast cell line lacks the enzyme hypoxanthine guanine phosphoribosyl transferase (HGPRT; also known as inosinic acid pyrophosphorylase) and so cannot utilise hypoxanthine as a precursor for nucleic acids. Chick embryo erythrocytes contain this enzyme although it is lost with the cytoplasm during cell fusion. When Harris and Cook (1969) used autoradiography to follow the incorporation of hypoxanthine by hybrid cells, they found that the heterocaryons are unable to utilise the nucleotide for the first three days after fusion, but begin to do so upon the fourth day, their activity increasing subsequently. The ability of the hybrid cells to produce the enzyme correlated with the development of nucleoli, which shows a sharp rise on the fourth day. Cook (1970) showed that the

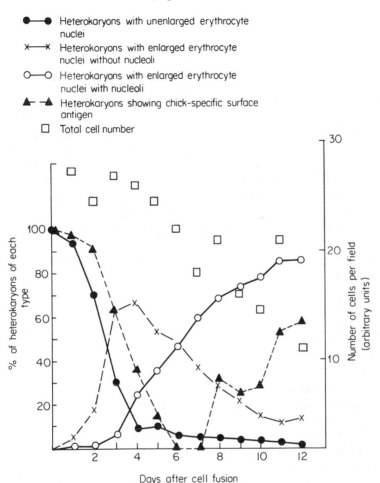

●——● Heterokaryons with unenlarged erythrocyte nuclei

×——× Heterokaryons with enlarged erythrocyte nuclei without nucleoli

○——○ Heterokaryons with enlarged erythrocyte nuclei with nucleoli

▲– –▲ Heterokaryons showing chick-specific surface antigen

☐ Total cell number

Figure 7.5: reappearance of hen specific antigens on the surface of heterocaryons following fusion between irradiated mouse fibroblasts and adult hen erythrocytes. The proportion of heterocaryons which gain hen specific antigens is closely related to the number of cells which possess erythrocyte nuclei displaying nucleoli. Data of Harris et al. (1969).

enzyme synthesized in the hybrids has the electrophoretic mobility characteristic of the hen and not of the mouse; so its synthesis represents an activity specified by the erythrocyte nucleus and not a reversion in the mouse genome (see also page 413).

The kinetics of appearance of this enzyme are similar to those of the production of surface antigens. Since these proteins are unrelated, it is probable that these kinetics are common to all genes expressed in the reactivating erythrocyte nucleus. The correlation between the appearance in the cytoplasm of hen specific proteins and the reconstruction of the nucleolus suggests that messengers synthesized in the erythrocyte nucleus—synthesis of Hn RNA is extensive by 48 hours after fusion—cannot be transported to the cytoplasm until the nucleolus forms. Although the nucleolus is itself responsible for the synthesis only of ribosomal RNA, this model implies that it may be responsible for the transport from nucleus to cytoplasm of all RNA.

By suppressing synthesis of RNA in the Hela or mouse nuclei of hybrids with erythrocytes, it is possible to follow by autoradiography the ability of the erythrocyte nucleus to synthesize and transport RNA to the cytoplasm. Ultraviolet irradiation of the mouse A9 or the Hela nucleus greatly reduces the labelling of the cytoplasm in the parent cells. Hybrid cells which contain only erythrocyte nuclei lacking nucleoli show the same reduced incorporation over the cytoplasm after irradiation of the mouse or human nucleus as the control A9 or Hela cells (although the erythrocyte nuclei themselves are labelled).

Figure 7.6 compares an unirradiated heterocaryon in which an A9 nucleus and an erythrocyte nucleus incorporate labelled uridine, and the cytoplasm becomes labelled, with a similar cell in which the A9 nucleus has been irradiated; this reveals that the erythrocyte nucleus, which lacks nucleoli, can utilise the labelled precursor but cannot transport RNA to the cytoplasm. However, in hybrids in which the erythrocyte nuclei have developed nucleoli, labelling of the cytoplasm continues even after inactivation of the A9 or Hela nucleus (part c). This supports the idea that although a reactivating erythrocyte nucleus begins to synthesize RNA very soon after cell fusion, it can transport messengers to the cytoplasm only after it has developed a nucleolus (for review see Harris, 1970).

Somatic mononucleate cells can also be examined by this technique. Sidebottom and Harris (1969) used a small microbeam to irradiate either the nucleolus alone or an equivalent area of the nucleoplasm of Hela cells; with a larger ultraviolet microbeam, the entire nucleus can be irradiated. After exposing the irradiated cells to labelled precursors, the number of grains over the nucleoplasm, nucleolus and cytoplasm reveals the effect of the irradiation. Irradiation of the nucleolus alone or the whole nucleus reduces labelling of the cytoplasm to some 10% of the control level of unirradiated cells; but irradiation of the nucleoplasm alone reduces labelling by only 50%. Inactivation of the nucleolus therefore specifically prevents all RNA transport from nucleus to cytoplasm.

Irradiation of the nucleolus has also been used to demonstrate that the temporal correlation between nucleolar formation and expression of specific gene functions in the reactivating erythrocyte represents a causal relationship. If erythrocyte gene expression depends upon continued synthesis of messenger RNA and its transport via the nucleolus to the cytoplasm, the production of proteins coded by the erythrocyte genome should decline after irradiation

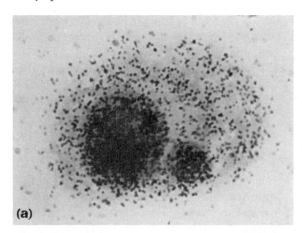

(a)

Figure 7.6: RNA synthesis in heterocaryons between mouse A9 fibroblasts and hen erythrocytes. (a) A control which has not been irradiated. Both nuclei are labelled by H³-uridine and extensive transport to the cytoplasm takes place. (b) Effect of irradiating the A9 nucleus before the erythrocyte nucleus develops a nucleolus. Although the erythrocyte nucleus synthesizes RNA, it cannot transport it to the cytoplasm because of the absence of the nucleolus. (c) Control in which the A9 nucleus is irradiated after the erythrocyte nucleus has developed a nucleolus. Transport by the erythrocyte nucleus of RNA to the cytoplasm can take place via its nucleolus. Photographs kindly provided by Professor Henry Harris.

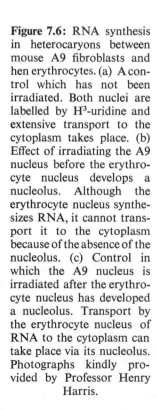

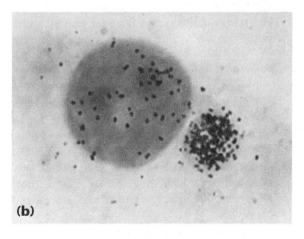

(b)

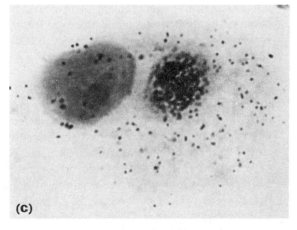

(c)

of its nucleolus. This prediction has been tested by Deak, Sidebottom and Harris (1972). Hybrid cells in which the only nucleolus of the erythrocyte nucleus has been irradiated, progressively lose their ability to incorporate hypoxanthine into RNA; by 4–5 days after irradiation their abilities are reduced to the level of the A9 control cells. In cells in which the erythrocyte nucleus has formed two nucleoli, it is possible to irradiate only one of them; in this case the cells remain able to utilise hypoxanthine. Analogous results are obtained when the synthesis of hen surface antigens or sensitivity to destruction by diphtheria toxin (a characteristic of chick but not of mouse cells) is followed.

The similar behaviour of proteins coded by genes at many different chromosome loci reinforces the conclusion that the nucleolus is responsible for transferring all messenger RNAs to the cytoplasm. How this function may be related to the synthesis and transport of ribosomal RNA molecules is not known; that transport of messengers is independent of ribosomal RNA synthesis and transport is suggested by the continued synthesis of Hn RNA and transport of mRNA from the nucleus when nucleolar synthesis of rRNA is inhibited by low doses of actinomycin (see page 248).

Molecular Reconstruction of Nucleoplasm and Nucleolus

Immediately after fusion between an erythrocyte and a Hela cell, the red blood cell nucleus displays about the same dimensions as those observed in its native state. But Harris (1967) observed that the inert nucleus enlarges under the influence of Hela cytoplasm. Enlargement usually begins within 24 hours and by the third day unenlarged nuclei are rare. The extent of RNA synthesis in the reactivating erythrocyte nucleus—judged by autoradiography of hybrid cells incubated with labelled precursors—correlates well with its cross sectional area (measured as a representation of volume), as shown in figure 7.7.

Irradiation of the red blood cell nucleus with ultraviolet before fusion does not prevent its subsequent enlargement in Hela cytoplasm. Since the irradiation abolishes its ability to resume synthesis of RNA, this implies that enlargement is not a consequence of increased activity in RNA synthesis. This suggests that enlargement may be the cause of, or at least a prerequisite for, RNA synthesis (see also Bolund et al., 1969a).

Cytochemical studies of reactivation have been extended by Bolund, Ringertz and Harris (1969b). The increase in nuclear volume which takes place during reactivation of erythrocyte nuclei in Hela cytoplasm is accompanied by an increase in dry mass (measured by microinterferometry). The largest erythrocyte nuclei increase their mass some five times during the first 41 hours after fusion. Some of this increase takes place during the first 16–20 hours after fusion, whilst the erythrocyte nuclei remain in the G1 phase characteristic of the red blood cell. By 24 hours after fusion, the feulgen values of the nuclei increase as DNA is replicated; and by 41–47 hours many of the erythrocyte nuclei show G2 values for their reaction with feulgen. This supports the conclusion that autoradiographic observations of the incorporation of H³-thymidine into the nucleus reflect complete replication of its genetic material.

At an early stage in reactivation, the ability of the DNA of erythrocyte nuclei to bind the intercalating dye acridine orange increases. This occurs

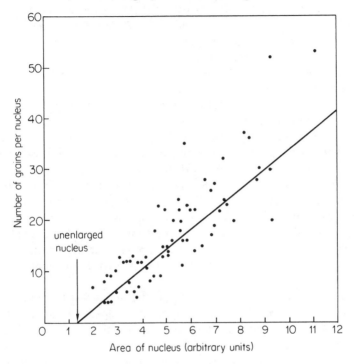

Figure 7.7: relationship between area of reactivating erythrocyte nucleus in Hela cytoplasm and extent of RNA synthesis measured by incorporation of H³-uridine.

before replication, and so may reflect a dispersion of their chromatin. During reactivation, the erythrocyte DNA also becomes more susceptible to denaturation, also consistent with the idea that an unfolding of its compact structure may be necessary for its replication and for activation of repressed genes. (Similar changes seem to take place in lymphocyte chromatin when the cells are stimulated into division by phytohaemoglutinins.) The early reconstruction of the nucleus which takes place before replication of its DNA therefore involves uptake of proteins and dispersion of chromatin.

A similar series of changes in the structure of chromatin has been observed when epithelial kidney cells from 14-day mice are seeded into culture. Their compact nuclei with condensed chromatin and relatively low nuclear protein content display dispersion of the chromatin structure and uptake of proteins. Auer (1972) and Auer et al. (1973) observed that first the phosphate groups of DNA acquire an increased ability to bind the dyes acridine orange and ethidium bromide and the staining with fast green or bromophenol of histone groups bound to DNA decreases. Chromatin also displays increased susceptibility to heat denaturation.

These changes suggest that the structure of chromatin becomes less compact so that DNA is less tightly bound by histones. The non histone protein content of the nucleus increases greatly, although it is not clear precisely how the entry of non histone proteins is related to changes in chromatin structure. Auer and Zetterberg (1973) observed that the rate of nuclear RNA synthesis increases with the nuclear content of proteins and is not directly correlated with the dispersion of chromatin structure, which although necessary may therefore be insufficient in itself for gene activation.

A common sequence of events therefore takes place in reactivation of inert nuclei, whether under stimulation of transplantation to egg cytoplasm (as in Xenopus), hormonal action (as with lymphocytes), or upon fusion with a different species of cell which has a greater range of synthetic activities. The dormant nucleus tends always to be rather small, to have comparatively condensed chromatin and to lack, or have only one or two, nucleoli. Before or concomitant with reaction, these nuclei must enlarge and allow their chromatin to achieve a more dispersed state. We may suppose that this reflects topological demands for the state of the template necessary for replication and transcription.

Reversal of differentiation in Xenopus depends upon the entry into the nucleus of proteins of the egg cytoplasm. A comparable interaction is responsible for the reactivation of erythrocyte nuclei in Hela cytoplasm. The extensive increase in dry weight of the nucleus during the first 48 hours after fusion represents uptake of proteins from Hela cytoplasm; most and probably all of these proteins must be human in derivation since at this time the erythrocyte nucleus has not yet started to direct the synthesis of hen specific proteins. The cytoplasmic proteins which enter the nucleus to change its state of differentiation therefore appear to be specific for neither the species nor the cell phenotype. Viewed within the perspective of the general mobility of movement of proteins between nucleus and cytoplasm in tissue culture cells (see page 72), these results are in accord with the concept that nuclear proteins achieve an equilibrium with those of the cytoplasm which establishes the genetic activity of the nucleus.

That the progress of reactivation depends upon the extent of uptake of proteins by the reactivating nucleus is suggested by the studies of chick erythrocyte nuclear reactivation in rat epithelial cells shown in figure 7.8. When Carlsson et al. (1973) prepared heterocaryons with proportions of the two species of nuclei varying from 2 rat:1 chick to 1 rat:4 chick, they found

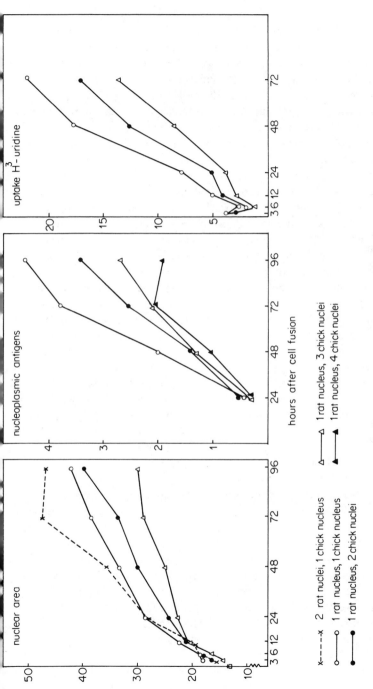

Figure 7.8: reactivation of chick erythrocyte nuclei in heterocaryons with rat epithelial cells. The nuclear area of the reactivating chick erythrocyte nuclei, their uptake of rat nucleoplasmic antigens, and their incorporation into RNA of H³-uridine all depend in a similar manner upon the ratio of chick:rat nuclei in the heterocaryon. The greater the proportion of chick nuclei, the slower their rate of reactivation. This suggests that reactivation depends upon uptake of molecules available in limiting amounts in the rat cytoplasm. Data of Carlsson et al. (1973).

that the rate of reactivation depends upon the numbers of chick nuclei in the cell; the greater the number of chick nuclei, the slower is the rate of reactivation measured by increase in nuclear area, uptake of nucleoplasmic antigens characteristic of the rat, or incorporation of H^3-uridine into RNA. These experiments therefore suggest that the chick nuclei in a common cytoplasm compete with each other for uptake of the molecules upon which reactivation depends; the extent of reactivation of each nucleus depends upon its success in obtaining these molecules from the cytoplasm in which it finds itself.

The distribution of human nuclear proteins in chick erythrocyte–Hela cell fusions has been followed by Ringertz et al. (1971), who utilised anti-nuclear antibodies produced by patients with auto-immune diseases. These antibodies can be tagged with fluorescent groups so that antibodies to human nucleoli fluoresce only over the nucleolus, those to the nucleoplasm react principally with this area (although also showing some response to the nucleolus); antibodies to human cytoplasm reveal Hela cells as fluorescent cytoplasms in which the nuclei appear as dark holes. Chick cells react only very weakly or not at all with these antibody preparations.

Antisera to human nucleoli and nucleoplasm react increasingly strongly with the chick nuclei as they reactivate. Cytoplasmic antigens remain unreactive. As nucleoli form in the chick nuclei, they react first with human nucleolar antigens and later during reconstruction with antibodies prepared against chick nucleoli. The Hela nuclei of the hybrid cells at first show only a weak reaction with the chick specific antigens against the nucleolus, but as the chick nuclei grow in size and develop nucleoli, the Hela nucleoli become as reactive as those in the erythrocyte nuclei. The intensity with which Hela nucleoli react with antigens against chick nucleoli depends upon the ratio of Hela to chick nuclei in the cell; the greater the number of chick nuclei, the higher the reaction of all nuclei to chick specific antigens. Within individual hybrid cells, the Hela nuclei closest to the chick nuclei show the strongest reaction with chick specific antigens.

These results imply that human nucleoplasmic and nucleolar proteins move into chick nuclei as they start to reactivate. The failure of Hela cytoplasmic proteins to enter the chick nuclei shows that their enlargement does not depend upon a passive swelling process in which human proteins at random enter the nucleus—reactivation is specific and only appropriate proteins are taken up and concentrated in the appropriate regions of the chick nucleus. The observation that chick nucleoli react first to the human antigens and only later during development to the chick antigens suggests the the nucleolus in the chick erythrocyte is initially reconstructed with human nucleolar proteins. Only later, when its formation has enabled the chick genome to direct protein synthesis, are chick nucleolar proteins synthesized.

The appearance of the chick nucleolar proteins in both chick and Hela nuclei of hybrid cells suggests that once synthesized in the cytoplasm they may be utilised by any nucleus present, of either chick or human species. The effect of the relative numbers and positions of the chick and Hela nuclei in their common cytoplasm suggests that a gradient of nucleolar proteins may exist in the cell from the nuclei coding for them. The structural components of the nucleoli of chick and human cells must be sufficiently similar to replace each other and to form functional nucleoli consisting in part of components from each species. And since at least the first chick messengers must be translated exclusively on pre-existing Hela cytoplasmic ribosomes, the translation as well as the transport apparatus must lack species specificity.

Genetic Complement of Hybrid Cells

Fusion of Cells in Different Stages of the Cycle

When a population of Hela cells grows in suspension, about 30–40% of the cells are in S phase at any given moment. Johnson and Harris (1969a) found that the proportion of labelled nuclei is the same in the homocaryons immediately after fusion, which implies that there is no selection for fusion of cells in any particular stage of the cell cycle. At one hour after fusion, cells display a high frequency of asynchronous labelling when incubated with H^3-thymidine to determine whether none, some, or all of the nuclei in the hybrid are in S phase; fusion between cells therefore appears to be random so that each nucleus in the hybrid retains its DNA in the state characteristic of the cell from which it has been derived. Similar results are obtained in hybrids with 2, 3 or 4 nuclei.

Hybrid cells show an increasing tendency to establish synchrony in nuclear replication and division after their formation. Synchrony in nuclear replication begins to be imposed during the first day and by the third day is appreciable in binucleate, trinucleate and tetranucleate cells. The imposition of synchrony can also be observed by following mitosis. At 2 hours after fusion, more than 70% of the cells which contain a mitotic nucleus are asynchronous so that only some of the nuclei in the cell have entered mitosis. By 23 hours and 33 hours after fusion, only 20% of the cells show asynchronous mitosis and in the majority all nuclei divide simultaneously; by 50 hours after fusion the proportion of cells with asynchronous nuclei has declined to 16·5%. With increasing incubation, homocaryons therefore establish improved synchrony in both nuclear DNA synthesis and mitosis; as nuclei continue to reside in common cytoplasm they become increasingly likely to follow simultaneous division cycles, presumably due to their exposure to the same control signals.

During the first 44 hours of fusion between a hen erythrocyte and a Hela cell, the labelling of the Hela nuclei in heterocaryons is maintained at the level characteristic of Hela cells. Johnson and Harris (1969b) therefore noted that the presence of the erythrocyte nuclei can have little effect on the replication of the Hela nuclei. The same result is found in hybrid cells containing one Hela nucleus and several erythrocyte nuclei. Within the first 20 hours after fusion, about 30% of the erythrocyte nuclei start to replicate their DNA; they display considerable synchrony with the Hela nuclei, largely replicating their DNA only when the Hela nuclei are also in S phase. This suggests that the same component of the hybrid cell induces DNA synthesis in both human and hen nuclei, reinforcing the conclusion that the resumption of synthetic activities in the erythrocyte nucleus is a specific response to control signals of the Hela cell which lack species specificity.

Fusion of cells in different stages of their cycles shows that the state of a nucleus may be influenced by the states of the other nuclei present in the same cytoplasm. Rao and Johnson (1970) have fused G1 with S phase Hela cells, G1 with G2 cells, and S phase with G2 cells.

That the effects on the nuclei of these heterophasic fusions are specific is shown by the absence of any change in the states of nuclei subjected to homophasic fusions in which both cells are in the same stage of the cycle. By lightly prelabelling only one of the parental populations of the Hela cells, the two classes of nuclei in the hybrid can be distinguished. When a large dose of H^3-thymidine and colcemid is added after fusion, its density of labelling is great enough to be distinct from the light prelabel which distinguishes the two classes of parent nuclei.

In binucleate cells, fusion of S with G1 nuclei induces DNA synthesis more rapidly in the G1 nucleus. In multinucleate cells, the ratio of nuclei in advanced to early stages of the cycle determines their relative activities. In G1–S phase fusions, the time taken for half of the G1 nuclei to incorporate a label of H^3 thymidine is:

G1 controls	10 hours
2G1:S	3
G1:S	1·75
G1:2S	1·5

Synthesis of DNA therefore appears to be under a positive control in which inducers synthesized by S phase nuclei may hasten the entry into replication of G1 nuclei. Fusions of G2 with G1 nuclei do not induce DNA synthesis, so that the signals which activate replication do not remain in the cell after the termination of S phase.

In binucleate hybrid cells, fusion of S with G2 causes the S phase nucleus to enter mitosis more rapidly than usual; there is no induction of DNA synthesis in the G2 nucleus. The G2 nucleus is not receptive to signals to replicate DNA, for it is impossible to induce DNA synthesis in it even by the introduction of many S phase nuclei. In multinucleate hybrid cells, the initiation of mitosis is dose dependent; the greater the ratio of G2 to S phase nuclei, the sooner the S phase nuclei enter mitosis. In fusions involving several nuclei, the time taken to reach a mitotic index of 50% is:

4S nuclei	14·2 hours
G2:S	13·3
2G2:S	11·6
3G2:S	10·7
4G2	8·5

Mitosis may therefore result from the accumulation of an inducer in G2 cells which is also able to act upon S phase nuclei. This interpretation is supported by the results of fusions between G1 and G2 cells, in which the G1 nucleus enters DNA synthesis at the usual time but the entry into mitosis of the G2 nucleus is delayed; the presence of G1 nuclei may lower the concentration in the G2 nuclei of a mitotic inducer and therefore delay accumulation of a critical amount. Synchrony in cell fusion has been reviewed by Johnson and Rao (1971).

That the induction of DNA synthesis is not species specific but the duration of S phase is a characteristic of each nucleus is suggested by fusions between synchronized populations of Hela cells and mouse or Chinese hamster cells. Each of the parent cell populations spends a characteristic time in G1 and S phases: Hela cells display a G1 of 7–11 hours and an S phase of 7·5 hours; mouse cells a G1 phase of 1·5–3·5 hours and an S period of almost 9 hours; and Chinese hamster cells a G1 of 2–2·5 hours and an S phase of 5·5 hours. In hybrids produced by fusion together of mitotic Hela cells, both nuclei initiate DNA synthesis together after the usual G1 interval. But in fusions between mitotic Hela and mitotic mouse or Chinese hamster cells, both nuclei initiate S phase after a G1 period 4–5 hours shorter than that usually displayed by the Hela cell; each nucleus continues S phase, however, for its characteristic length of time. Signals lacking species specificity must therefore be synthesized by the mouse or Chinese hamster component of the hybrid and induce S phase in both species of nucleus; the period required for replication, however, must depend upon the characteristic organization of each chromatin and is not influenced by cytoplasmic signals.

Premature Condensation of Chromosomes

Mitotic asynchrony occurs only rarely in G1–S and G1–G2 fusions, perhaps because the nuclei can reach an equilibrium in which division takes place at a time appropriate for both. Johnson and Rao (1970) observed that in G2–S fusions, however, the G2 nucleus may enter mitosis before the nucleus in S phase is ready; this causes a premature condensation of the chromosomes in the S phase nucleus, in effect an atypical mitosis without a mitotic spindle. Synthesis of DNA is reduced but not halted completely in the condensed chromosomes and they cease RNA transcription; the morphological changes which take place parallel those of mitosis itself.

Premature chromosome condensation can be induced in all interphase cells by fusion with mitotic cells. If the chromosomes of G1 or G2 nuclei are prematurely condensed by such fusion, they display single and double chromatids respectively. Induction of condensation in S phase nuclei, however, yields unevenly condensed chromatin with patches of large and small dispersed fragments separating condensed regions. The induction of premature chromosome condensation in a hybrid cell depends upon the ratio of mitotic to non-mitotic nuclei; this dosage effect represents a titration of the inducer molecules which must be produced by G2 and M phase cells to cause condensation.

No species specificity is apparent in the induction of premature chromosome condensation. Johnson, Rao and Hughes (1970) found that when a population of cultured Chinese hamster ovary cells is fused with mitotic Hela cells, it displays the same three classes of prematurely condensed interphase chromosomes generated by homocaryotic fusions of Hela cells. The mitotic inducers of the Hela cell are also effective upon fusion with Xenopus or mosquito cells and with non-dividing cells such as lymphocytes or erythrocytes. The nature of the molecules which cause chromosome condensation for mitosis therefore appear to be independent of both species and cell phenotype and must interact with features of chromatin present in all cells.

Whatever factors induce chromosome condensation are produced and remain in mitotic cells for only a short time. Rao and Johnson (1972) found that when Hela cells are blocked in mitosis with colcemid before they are fused with an unsynchronized population, the proportion of the interphase nuclei which suffer premature chromosome condensation in the mitotic–interphase hybrids depends upon the length of time for which the mitotic cells have remained in colcemid. Freshly collected mitotic cells maintained in colcemid for 4 hours induce premature chromosome condensation in 88% of the resulting hybrids; but if the mitotic cells are incubated in the colcemid for another 20 hours before fusion, the proportion of hybrid cells with prematurely condensed chromosomes is only 10%. As cells are maintained in mitosis they therefore lose their inducing factors. This suggests that the factors are synthesized in late G2 and the beginning of mitosis and are not stable but decay as the cell passes through division, presumably since mitotic cells would otherwise usually be unable to disperse their condensed chromosomes.

By prelabelling the chromosomes of an interphase population of cells before fusion with mitotic cells, their fate can be followed during growth of the subsequent hybrid cells. In the first cell cycle after fusion, the interphase chromosomes suffer premature condensation to give the appearance dictated by their state prior to fusion. The prematurely condensed chromosomes induced in all interphase nuclei suffer the same fate; they are not integrated into the mitotic spindle and at the end of mitosis are randomly distributed to the daughter cells. (In cells in which the interphase chromosome set fails to suffer premature condensation, the mitotic set of chromosomes is unable to complete division and instead forms inactive micronuclei.) In the second cycle of hybrid cells which have suffered loss of the interphase set by premature condensation in the first cell cycle, some of the chromosomes are labelled. This implies that some of the prematurely condensed chromosomes have survived the first mitosis and are included in the nucleus; during subsequent cell cycles they display the same behaviour as the chromosomes of the other parental set.

By hybridizing populations of *glyA* and *glyB* mutants of Chinese hamster ovary cells, cells containing both $glyA^+$ and $glyB^+$ genes can be generated. In a suitable medium, only hybrid cells containing both these genes, that is one from each parent cell, can survive. When mitotic and interphase cells are fused, there is a low rate of survival in selective medium compared with that when cells in the same phase are fused. This implies that the formation of prematurely condensed chromosomes, with the consequent loss of one parental set from the hybrid cells, causes a low rate of survival by making unavailable one of the genes required for growth in the selective medium. But in those cells which survive mitotic–interphase fusion, genes provided by both parents must be expressed; it is therefore possible for chromosomes to survive premature condensation and to transcribe their genes in subsequent cell generations. Cells whose G1 or G2 chromosomes are prematurely condensed have a better chance of survival than cells with prematurely condensed S phase chromosomes, consistent with the view that chromosomes may suffer structural changes during replication of their DNA.

That genes may survive premature condensation of the chromosomes on which they are carried and may subsequently be expressed in an alien cell is suggested by experiments on the fusion of chick embryo erythrocytes with A9 mouse fibroblasts. After fusion, the chick chromosome set often suffers a premature condensation when the mouse nucleus passes through mitosis. However, the A9 mouse cells lack the enzyme hypoxanthine guanine phosphoribosyl transferase, also known as inosinic acid pyrophosphorylase, and are therefore unable to grow on HAT medium in which this enzyme is essential for survival. Schwartz, Cook and Harris (1971) found that when the hybrid cells of A9–erythrocyte fusions are incubated in HAT medium, most of them die; but some clones resistant to HAT appear within 2–3 weeks after cell fusion. These clones contain the chick enzyme HGPRT, identified as such by its characteristic electrophoretic mobility.

In spite of the presence of the chick enzyme, no chick chromosomes can be seen in the hybrid cells, even taking into account the many small "dot" chromosomes of the chick chromosome set. One explanation for their absence might be that chick genetic material has in some way been integrated into the mouse chromosome set. In this case, the chick gene should be retained in the population even after the selective pressure for the enzyme which it specifies is removed. This prediction can be tested by comparing the rate of loss of HGPRT from the hybrid cells with that of L cells, the mouse parental line whose loss of this function gave rise to the A9 variant.

After culturing the two populations of cells in HAT medium for 4 months to select for retention of the gene for HGPRT, Schwartz et al. transferred the hybrid cells and the L cells to a neutral medium for 6 weeks. The number of clones which have lost the enzyme can be tested by transfer to a medium containing 8-azaguanine; cells which have HGPRT activity incorporate this analogue into their nucleic acids and die, but those lacking the enzyme are resistant and survive. The frequency of production of cells lacking the enzyme is 10^{-6} in L cells, but 20% in the hybrid population. The hybrid cells must therefore retain the chick gene in an unstable state under selective pressure but lose it readily when the pressure is removed.

Genetic material of the erythrocyte chromosomes which have suffered premature condensation is probably incorporated into the mouse nucleus during its post mitotic reconstitution. In cell hybrids in which whole chromosomes are included in the reconstituting nucleus, the genetic information which they carry appears to be stably expressed in subsequent generations. But the chick erythrocyte genes are included in the mouse nucleus in only an unstable state, perhaps because the erythrocyte chromosomes have been fragmented during their premature condensation into material which can in some way associate with mouse chromosomes but which does not behave as an integral component of the nucleus. Since no chick specific antigens can be detected on the surface of the hybrid cells, and since the genes coding for these proteins are widespread in the chromosome set, it is likely that the amount of chick genetic information retained in the mouse nucleus is quite small.

Genetic Mapping in Somatic Cell Hybrids

Two classes of problem have prevented mapping of the mammalian genome with the resolution of that possible for bacteria: the lack of mutants and the difficulty of obtaining recombinants. It is in general more difficult to induce and isolate mutants in particular genes in multicellular than in unicellular organisms, for it is not possible to clone large numbers of identical genotypes and to isolate conditional lethal mutants by the use of appropriate selective techniques—mammalian organisms would in general suffer fatal damage from such treatment. And the long generation times of mammals and the small numbers of progeny of each mating hinder mapping of those mutants which have been identified; in man, selective crosses are not possible and genetic information must be derived solely from familial studies.

Use of cell cultures allows some of the techniques of bacterial genetics to be applied to eucaryotic cells, for mutants may be induced and selected by establishing appropriate conditions of growth. Of course, one important limitation is that only those genes expressed in the cultured cell—in general non-differentiated functions—can be investigated in this way. One problem common to all attempts to map mammalian genomes has been that, with the exception of the X chromosome, correlation between linkage groups and chromosomes is difficult to establish in species in which not all chromosomes are morphologically distinguishable; the use of translocations to identify chromosomes suffers from the same difficulties encountered in obtaining genetic recombinants. However, the discovery that specific banding patterns may be generated in each chromosome by quinacrine or Giemsa staining may overcome this problem; each chromosome of the human complement, for example, can be identified by its fluorescent bands in the metaphase condition.

Obtaining a genetic map of the chromosomes of cultured cells depends first upon the induction and isolation of mutants. Mutant cells deficient in particular enzyme activities have been identified by selective techniques which rely upon drug resistance or nutritional deficiencies. Drug resistant cell lines were amongst the first mutant classes isolated. By exposing mouse L cells to 8-azaguanine, Littlefield (1964) isolated the A9 mutant which lacks the enzyme hypoxanthine-guanine phosphoribosyl-transferase (HGPRT); cells possessing this enzyme activity incorporate 8-azaguanine into DNA and therefore die, but mutants lacking it are resistant to the drug and therefore survive.

When plated on HAT (hypoxanthine aminopterin thymidine) medium, however, cells are prevented by the aminopterin from synthesising purines and pyrimidines; they can therefore survive only by utilising the bases provided in the medium. Hypoxanthine must be utilised via the pathway catalysed by HGPRT; cells lacking the HGPRT enzyme activity therefore die on HAT medium. It is therefore possible to select cells for the absence of this enzyme by growth on 8-azaguanine, or for its presence by growth on HAT, as is illustrated in figure 7.9.

Similar protocols have been used for selecting for or against cells possessing the enzymes thymidine kinase (TK) or adenine phosphoribosyl-transferase (APRT). Cells possessing thymidine kinase activity can incorporate the analogue bromodeoxyuridine (BUdR) into their DNA; this is fatal to the cell so growth on BUdR allows only cells lacking thymidine kinase to survive. Cells which possess thymidine kinase activity may be selected by plating on HAT, when the presence of aminopterin makes it essential for the cells to utilise the thymidine of the medium via the pathway catalysed by the kinase enzyme. Lines of mouse L cells lacking thymidine kinase have been isolated by this means.

Cells lacking APRT have been isolated by growth in the presence of the analogues 2,6-diaminopurine or 2-fluoroadenine whose incorporation into DNA is catalysed by the enzyme. Kusano, Long and Green (1971) found that the antibiotic alanosine prevents the conversion of IMP to AMP so that

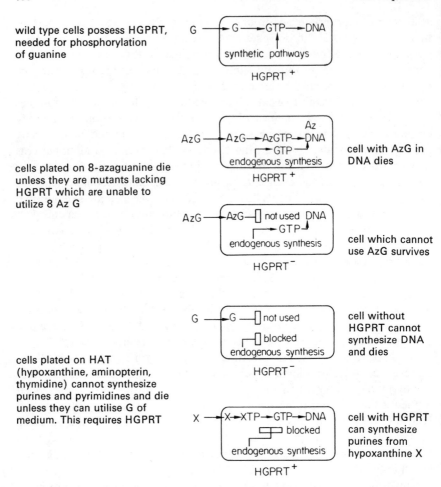

Figure 7.9: selection of cells for deficiency in HGPRT by growth on 8-azaguanine medium, or for presence of HGPRT by growth on HAT medium.

in its presence cells grown on adenine must utilise APRT to produce AMP if they are to survive.

Nutritionally deficient mutants of Chinese hamster cells which lack an enzyme activity have been isolated by their failure to grow when the medium is not supplemented with the metabolite they are unable to produce; the protocol developed by Puck and Kao (1967) is illustrated in figure 7.10. This constitutes an adaptation of a method for isolating bacterial mutants in which a population of cells is mutated and placed on a medium in which the mutant cells lacking some metabolic pathway cannot grow; cells possessing the enzymes of the pathway, however, can grow in spite of the absence of the

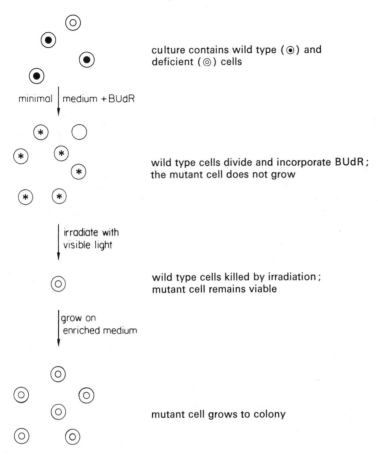

Figure 7.10: protocol for selection of nutritional mutants. Cells lacking ability to utilise some growth factor cannot grow on minimal medium and are therefore immune from the effects of incorporation of BUdR and irradiation with visible light. Method of Kao and Puck (1968).

metabolite omitted from the selective medium and are killed by application of an agent which discriminates between dividing and non-dividing cells.

Mammalian cells which grow and divide must replicate their DNA and may therefore be killed selectively by addition of BUdR and irradiation with near visible light (which is lethal to cells which have incorporated the analogue into DNA). The surviving cells may then be plated on an enriched medium containing the metabolite needed by the mutants and may be cloned and characterized. By subjecting cells previously mutated with ethane methane sulfate or nitrosoguanidine to such protocols, Kao and Puck (1968) isolated several mutants in various metabolic pathways, including those unable to synthesize glycine, proline or inositol.

Patients with inherited diseases of metabolism provide a source of human cells lacking particular enzymes; mapping of these mutants is rarely possible except for those located on the X chromosome, whose characteristic pattern of inheritance is readily revealed from familial studies.

Identification of the chromosomes which carry genes coding for particular enzymes has been made possible by a useful property of hybrids formed between cells of different species; cells in which the two parent nuclei have fused into a joint nucleus containing both parental sets of chromosomes may subsequently preferentially lose the chromosomes of one parent. The characteristics of this loss vary with the species. Weiss and Ephrussi (1966) first noted that one month after fusion between mouse and rat cells the number of chromosomes is much lower than the total of the two parent complements. The principal loss of chromosomes seems to take place soon after fusion to yield stable hybrids which retain their remaining chromosomes.

When human and mouse cells are fused, hybrids which form a single joint nucleus at first appear to possess all the chromosomes of both parents. Within only a few generations, however, many of the human chromosomes are lost although the hybrids retain all or nearly all of the mouse chromosomes. The results of Matsuya, Green and Basilico (1968) and Nabholz, Miggiano and Bodmer (1969) show that when the hybrids are first examined a few generations after fusion, they have already lost many of their human chromosomes. Further human chromosomes are gradually eliminated from the nucleus during subsequent cycles of division so that the progeny eventually contain chromosomes derived largely or entirely from the mouse parent alone. We do not know why the human chromosome complement is preferentially lost.

Preferential chromosome loss is not a characteristic of all inter-species cell fusions, for Pontecorvo (1971) observed that hybrids between Chinese hamster and mouse cells eliminate the chromosomes of neither parent. It would of course be useful to direct the elimination of chromosomes in hybrids so that those of either parent may be lost; and irradiation with X-rays or γ-rays of either the Chinese hamster or mouse cell parent line in this cross leads to preferential loss of its chromosomes after fusion.

By correlating the loss or retention of particular chromosomes with the disappearance or maintenance of particular gene functions, it is possible to identify the chromosomes which carry certain genes. Because human chromosomes are preferentially lost from hybrids formed with mouse or Chinese hamster cells, this technique has proved particularly useful for locating genes in the human complement. A human cell which possesses the gene coding for some active enzyme is fused with a cell of another species which lacks this catalytic function. The hybrid cells are then transferred to a medium in which they can survive only if they possess this enzyme activity.

This establishes a selection system for cells carrying at least one copy of the gene coding for the essential human enzyme. Any hybrids which lose all their copies of the human chromosome carrying this gene therefore die; all the hybrids which survive on the selective medium must contain at least one copy of the human gene and therefore of the homologue which carries it. As increasing numbers of human chromosomes are lost from the hybrid cells in successive growth cycles, the survivors come to retain only the single human chromosomes carrying the gene necessary for survival, or at least only a very small number of human chromosomes including this one. Identification of the common human chromosome retained by all survivors thus locates the gene in the human complement. By mutating some function of mouse or Chinese hamster cells, it is therefore possible to map the corresponding locus in the human genome.

The gene coding for thymidine kinase in the human genome was the first to be located by this technique. Weiss and Green (1967) fused human diploid fibroblasts which synthesize thymidine kinase with a mouse cell line which lacks the enzyme. Human–mouse hybrids were grown for four days in a standard medium and then transferred to HAT medium in which thymidine kinase activity is essential to allow the cells to utilise thymidine. The mouse parent cells die in this medium and the human parents grow only rather poorly; hybrid cells which can synthesize thymidine kinase grow well.

In hybrid cells examined after 20 generations, all or nearly all the mouse chromosomes were present, but only a few (2–15) of the human chromosomes remained in the various populations. But all cells grown in HAT contain at least one human chromosome of the E group. Cells which lose the thymidine kinase gene occur with high frequency in each generation; although they die in HAT medium they can be selected by growth in medium containing BUdR, in which cells possessing the enzyme die. Chromosomes of group E are rare in hybrid cells grown in BUdR. These results therefore suggest that the thymidine kinase gene is located on one of the chromosomes of this group.

The number of human chromosomes in the man–mouse hybrids falls steadily during successive growth cycles. Matsuya, Green and Basilico (1968) obtained the results shown in figure 7.11; after 18 generations, when the cells were first examined, this population showed 10·2 human chromosomes per hybrid cell. The human chromosomes continue to be lost and by 80 generations only 3 may be present; the value may fall to zero when the cells are grown on standard medium so that the selective pressure to retain the human thymidine kinase gene is removed.

The presence of human chromosomes in the hybrids influences their growth rate. The mouse parental cells have a doubling time of 15 hours; the human diploid fibroblasts used to provide the other parent grow somewhat more slowly. The rate of growth of the hybrids increases with the loss of human chromosomes. Hybrids with 10 human chromosomes double in 35–60 hours; the doubling time reaches 25–30 hours when the human chromosome complement falls below 4. Hybrids which retain only the human chromosome carrying the thymidine kinase gene double in 23 hours; cells which have lost all the human chromosomes double in 16–19 hours. The effect of the human chromosomes on growth may be connected with the reason for their preferential loss.

Because fusion experiments may induce reversion of mutations in the defective parent, it is particularly important to characterize the enzyme activity in the hybrid. Migeon, Smith and Leddy (1969) confirmed by analysis of

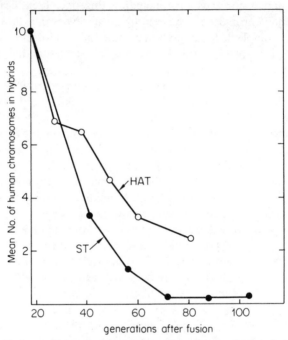

Figure 7.11: loss of human chromosomes from human-mouse hybrid cells. On standard medium (ST) all human chromosomes may be lost in 80 generations. On HAT medium, some human chromosomes must be retained so that the hybrid possesses thymidine kinase (absent from the mouse parent) necessary for survival; only 2–3 human chromosomes are retained after 80 generations. Data of Matsuya, Green and Basilico (1968).

electrophoretic mobility and heat sensitivity that the thymidine kinase activity in hybrid cells is that coded by the human genome. Miller et al. (1971c) utilised identification of chromosomes by their fluorescent banding patterns to confirm that the E group human chromosome retained in the hybrids is number 17. Green et al. (1971) found that although many of the hybrids surviving on HAT retain only a single copy of chromosome 17, others may have upto six, presumably as the result of non-disjunction of chromosome 17 at mitosis; although one copy of the thymidine kinase gene is clearly sufficient to alleviate the selection pressure, it is possible that in some circumstances it may be advantageous for the cell to possess further copies.

The mutations responsible for loss of thymidine kinase and HGPRT in the mouse cell lines used for hybridization have not been defined but behave as deletions in failing to produce revertants possessing the enzymes. That the HGPRT deficiency does not result from a deletion, however, has been shown by more recent experiments in which revertants were obtained during cell hybridization experiments. Watson et al. (1972) fused a culture of mouse A9 cells with human lymphocytes and then transferred the population to HAT medium on which only hybrids possessing the HGPRT enzyme activity provided by the human parent should survive. But three of the surviving cell lines possessed no human chromosomes and synthesized the HGPRT enzyme specified by the mouse genome. This implies that reversion is in some way promoted by exposure of the cells to fusion.

Using both A9 cells and RAG cells (a variant of the Balb/c mouse adrenal-carcinoma selected for lack of HGPRT), Shin et al. (1973) found that exposure to two of the stages of the fusion protocol induces reversions: bringing the cells to a high (10^7/ml) concentration and exposure to heat shock. All the revertants synthesize the mouse enzyme; we do not know the nature of the original mutation or why its reversion should become possible only under these conditions. The reversion frequency of the A9 cells is $1{\cdot}8 \times 10^{-8}$ and that of the RAG cells is 9×10^{-8}.

Cells which survive hybridization and grow on selective medium may therefore possess a reversion of the mutation in one parent instead of an active gene provided by the other parent. Experiments to identify the chromosome responsible for specifying some enzyme essential for survival under adverse conditions must therefore utilise only enzymes which differ in the two parental species; this means that by characterizing the active protein it is possible to rely upon cells which survive because of their possession of an alien (human) chromosome and not because of reversion of the original mutation.

Fusion between cells of different species is most powerful as a technique for gene mapping when a selective pressure can be established for loss or retention of some enzyme activity. But correlations between chromosomes and genes can also be made for any enzyme coded by one species which is distinguish-able—for example by its electrophoretic mobility—from that of the other species. By screening cell populations to identify those lacking and possessing the human enzyme, it is possible to deduce from the remaining human chromo-somes which specifies the enzyme.

By following the abilities of hybrid cells to synthesize two enzymes of the human genome, it is possible to establish linkage groups. If two enzymes are always retained together or lost together, they must be coded by genes which lie on the same chromosome. Although it is possible simply to screen hybrids for pairs of human enzymes, a more productive technique is to grow clones under selective conditions demanding the retention of one enzyme and follow the retention or loss of other, unselected enzyme markers. Nabholz, Miggiano and Bodmer (1969) used this approach to screen human–mouse hybrids grown on HAT medium—which must therefore retain HGPRT—for their ability also to synthesize human glucose-6-phosphate dehydrogenase (G6P DH) and the lactate dehydrogenase polypeptides, LDH A and LDH B.

By 40–60 days after fusion, the hybrids contained only 4–14 chromosomes in excess of the parental mouse complement. All the hybrids surviving on HAT retained the human HGPRT enzyme and also the human G6P DH. This correlation suggests that both enzymes are coded by genes on one chromosome. Since studies of human families have shown that G6P DH is carried by the X chromosome, this implies that HGPRT must also be sex linked, a conclusion which is consistent with results obtained from patients with inherited metabolic diseases in HGPRT. The activities of LDH A and LDH B, by contrast, vary very considerably in the different clones, showing no correlation with the retention of HGPRT and G6P DH or with each other; the genes for these two polypeptides must therefore be carried on different autosomes of the human complement.

An exception to the rule that genes on the same chromosome are jointly inherited may result when a chromosome break and/or translocation takes place in the hybrid cell. Miller et al. (1971d) observed the mitotic separation of linked loci in hybrids between human and mouse A9 cells; 47 out of 105 clones grown on HAT lacked G6P DH in spite of its usual linkage with HGPRT. This implies that a break must have taken place in the X chromosome to separate the two genes so that only the part carrying HGPRT is selected for retention.

Another gene located on the X chromosome codes for phosphoglycerate kinase (PGK). Grzeschek et al. (1972) examined its retention in hybrids derived from fusion of mouse A9 or Syrian hamster TG2 cells lacking HGPRT with human diploid fibroblasts derived from a heterozygote for a translocation between the X chromosome and chromosome 14 of the D group; the translocation involves breakage of the X chromosome into its two arms, the short arm forming an independent Xq chromosome and the long arm becoming attached to the end of chromosome 14 to yield the large chromosome t(14q, Xq). Cells of the human parent (KOP) possess one copy each of chromosome 14, the X chromosome, the Xq short arm, and the t(14q, Xq) translocation which carries the long arm of the X chromosome. Because the normal X chromosome provides the inactive sex chromatin of these cells—and is not reactivated by cell fusion—all active sex-linked genes of the human complement must be located on one of the two translocation chromosomes, Xq and t(14q, Xq).

All the hybrid cells grown on HAT medium retained both the HGPRT enzyme essential for survival and also the human G6P DH enzyme; some but not all clones also retained PGK of human origin. The translocation chromosome t(14q, Xq) can be identified by its characteristic fluorescent banding pattern and proved to be present in all clones synthesising human PGK but absent in all clones lacking this enzyme; the gene for PGK must therefore be carried on the long arm of the X chromosome. The other translocation chromosome, t(14q, Xq) cannot always be unequivocally distinguished from the chromosomes of the non-human parent but appears to be present in at least some clones; it is therefore likely that the genes coding for HGPRT and G6P DH are carried on the short arm of the X chromosome— although sufficiently far apart to be separated by the break identified by Miller et al. (1971d)—and that Xq is present in all clones.

One possible qualification on the use of translocations to allow separation of linked genes and thus to identify the chromosome region carrying some gene is that their inheritance may not always be independent. Only a small proportion of hybrid clones lost the t(14q, Xq) chromosome carrying the gene for PGK; most clones retained this enzyme during growth in HAT as well as the essential HGPRT and the linked G6P DH. Upon transfer to medium containing 8-azaguanine, most cells lost human PGK as well as the HGPRT and G6P DH enzymes; only some clones lost HGPRT and G6P DH but retained PGK.

Although only the Xq homologue is implicated in responding to the demands of the selective medium, during growth under selective conditions both parts of the X chromosome therefore tend to show the same behaviour. Ruddle et al. (1971) also found that KOP–mouse hybrids retain all three sex linked human enzymes when they are transferred into HAT medium directly after fusion. Grzeschek et al. (1972) noted, however, that separation in some clones is achieved when the hybrid cells are first cloned in neutral medium and later transferred to selective medium to allow the survival of only those with the essential HGPRT gene. They suggested that chromosome loss may not always be random and that cells may retain both parts of the X chromosome after its translocation, although no reason is apparent for this relationship.

Another alteration in the usual response of the cells to selective pressure was revealed by the inviability of all derivatives of one hybrid clone in 8-azaguanine; this might be caused by translocation of the lethal HGPRT gene to a mouse chromosome whose loss would render the cell inviable. In this situation, a cell line cannot survive in 8-azaguanine, for it must either retain the HGPRT gene and die by incorporating the analogue (probably its usual fate) or it must lose a mouse chromosome and with it genes coding for other proteins essential to the survival of the cell.

Starch gel electrophoresis of seventeen enzymes of human KOP–mouse RAG cell hybrids was used by Ruddle et al. (1971) to establish linkage relationships. Only the three human enzyme markers HGPRT, G6P DH and PGK are invariably retained during growth on HAT medium; the other fourteen human enzymes studied must therefore be coded by autosomal genes. The autosomal coded enzymes present in the various clones show no pairwise correlations of inheritance; all are independently retained or lost as the hybrids lose human chromosomes. All fourteen genes must therefore be located on different chromosomes.

Nutritional mutants of Chinese hamster cells have been used by Kao and Puck (1970) to investigate whether human genes providing the missing functions in hybrids are carried on the same or different chromosomes. Extensive chromosome loss commences in the Chinese hamster–human hybrids immediately after fusion and within a few days the number of chromosomes in each hybrid cell falls to a value close to that of the Chinese hamster parent. The Chinese hamster parents have a generation time of about 12 hours; the human fibroblast parents display a doubling time of about 22 hours. The hybrids initially grow slowly but reduce their doubling time as they lose further human chromosomes; the increase in growth rate with loss of human chromosomes helps establish an automatic selection system in which cells with fewer human chromosomes leave more progeny.

After Chinese hamster cells whose genomes have two metabolic deficiencies have been fused with normal human cells, the hybrids can be grown on a medium containing only one of the two essential metabolites. Survivors must therefore retain the human chromosome carrying the gene which allows the cells to synthesize the metabolite omitted from the medium. After growth for long enough to allow all other human chromosomes to be eliminated, the cells are transferred to medium lacking the second metabolite. If they can grow in this medium, the human gene providing the missing function must be carried on the same chromosome which bears the gene necessary to alleviate the first deficiency. If cells retaining the first gene cannot grow under conditions used to define the second deficiency, the two genes must be located on different chromosomes.

Using the mutants which have been isolated in mouse and Chinese hamster cells, hybridization with human cells has allowed some 30 genes to be assigned to 16 of the human chromosomes. The assignment of genes to human chromosomes has been reviewed in detail by Ruddle (1973). An encouraging feature of the large number of chromosomes which has now been identified as carrying a specific gene is that further tests for the presence of other genes on any chromosome can be made simply by testing whether an unmapped enzyme marker is always inherited together with one known to be coded by the chromosome. Cytological identification of the chromosome common to all cells possessing the enzyme thus becomes necessary only as a check that no translocation or other rearrangement of the genetic material has occurred.

References

Abelson H.T. and Penman S. (1972). Messenger RNA formation: resistance to inhibition of 3' deoxycytidine. *Biochem. Biophys. Acta*, **277**, 129–133.
Abuelo J.G. and Moore D.E. (1969). The human chromosome. Electron microscopic observations on chromatin fibre organization. *J. Cell Biol.*, **41**, 73–90.
Adesnik M. and Darnell J.E. (1972). Biogenesis and characterization of histone mRNA in Hela cells. *J. Mol. Biol.*, **67**, 397–406.
Adesnik M., Salditt M., Thomas W. and Darnell J.S. (1972). Evidence that mRNA molecules (except histone mRNA) contain poly-A sequences and that the poly-A has a nuclear function. *J. Mol. Biol.*, **71**, 21–30.
Alfert M. (1958). Variations in cytochemical properties of cell nuclei. *Exp. Cell Res.*, *suppl.* **6**, 227–235.
Alfert M. and Das N.K. (1969). Evidence for control of the rate of nuclear DNA synthesis by the nuclear membrane in eucaryotic cells. *Proc. Nat. Acad. Sci.*, **63**, 123–128.
Allfrey V.G. (1971). Functional and metabolic aspects of DNA-associated proteins. In Phillips D.M.P. (Ed.), *Histones and Nucleohistones*. Plenum Press, New York, pp 241–294.
Allfrey V.G. and Mirsky A.E. (1963). Mechanisms of synthesis and control of protein and RNA synthesis in the cell nucleus. *Cold Spring Harbor Symp. Quant. Biol.*, **28**, 247–272.
Aloni Y., Hatlen L.E. and Attardi G. (1971). Studies of fractionated Hela cell metaphase chromosomes. II. Chromosomal distribution of sites for tRNA and 5S RNA. *J. Mol. Biol.*, **56**, 555–563.
Amaldi F. and Attardi G. (1968). Partial sequence analysis of ribosomal RNA from Hela cells. I. Oligonucleotide pattern of 28S and 18S RNA after pancreatic ribonuclease digestion. *J. Mol. Biol.*, **33**, 737–756.
Appels R. and Wells J.R.E. (1972). Synthesis and turnover of DNA-bound histone during maturation of avian red blood cells. *J. Mol. Biol.*, **70**, 425–434.
Appels R., Wells J.R.E. and Williams A.F. (1972). Characterization of DNA bound to histones in the cells of the avian erythropoietic series. *J. Cell Sci.*, **10**, 47–60.
Arms K. (1968). Cytonucleoproteins in cleaving eggs of X. laevis. *J. Emb. Exp. Morph.*, **20**, 367–374.
Arrhigi F.E. and Hsu T.C. (1971). Localization of heterochromatin in human chromosomes. *Cytogenetics*, **10**, 81–86.
Arrhigi F.R., Hsu T.C., Saunders P. and Saunders G.F. (1970). Localization of repetitive DNA in the chromosomes of M. agrestis by means of in situ hybridization. *Chromosoma*, **32**, 224–236.
Artman M. and Roth J.S. (1971). Chromosomal RNA: an artefact of preparation? *J. Mol. Biol.*, **60**, 291–302.
Ashburner M. (1967). Autosomal puffing patterns in a laboratory stock of D. melanogaster. *Chromosoma*, **21**, 398–428.

417

Ashburner M. (1969a). The X chromosome puffing pattern of D. melanogaster and D. simulans. *Chromosoma*, **27**, 47–63.

Ashburner M. (1969b). Patterns of puffing activity in the salivary gland chromosomes of Drosophila. IV. Variability of puffing patterns. *Chromosoma*, **27**, 156–177.

Ashburner M. (1970). Formation and structure of polytene chromosomes during insect development. *Adv. Insect Physiol.*, **7**, 1–95.

Ashburner M. (1972). Ecdysone induction of puffing in polytene chromosomes of D. melanogaster. Effects of inhibitors of RNA synthesis. *Exp. Cell Res.*, **71**, 433–440.

Auer G. (1972). Nuclear protein content and DNA-histone interaction. *Exp. Cell Res.*, **75**, 231–236.

Auer G., Moore G.P.M., Ringertz N.R. and Zetterberg A. (1973). DNA dependent RNA synthesis in nuclear chromatin of fixed cells. Relationship between dye binding properties, nuclear protein content and RNA polymerase activity. *Exp. Cell Res.*, **76**, 229–233.

Auer G. and Zetterberg A. (1972). The role of nuclear proteins in RNA synthesis. *Exp. Cell Res.*, **75**, 245–253.

Auer G., Zetterberg A. and Foley G.E. (1973). The relationship of DNA synthesis to protein accumulation in the cell nucleus. *J. Cell Physiol.*, **76**, 357–364.

Aula P. and Saksela E. (1972). Comparison of areas of quinacrine mustard fluorescence and modified Giemsa staining in human metaphase chromosomes. *Exp. Cell Res.*, **71**, 161–167.

Aviv H. and Leder P. (1972). Purification of biologically active globin mRNA by chromatography on oligothymidylic acid cellulose. *Proc. Nat. Acad. Sci.*, **69**, 1408–1412.

Bahr G.F. and Golomb H.M. (1971). Karyotyping of single human chromosomes from dry mass determined by electron microscopy. *Proc. Nat. Acad. Sci.*, **68**, 726–730.

Bajer A. and Mole-Bajer J. (1970). Architecture and function of the mitotic spindle. *Adv. Cell Mol. Biol.*, **1**, 213–267.

Bajer A.S. and Mole-Bajer J. (1972). Spindle dynamics and chromosome movements. *Int. Rev. Cytol.*, *suppl.* **3**.

Balhorn R., Bordwell J., Sellers L., Granner D. and Chalkley R. (1972). Histone phosphorylation and DNA synthesis are linked in synchronous cultures of HTC cells. *Biochem. Biophys. Res. Commun.*, **46**, 1326–1333.

Balhorn R., Chalkley R. and Granner D. (1972). Lysine rich phosphorylation. A positive correlation with cell replication. *Biochem.*, **11**, 1094–1098.

Balhorn R., Oliver D., Hohmann P., Chalkley R. and Granner D. (1972). Turnover of DNA, histones and lysine rich phosphate in hepatoma tissue culture cells. *Biochem.*, **11**, 3915–1920.

Baltimore D. and Huang A.S. (1970). Interaction of Hela cell proteins with RNA. *J. Mol. Biol.*, **47**, 263–273.

Barnicott N.A. (1966). A note on the structure of spindle fibres. *J. Cell Sci.*, **1**, 217–222.

Barnicott N.A. (1967). A study of newt mitotic chromosomes by negative staining. *J. Cell. Biol.*, **32**, 585–602.

Barr G.C. and Butler J.A.V. (1963). Histones and gene function. *Nature*, **199**, 1170–1172.

Bartley J.A. and Chalkley R. (1972). The binding of DNA and histone in native nucleohistone. *J. Biol. Chem.*, **247**, 3647–3655.

Bartley J. and Chalkley R. (1973). An approach to the structure of native nucleohistone. *Biochem.*, **12**, 468–474.

Baserga R. (1968). Biochemistry of the cell cycle: a review. *Cell Tissue Kinetics*, **1**, 167–191.

Baserga R. and Wiebel F. (1969). The cell cycle of mammalian cells. *Int. Rev. Exp. Path.*, **7**, 1–31.

Baxter J.D., Rousseau G.G., Benson M.C., Garcea R.L., Ito J. and Tomkins G.M. (1972). Role of DNA and specific cytoplasmic receptors in glucocorticoid cells. *Proc. Nat. Acad. Sci.*, **69**, 1892–1896.

Becker H.J. (1959). Die puffs der speicheldrusenchromosomen von D. melanogaster. I. Beobachtungen zum verhalten des puffmasters im normalsta, und bei zwei mutanten, giant und lethal giant leavae. *Chromosoma*, **10**, 654–678.

Becker H.J. (1962a). Die auslosung der puffbildung, ihre spezifitat und ihre beziehung zur funktion der ringdruse. *Chromosoma*, **13**, 341–384.

Becker H.J. (1962b). Stadienspezifische genaktivierung in speicheldrusen nach transplantation bei D. melanogaster. *Zool. Anz. Suppl.*, **25**, 92–101.

Beerman W. (1952). Chromerenkonstanz und spezifische modifikationen der chromosmenstruktur in der entwicklung und organdifferenzierung von C. tentans. *Chromosoma*, **5**, 139–198.

Beerman W. (1964). Control of differentiation at the chromosomal level. *J. Exp. Zool.*, **157**, 49–62.

Beerman W. (1965). Differentiation at the level of the chromosomes. In Beerman W. (Ed.), *Cell Differentiation and morphogenesis*. North Holland, Amsterdam, pp 24–54.

Beerman W. (1971). Effect of α-amanitin on puffing and intranuclear RNA synthesis in Chrinomus salivary glands. *Chromosoma*, **34**, 152–167.

Beerman W. (1973). Directed changes in the pattern of Balbiani ring puffing in Chironomus: effects of a sugar treatment. *Chromosoma*, **41**, 297–326.

Beerman W. and Bahr G.R. (1954). The submicroscopic structure of the Balbiani ring. *Exp. Cell Res.*, **6**, 195–201.

Beerman W. and Pelling C. (1965). H³ thymidin markierung einzelner chromatiden in Riesenchromosomen. *Chromosoma*, **16**, 1–21.

Bekhor I., Bonner J. and Dahmus G.C. (1969). Hybridization of chromosomal RNA to native DNA. *Proc. Nat. Acad. Sci.*, **62**, 271–277.

Bekhor I., Kung G.M. and Bonner J. (1969). Sequence specific interaction of DNA and chromosomal protein. *J. Mol. Biol.*, **39**, 351–364.

Bendich A.J. and McCarthy B.J. (1970). Ribosomal RNA homologies among distantly related organisms. *Proc. Nat. Acad. Sci.*, **65**, 349–356.

Berendes H.D. (1965a). Salivary gland functions and chromosomal puffing patterns in D. hydei. *Chromosoma*, **17**, 35–77.

Berendes H.D. (1965b). The induction of changes in chromosomal activity in different polytene types of cell in D. hydei. *Develop. Biol.*, **11**, 371–384.

Berendes H.D. (1966). Gene activities in the Malpighian tubules of D. hydei at different stages. *J. Exp. Zool.*, **162**, 209–218.

Berendes H.D. (1967). The hormone ecdysone as effector of specific changes in the pattern of gene activities of D. hydei. *Chromosoma*, **22**, 274–293.

Berendes H.D. (1968). Factors involved in the expression of gene activity in polytene chromosomes. *Chromosoma*, **24**, 418–437.

Berendes H.D. (1970). Polytene chromosome structure at the submicroscopic level. I. A map of region X, 1–4E of D. melanogaster. *Chromosoma*, **29**, 118–130.

Berendes H.D. and Keyl H.G. (1967). Distribution of DNA in heterochromatin and euchromatin of polytene nuclei of D. hydei. *Genetics*, **57**, 1–13.

Berlowitz L. (1965). Correlation of genetic activity, heterochromatinization and RNA metabolism in mammals. *Proc. Nat. Acad. Sci.*, **53**, 68–73.

Berns A.J.M., Strous G.J.A.M. and Bloemendal H. (1972). Heterologous in vitro synthesis of lens α-crystallin polypeptide. *Nature New Biol.*, **236**, 7–9.

Berns A.J.M., Van Kraaikamp M., Bloemendal H. and Lane C.D. (1972). Calf crystallin synthesis in frog cells: the translation of lens cell 14S RNA in oocytes. *Proc. Nat. Acad. Sci.*, **69**, 1606–1609.

Beutler E., Yeh M. and Fairbanks V.F. (1962). The normal human female as a mosaic of X-chromosome activity: studies using the gene for G-6-PD deficiency as a marker. *Proc. Nat. Acad. Sci.*, **48**, 9–16.

Bhorjee J.S. and Pederson T. (1972). Non histone chromosomal proteins in synchronized Hela cells. *Proc. Nat. Acad. Sci.*, **69**, 3345–3349.

Bibring T. and Baxendall J. (1968). Mitotic apparatus: the selective extraction of protein with mild acid. *Science*, **161**, 377–378.

Bick M.D., Huang H.L. and Thomas C.A.jr. (1973). Stability and fine structure of eucaryotic DNA rings in formamide. *J. Mol. Biol.*, **77**, 75–84.

Billeter M.A. and Hindley J. (1972). A study of the quantitative variation of histones and their relationship to RNA synthesis during erythropoiesis in the adult chicken. *Europ. J. Biochem.*, **28**, 451–462.

Birnboim H.C. and Coakley B.V. (1971). Adenylate rich oligonucleotides of ribosomal and ribosomal precursor RNA from Hela cells. *Biochem. Biophys. Res. Commun.*, **42**, 1169–1176.

Birnsteil M.L., Chipchase M.I.H. and Hyde B.B. (1966). The nucleolus as a source of ribosomes. *Biochim. Biophys. Acta*, **76**, 454–462.

Birnsteil M.L., Sells B.H. and Purdom I.F. (1972). Kinetic complexity of RNA molecules. *J. Mol. Biol.*, **63**, 21–39.

Birnsteil M., Spiers J., Purdom I., Jones K. and Loening U.E. (1968). Properties and composition of the isolated ribosomal DNA satellite of X. laevis. *Nature*, **219**, 454–463.

Bishop J.O., Pemberton R. and Baglioni C. (1972). Reiteration frequency of haemoglobin genes in the duck. *Nature New Biol.*, **235**, 231–234.

Bishop J.O. and Rosbash M. (1973). Reiteration frequency of duck haemoglobin genes. *Nature New Biol.*, **241**, 204–207.

Blatti S.P., Ingles C.J., Lindell T.J., Morris P.W., Weaver R.F., Weinberg F. and Rutter W.J. (1970). Structure and regulatory properties of eucaryotic RNA polymerase. *Cold Spring Harbor Symp. Quant. Biol.*, **35**, 649–658.

Blobel G. (1972). Protein tightly bound to globin mRNA. *Biochem. Biophys. Res. Commun.*, **47**, 88–95.

Blobel G. (1973). A protein of molecular weight 78,000 bound to a polyadenylate region of eucaryotic messenger RNA. *Proc. Nat. Acad. Sci.*, **70**, 924–928.

Bloch D.P. and Goodman G.C. (1955). A microphotometric study of the synthesis of DNA and nuclear histone. II. Evidence of differences in the DNA-protein complex of rapidly proliferating and non-dividing cells. *J. Biochem. Biophys. Cytol.*, **1**, 17–28.

Bloch D.P., MacQuigg R.A., Brack S.D. and Wu J.R. (1967). The synthesis of DNA and histone in the onion root meristem. *J. Cell Biol.*, **33**, 451–467.

Blumenfeld M. and Forrest H.S. (1971). Is Drosophila dAT on the Y chromosome? *Proc. Nat. Acad. Sci.*, **68**, 3145–3149.

Bobrow M., Pearson P.L. and Collacott H.E.A.C. (1971). Paranucleolar position of the human Y chromosome in interphase nuclei. *Nature*, **232**, 556–557.

Bolton E.T. and McCarthy B.J. (1962). A general method for the isolation of RNA complementary to DNA. *Proc. Nat. Acad. Sci.*, **48**, 1390–1397.

Bolund L., Darzynkiewicz Z. and Ringertz N.R. (1969a). Growth of the erythrocyte nuclei undergoing reactivation in heterocaryons. *Exp. Cell Res.*, **56**, 406–410.

Bolund L., Ringertz N.R. and Harris H. (1969b). Changes in the cytochemical properties of erythrocyte nuclei reactivated by cell fusion. *J. Cell Sci.*, **4**, 71–87.

Bonner J., Dahmus M.E., Fambrough D., Huang R.C., Marushige K. and Tuan D.Y.H. (1968a). The biology of isolated chromatin. *Science*, **159**, 47–56.

Bonner J., Chalkley G.R., Dahmus M., Fambrough D., Fujimura F., Huang R.C., Huberman J., Jensen R., Marushige K., Ohlenbusch H., Olivera B. and Widholm J. (1968b). Isolation and characterization of chromosomal nucleoproteins. *Methods Exymol.*, **12B**, 3–64.

Bonner J. and Huang R.C.C. (1966). Methodology for the study of the template activity of chromosomal nucleohistone. *Biochem. Biophys. Res. Commun.*, **22**, 211–217.

Bonner J. and Wu J.R. (1973). A proposal for the structure of the Drosophila genome. *Proc. Nat. Acad. Sci.*, **70**, 535–537.

Borisy G.G. and Taylor E.W. (1967a). The mechanism of action of colchicine. Binding of colchicine-H^3 to cellular protein. *J. Cell Biol.*, **64**, 525–533.

Borisy G.G. and Taylor E.W. (1967b). The mechanism of action of colchicine. Colchicine binding to sea urchin eggs and the mitotic apparatus. *J. Cell Biol.*, **34**, 534–548.

Borun T., Scharff M.D. and Robbins E. (1967). Rapidly labelled, polyribosome associated RNA having the properties of histone messenger. *Proc. Nat. Acad. Sci.*, **58**, 1977–1983.

Bostock C.J. and Prescott D.M. (1971a). Buoyant density of DNA synthesized at different stages of the S phase of mouse L cells. *Exp. Cell Res.*, **64**, 267–274.

Bostock C.J. and Prescott D.M. (1971b). Buoyant density of DNA synthesized at different stages of the S phase in Chinese hamster cells. *Exp. Cell Res.*, **64**, 481–484.

Bostock C.J. and Prescott D.M. (1971c). Shift in buoyant density of DNA during the synthetic period and its relation to euchromatin and heterochromatin in mammalian cells. *J. Mol. Biol.*, **60**, 151–162.

Bostock C.J., Prescott D.M. and Hatch F.T. (1972). Timing of replication of the satellite and main band DNAs in cells of the kangaroo rat. *Exp. Cell Res.*, **74**, 487–495.

Boublik M., Bradbury E.M. and Crane-Robinson C. (1970a). An investigation of the conformational changes of histones f1 and f2a1 by proton magnetic resonance spectroscopy. *Europ. J. Biochem.*, **14**, 486–497.

Boublik M., Bradbury E.M., Crane-Robinson C. and Johns E.W. (1970b). An investigation of the conformational changes of histone f2B by high resolution nuclear magnetic resonance. *Europ. J. Biochem.*, **17**, 151–159.

Bradbury E.M., Cary P.D., Crane-Robinson C., Riches P.L. and Johns E.W. (1972b). Nuclear magnetic resonance and optical spectroscopic studies of conformation and interactions in the cleaved halves of histone f2b. *Europ. J. Biochem.*, **26**, 482–489.

Bradbury E.M. and Crane-Robinson C. (1971). Physical and conformational studies of histones and nucleohistones. In Phillips D.M.P. (Ed.), *Histones and Nucleohistones*. Plenum Press, New York, pp 83–135.

Bradbury E.M., Crane-Robinson C., Goldman H., Rattle H.W.E. and Stephens R.M. (1967). Spectroscopic studies of the conformations of histones and protamine. *J. Mol. Biol.*, **29**, 507–523.

Bradbury E., Molgaard H.V., Stephens R.M., Bolund L. and Johns E.W. (1972a). X-ray studies of nucleoproteins depleted of lysine rich histones. *Europ. J. Biochem.*, **31**, 474–482.

Bradbury E.M. and Rattle H.W.E. (1972). Simple computer aided approach for the analyses of the nuclear magnetic resonance spectra of histone fractions f1, f2a1, f2b, cleaved halves of f2b and f2b-DNA. *Europ. J. Biochem.*, **27**, 270–281.

Bram S. (1971). The secondary structure of DNA in solution and in nucleohistone. *J. Mol. Biol.*, **58**, 277–288.

Bram S. and Beeman W.W. (1971). On the structure of nucleohistone. *J. Mol. Biol.*, **55**, 311–324.

Brasch K., Setterfield G. and Neelin J.M. (1972). Effects of sequential extraction of histone proteins on structural organization of avian erythrocyte and liver nuclei. *Exp. Cell Res.*, **74**, 27–41.

Bray G. and Brent T.P. (1972). Deoxynucleoside triphosphate pool fluctuations during the mammalian cell cycle. *Biochim. Biophys. Acta*, **269**, 184–191.

Breindl M. and Gallwitz D. (1973). Identification of histone mRNA from Hela cells. Appearance of histone mRNA in the cytoplasm and its translation in a rabbit reticulocyte cell free system. *Europ. J. Biochem.*, **32**, 381–391.

Brent T.P., Butler J.A.V. and Crathorn A.R. (1965). Variations in phosphokinase activities during the cell cycle in synchronous populations of Hela cells. *Nature*, **207**, 176–177.

Bridges C.B. (1935). Salivary gland chromosome maps with a key to the binding of the chromosomes of D. melanogaster. *J. Hered.*, **26**, 60–64.

Bridges C.B. (1938). A revised map of the salivary gland X chromosome of D. melanogaster. *J. Hered.*, **29**, 11–13.

Briggs R. and King T.J. (1960). Nuclear transplantation studies on the early gastrula R. pipiens. *Develop. Biol.*, **2**, 252–270.

Britten R.J. and Davidson E.H. (1969). Gene regulation for higher cells: a theory. *Science*, **165**, 349–357.

Britten R.J. and Davidson E.H. (1971). Repetitive and non repetitive DNA sequences and a speculation on the origins of evolutionary novelty. *Quart. Rev. Biol.*, **46**, 111–133.

Britten R.J. and Kohne D.E. (1968). Repeated sequences in DNA. *Science*, **161**, 529–540.

Britten R.J. and Kohne D.E. (1969a). Repetition of nucleotide sequences in chromosomal DNA. In Lima-de-Faria A. (Ed.), *Handbook of Molecular Cytology*. North Holland, Amsterdam, pp 21–36.

Britten R.J. and Kohne D.E. (1969b). Implications of repeated nucleotide sequences. In Lima-de-Faria A. (Ed.), *Handbook of Molecular Cytology*. North Holland, Amsterdam, pp 37–51.

Brown D.D. and Blackler A.W. (1972). Gene amplification proceeds by a chromosome copy mechanism. *J. Mol. Biol.*, **63**, 75–84.

Brown D.D. and Dawid I. (1968). Specific gene amplification in oocytes. *Science*, **160**, 272–280.

Brown D.D. and Gurdon J.B. (1964). Absence of ribosomal RNA synthesis in the anucleolate mutant of X. laevis. *Proc. Nat. Acad. Sci.*, **51**, 139–146.

Brown D.D. and Weber C.S. (1968a). Gene linkage by RNA–DNA hybridization. I. DNA sequences homologous to 4S RNA, 5S RNA and rRNA. *J. Mol. Biol.*, **34**, 661–680.

Brown D.D. and Weber C.S. (1968b). Gene linkage by RNA–DNA hybridization. II. Arrangement of the redundant gene sequences for 21S and 18S rRNA. *J. Mol. Biol.*, **34**, 681–698.

Brown D.D., Wensink P.C. and Jordan E. (1971). Position and characteristics of 5S DNA from X. laevis. *Proc. Nat. Acad. Sci.*, **68**, 3175–3179.

Brown I.R. and Church R.B. (1971). RNA transcription from non repetitive DNA in the mouse. *Biochem. Biophys. Res. Commun.*, **42**, 850–856.

Brown I.R. and Church R.B. (1972). Transcription of non repeated DNA during mouse and rabbit development. *Develop. Biol.*, **29**, 73–84.

Brown S.W. (1966). Heterochromatin. *Science*, **151**, 417–425.

Brown S.W. and Chandra H.S. (1973). Inactivation system of the mammalian X chromosome. *Proc. Nat. Acad. Sci.*, **70**, 195–199.

Burkholder G.D., Okada T.A. and Comings D.E. (1972). Whole mount electron microscopy of metaphase I chromosomes and microtubules from mouse oocytes. *Exp. Cell Res.*, **75**, 497–511.

Burr H. and Lingrel J.B. (1971). Poly-A sequences at the 3′ termini of rabbit globin mRNAs. *Nature New Biol.*, **233**, 41–43.

Bustin M. and Cole R.D. (1969a). A study of the multiplicity of lysine-rich histones. *J. Biol. Chem.*, **244**, 5286–5290.

Bustin M. and Cole R.D. (1969b). Bisection of a lysine rich histone by N-bromo-succinimide. *J. Biol. Chem.*, **244**, 5291–5294.

Bustin M. and Cole R.D. (1970). Regions of high and low cationic charge in a lysine rich histone. *J. Biol. Chem.*, **245**, 1458–1466.

Bustin M. and Stollar R.D. (1972). Immunochemical specificity in lysine rich subfractions. *J. Biol. Chem.*, **247**, 5716–5721.

Butler J.A.V. and Chipperfield A.R. (1967). Inhibition of RNA polymerase by histones. *Nature*, **215**, 1188–1189.

Butterworth P.H.W., Cox R.F. and Chesterton C.J. (1971). Transcription of mammalian chromatin by mammalian DNA-dependent RNA polymerase. *Europ. J. Biochem.*, **23**, 229–241.

Byers T.J., Platt D.B. and Goldstein L. (1963). The cytonucleoproteins of Amoeba. I. Chemical properties and intracellular distribution. *J. Cell Biol.*, **19**, 453–466.

Cairns J. (1966). Autoradiography of Hela cell DNA. *J. Mol. Biol.*, **15**, 372–373.

Callan H.G. (1967). The organization of genetic units in chromosomes. *J. Cell Sci.*, **2**, 1–8.

Callan H.G. and Lloyd L. (1960). Lampbrush chromosomes of crested newts. *Phil. Trans. Roy. Soc. B.*, **243**, 135–219.

Callan H.G. and McGregor H.C. (1958). Action of DNAase on lampbrush chromosomes. *Nature*, **181**, 1479–1480.

Candido E.P.M. and Dixon G.H. (1972a). Acetylation of trout testis histones in vivo. Site of the modification in histone IIb1. *J. Biol. Chem.*, **247**, 3868–3873.

Candido E.P.M. and Dixon G.H. (1972b). Amino terminal sequences and sites of in vivo acetylation of trout testis histones III and IIb2. *Proc. Nat. Acad. Sci.*, **69**, 2015–2019.

Candido E.P.M. and Dixon G.H. (1972c). Acetylation of histones in different cell types from developing trout testis. *J. Biol. Chem.*, **247**, 5506–5510.

Carlsson S.A., Moore G.P.M. and Ringertz N.R. (1973). Nucleocytoplasmic protein migration during the activation of chick erythrocyte nuclei in heterocaryons. *Exp. Cell Res.*, **76**, 234–241.

Caspersson T., Zech L. and Johansson C. (1970). Analysis of human metaphase chromosome sets by aid of DNA-binding fluorescent agents. *Exp. Cell Res.*, **62**, 490–492.

Caspersson T., De La Chappelle A., Schroder J. and Zech L. (1972). Quinacrine fluorescence of metaphase chromosomes. Identical patterns in different tissues. *Exp. Cell Res.*, **72**, 56–59.

Caston J.D. and Jones P.H. (1972). Synthesis and processing of high molecular weight by nuclei isolated from embryos of R. pipiens. *J. Mol. Biol.*, **69**, 19–38.

Cattanach B.M. and Isaacson J.H. (1965). Genetic control over the inactivation of autosomal genes attached to the X chromosome. *Z. Verebungsl.*, **96**, 313–323.

Cattanach B.M. and Isaacson J.H. (1967). Controlling elements in the mouse X chromosome. *Genetics*, **57**, 331–346.

Cattanach B.M., Perez J.N. and Pollard C.E. (1970). Controlling elements in the mouse X chromosome. II. Location in the linkage map. *Genetic Res.*, **15**, 183–195.

Cattanach B.M., Pollard C.E. and Perez J.N. (1969). Controlling elements in the mouse X chromosome. I. Interaction with the X-linked genes. *Genetic Res.*, **14**, 223–235.

Cattanach B.M. and Williams C.E. (1972). Evidence of non random X chromosome activity in the mouse. *Genetic Res.*, **19**, 229–240.

Chalkley G.R. and Jensen R.H. (1968). A study of the structure of isolated chromatin. *Biochem.*, **7**, 4380–4387.

Chambon P., Gissinger F., Mandel J.L., Kedinger C., Gniazdowski M. and Meihlac M. (1970). Purification and properties of calf thymus DNA dependent RNA polymerases A and B. *Cold Spring Harbor Symp. Quant. Biol.*, **35**, 693–708.

Chan V.L., Whitmore G.F. and Siminovitch L. (1972). Mammalian cells with altered forms of RNA polymerase II. *Proc. Nat. Acad. Sci.*, **69**, 3119–3123.

Chang L.M.S., Brown M. and Bollum F.H. (1973). Induction of DNA polymerase in mouse L cells. *J. Mol. Biol.*, **74**, 1–8.

Chantrenne H., Burny A. and Marbaix G. (1967). The search for mRNA of haemoglobin. *Prog. Nuc. Acid Res.*, **7**, 173–194.

Chesterton C.J. and Butterworth P.H.W. (1971). Selective extraction of form I DNA dependent RNA polymerase from rat liver nuclei and its separation into two species. *Europ. J. Biochem.*, **19**, 232–241.

Choi Y.C. and Busch H. (1970). Studies on the 5′ terminal and alkali resistant dinucleotides of nucleolar high molecular weight RNA. *J. Biol. Chem.*, **245**, 1954–1961.

Church R.B. and McCarthy B.J. (1967a). RNA synthesis in regenerating and embryonic liver. I. The synthesis of new species of RNA during regeneration of mouse liver after partial hepatectomy. *J. Mol. Biol.*, **23**, 459–475.

Church R.B. and McCarthy B.J. (1967b). RNA synthesis in regenerating and embryonic liver. II. The synthesis of RNA during embryonic liver development and its relationship to regenerating liver. *J. Mol. Biol.*, **23**, 477–486.

Church R.B. and McCarthy B.J. (1967c). Changes in nuclear and cytoplasmic RNA in regenerating liver. *Proc. Nat. Acad. Sci.*, **58**, 1548–1555.

Church R.B. and McCarthy B.J. (1968). The interpretation of DNA/RNA hybridization studies with mammalian nucleic acids. *Biochem. Genetics*, **2**, 55–73.

Clark R.J. and Felsenfeld G. (1971). Structure of chromatin. *Nature New Biol.*, **229**, 101–105.

Clever U. (1966). Puffing in giant chromosomes of Diptera and the mechanism of its control. In Bonner J. and T'so (Eds.), *The Nucleohistones*. Holden-Day, San Francisco, pp 317–331.

Clever U. (1969). The formation of secretion in the salivary glands of Chrinomus. *Exp. Cell Res.*, **55**, 317–322.

Clever U. and Karlson P. (1960). Induktion von puff-veranderungen in den speicheldrusenchromosemen von C. tentans durch ecdysone. *Exp. Cell Res.*, **20**, 623–626.

Clever U. and Storbeck I. (1970). Chromosome activity and cell function in polytenic cells. IV. Polyribosomes and their sensitivity to actinomycin. *Biochim. Biophys. Acta.*, **217**, 108–119.

Clever U., Storbeck I. and Romball C.G. (1969). Chromosome activity and cell function in polytene cells. I. Protein synthesis at various stages of larval development. *Exp. Cell Res.*, **55**, 306–316.

Cohen M.M. and Ratazzi M.C. (1971). Cytological and biochemical correlation of late X-chromosomal replication and gene inactivation in the mule. *Proc. Nat. Acad. Sci.*, **68**, 544–548.
Comings D.E. (1967a). The duration of replication of the inactive X chromosome in humans based on the persistence of the heterochromatic sex nody during DNA synthesis. *Cytogenetics*, **6**, 20–37.
Comings D.E. (1967b). Sex chromatin, nuclear size and the cell cycle. *Cytogenetics*, **6**, 120–144.
Comings D.E. (1968). The rationale for an ordered arrangement of chromatin in the interphase nucleus. *Amer. J. Hum. Gen.*, **20**, 440–460.
Comings D.E., Avelino E., Okada T.A. and Wyandt H.E. (1973). The mechanism of C and G banding of chromosomes. *Exp. Cell. Res.*, **77**, 469–493.
Comings D.E. and Kakefuda T. (1968). Initiation of DNA replication at the nuclear membrane in human cells. *J. Mol. Biol.*, **33**, 225–229.
Comings D.E. and Mattoccia E. (1972). DNA of mammalian and avian heterochromatin. *Exp. Cell Res.*, **71**, 113–131.
Comings D.E. and Okada T.A. (1970a). Whole mount electron microscopy of the centromere region of metacentric and telocentric mammalian chromosomes. *Cytogenetics*, **9**, 436–449.
Comings D.E. and Okada T.A. (1970b). Do half chromatids exist? *Cytogenetics*, **9**, 450–459.
Comings D.E. and Okada T.A. (1971a). Triple chromosome pairing in triploid chickens. *Nature*, **231**, 119–121.
Comings D.E. and Okada T.A. (1971b). Whole mount electron microscopy of human meiotic chromosomes. *Exp. Cell Res.*, **65**, 99–103.
Comings D.E. and Okada T.A. (1971c). Fine structure of the synaptonemal complex. *Exp. Cell Res.*, **65**, 104–119.
Comings D.E. and Okada T.A. (1972). Architecture of meiotic cells and mechanisms of chromosome pairing. *Adv. Cell Mol. Biol.*, **2**, 310–384.
Comings D.E. and Okada T.A. (1973). DNA replication and the nuclear membrane. *J. Mol. Biol.*, **74**, 609–618.
Cook P.R. (1970). Species specificity of an enzyme determined by an erythrocyte nucleus in an interspecific hybrid cell. *J. Cell Sci.*, **7**, 1–4.
Cooper D.W. (1971). Directed genetic change model for X chromosome inactivation in eutherian mammals. *Nature*, **230**, 292–294.
Cooper K.W. (1959). Cytogenetic analysis of major heterochromatic elements (especially Xh and Y) in D. melanogaster and the theory of heterochromatin. *Chromosoma*, **10**, 535–588.
Corneo G., Ginelli E. and Polli E. (1970). Different satellite DNAs of guinea pig and ox. *Biochem.*, **9**, 1565–1570.
Corneo G., Ginelli E. and Polli E. (1971). Renaturation properties and localization in the heterochromatin of human satellite DNAs. *Biochim. Biophys. Acta*, **247**, 528–534.
Corneo G., Ginelli E., Soave C. and Bernadi C. (1968). Isolation and characterization of mouse and guinea pig satellite DNA. *Biochem.*, **7**, 4373–4379.
Craig A.P. and Shaw M.W. (1971). Autoradiographic studies on the human Y chromosome. *Chromosoma*, **32**, 364–377.
Crick F.H.C. (1971). General model for chromosomes of higher organisms. *Nature*, **234**, 25–27.

Dahmus M.E. and McConnell D.J. (1969). Chromosomal RNA of non histone protein of rat liver. *Biochem.*, **8**, 1524–1534.

Daneholt B. (1972). Giant RNA transcript in a Balbiani ring. *Nature*, **240**, 229–232.

Daneholt B. and Edstrom J.E. (1967). The content of DNA in individual polytene chromosomes of Chironomus tentans. *Cytogenetics*, **6**, 350–356.

Daneholt B., Edstrom J.E., Egyhazi E., Lambert B. and Ringborg U. (1969a). Physicochemical properties of chromosomal RNA in C. tentans polytene chromosomes. *Chromosoma*, **28**, 379–398.

Daneholt B., Edstrom J.E., Egyhazi E., Lambert B. and Ringborg U. (1969b). Chromosomal RNA synthesis in polytene chromosomes of C. tentans. *Chromosoma*, **28**, 399–417.

Daneholt B., Edstrom J.E., Egyhazi E., Lambert B. and Ringborg U. (1969c). RNA synthesis in a Balbiani ring in C. tentans salivary gland nuclei. *Chromosoma*, **28**, 418–429.

Daneholt B. and Hosick H. (1973). Evidence for transport of 75S RNA from a discrete chromosome region via nuclear sap to cytoplasm in C. tentans. *Proc. Nat. Acad. Sci.*, **70**, 442–446.

Daneholt B. and Svedhem L. (1971). Differential representation of chromosomal Hn RNA in nuclear sap. *Exp. Cell Res.*, **67**, 263–272.

Darnell J.E., Jelinek W.R. and Molloy G.R. (1973). Biogenesis of mRNA: genetic regulation in mammalian cells. *Science*, **181**, 1215–1221.

Darnell J.E., Pagoulatos G.N., Lindberg U. and Balint R. (1970). Studies on the relationship of mRNA to heterogeneous nuclear RNA in mammalian cells. *Cold Spring Harbor Symp. Quant. Biol.*, **35**, 555–560.

Darnell J.E., Philipson L., Wall R. and Adesnik M. (1971). Polyadenylic acid sequences: role in conversion of nuclear RNA into mRNA. *Science*, **174**, 507–510.

Darnell J.E., Wall R. and Tushinski R.J. (1971). An adenylic acid rich sequence in mRNA of Hela cells and its possible relationship to reiterated sites in DNA. *Proc. Nat. Acad. Sci.*, **68**, 1321–1325.

Das N.K., Micou-Eastwood J., Ramamurthy G. and Alfert M. (1970). Sites of synthesis and processing of rRNA precursors within the nucleus of Urechis caupo eggs. *Proc. Nat. Acad. Sci.*, **67**, 968–975.

Davidson D. (1964). RNA synthesis in roots of Vicia faba. *Exp. Cell Res.*, **35**, 317–325.

Davidson E.H. and Britten R.J. (1971). Note on the control of gene expression during development. *J. Theor. Biol.*, **32**, 123–130.

Davidson E.H. and Britten R.J. (1973). Organization, transcription and regulation in the animal genome. *Quart. Rev. Biol.*, **48**, in press.

Davidson E.H., Graham D.E., Neufeld B.R., Chamberlin M.E., Amenson C.S., Hough B.R. and Britten R.J. (1973). Arrangement and characterization of repetitive sequence elements in animal DNAs. *Cold Spring Harbor Symp. Quant. Biol.*, **38**, 293–301

Davidson E. H. & Hough B. R. (1971). Genetic information in oocyte RNA. *J. Mol. Biol.*, **56**, 491–506.

Davidson E. H., Hough B. R., Amenson C. S. & Britten R. J. (1973). General interspersion of repetitive with non repetitive sequence elements in the DNA of Xenopus. *J. Mol. Biol.*, **77**, 1–24.

Davis B.K. (1971). Genetic analysis of a meiotic mutant resulting in precocious sister-centromere separation in D. melanogaster. *Molec. Gen. Genet.*, **113**, 251–272.

Deak I., Sidebottom E. and Harris H. (1972). Further experiments on the role of the nucleolus in the expression of structural genes. *J. Cell Sci.*, **11**, 379–392.

Defendi V. and Manson L.A. (1963). Analysis of the life cycle in mammalian cells. *Nature*, **198**, 359–361.

DeLange R.J., Fambrough D.M., Smith E.L. and Bonner J. (1969a). Calf and thymus pea histone IV. II. The complete amino acid sequence of calf thymus histone IV: presence of ε-N-acetyllysine. *J. Biol. Chem.*, **244**, 319–334.

DeLange R.J., Fambrough D.M., Smith I.L. and Bonner J. (1969b). Calf and pea histone IV. III. Complete amino acid sequence of pea seedling histone IV: comparison with the homologous calf thymus histone. *J. Biol. Chem.*, **244**, 5669–5679.

DeLange R.J., Hooper J.A. and Smith E.L. (1972). Complete amino acid sequence of calf thymus histone III. *Proc. Nat. Acad. Sci.*, **69**, 882–884.

DeLange R.J., Hooper J.A. and Smith E.L. (1973). Sequence studies on the cyanogen bromide peptides: complete amino acid sequence of calf thymus histone III. *J. Biol. Chem.*, **248**, 3261–3274.

DeLange R.J. and Smith E.L. (1971). Histones: structure and function. *Ann. Rev. Biochem.*, **40**, 279–314.

Denhardt D.T. (1966). A membrane filter technique for the detection of complementary DNA. *Biochem. Biophys. Res. Commun.*, **23**, 641–646.

Dev V.G., Warburton D., Miller O.J., Miller D.A., Erlanger B.F. and Beiser S.M. (1972). Consistent pattern of binding of anti-adenosine antibodies to human metaphase chromosomes. *Exp. Cell Res.*, **74**, 288–293.

Dick C. and Johns E.W. (1969). A quantitative comparison of histones from immature and mature erythroid cells of the duck. *Biochim. Biophys. Acta*, **175**, 414–418.

Dickson E., Boyd J.B. and Laird C.D. (1971). Sequence diversity of polytene chromosome DNA from D. hydei. *J. Mol. Biol.*, **61**, 615–628.

Dina D., Crippa M. and Beccari E. (1973). Hybridization properties and sequence arrangement in a population of mRNAs. *Nature New Biol.*, **242**, 101–105.

Dixon G.H. (1972). The basic proteins of trout testis chromatin: aspects of their synthesis, post synthetic modifications and binding to DNA. *Karolinska Symp. Res. Methods Reproductive Biology*, **5**, 128–154.

Dixon G.H. and Smith M. (1968). Nucleic acids and protamine in salmon testis. *Prog. Nucleic Acid Res.*, **8**, 9–34.

Doty P., Marmur J., Eigner J. and Schildkraut C. (1960). Strands separation and specific recombination in DNAs: physical studies. *Proc. Nat. Acad. Sci.*, **46**, 461–476.

Doyle D. and Lauffer H. (1969). Requirements of RNA synthesis for the formation of salivary gland specific proteins in larval C. tentans. *Exp. Cell Res.*, **57**, 205–210.

DuPraw E.J. (1965a). Macromolecular organization of nuclei and chromosomes: a folded fibre model based on whole mount electron microscopy. *Nature*, **206**, 338–343.

DuPraw E.J. (1965b). The organization of nuclei and chromosomes in honeybee embryonic cells. *Proc. Nat. Acad. Sci.*, **53**, 161–165.

DuPraw E.J. (1966). Evidence for a folded fibre organization in human chromosomes. *Nature*, **209**, 577–581.

DuPraw E.J. (1968). *Cell and Molecular Biology*. Academic Press, New York.

DuPraw E.J. (1970). *DNA and chromosomes*. Holt, Rinehart and Winston, New York.

DuPraw E.J. (1972). Stages in chromosome evolution: the chromatid twins and how they grew. *Brookhaven Symp. Biol.*, **23**, 230–249.

DuPraw E.J. and Rae P.M.M. (1966). Polytene chromosome structure in relation to the folded fibre concept. *Nature*, **212**, 598–600.

DuPraw E.J. and Bahr G.F. (1969). The arrangement of DNA in human chromosomes as investigated by quantitative electron microscopy. *Acta Cytol.*, **13**, 188–205.

Edmonds M. and Caramela M.G. (1969). The isolation and characterization of AMP rich polynucleotide synthesized by Ehrlich ascites cells. *J. Biol. Chem.*, **244**, 1314–1324.

Edmonds M., Vaughan M.H. and Nakazoto H. (1971). Polyadenylic acid sequences in the heterogeneous nuclear RNA and rapidly labelled polyribosomal RNA of Hela cells: possible evidence for a precursor relationship. *Proc. Nat. Acad. Sci.*, **68**, 1336–1340.

Edstrom J.E. and Daneholt B. (1967). Sedimentation properties of the newly synthesized RNA from isolated nuclear components of C. tentans salivary gland cells. *J. Mol. Biol.*, **28**, 331–343.

Edwards L.J. and Hnilica L.S. (1968). The specificity of histones in nucleated erythrocytes. *Experientia*, **24**, 228–229.

Eicher E.M. (1970). X-autosome translocation in the mouse: total inactivation versus partial inactivation of the X chromosome. *Advances Genetics*, **15**, 175–259.

Elgin S.C.R. and Bonner J. (1970). Limited heterogeneity at the major non histone chromosomal proteins. *Biochem.*, **9**, 4440–4448.

Elgin S.C.R. and Bonner J. (1972). Partial fractionation and chemical character-ization of the major non histone chromosomal proteins. *Biochem.*, **11**, 772–781.

Ellgaard E.G. and Clever U. (1971). RNA metabolism during puff induction in D. melanogaster. *Chromosoma*, **36**, 60–78.

Ellison J.R. and Barr H.J. (1972). Quinacrine fluorescence of specific chromosome regions. Late replication and high A:T content in Samoaia leonensis. *Chromosoma*, **36**, 375–390.

Enger M.D. and Tobey R.A. (1969). RNA synthesis in Chinese hamster cells. II. Increase in rate of RNA synthesis during G1. *J. Cell Biol.*, **42**, 308–315.

Evans H.J. (1964). Uptake of H^3 thymidine and patterns of replication in nuclei and chromosomes of Vicia faba. *Exp. Cell Res.*, **35**, 381–393.

Evans H.J., Ford C., Lyon M. and Gray J. (1965). DNA replication and genetic expression in female mice with morphologically distinguishable X chromosomes. *Nature*, **206**, 900–903.

Evans M.J. and Lingrel J.B. (1969a). Haemoglobin mRNA. Distribution of the 9S RNA in polysomes of different sizes. *Biochem.*, **8**, 829–831.

Evans M.J. and Lingrel J.B. (1969b). Haemoglobin mRNA. Synthesis of 9S and rRNA during erythroid development. *Biochem.*, **8**, 3000–3005.

Fakan S., Turner G.N., Pagano J.S. and Hancock R. (1972). Sites of replication of chromosomal DNA in a eucaryotic cell. *Proc. Nat. Acad. Sci.*, **69**, 2300–2305.

Fambrough D.M. and Bonner J. (1966). On the similarity of plant and animal histones. *Biochem.*, **5**, 2563–2570.

Fambrough D.M. and Bonner J. (1968). Sequence homology and role of cysteine in plant and animal arginine rich histones. *J. Biol. Chem.*, **243**, 4434–4439.

Fan H. and Penman S. (1970a). Mitochondrial RNA synthesis during mitosis. *Science*, **168**, 135–138.

Fan H. and Penman S. (1970b). Regulation of protein synthesis in mammalian cells. II. Inhibition of protein synthesis at the level of initiation during mitosis. *J. Mol. Biol.*, **50**, 655–670.

Fan H. and Penman S. (1971). Regulation of synthesis and processing of nucleolar components in metaphase arrested cells. *J. Mol. Biol.*, **59**, 27–42.

Fansler B. and Loeb L.A. (1972). Sea urchin nuclear DNA polymerase. IV. Revers-ible association of DNA polymerase with nuclei during the cell cycle. *Exp. Cell Res.*, **75**, 433–441.

Fawcett D.W. (1956). The fine structure of chromosomes in the meiotic prophase of vertebrate spermatocytes. *J. Biochem. Biophys. Cytol.*, **2**, 403–406.

Flamm W.G., Bernheim N.J. and Brubacker P.E. (1971). Density gradient analysis of newly replication DNA from synchronized mouse lymphoma cells. *Exp. Cell Res.*, **64**, 97–104.

Flamm W.G., Walker P.M.B. and McCallum M. (1969). Some properties of the single strands isolated from the DNA of the nuclear satellite of the mouse (mus musculus). *J. Mol. Biol.*, **40**, 423–443.

Forer A. (1969). Chromosome movements during cell division. In Lima-de-Faria A. (Ed.), *Handbook of Molecular Cytology*. North Holland, Amsterdam, pp 553–604.

Fox T.O. and Pardee A.B. (1971). Proteins made in the mammalian cell cycle. *J. Biol. Chem.*, **246**, 6159–6165.

Freese E. (1958). The arrangement of DNA in the chromosome. *Cold Spring Harbor Symp. Quant. Biol.*, **23**, 13–18.

Frenster J.H. (1969). Biochemistry and molecular biophysics of heterochromatin and euchromatin. In Lima-de-Faria A. (Ed.), *Handbook of Molecular Cytology*. North Holland, Amsterdam, pp 251–276.

Friedman D.L. (1970). DNA polymerase from Hela cell nuclei: levels of activity during a synchronized cell cycle. *Biochem. Biophys. Res. Commun.*, **39**, 100–109.

Friedman D.L. and Mueller G.C. (1968). A nuclear system for DNA replication from synchronized Hela cells. *Biochim. Biophys. Acta*, **161**, 455–468.

Fujiwara Y. (1967). Role of RNA synthesis in DNA replication of synchronized populations of cultured mammalian cells. *J. Cell Physiol.*, **70**, 291–300.

Galau G.A., Britten R.J. and Davidson E.H. (1974). A measurement of the sequence complexity of polysomal mRNA in sea urchin embryos. *Cell*, **2**, 9–21.

Gall J.G. (1959). Macromolecular duplication in the ciliated protozoan Euplotes. *J. Biochem. Biophys. Cytol.*, **5**, 295–308.

Gall J.G. (1963). Kinetics of DNAase action on chromosomes. *Nature*, **198**, 36–38.

Gall J.G. (1968). Differential synthesis of the genes for rRNA during amphibian oogenesis. *Proc. Nat. Acad. Sci.*, **60**, 553–560.

Gall J.G. (1969). The genes for ribosomal RNA during oogenesis. *Genetics*, **61**, suppl. 1, 121–132.

Gall J.G., Cohen E.H. and Polan M.L. (1971). Repetitive DNA sequences in Drosophila. *Chromosoma*, **33**, 319–344.

Gall J.G. and Pardue M.L. (1969). Formation and detection of RNA–DNA hybrid molecules in cytological preparations. *Proc. Nat. Acad. Sci.*, **63**, 378–383.

Galton M. and Holt S.F. (1964). DNA replication patterns of the sex chromosomes in somatic cells of the Syrian hamster. *Cytogenetics*, **3**, 97–111.

Ganner E. and Evans H.J. (1971). The relationship between patterns of DNA replication and of quinacrine fluorescence in the human chromosome complement. *Chromosoma*, **35**, 326–341.

Gardner R.L. and Lyon M.F. (1971). X chromosome inactivation studied by injection of a single cell into the mouse blastocyst. *Nature*, **231**, 383–386.

Gaskill P. and Kabat D. (1971). Unexpectedly large size of globin mRNA. *Proc. Nat. Acad. Sci.*, **68**, 72–75.

Gavosto F., Pegora L., Masera P. and Rovera G. (1968). Late DNA replication pattern in human haemapoietic cells. A comparative investigation using a high resolution quantitative autoradiography. *Exp. Cell Res.*, **49**, 340–358.

Gelderman A., Rake A. and Britten R.J. (1971). Transcription of non repeated DNA in neonatal and fetal mice. *Proc. Nat. Acad. Sci.*, **68**, 172–176.

Georgiev G.P. (1969). On the structural organization of operon and the regulation of RNA synthesis in animal cells. *J. Theor. Biol.*, **25**, 473–490.

Georgiev G.P., Ryskov A.P., Coutelle C., Mantieva V.L. and Avakyan E.R. (1972). On the structure of transcriptional unit in mammalian cells. *Biochim. Biophys. Acta*, **259**, 259–282.

Giacomoni D. and Finkel D. (1972). Time of duplication of ribosomal RNA cistrons in a cell line of Potodous tridactylis (rat kangaroo). *J. Mol. Biol.*, **70**, 725–728.

Gilbert C.W., Muldal S. and Lathja L.G. (1965). Rate of chromosome duplication at the end of the DNA S period in human blood cells. *Nature*, **208**, 159–161.

Gillespie D., Marshall S. and Gallo R.C. (1972). RNA of RNA tumour viruses contains poly-A. *Nature New Biol.*, **236**, 227–231.

Gillespie D. and Spigelman S. (1965). A quantitative assay for RNA–DNA hybrids with DNA immobilised on a membrane. *J. Mol. Biol.*, **12**, 829–842.

Gilmour R.S. and Paul J. (1969). RNA transcribed from reconstituted nucleoprotein is similar to natural RNA. *J. Mol. Biol.*, **40**, 137–140.

Girard M., Latham H., Penman S. and Darnell J.E. (1965). Entrance of newly formed mRNA and ribosome subunits into Hela cell cytoplasm. *J. Mol. Biol.*, **11**, 187–201.

Goldberg R.B., Galau G.A., Britten R.J. and Davidson E.H. (1973). Nonrepetitive DNA sequence representation in sea urchin embryo messenger RNA. *Proc. Nat. Acad. Sci. USA*, **70**, 3516–3520.

Goldstein L. and Prescott D.M. (1967). Proteins in nucleocytoplasmic interactions. I. The fundamental characteristic of the rapidly migrating proteins and the slow turnover proteins of the Amoeba proteus nucleus. *J. Cell Biol.*, **33**, 637–644.

Goldstein L. and Prescott D.M. (1968). Proteins in nucleocytoplasmic interactions. II. Turnover and changes in nuclear protein distribution with time and free growth. *J. Cell Biol.*, **36**, 53–61.

Gorovsky M.A. and Woodard J. (1967). Histone content of chromosomal loci active and inactive in RNA synthesis. *J. Cell Biol.*, **33**, 723–727.

Gorski J., Toft D., Shyamala G., Smith D. and Notides A. (1968). Hormone receptors: studies on the interaction of estrogen with the nucleus. *Rec. Prog. Horm. Res.*, **24**, 45–72.

Graham C.F., Arms K. and Gurdon J.B. (1966). The induction of DNA synthesis by frog egg cytoplasm. *Develop. Biol.*, **14**, 349–381.

Graham C.F. and Morgan R.W. (1966). Changes in the cell cycle during early amphibian development. *Develop. Biol.*, **14**, 439–460.

Graham D.E., Neufeld B.R., Davidson E.H. and Britten R.J. (1974). Interspersion of repetitive and non-repetitive DNA sequences in the sea urchin genome. *Cell*, **1**, 127–137.

Granboulan N. and Scherrer K. (1969). Visualization in the electron microscope and size of RNA from animal cells. *Europ. J. Biochem.*, **9**, 1–20.

Green H., Wang R., Kehinde O. and Meuth M. (1971). Multiple human TK. chromosomes in human–mouse somatic cell hybrids. *Nature New Biol.*, **234**, 138–140.

Green M. and Cartas S. (1972). The genome of RNA tumour viruses contains poly-A sequences. *Proc. Nat. Acad. Sci.*, **69**, 791–794.

Greenaway P.J. and Murray K. (1971). Heterogeneity and polymorphism in chicken erythrocyte histone fraction V. *Nature New Biol.*, **229**, 233–238.

Greenberg H. and Penman S. (1966). Methylation and processing of rRNA in Hela cells. *J. Mol. Biol.*, **21**, 527–535.

Greenberg J.R. (1972). High stability of messenger RNA in growing cultured cells. *Nature*, **242**, 102–104.

Greenberg J.R. and Perry R.P. (1971). Hybridization properties of DNA sequences directing the synthesis of mRNA and heterogeneous nuclear RNA. *J. Cell Biol.*, **50**, 774–787.

Greenberg J.R. and Perry R.P. (1972). Relative occurrence of poly-A sequences in messenger and heterogeneous nuclear RNA of L cells as determined by poly-U hydroxyapatite chromatography. *J. Mol. Biol.*, **72**, 91–98.

Grossbach U. (1969). Chromosome aktivitat und biochemische zelldifferentzierung in der speicheldrusen von Camptochironomus. *Chromosoma*, **28**, 136–187.

Grouse L., Chilton M.D. and McCarthy B.J. (1972). Hybridization of RNA with unique sequences of mouse DNA. *Biochem.*, **11**, 798–805.

Grzezchik K.H., Allerdice P.W., Grzezchik A., Opitz J.M., Miller O.J., and Siniscalco M. (1972). Cytological mapping of human X-linked genes by use of somatic cell hybrids involving an X-autosome translocation. *Proc. Nat. Acad. Sci.*, **69**, 69–73.

Gurdon J.B. (1960). Factors responsible for the abnormal development of embryos obtained by nuclear transplantation in X. laevis. *J. Emb. Exp. Morph.*, **8**, 327–340.

Gurdon J.B. (1962a). Adult frogs derived from the nuclei of single somatic cells. *Develop. Biol.*, **4**, 256–273.

Gurdon J.B. (1962b). The transplantation of nuclei between two species of Xenopus. *Develop. Biol.*, **5**, 68–73.

Gurdon J.B. (1962c). The developmental capacity of nuclei taken from intestinal epithelian cells of feeding tadpoles. *J. Emb. Exp. Morph.*, **10**, 622–640.

Gurdon J.B. (1963). Nuclear transplantation in amphibia and the importance of stable nuclear changes in promoting cellular differentiation. *Quart. Rev. Biol.*, **38**, 54–78.

Gurdon J.B. (1964). The transplantation of living cell nuclei. *Adv. Morph.*, **4**, 1–43.

Gurdon J.B. (1968). Changes in somatic cell nuclei inserted into growing and maturing amphibian oocytes. *J. Emb. Exp. Morph.*, **20**, 401–414.

Gurdon J.B. (1968). Nuclear transplantation and the control of gene activity in animal development. *Proc. Roy. Soc. B*, **176**, 303–314.

Gurdon J.B., Birnsteil M.L. and Speight V.A. (1969). The replication of purified DNA introduced into living egg cytoplasm. *Biochim. Biophys. Acta.*, **174**, 614–628.

Gurdon J.B. and Brown D.D. (1965). Cytoplasmic regulation of RNA synthesis and nucleolus formation in developing embryos of X. laevis. *J. Mol. Biol.*, **12**, 27–35.

Gurdon J.B. and Laskey R.A. (1970). The transplantation of nuclei from single cultured eggs into enucleated frogs' eggs. *J. Emb. Exp. Morph.*, **24**, 227–248.

Gurdon J.B. and Speight V.A. (1969). The appearance of cytoplasmic DNA polymerase activity during the maturation of amphibian oocytes into eggs. *Exp. Cell Res.*, **55**, 253–256.

Gurdon J.B. and Woodland H.R. (1970). On the long term control of nuclear activity during cell differentiation. *Current Topics Devel. Biol.*, **5**, 39–70.

Gurley L.R., Enger M.D. and Walters R.A. (1973). The nature of histone f1 isolated from polysomes. *Biochem.*, **12**, 237–246.

Gurley L.R., Walters R.A. and Tobey R.A. (1973). Histone phosphorylation in late interphase and mitosis. *Biochem. Biophys. Res. Commun.*, **50**, 744–750.

Hallberg R.L. and Brown D.D. (1969). Coordinated synthesis of some ribosomal proteins and ribosomal RNA in embryos of X. laevis. *J. Mol. Biol.*, **46**, 393–411.

Hamerton J.L., Richardson B.J., Gee P.A., Allen W.R. and Short R.V. (1971). Non random X chromosome expression in female mules and hinnies. *Nature*, **232**, 312–315.

Hammerling J. (1953). Nucleocytoplasmic relationships in the development of Acetabularia. *Int. Rev. Cytol.*, **2**, 475–498.

Hammerling J. (1963). Nucleocytoplasmic interactions in Acetabularia and other cells. *Ann. Rev. Plant. Physiol.*, **14**, 65–92.

Hancock R. (1969). Conservation of histones in chromatin during growth and mitosis in vitro. *J. Mol. Biol.*, **40**, 457–466.

Harris H. (1963). Nuclear RNA. *Prog. Nucleic Acid Res.*, **2**, 20–60.

Harris H. (1967). The reactivation of the red cell nucleus. *J. Cell Sci.*, **2**, 23–32.

Harris H. (1970). *Nucleus and Cytoplasm*, 2nd edition. Oxford University Press, Oxford.

Harris H. and Cook P.R. (1969). Synthesis of an enzyme determined by an erythrocyte nucleus in a hybrid cell. *J. Cell Sci.*, **5**, 121–134.

Harris H., Sidebottom E., Grace D.M. and Bramwell M.E. (1969). The expression of genetic information: a study with hybrid animal cells. *J. Cell Sci.*, **4**, 499–526.

Harris H., Watkins J.F., Ford C.E. and Schoefl G.I. (1966). Artificial heterocaryons of animal cells from different species. *J. Cell Sci.*, **1**, 1–30.

Harrison P.R., Hell A., Birnie G. and Paul J. (1972). Evidence for single copies of globin genes in the mouse genome. *Nature*, **239**, 219–221.

Hatlen L.E. and Attardi G. (1971). Proportion of the Hela cell genome complementary to tRNA and 5S RNA. *J. Mol. Biol.*, **56**, 535–554.

Hay E.D. and Revel J.P. (1963). The fine structure of the DNP component of the nucleus. *J. Cell Biol.*, **16**, 29–51.

Heitz E. (1928). Das heterochromatin der moose. *Jb. Wiss. Bot.*, **69**, 762–818.

Heitz E. and Bauer H. (1933). Beweise fur die chromosomennatur der kernschleifen in den knauelkernen von Bibio hortulanus. *L.Z. Zellforsch.*, **17**, 67–82.

Henderson A. and Ritossa F. (1970). On the inheritance of rDNA of magnified bobbed loci in D. melanogaster. *Genetics*, **66**, 463–473.

Henderson S.A. (1969). Chromosome pairing, chiasmata and crossing over. In Lima-de-Faria A. (Ed.), *Handbook of Molecular Cytology*. North Holland, Amsterdam, pp 326–360.

Hennig W. (1972a). Highly repetitive DNA sequences in the genome of D. hydei. I. Preferential localization in the X chromosomal heterochromatin. *J. Mol. Biol.*, **71**, 407–417.

Hennig W. (1972b). Highly repetitive DNA sequences in the genome of D. hydei. II. Occurrence in polytene tissues. *J. Mol. Biol.*, **71**, 419–431.

Hennig W., Hennig I. and Stein H. (1970). Repeated sequences in the DNA of Drosophila and their localization in giant chromosomes. *Chromosoma*, **32**, 31–63.

Hennig W. and Meer B. (1971). Reduced polyteny of rRNA cistrons in giant chromosomes of D. hydei. *Nature New Biol.*, **233**, 70–72.

Hennig W. and Walker P.M.B. (1970). Variations in the DNA from two rodent families (cricetidae and muridae). *Nature*, **225**, 915–919.

Heyden H.W. Von. and Zachau H.G. (1971). Characterization of RNA in fractions of calf thymus chromatin. *Biochim. Biophys. Acta*, **232**, 651–660.

Highfield D.P. and Dewey W.C. (1972). Inhibition of DNA synthesis in synchronized Chinese hamster cells treated in G1 or early S phase with cycloheximide or puromycin. *Exp. Cell Res.*, **75**, 314–320.

Hodge L.D., Robbins E. and Scharff M.D. (1969). Persistence of mRNA through mitosis in Hela cells. *J. Cell Biol.*, **40**, 497–507.

Holmes D.W., Mayfield J.E., Sander G. and Bonner J. (1972). Chromosomal RNA: its properties. *Science*, **177**, 72–74.

Holt T.K.H. (1970). Local protein accumulation during gene activation. I. Quantitative measurements on dye binding capacity at subsequent stages of puff formation in D. hydei. *Chromosoma*, **32**, 64–78.

Holt T.K.H. (1971). Local protein accumulation during gene activation. II. Interferometric measurements of the amount of solid material in temperature induced puffs of D. hydei. *Chromosoma*, **32**, 428–435.

Hook E.B. and Brustman L.D. (1971). Evidence for selective differences between cells with an active horse X chromosome and cells with an active donkey X chromosome in the female mule. *Nature*, **232**, 349–350.

Hooper J.A., Smith E.L., Sommer K.R. and Chalkley R. (1973). Amino acid sequence of histone III of the testes of the carp, Letiobus bubalus. *J. Biol. Chem.*, **248**, 3275–3279.

Hori T.A. and Lark K.G. (1973). Effect of puromycin on DNA replication in Chinese hamster cells. *J. Mol. Biol.*, **77**, 391–404.

Hotta Y. and Stern H. (1971). Analysis of DNA synthesis during meiotic prophase in Lilium. *J. Mol. Biol.*, **55**, 337–356.

Hough B.R. and Davidson E.H. (1972). Studies on the repetitive sequence transcripts of Xenopus oocytes. *J. Mol. Biol.*, **70**, 491–510.

Howard A. and Pelc S. (1953). Synthesis of DNA in normal and irradiated cells and its relation to chromosome breakage. *Heredity*, **6**, *suppl*, 261–273.

Hsu T.C. (1964). Mammalian chromosomes in vitro. XVIII. DNA replication sequence in the Chinese hamster. *J. Cell Biol.*, **23**, 53–62.

Hsu T.C. and Arrhigi F.E. (1971). Distribution of constitutive heterochromatin in mammalian chromosomes. *Chromosoma*, **34**, 243–253.

Hsu T.C., Cooper J.E.K., Mace M.L.jr. and Brinkley B.R. (1971). Arrangement of centromeres in mouse cells. *Chromosoma* **34**, 73–87.

Huang R.C.C. and Bonner J. (1965). Histone bound RNA, a component of native nucleohistone. *Proc. Nat. Acad. Sci.*, **54**, 960–967.

Huang R.C.C., Bonner J. and Murray K. (1964). Physical and biological properties of soluble nucleohistones. *J. Mol. Biol.*, **8**, 54–64.

Huang R.C.C. and Huang P.C. Effect of protein bound RNA associated with chick embryo chromatin on template specificity of the chromatin. *J. Mol. Biol.*, **39**, 365–378.

Huberman J.A. (1973). Structure of chromosome fibres and chromosomes. *Ann. Rev. Biochem.*, **42**, 355–378.

Huberman J.A. and Attardi G. (1966). Isolation of metaphase chromosomes from Hela cells. *J. Cell Biol.*, **31**, 95–105.

Huberman J.A. and Attardi G. (1967). Studies of fractionated Hela cell metaphase chromosomes. I. The chromosomal distribution of DNA complementary to 28S and 18S rRNA and the cytoplasmic messenger RNA. *J. Mol. Biol.*, **29**, 487–505.

Huberman J.A. and Riggs D.A. (1968). On the mechanism of replication in mammalian chromosomes. *J. Mol. Biol.*, **32**, 327–341.

Huberman J.A., Tsai A. and Deich R.A. (1973). DNA replication within nuclei of mammalian cells. *Nature*, **241**, 32–36.

Imaizumi T., Diggelmann H. and Scherrer K. (1973). Demonstration of globin messenger sequences in giant nuclear precursors of messenger RNA of avian erythroblasts. *Proc. Nat. Acad. Sci.*, **70**, 1122–1126.

Inagaki A. and Busch H. (1972). Marker nucleotides for non ribosomal spacer elements of preribosomal RNA. *Biochem. Biophys. Res. Commun.*, **49**, 1398–1406.

Infante A.A. and Nemer M. (1968). Heterogeneous RNP in the cytoplasm of sea urchin embryos. *J. Mol. Biol.*, **32**, 543–565.

Inoue S. (1953). Polarization optical studies of the mitotic spindle. I. The demonstration of spindle fibres in living cells. *Chromosoma*, **5**, 487–500.

Inoue S. (1964). Organization and function of the mitotic spindle. In Allen R.D. and Kamiya N. (Eds.), *Primitive motile systems in cell biology*. Academic Press, New York, pp 549–594.

Iwai K., Ishikawa K. and Hayashi H. (1970). Amino acid sequence of slightly lysine rich histone. *Nature*, **226**, 1056–1058.

Jacobs S.T., Muecke W., Sajdel E.M. and Munro H.N. (1970). Evidence for extranucleolar control of RNA synthesis in the nucleolus. *Biochem. Biophys. Res. Commun.*, **40**, 334–342.

Jacobs-Lorena M. and Baglioni C. (1972). mRNA for globin in the postribosomal supernatant of rabbit reticulocytes. *Proc. Nat. Acad. Sci.*, **69**, 1425–1428.

Jacobs-Lorena M., Baglioni C. and Borun T.W. (1972). Translation of messenger RNA for histones from Hela cells by a cell free extract from mouse ascites tumour. *Proc. Nat. Acad. Sci.*, **69**, 2095–2099.

Jacobson R.A. and Bonner J. (1971). Studies of the chromosomal RNA and of the chromosomal RNA binding protein of higher organisms. *Arch. Biochem. Biophys.*, **146**, 557–563.

Jakob K.M. (1972). RNA synthesis during the DNA synthetic period of the first cell cycle in the root meristem of germinating Vicia faba. *Exp. Cell Res.*, **72**, 370–376.

Jeanteur P., Amaldi F. and Attardi G. (1968). Partial sequence analysis of ribosomal RNA from Hela cells. II. Evidence for sequences of non ribosomal type in 45S and 32S rRNA precursors. *J. Mol. Biol.*, **33**, 757–776.

Jeanteur P. and Attardi G. (1969). Relationship between Hela cell rRNA and its precursors studied by high resolution RNA–DNA hybridization. *J. Mol. Biol.*, **45**, 305–324.

Jelinek W., Adesnik M., Salditt M., Sheiness D., Wall R., Molloy G., Philipson L. and Darnell J.E. (1973). Further evidence on the nuclear origin and transfer to the cytoplasm of poly-A sequences in mammalian cell RNA. *J. Mol. Biol.*, **75**, 515–532.

Jelinek W. and Darnell J.E. (1972). Double stranded regions on heterogeneous nuclear RNA from Hela cells. *Proc. Nat. Acad. Sci.*, **69**, 2537–2541.

Jensen E.V., Suzuki T., Kawashima T., Stumpf W.E., Jungblatt P.W. and De Sombre P.R. (1968). A two step mechanism for the interaction of estradiol with rat uterus. *Proc. Nat. Acad. Sci.*, **59**, 632–638.

John H.A., Birnsteil M.L. and Jones K.W. (1969). RNA–DNA hybrids at the cytological level. *Nature*, **223**, 582–587.

Johns E.W. (1964). Preparative methods for histone fractions from calf thymus. *Biochem. J.*, **92**, 55–59.

Johns E.W. (1971). The preparation and characterization of histones. In Phillips D.M. (Ed.), *Histones and Nucleohistones*. Plenum Press, New York, pp 2–46.

Johns E.W. and Butler J.A.V. (1962). Further fractionation of histones from calf thymus. *Biochem. J.*, **82**, 15–18.

Johns E.W. and Diggle J.H. (1969). A method for the large scale preparation of the avian specific histone f2c. *Europ. J. Biochem.*, **11**, 495–498.

Johns E.W. and Forrester S. (1969). Interactions between the lysine rich histone f1 and DNA. *Biochem. J.*, **111**, 371–374.

Johns E.W. and Hoare T.A. (1970). Histones and gene control. *Nature*, **226**, 650–651.

Johnson R.T. and Harris H. (1969a). DNA synthesis and mitosis in fused cells. I. Hela homocaryons. *J. Cell Sci.*, **5**, 603–624.

Johnson R.T. and Harris H. (1969b). DNA synthesis and mitosis in fused cells. II. Hela–chick erythrocyte heterocaryons. *J. Cell Sci.*, **5**, 625–644.

Johnson R.T. and Rao P.N. (1970). Mammalian cell fusion: induction of premature chromosome condensation in interphase nuclei. *Nature*, **226**, 717–722.

Johnson R.T. and Rao P.N. (1971). Nucleocytoplasmic interactions in the achievement of nuclear synchrony in DNA synthesis and mitosis in multinucleate cells. *Biol. Rev.*, **46**, 97–155.

Johnson R.T., Rao P.N. and Hughes H.D. (1970). Mammalian cell fusion. III. A Hela cell inducer of premature chromosome condensation active in cells from a variety of animal species. *J. Cell Physiol.*, **76**, 151–158.

Johnston R.E. and Bose H.R. (1972). Correlation of mRNA function with adenylate-rich segments in the genomes of single stranded RNA viruses. *Proc. Nat. Acad. Sci.*, **69**, 1514–1516.

Jokelainen P.T. (1967). The ultrastructure and spatial organization of the metaphase kinetochore in mitotic rat cells. *J. Ult. Res.*, **19**, 19–44.

Joklik W.K. and Becker Y. (1965). Studies on the genesis of polyribosomes. II. The association of nascent messenger RNA with the 40S subribosomal particle. *J. Mol. Biol.*, **13**, 496–510.

Jones K.W. (1970). Chromosomal and nuclear location of mouse satellite DNA in individual cells. *Nature*, **225**, 912–915.

Jones K.W. and Corneo G. (1971). Location of satellite and homogeneous DNA sequences on human chromosomes. *Nature New Biol.*, **233**, 268–271.

Jones K.W. and Robertson F.W. (1970). Localization of reiterated nucleotide sequences in Drosophila or mouse by in situ hybridization of complementary RNA. *Chromosoma*, **31**, 331–345.

Kacian D.L., Spiegelman S., Bank A., Terada M., Metafora S., Dow L. and Marks P.A. (1972). In vitro synthesis of DNA components of human genes for globins. *Nature New Biol.*, **235**, 167–169.

Kane R.E. (1967). The mitotic apparatus. Identification of the major soluble component of the glycerol isolated mitotic apparatus. *J. Biol. Chem.*, **32**, 243–253.

Kao F.T. and Puck T.T. (1968). Genetics of somatic mammalian cells. VIII. Induction and isolation of nutritional mutants in Chinese hamster cells. *Proc. Nat. Acad. Sci.*, **60**, 1275–1281.

Kao F.T. and Puck T.T. Genetics of somatic mammalian cells: linkage studies with human–Chinese hamster cell hybrids. *Nature*, **228**, 329–333.

Kates J. and Beeson J. (1970). RNA synthesis in vaccinia virus. II. Synthesis of poly-A. *J. Mol. Biol.*, **50**, 19–34.

Kato H. and Morikawa K. (1972). Factors involved in the production of banded structures in mammalian chromosomes. *Chromosoma*, **38**, 105–120.

Kato H. and Yoshida T.H. (1972). Banding patterns of Chinese hamster chromosomes revealed by new techniques. *Chromosoma*, **36**, 272–280.

Kavenoff R. and Ziman B.H. (1973). Chromosome sized DNA molecules from Drosophila. *Chromosoma*, **41**, 1–28.

Kedes L.H. and Birnsteil M.L. (1971). Reiteration and clustering of DNA sequences complementary to histone mRNA. *Nature*, **230**, 165–169.

Kedinger C., Nuret P. and Chambon P. (1971). Structural evidence for two α-amanitin sensitive RNA polymerases in calf thymus. *FEBS Lett.*, **15**, 169–174.

Keshgegian A. and Furth J.J. (1972). Comparison of transcription of chromatin by calf thymus and E.coli RNA polymerase. *Biochem. Biophys. Res. Commun.*, **48**, 757–763.

Keyl H.G. (1965). A demonstrable local and geometric increase in the chromosomal DNA of Chironomus. *Experientia*, **21**, 191–199.

Keyl H.G. and Pelling C. (1963). Differentielle DNS-replikation in den speicheldrusen chromosomen von Chironomus thummi. *Chromosoma*, **14**, 347–359.

Kiefer B., Sakai H., Solari A.J. and Mazia D. (1966). The molecular unit of the microtubules of the mitotic apparatus. *J. Mol. Biol.*, **29**, 75–79.

Kihlman B.A. (1971). Molecular mechanisms of chromosome breakage and rejoining. *Adv. Cell Mol. Biol.*, **1**, 59–108.

Kihlman B.A. and Hartley B. (1967). Subchromatid exchanges and the folded fibre model of chromosome structure. *Hereditas*, **57**, 289–294.

Killander D. and Zetterberg A. (1965a). Quantitative cytochemical studies on interphase growth. I. Determination of DNA, RNA and mass content of age determined mouse fibroblasts in vitro and of intercellular variations in generation time. *Exp. Cell Res.*, **38**, 272–284.

Killander D. and Zetterberg A. (1965b). A quantitative cytochemical investigation of the relationship between cell mass and initiation of DNA synthesis in mouse fibroblasts in vitro. *Exp. Cell Res.*, **40**, 12–20.

King R.C. and Akai H. (1971). Spermatogenesis in Bombyx mori. II. The ultra-structure of synapsed bivalents. *J. Morph.*, **134**, 181–194.

King T.J. and Briggs R. (1956). Serial transplantation of embryonic nuclei. *Cold Spring Harbor Symp. Quant. Biol.*, **21**, 271–290.

Kinkade J.M.jr. (1969). Qualitative differences and quantitative tissue differences in the distribution of lysine rich histones. *J. Biol. Chem.*, **244**, 3375–3386.

Kinkade J.M.jr. and Cole R.D. (1966). A structural comparison of different lysine rich histones of calf thymus. *J. Biol. Chem.*, **241**, 5798–5805.

Kit S. (1961). Equilibrium centrifugation in density gradients of DNA preparations from animal tissues. *J. Mol. Biol.*, **3**, 711–716.

Kleiman L. and Huang R.C.C. (1972). Reconstitution of chromatin. The sequential binding of histones to DNA in the presence of salt and urea. *J. Mol. Biol.*, **64**, 1–8.

Kleinsmith L.J. and Allfrey V.G. (1969a). Isolation and characterization of a phosphoprotein fraction from calf thymus nuclei. *Biochim. Biophys. Acta*, **175**, 123–135.

Kleinsmith L.J., Heidema J. and Carroll A. (1970). Specific binding of rat liver nuclear proteins to DNA. *Nature*, **226**, 1025–1027.

Klevecz R.R. and Stubblefield E. (1967). RNA synthesis in relation to DNA replication in synchronized Chinese hamster cell cultures. *J. Exp. Zool.*, **165**, 259–268.

Knight L.A. and Luzzatti L. (1973). Replication pattern of the X and Y chromosomes in partially synchronized human lymphocyte cultures. *Chromosoma*, **40**, 153–166.

Knowland J. and Miller L. (1970). Reduction of ribosomal RNA synthesis and ribosomal RNA genes in a mutant of X. laevis which organises only a partial nucleolus. I. Ribosomal RNA synthesis in embryos of different nucleolar types. *J. Mol. Biol.*, **53**, 321–328.

Kohne D.E. (1970). Evolution of higher organism DNA. *Quart. Rev. Biophys.*, **3**, 327–375.

Kram R., Botchasn M. and Hearst J.E. (1972). Arrangement of the highly reiterated DNA sequences in the centric heterochromatin of D. melanogaster. Evidence for interspersed spacer DNA. *J. Mol. Biol.*, **64**, 103–119.

Kuechler E. and Rich A. (1969). Two rapidly labelled RNA species in the polysomes of antibody producing lymphoid tissue. *Proc. Nat. Acad. Sci.*, **63**, 520–527.

Kumar A. and Lindberg U. (1972). Characterization of messenger ribonucleoprotein and mRNA from KB cells. *Proc. Nat. Acad. Sci.*, **69**, 681–685.

Kumar A. and Warner J.R. (1972). Characterization of ribosomal precursor particles from Hela cell nucleoli. *J. Mol. Biol.*, **63**, 233–246.

Kurnit D.M., Schildkraut C.L. and Maio J.J. (1972). Single stranded interactions of mouse satellite DNA. *Biochim. Biophys. Acta*, **259**, 297–312.

Kusano T., Long C. and Green H. (1971). A new reduced human–mouse somatic cell hybrid containing the human gene for adenine phosphoribosyl transferase. *Proc. Nat. Acad. Sci.*, **68**, 82–86.

Kwan S.W. and Brawerman G. (1972). A particle associated with the poly-A segment in mammalian mRNA. *Proc. Nat. Acad. Sci.*, **69**, 3247–3250.

Labrie F. (1969). Isolation of an RNA with the properties of haemoglobin messenger. *Nature*, **221**, 1217–1222.

Laird C.D. (1971). Chromatid structure: relationship between DNA content and nucleotide sequence diversity. *Chromosoma*, **32**, 378–406.

Laird C.D. and McCarthy B.J. (1968a). Magnitude of interspecific nucleotide sequence variability in Drosophila. *Genetics*, **60**, 303–322.

Laird C.D. and McCarthy B.J. (1968b). Nucleotide sequence homology within the genome of D. melanogaster. *Genetics*, **60**, 323–334.

Laird C.D. and McCarthy B.J. (1969). Molecular characterization of the Drosophila genome. *Genetics*, **63**, 865–882.

Lake R.S. and Salzman N.P. (1972). Occurrence and properties of a chromatin associated f1-histone phosphokinase in mitotic Chinese hamster cells. *Biochem.*, **11**, 4817–4825.

Lam D.M.K. and Bruce W.R. (1971). The biosynthesis of protamine during spermatogenesis of the mouse: extraction, partial characterization and site of synthesis. *J. Cell Physiol.*, **78**, 13–24.

Lambert B. (1972). Repeated DNA sequences in a Balbiani ring. *J. Mol. Biol.*, **72**, 65–76.

Lambert B., Egyhazi E., Daneholt B. and Ringborg U. (1973a). Quantitative microassay for RNA/DNA hybrids in the study of nucleolar RNA from C. tentans salivary gland cells. *Exp. Cell Res.*, **76**, 369–380.

Lambert B., Daneholt B., Edstrom J.E., Egyhazi E. and Ringborg U. (1973b). Comparison between chromosomal and nuclear sap RNA from C. tentans salivary glands by RNA/DNA hybridization. *Exp. Cell Res.*, **76**, 381–390.

Lambert B., Wieslander L., Daneholt B., Egyhazi E. and Ringborg U. (1972). In situ demonstration of DNA hybridizing with chromosomal and nuclear sap RNA in C. tentans. *J. Cell Biol.*, **53**, 407–418.

Lampert F. (1971). Coiled supercoiled DNA in critical point dried and thin sectioned human chromosome fibres. *Nature New Biol.*, **234**, 187–188.

Lane C.D., Gregory C.M. and Morel C. (1973). Duck haemoglobin synthesis in frog cells. The translation and assay of reticulocyte 9S RNA in oocytes of X. laevis. *Europ. J. Biochem.*, **34**, 219–227.

Lane C.D., Marbaix G. and Gurdon J.B. (1971). Rabbit haemoglobin synthesis in frog cells: the translation of reticulocyte 9S RNA in frog oocytes. *J. Mol. Biol.*, **61**, 73–92.

Laskowski M.sr. (1972). The poly-dAT of crab. *Prog. Nucleic Acid Res. Mol. Biol.*, **12**, 161–188.

Latham H. and Darnell J.E. (1965). Distribution of mRNA in the cytoplasmic polyribosomes of the Hela cell. *J. Mol. Biol.*, **14**, 1–12.

Laycock D.G. and Hunt J.A. (1969). Synthesis of rabbit globin by a bacterial cell free system. *Nature*, **221**, 1118–1122.

Lebleu B., Marbaix G., Huez G., Temmerman J., Burny A. and Chantrenne H. (1971). Characterization of the messenger ribonucleoprotein released from reticulocyte polysomes by EDTA treatment. *Europ. J. Biochem.*, **19**, 264–269.

Ledbetter M.C. and Porter K.R. (1963). A microtubule in plant cell fine structure. *J. Cell Biol.*, **19**, 239–250.

Lee C.S. and Thomas C.A.jr. (1973). Formation of rings from Drosophila DNA fragments. *J. Mol. Biol.*, **77**, 25–56.

Lee J.C. and Yunis J.J. (1971a). A developmental study of constitutive heterochromatin in M. agrestis. *Chromosoma*, **32**, 237–250.

Lee J.C. and Yunis J.J. (1971b). Cytological variations in the constitutive heterochromatin of M. agrestis. *Chromosoma*, **35**, 117–124.

Lee S.L. and Brawerman G. (1971). Pulse labelled RNA complexes released by dissociation of rat liver polysomes. *Biochem.*, **10**, 510–516.

Lee S.Y., Mendecki J. and Brawerman G. (1971). A polynucleotide segment rich in adenylic acid in the rapidly labelled RNA component of mouse sarcoma 180 ascites cells. *Proc. Nat. Acad. Sci.*, **68**, 1331–1335.

Lefevre G.jr. (1971). Salivary gland chromosome bands and the frequency of crossing over in D. melanogaster. *Genetics*, **67**, 497–513.

Legname C. and Goldstein L. (1972). Proteins in nucleocytoplasmic interactions. VI. Is there an artefact responsible for the observed shuttling of proteins between cytoplasm and nucleus in Amoeba proteus. *Exp. Cell Res.*, **75**, 111–121.

Lentfer D. and Lezius A.G. (1972). Mouse myeloma RNA polymerase B. Template specificities and the role of a transcription-stimulation factor. *Europ. J. Biochem.*, **30**, 278–284.

Lewin S. (1973). *Water displacement and its control of biochemical reactions.* Academic Press, London.

Lewis E.B. (1950). The phenomenon of position effect. *Adv. Gen.*, **3**, 75–115.

Lezzi M. (1965). Die wirkung von DNAase auf isolierte polytan-chromosomen. *Esp. Cell Res.*, **39**, 289–202.

Li H.J. and Bonner J. (1971). Interaction of histone half molecules with DNA. *Biochem.*, **10**, 1461–1470.

Li H.J., Chang C. and Weiskopf M. (1972). Polylysine binding to histone bound regions in chromatin. *Biochem. Biophys. Res. Commun.*, **47**, 883–888.

Li H.J., Chang C. and Weiskopf M. (1973). Helix coil transition in nucleoprotein chromatin structures. *Biochem.*, **12**, 1763–1772.

Lim L. and Canellakis E.S. (1970). Adenine rich polymer associated with rabbit reticulocyte mRNA. *Nature*, **227**, 710–712.

Lima-de-Faria A. (1959). Differential uptake of tritiated thymidine into heter- and euchromatin in Melanoplus and Secale. *J. Biochem. Biophys. Cytol.*, **6**, 457–466.

Lima-de-Faria A. (1969). DNA replication and gene amplification in heterochromatin. In Lima-de-Faria A. (Ed.), *Handbook of Molecular Cytology.* North Holland, Amsterdam, pp 277–325.

Lima-de-Faria A. and Jaworska H. (1968). Late DNA synthesis in heterochromatin. *Nature*, **217**, 138–142.

Lindell T.J., Weinberg F., Morris P.W., Roeder R.G. and Rutter W.J. (1970). Specific inhibition of nuclear RNA polymerase by α-amanitin. *Science*, **170**, 447–448.

Littau V.C., Burdick J.C., Allfrey C.F. and Mirsky E.A. (1965). The role of histones in the maintenance of chromatin structure. *Proc. Nat. Acad. Sci.*, **54**, 1204–1212.

Littlefield J.W. (1964). Three degrees of guanylic acid–inosinic acid pyrophosphorylase deficiency in mouse fibroblasts. *Nature*, **203**, 1142–1144.

Littlefield J.W. (1966). The periodic synthesis of thymidine kinase in mouse fibroblasts. *Biochim. Biophys. Acta*, **114**, 398–403.

Littlefield J.W., McGovern A.P. and Margeson K.B. (1963). Changes in the distribution of polymerase activity during DNA synthesis in mouse fibroblasts. *Proc. Nat. Acad. Sci.*, **49**, 102–107.

Lockard R.E. and Lingrel J.B. (1969). The synthesis of mouse haemoglobin β chains in a rabbit reticulocyte cell free system programmes with mouse reticulocyte 9S RNA. *Biochem. Biophys. Res. Commun.*, **37**, 204–212.

Lodish H.F. and Jacobsen M. (1972). Equal rates of translation and termination of α and β globin chains. *J. Biol. Chem.*, **247**, 3622–3629.

Loeb L.A., Ewald J.L. and Agarwal S.S. (1970). DNA polymerase and DNA replication during lymphocyte transformation. *Cancer Res.*, **30**, 2514–2520.

Loening U.E., Jones K.W. and Birnsteil M.L. (1969). Properties of the rRNA precursor in X. laevis: comparison to the precursor in mammals and in plants. *J. Mol. Biol.*, **45**, 353–366.

Louie A.J. and Dixon G.H. (1972a). Synthesis, acetylation and phosphorylation of histone IV and its binding to DNA during spermatogenesis in trout. *Proc. Nat. Acad. Sci.*, **69**, 1975–1979.

Louie A.J. and Dixon G.H. (1972b). Trout testis cells. I. Characterization by DNA and protein analysis of cells separated by velocity sedimentation. *J. Biol. Chem.*, **247**, 5490–5497.

Louie A.J. and Dixon G.H. (1972c). Kinetics of enzymatic modification of the protamines and a proposal for their binding to chromatin. *J. Biol. Chem.*, **247**, 7962–7968.
Louie A.J., Sung M.T. and Dixon G.H. (1973). Modification of histones during spermatogenesis in trout. III. Levels of phosphohistone species and kinetics of phosphorylation of histone IIb1. *J. Biol. Chem.*, **248**, 3335–3341.
Lukanidin E.M., Zalmanzon E.S., Komaromi L., Samarina O.P. and Georgiev G.P. (1972). Structure and function of informomers. *Nature New Biol.*, **238**, 193–197.
Luykx P. (1970). Cellular mechanisms of chromosome distribution. *Int. Rev. Cytol.*, suppl. 2.
Lyon M.F. (1961). Gene action in the X chromosome of the mouse. *Nature*, **190**, 372–373.
Lyon M.F. (1962). Sex chromatin and gene action in the mammalian X chromosome. *Amer. J. Hum. Genet.*, **14**, 135–148.
Lyon M.F. (1964). Lack of evidence that inactivation of the mouse X chromosome is incomplete. *Genet. Res.*, **8**, 197–203.
Lyon M.F. (1971). Possible mechanisms of X chromosome inactivation. *Nature New Biol.*, **232**, 229–232.
Lyon M.F. (1972). X chromosome inactivation and developmental patterns in mammals. *Biol. Rev.*, **47**, 1–35.
Lyon M.F., Searle A.G., Ford C.E. and Ohno S. (1964). A mouse translocation suppressing sex linked variegation. *Cytogenetics*, **3**, 306–323.

Mach B., Faust C. and Vassalli P. (1973). Purification of 14S mRNA of immunoglobulin light chain that codes for a possible light chain precursor. *Proc. Nat. Acad. Sci.*, **70**, 451–455.
Maden B.E.H., Salim M. and Summers D.F. (1972). Maturation pathway for rRNA in the Hela cell nucleus. *Nature New Biol.*, **237**, 5–9.
Maio J.J. (1971). DNA strand reassociation and polyribonucleotide binding in the African green monkey, Ceropithecus aethiops. *J. Mol. Biol.*, **56**, 579–586.
Maio J.J. and Schildkraut C.L. (1969). Isolated mammalian metaphase chromosomes. II. Fractionated chromosomes of mouse and Chinese hamster cells. *J. Mol. Biol.*, **40**, 203–216.
Marbaix G. and Gurdon J.B. (1972). The effect of reticulocyte ribosome "factors" on the translation of haemoglobin messenger. *Biochim. Biophys. Acta*, **281**, 86–92.
Marbaix G. Lane C.D. (1972). Rabbit haemoglobin synthesis in frog cells. II. Further characterization of the products of translation of reticulocyte 9S RNA. *J. Mol. Biol.*, **67**, 517–524.
Margulis L. (1973). Colchicine sensitive microtubules. *Int. Rev. Cytol.*, **34**, 333–362.
Marin G. and Prescott D.M. (1964). The frequency of sister chromatid exchanges following exposure to varying doses of H³-thymidine or X-rays. *J. Cell Biol.*, **21**, 159–167.
Marmur J. and Doty P. (1959). Heterogeneity in DNA. I. Dependence on composition of the configurational stability of DNAs. *Nature*, **183**, 1427–1428.
Marmur J. and Doty P. (1961). Thermal renaturation of DNAs. *J. Mol. Biol.*, **3**, 585–594.
Marmur J., Rownd R. and Shildkraut C.L. (1963). Denaturation and renaturation of DNA. *Prog. Nucleic Acid Res.*, **1**, 232–300.
Martin M.A. and Hoyer B.H. (1966). Thermal stabilities and species specificities of reannealed animal DNAs. *Biochem.*, **5**, 2706–2713.
Marzluff W.F.jr. and McCarty K.S. (1972). Structural studies of calf thymus f3 histone. II. Occurrence of phosphoserine and ε-N-acetyllysine in thermolysin peptides. *Biochem.*, **11**, 2677–2681.

Marzluff W.F.jr. Miller D.M. and McCarty K.S. (1972). Occurrence of ε-N-acetyl-lysine in calf thymus histone f2b. *Arch. Biochem. Biophys.*, **152**, 472–474.

Matsuya Y., Green H. and Basilico C. (1968). Properties and uses of human mouse hybrid cell lines. *Nature*, **220**, 1199–1202.

Matthews M.B. (1972). Further studies on the translation of globin mRNA and EMP virus RNA in a cell free system from Krebs II ascites cells. *Biochim. Biophys. Acta*, **272**, 108–118.

Matthews M.B., Osborn M., Berns A.J.M. and Bloemendal H. (1972). Translation of two mRNAs from lens in a cell free system from Krebs II ascites cells. *Nature New Biol.*, **236**, 5–6.

Maul G.G. and Hamilton T.H. (1967). The intranuclear localization of two DNA dependent RNA polymerase activities. *Proc. Nat. Acad. Sci.*, **57**, 1371–1378.

Maurer H.R. and Chalkley G.R. (1967). Some properties of a nuclear binding site of estradiol. *J. Mol. Biol.*, **27**, 431–441.

Mayfield J.E. and Bonner J. (1971). Tissue differences in rat chromosomal RNA. *Proc. Nat. Acad. Sci.*, **68**, 2652–2655.

Mayfield J.E. and Bonner J. (1972). A partial sequence of nuclear events in regenerating rat liver. *Proc. Nat. Acad. Sci.*, **69**, 7–10.

Mazia D. (1961). Mitosis and the physiology of cell division. In Brachet J. and Mirsky A.E. (Eds.), *The Cell*, volume 3, pp 77–412.

Mazia D. (1963). Synthetic activities leading to mitosis. *J. Cell Comp. Physiol.*, **62**, *suppl.*, **1**, 123–140.

Mazia D. and Dan K. (1952). The isolation and biochemical characterization of the mitotic apparatus of dividing cells. *Proc. Nat. Acad. Sci.*, **38**, 826–738.

Mazia D., Harris P.J. and Bibring T. (1960). The multiplicity of the mitotic centres and the time course of their duplication and separation. *J. Biochem. Biophys. Cytol.*, **7**, 1–20.

Mazrimas J.A. and Hatch F.T. (1972). A possible relationship between satellite DNA and the evolution of kangaroo rat species (genus Dipodomys). *Nature New Biol.*, **240**, 102–105.

McCarthy B.J. (1967). The arrangement of base sequences in DNA. *Bact. Rev.*, **31**, 215–229.

McCarthy B.J. (1969). The evolution of base sequences in nucleic acids. In Lima-de-Faria (Ed.), *Handbook of Molecular Cytology*. North Holland, Amsterdam, pp 3–20.

McCarthy B.J. and Church R.B. (1970). The specificity of molecular hybridization reactions. *Ann. Rev. Biochem.*, **39**, 131–150.

McCarthy B.J. and McConaughy B.L. (1968). DNA/DNA duplex formation and the incidence of partially related base sequences in DNA. *Biochem. Genet.*, **2**, 37–53.

McClintock B. (1965). The control of gene action in maize. *Brookhaven Symp. Biol.*, **18**, 162–184.

McConaughy B.L., Laird C.D. and McCarthy B.J. (1969). Nucleic acid reassociation in formamide. *Biochem.*, **8**, 3289–3294.

McConaughy B.L. and McCarthy B.J. (1970a). The specificity of interactions between oligonucleotides and denatured DNA. *Biochem. Genet.*, **4**, 409–424.

McConaughy B.L. and McCarthy B.J. (1970b). The extent of base sequence divergence among the DNAs of various rodents. *Biochem. Genet.*, **4**, 425–446.

McConaughy B.L. and McCarthy B.J. (1972). Fractionation of chromatin by thermal chromatography. *Biochem.*, **11**, 998–1002.

McConkey E.H. and Hopkins J.W. (1964). The relationship of the nucleolus to the synthesis of rRNA in Hela cells. *Proc. Nat. Acad. Sci.*, **51**, 1197–1204.

McConkey E.H. and Hopkins J.W. (1965). Subribosomal particles and the transport of messenger RNA in Hela cells. *J. Mol. Biol.*, **14**, 257–270.

McGillivray A.J., Carroll D. and Paul J. (1971). The heterogeneity of the non histone chromatin proteins from mouse tissues. *FEBS Lett.*, **13**, 204–207.

McGillivray A.J., Cameron A., Krauze R.J., Rickwood D. and Paul J. (1972). The non histone proteins of chromatin. Their isolation and composition in a number of tissues. *Biochim. Biophys. Acta*, **277**, 384–402.

Means A.R., Comstock J.P., Rosenfeld G.G. and O'Malley B.W. (1972). Ovalbumin mRNA of chick oviduct: partial characterization, oestrogen dependence, and translation in vitro. *Proc. Nat. Acad. Sci.*, **69**, 1146–1150.

Melli M. and Bishop J. (1969). Hybridization between rat liver DNA and complementary RNA. *J. Mol. Biol.*, **40**, 117–136.

Melli M., Whitfield C., Rao K.V., Richardson M. and Bishop J.O. (1971). DNA–RNA hybridization in great excess. *Nature New Biol.*, **231**, 8–12.

Mendecki J., Lee S.Y. and Brawerman G. (1972). Characteristics of the poly-A segment associated with mRNA in mouse sarcoma 180 ascites cells. *Biochem.*, **11**, 792–798.

Merriam R.W. (1969). Movement of cytoplasmic proteins into nuclei induced to enlarge and initiate DNA or RNA syntheis. *J. Cell Science*, **5**, 333–350.

Meselson M. and Stahl F.W. (1958). The replication of DNA in E.coli. *Proc. Nat. Acad. Sci.*, **44**, 671–682.

Meselson M., Stahl F.W. and Vinograd J. (1957). Equilibrium sedimentation of macromolecules in density gradients. *Proc. Nat. Acad. Sci.*, **43**, 581–588.

Migeon B.R. (1972). Stability of X chromosome inactivation in human somatic cells. *Nature*, **239**, 87–89.

Migeon B.R., Der Kaloustian V.M., Nyhan W.L., Young W.J. and Childs B. (1968). X linked HGPRT deficiency: heterozygote has two clonal populations. *Science*, **160**, 425–427.

Migeon B.R., Smith S.W. and Leddy C.L. (1969). The nature of thymidine kinase in the human-mouse hybrid cell. *Biochem. Genet.*, **3**, 583–590.

Mill D.A., Allerdice P.W., Miller O.J. and Breg W.R. (1971a). Quanacrine fluorescence patterns of human D group chromosomes. *Nature*, **232**, 24–27.

Miller D.A., Kouri R.E., Dev V.G., Grewal M.S., Hutton J.J. and Miller O.J. (1971b). Assignment of four linkage groups to chromosomes in Mus musculus and a cytogenetic method for locating their centromeric ends. *Proc. Nat. Acad. Sci.*, **68**, 2699–2702.

Miller G., Berlowitz L. and Regelson W. (1971). Chromatin and histones in mealy bug explants: activation and decondensation of facultative heterochromatin by a synthetic polyanion. *Chromosoma*, **32**, 251–261.

Miller O.J., Allerdice P.W., Miller D.A., Breg W.R. and Migeon B.R. (1971c). Human thymidine kinase gene locus: assignment to chromosome 17 in a hybrid of man and mouse cells. *Science*, **173**, 244–245.

Miller O.J., Cook P.R., Khan P.M., Shin S. and Siniscalco M. (1971d). Mitotic separation of two human X-linked genes in man–mouse somatic cell hybrids. *Proc. Nat. Acad. Sci.*, **68**, 116–120.

Miller O.J., Miller D.A., Kouri R.E., Allerdice P.W., Dev V.G., Grewal M.S. and Hutton J.J. (1971e). Identification of the mouse karyotype by quinacrine fluorescence and tentative assignment of seven linkage groups. *Proc. Nat. Acad. Sci.*, **68**, 1530–1533.

Mill O.L. (1965). Fine structure of lampbrush chromosomes. *National Cancer Institute Monograph*, **18**, 79–100.

Miller O.L.jr. and Beatty B.R. (1969a). Visualization of nucleolar genes. *Science*, **164**, 955–957.

Miller O.L.jr. and Beatty B.R. (1969b). Extrachromosomal nucleolar genes in amphibian oocytes. *Genetics*, **61**, 133–143.

Miller O.L.jr. and Beatty B.R. (1969c). Portrait of a gene. *J. Cell Physiol.*, **74**, *suppl.*, **1**, 225–230.

Mirsky A.E. (1971). The structure of chromatin. *Proc. Nat. Acad. Sci.*, **68**, 2945–2948.

Mirsky A.E. and Ris H. (1951). The composition and structure of isolated chromosomes. *J. Gen. Physiol.*, **34**, 475–492.

Mirsky A.E. and Silverman B. (1972). Blocking by histones of accessibility to DNA in chromatin. *Proc. Nat. Acad. Sci.*, **69**, 2115–2119.

Mirshk A.E., Silverman B. and Panda N.C. (1972). Blocking by histones of accessibility to DNA in chromatin: addition to histones. *Proc. Nat. Acad. Sci.*, **69**, 3243–3246.

Mitchison J.M. (1969). Enzyme synthesis in synchronous cultures. *Science*, **165**, 657–663.

Mitchison J.M. (1971). *The Biology of the Cell Cycle.* Cambridge University Press, Cambridge.

Mittwoch U. (1967). Barr bodies in relation to DNA values and nuclear size in cultured human cells. *Cytogenetics*, **6**, 38–50.

Mizuno N.S., Stoops C.E. and Peiffer R.L. (1971). Nature of the DNA associated with the nuclear envelope of regenerating liver. *J. Mol. Biol.*, **59**, 517–525.

Moav B. and Nemer M. (1971). Histone synthesis: assignment to a special class of polysomes in sea urchin embryos. *Biochem.*, **10**, 881–888.

Moens P.B. (1968). The structure and function of the synaptonemal complex in Lilium longiflorum sporocytes. *Chromosoma*, **23**, 418–426.

Moens P.B. (1969a). The fine structure of meiotic chromosome pairing in the triploid, Lilium tigranum. *J. Cell Biol.*, **42**, 272–279.

Moens P.B. (1969b). Multiple core complex in grasshopper spermatocytes and spermatids. *J. Cell Biol.*, **42**, 542–551.

Molloy G.R. and Darnell J.E. (1973). Characterization of poly-A regions and the adjacent nucleotides in heterogeneous nuclear RNA and mRNA from Hela cells. *Biochem.*, **12**, 2324–2330.

Molloy G.R., Thomas W.L. and Darnell J.E. (1972). Occurrence of uridylate rich oligonucleotide regions in heterogeneous nuclear RNA of Hela cells. *Proc. Nat. Acad. Sci.*, **69**, 3684–3688.

Monesi V. (1969). DNA, RNA and protein synthesis during the mitotic cell cycle. In Lima-de-Faria A. (Ed.), *Handbook of Molecular Cytology*. North Holland, Amsterdam, pp 472–499.

Moore G.P.M. and Ringertz N.R. (1973). Localization of DNA-dependent RNA polymerase activities in fixed human fibroblasts by autoradiography. *Exp. Cell Res.*, **76**, 223–228.

Morrison M.R., Paul J. and Williamson R. (1972). The DNA–RNA hybridization of H^3 methylated mouse globin mRNA in conditions of RNA excess. *Europ. J. Biochem.*, **27**, 1–9.

Moses M.J. (1956). Studies on nuclei using correlated cytochemical, light, and electron microscope techniques. *J. Biochem. Biophys. Cytol.*, **2**, *suppl.*, 397–406.

Moses M.J. (1958). The relation between the axial complex of meiotic prophase chromosomes and chromosome pairing in a Salamander (Plethodon cinercus). *J. Biochem. Biophys. Cytol.*, **4**, 633–638.

Moses M.J. (1968). Synaptonemal complex. *Ann. Rev. Genet.*, **2**, 363–412.

Moses M.J. and Coleman J.R. (1964). Structural patterns and the functional organization of chromosomes. In M. Locke (Ed.), *The role of chromosomes in development*. Academic Press, New York, pp 11–49.

Mueller G.C. and Kajiwara K. (1966a). Early and late replicating DNA complexes in Hela nuclei. *Biochim. Biophys. Acta*, **114**, 108–115.

Mueller G.C. and Kajiwara K. (1966b). Actinomycin D and p-fluorphenylalanine, inhibitors of nuclear replication in Hela cells. *Biochim. Biophys. Acta*, 119, 557–565.
Mueller G.C., Vonderhaar B., Kim U.H. and Mahieu M.L. (1971). Estrogen action: an inroad to cell biology. *Rec. Prog. Horm. Res.*, 28, 1–45.
Mukherjee B.B. and Milet R.G. (1972). Non random X chromosome inactivation—an artefact of cell selection. *Proc. Nat. Acad. Sci.*, 69, 37–39.
Mukherjee A.B. and Nitowsky H.M. (1972). Fluorescence of constitutive heterochromatin of M. agrestis. *Exp. Cell Res.*, 73, 248–251.
Mukherjee B.B. and Sinha A.K. (1964). Single X hypothesis: cytological evidence for random inactivation of X chromosomes in a female mule complement. *Proc. Nat. Acad. Sci.*, 51, 252–259.
Muller H.J. (1967). The genetic material as the initiator and the organising basis of life. In Brink R.A. (Ed.), *Heritage from Mendel*. University of Wisconsin Press, Madison, pp 419–447.
Murray K. (1965). The basic proteins of cell nuclei. *Ann. Rev. Biochem.*, 34, 209–246.
Murray K. (1969). Stepwise removal of histone from native deoxyribonucleoprotein by titration with acid at low temperature and some properties of the resulting partial nucleoproteins. *J. Mol. Biol.*, 39, 125–144.

Nabholz M., Miggiano V. and Bodmer W. (1969). Genetic analysis with human–mouse somatic cell hybrids. *Nature*, 223, 358–363.
Nesbitt M.N. and Francke U. (1973). A system of nomenclature for band patterns of mouse chromosomes. *Chromosoma*, 41, 145–158.
Nicklas R.B. (1967). Chromosome micromanipulation. II. Induced reorientation and the experimental control of segregation in mitosis. *Chromosoma*, 21, 17–50.
Nicklas R.B. (1972). Mitosis. *Adv. Cell Biol.*, 2, 225–298.
Nicklas R.B. and Koch C.A. (1972). Chromosome micromanipulation. IV. Polarized motions within the spindle and models for mitosis. *Chromosoma*, 39, 1–26.
Nicklas R.B. and Staehly C.A. (1967). Chromosome micromanipulation. I. The mechanics of chromosome attachment to the spindle. *Chromosoma*, 21, 1–16.
Nudel U., Lebleu B., Zehavi-Willner T. and Revel M. (1973). Messenger RNP and initiation factors in rabbit reticulocyte polyribosomes. *Europ. J. Biochem.*, 33, 314–322.
Nygaard A.P. and Hall B.D. (1963). A method for the detection of RNA–DNA complexes. *Biochem. Biophys. Res. Commun.*, 12, 98–104.
Nygaard A.P. and Hall B.D. (1964). Formation and properties of RNA–DNA complexes. *J. Mol. Biol.*, 9, 125–142.

O'Brien R.L., Sanyal A.B. and Stanton R.H. (1972). Association of DNA replication with the nuclear membrane of Hela cells. *Exp. Cell Res.*, 70, 106–112.
Ockey C.H. (1972). Distribution of DNA replicator sites in mammalian nuclei. II. Effects of prolonged inhibition of DNA synthesis. *Exp. Cell Res.*, 70, 203–213.
Ogawa Y., Quagliarotti G., Jordan J., Taylor C.W., Starbuck W.C. and Busch H. (1969). Structural analysis of the glycine rich arginine rich histone. III. Sequence of the amino terminal half of the molecule containing the modified lysine residues and the total sequence. *J. Biol. Chem.*, 244, 4387–4392.
Ohlenbusch H.H., Olivera B.M., Tuan D. and Davidson N. (1967). Selective dissociation of histones from calf thymus nucleoprotein. *J. Mol. Biol.*, 25, 299–315.
Ohno S. (1971). Simplicity of mammalian regulatory systems inferred by single gene determination of sex phenotypes. *Nature*, 234, 134–137.
Ohno S. and Cattanach B.M. (1962). Cytological study of an X chromosome translocation in Mus musculus. *Cytogenetics*, 1, 129–140.

Ohno S., Kaplan W.D. and Kinosita R. (1959). Formation of the sex chromatin by a single X chromosome in liver cells of Rattus norvegicus. *Exp. Cell Res.*, **18**, 415–418.

Ohnuki Y. (1968). Structure of chromosomes. I. Morphological studies of the spiral structure of human somatic chromosomes. *Chromosoma*, **25**, 402–428.

Ohta T. and Kimura M. (1971). Functional organization of genetic material as a product of molecular evelution. *Nature*, **233**, 118–119.

Oliver D., Balhorn R., Granner D. and Chalkley R. (1972). Molecular nature of f1 histone phosphorylation in cultured hepatoma cells. *Biochem.*, **11**, 3921–3925.

Oliver D.R. and Chalkley R. (1972a). An electrophoretic analysis of Drosophila histones. I. Isolation and identification. *Exp. Cell Res.*, **73**, 295–302.

Oliver D.R. and Chalkley R. (1972b). Comparison of larval and adult histone patterns in two species of Drosophila. *Exp. Cell Res.*, **73**, 303–310.

Olsnes S. (1971). Further studies on the protein bound to the mRNA in mammalian polysomes. *Europ. J. Biochem.*, **23**, 557–563.

O'Malley B.W., McGuire W.L., Kohler P.O. and Korenman S.G. (1969). Studies on the mechanism of steroid hormone regulation of synthesis of specific proteins. *Rec. Prog. Horm. Res.*, **25**, 105–153.

O'Malley B.W., Sherman M.R. and Toft D.O. (1970). Progesterone receptors in the cytoplasm and nucleus of chick oviduct target tissues. *Proc. Nat. Acad. Sci.*, **67**, 501–508.

O'Malley B.W., Toft D.O. and Sherman M.R. (1971). Progesterone binding components of chick oviduct. II. Nuclear components. *J. Biol. Chem.*, **246**, 1117–1122.

Pachmann U. and Rigler R. (1972). Quantum yield of acridines interacting with DNA of defined base sequence. *Exp. Cell Res.*, **72**, 602–608.

Packman S., Aviv H., Ross J. and Leder P. (1972). A comparison of globin genes in duck reticulocytes and liver cells. *Biochem. Biophys. Res. Commun.*, **49**, 813–819.

Pagoulatos G.N. and Darnell J.E. (1970a). A comparison of the heterogeneous nuclear RNA of Hela cells in different periods of the cell growth cycle. *J. Cell Biol.*, **44**, 476–483.

Pagoulatos G.N. and Darnell J.E. (1970b). Fractionation of heterogeneous nuclear RNA: rates of hybridization and chromosomal distribution of reiterated sequences. *J. Mol. Biol.*, **54**, 517–536.

Painter T.S. (1934). Salivary chromosomes and the attack on the gene. *J. Hered.*, **25**, 465–476.

Panyim S., Bilek D. and Chalkley R. (1971). An electrophoretic comparison of vertebrate histones. *J. Biol. Chem.*, **246**, 4206–4215.

Panyim S. and Chalkley R. (1969a). The heterogeneity of histones. I. A quantitative analysis of calf histones in very long polyacrylamide gels. *Biochem.*, **8**, 3972–3979.

Panyim S. and Chalkley R. (1969b). A new histone found only in mammalian tissues with little cell division. *Biochem. Biophys. Res. Commun.*, **37**, 1042–1048.

Panyim S., Sommer K.R. and Chalkley R. (1971). Oxidation of the cysteine containing histone f3. Detection of an evolutionary mutation in a conservative histone. *Biochem.*, **10**, 3911–3917.

Pardon J.F., Richards B.M., Skinner L.G. and Ockey G.H. (1973). X-ray diffraction from isolated metaphase chromosomes. *J. Mol. Biol.*, **76**, 267–270.

Pardon J.F. and Wilkins M.F.H. (1972). A supercoil model for nucleohistone. *J. Mol. Biol.*, **68**, 115–124.

Pardon J.F., Wilkins M.H.F. and Richards B.M. (1967). Superhelical model for nucleohistone. *Nature*, **215**, 508–509.

Pardue M.L. and Gall J.G. (1969). Molecular hybridization of radioactive DNA to the DNA of cytological preparations. *Proc. Nat. Acad. Sci.*, **64**, 600–604.

Pardue M.L. and Gall J.G. (1970). Chromosomal localization of mouse satellite DNA. *Science*, **168**, 1356–1358.

Parsons J.T. and McCarty K.S. (1968). Rapidly labelled mRNA-protein complex of rat liver nuclei. *J. Biol. Chem.*, **243**, 5377–5384.

Paul J. (1970). DNA masking in mammalian chromatin: a molecular mechanism for determination of cell type. *Curr. Topics Devel. Biol.*, **5**, 317–352.

Paul J. (1972). General theory of chromosome structure and gene activation in eucaryotes. *Nature*, **238**, 444–446.

Paul J. and Gilmour R.S. (1968). Organ specific restriction of transcription in mammalian chromatin. *J. Mol. Biol.*, **34**, 305–316.

Peacock W.J., Brutlag D., Goldring E., Appels R., Hinton C.W. and Lindsley D.L. (1973). *Cold Spring Harbor Symp. Quant. Biol.*, **38**, 405–416.

Pearson P.L., Bobrow M. and Vosa C.G. (1970). Technique for identifying Y chromosomes in human interphase nuclei. *Nature*, **226**, 78–80.

Pederson T. (1972). Chromatin structure and the cell cycle. *Proc. Nat. Acad. Sci.*, **69**, 2224–2228.

Pederson T. and Kumar A. (1971). Relationship between protein synthesis and ribosome assembly in Hela cells. *J. Mol. Biol.*, **61**, 655–668.

Pederson T. and Robbins E. (1972). Chromatin structure and the cell division cycle. Actinomycin binding in synchronized Hela cells. *J. Cell Biol.*, **55**, 322–327.

Pelling C. (1964). Ribonuklein saure synthese der reisenchromosemen. *Chromosoma*, **15**, 71–122.

Pemberton R.E. and Baglioni C. (1972). Duck haemoglobin mRNA contains a polynucleotide sequence rich in adenylic acid. *J. Mol. Biol.*, **65**, 531–536.

Penman S. (1966). RNA metabolism in the cell nucleus. *J. Mol. Biol.*, **17**, 117–130.

Penman S., Rosbash M. and Penman M. (1970). Messenger and heterogeneous nuclear RNA in Hela cells: differential inhibition by cordycepin. *Proc. Nat. Acad. Sci.*, **67**, 1878–1885.

Penman S., Smith I. and Holtzman E. (1966). Ribosomal RNA synthesis and processing in a particular site in the Hela cell nucleus. *Science*, **154**, 786–789.

Penman S., Vesco C. and Penman M. (1968). Localization and kinetics of formation of nuclear heterodisperse RNA, cytoplasmic heterodisperse RNA and polyribosome associated messenger RNA in Hela cells. *J. Mol. Biol.*, **34**, 49–69.

Perlman S., Abelson H.T. and Penman S. (1973). Mitochondrial protein synthesis: RNA with the properties of eucaryotic mRNA. *Proc. Nat. Acad. Sci.*, **70**, 350–353.

Perry R.P. (1967). The nucleolus and the synthesis of ribosomes. *Prog. Nucleic Acid Res.*, **6**, 219–257.

Perry R.P. (1969). Nucleoli: the cellular sites of ribosome production. In Lima-de-Faria A. (Ed.), *Handbook of Molecular Cytology*. North Holland, Amsterdam. pp 620–638.

Perry R.P., Chen T.T., Freed J.J., Greenberg J.R., Kelley D.E. and Tartof K.D. (1970). Evolution of the transcription unit of ribosomal RNA. *Proc. Nat. Acad. Sci.*, **65**, 609–616.

Perry R.P. and Kelley D.E. (1968). Messenger RNA–protein complexes and newly synthesized ribosomal subunits: analysis of free particles and components of polyribosomes. *J. Mol. Biol.*, **35**, 37–60.

Perry R.P. and Kelley D.E. (1970). Inhibition of RNA synthesis by actinomycin D: characteristic dose response of different RNA species. *J. Cell Physiol.*, **76**, 127–140.

Pfeiffer S.E. (1968). RNA synthesis in synchronously growing populations of Hela S3 cells. II. Rate of synthesis of individual RNA fractions. *J. Cell Pyhsiol.*, **71**, 95–104.

Pfeiffer S.E. and Tolmach L.J. (1968). RNA synthesis in synchronously growing populations of Hela S3 cells. I. Rate of total RNA synthesis and its relationship to DNA synthesis. *J. Cell Physiol.*, **71**, 77–94.

Pflueger O.H. and Yunis J.J. (1966). Late replication patterns of chromosomal DNA in somatic tissues of the Chinese hamster. *Exp. Cell Res.*, **44**, 413–420.

Philipson L., Wall R., Glickman R. and Darnell J.E. (1971). Addition of poly-adenylate sequences to virus specific RNA during adenovirus replication. *Proc. Nat. Acad. Sci.*, **68**, 2806–2809.

Phillips D.M.P. (1971). The primary structure of histones and protamines. In Phillips D.M.P. (Ed.), *Histones and Nucleohistones*. Plenum Press, New York, pp 47–84.

Phillips D.M.P. and Johns E.W. (1965). A fractionation of the histones of group f2a from calf thymus. *Biochem. J.*, **94**, 127–130.

Phillips S.G. (1972). Repopulation of the post mitotic nucleolus by preformed RNA. *J. Cell Biol.*, **53**, 611–623.

Pipkin J.L. and Larson D.A. (1972). Characterization of the "very" lysine rich histones of active and quiescent anther tissues of Hippeastratum belladonna. *Exp. Cell Res.*, **71**, 249–260.

Plaut W. (1969). On ordered DNA replication in polytene chromosomes. *Genetics*, **61**, *suppl.*, 239–244.

Plaut W., Nash D. and Fanning T. (1966). Ordered replication of DNA in polytene chromosomes of D. melanogaster. *J. Mol. Biol.*, **16**, 85–93.

Pogo B.G.T., Pogo A.O., Allfrey V.G. and Mirsky A.E. (1968). Changing patterns of histone acetylation and RNA synthesis in regeneration of the liver. *Proc. Nat. Acad. Sci.*, **59**, 1337–1344.

Pontecorvo G. (1971). Induction of directional chromosome elimination in somatic cell hybrids. *Nature*, **230**, 367–369.

Prescott D.M. (1963). Cellular sites of RNA synthesis. *Prog. Nucleic Acid Res.*, **3**, 33–57.

Prescott D.M. (1966). The synthesis of total macronuclear protein, histone and DNA during the cell cycle of Euplotes eurystomus. *J. Cell Biol.*, **31**, 1–9.

Prescott D.M. (1970). The structure and replication of eucaryotic chromosomes. *Adv. Cell Biol.*, **1**, 57–117.

Prescott D.M. and Bender M.A. (1962). Synthesis of RNA and protein during mitosis in mammalian tissue culture cells. *Exp. Cell Res.*, **26**, 260–268.

Prescott D.M. and Bender M.A. (1963a). Autoradiographic study of the chromatid distribution of labelled DNA in two types of mammalian cells in vitro. *Exp. Cell Res.*, **29**, 430–442.

Prescott D.M. and Bender M.A. (1963b). Synthesis and behaviour of nuclear proteins during the cell life cycle. *J. Cell Physiol.*, **62**, *suppl.*, 175–194.

Prestayko A.W., Lewis B.C. and Busch H. (1972). Endoribonuclease activity associated with nucleolar RNP particles from Novikoff hepatoma. *Biochim. Biophys. Acta*, **269**, 90–103.

Price R. and Penman S. (1972). A distinct RNA polymerase activity, synthesizing 5.5S, 5S and 4S RNA in nuclei from adenovirus 2 infected Hela cells. *J. Mol. Biol.*, **70**, 435–450.

Puck T.T. and Kao F.T. (1967). Genetics of somatic mammalian cells. V. Treatment with 5-BU and visible light for isolation of nutritionally defective mutants. *Proc. Nat. Acad. Sci.*, **58**, 1227–1234.

Pyeritz R.E. and Thomas C.A.jr. (1973). Regional organization of eucaryotic DNA sequences as studied by the formation of folded rings. *J. Mol. Biol.*, **77**, 57–74.

Quagliarotti G., Hidvegi E., Wikman J. and Busch H. (1970). Comparative hybridizations of nucleolar and rRNA with nucleolar DNA. *J. Biol. Chem.*, **245**, 1962–1969.

Rae P.M.M. (1966). Whole mount electron microscopy of Drosophila salivary chromosomes. *Nature*, **212**, 139–142.

Rae P.M.M. (1970). Chromosomal distribution of rapidly annealing DNA in D. melanogaster. *Proc. Nat. Acad. Sci.*, **67**, 1018–1025.

Rae P.M.M. (1972a). The interphase distribution of satellite DNA containing heterochromatin in mouse nuclei. *Chromosoma*, **39**, 443–456.

Rae P.M.M. (1972b). The distribution of repetitive DNA sequences in chromosomes. *Adv. Cell Mol. Biol.*, **2**, 109–150.

Rall S.C. and Cole R.D. (1971). Amino acid sequence and sequence variability of the amino terminal regions of lysine rich histones. *J. Biol. Chem.*, **246**, 7175–7190.

Randall J. and Disbrey C. (1965). Evidence for the presence of DNA at basal body sites in Tetrahymena pyriformis. *Proc. Roy. Soc. B*, **162**, 473–491.

Rao P.N. and Johnson R.T. (1970). Mammalian cell fusion: studies on the regulation of DNA synthesis and mitosis. *Nature*, **225**, 159–164.

Rao P.N. and Johnson R.T. (1972). Premature chromosome condensation: a mechanism for the elimination of chromosomes in virus fused cells. *J. Cell Sci.*, **10**, 495–514.

Rasch E.M., Barr H.J. and Rasch R.W. (1971). The DNA content of sperm of D. melanogaster. *Chromosoma*, **33**, 1–18.

Rasmussen P.S., Murray K. and Luck J.M. (1962). On the complexity of calf thymus histone. *Biochem.*, **1**, 79–89.

Rattner J.B. and Phillips S.G. (1973). Independence of centriole formation and DNA synthesis. *J. Cell Biol.*, **57**, 359–372.

Remington J.A. and Klevecz R.R. (1973). Families of repeating units in cultured hamster fibroblasts. *Exp. Cell Res.*, **76**, 410–418.

Richardson B.J., Czuppon A.B. and Sharman G.B. (1971). Inheritance of glucose-6-phosphate dehydrogenase variation in kangaroos. *Nature New Biol.*, **230**, 154–155.

Ringertz N.R., Carlsson S.A., Ege T. and Bolund L. (1971). Detection of human and chick nuclear antigens in nuclei of chick erythrocytes during reactivation in heterocaryons with Hela cells. *Proc. Nat. Acad. Sci.*, **68**, 3228–3232.

Ris H. (1956). A study of chromosomes with the electron microscope. *J. Biochem. Biophys. Cytol.*, **2**, *suppl*, 385–392.

Ris H. (1969). The molecular organization of chromosomes. In Lima-de-Faria A. (Ed.), *Handbook of Molecular Cytology*. North Holland, Amsterdam, pp 221–250.

Ris H. and Chandler B.L. (1963). The ultrastructure of genetic systems in procaryotes and eucaryotes. *Cold Spring Harbor Symp. Quant. Biol.*, **28**, 1–8.

Ritossa F.M., Malva C., Boncinelli E., Graziani F. and Polito L. (1971). The first steps of magnification of DNA complementary to ribosomal RNA in D. melanogaster. *Proc. Nat. Acad. Sci.*, **68**, 1580–1584.

Ritossa F.M. and Spiegelman S. (1965). Localization of DNA complementary to ribosomal RNA in the nucleolus organiser region of D. melanogaster. *Proc. Nat. Acad. Sci.*, **53**, 737–745.

Rizzoni M. and Palitti F. (1973). Regulatory mechanism of cell division. I. Colchicine induced endo-reduplication. *Exp. Cell Res.*, **77**, 450–458.

Robbins E. and Borun T. (1967). The cytoplasmic synthesis of histone in Hela cells and its temporal relationship to DNA replication. *Proc. Nat. Acad. Sci.*, **57**, 409–416.

Robbins E., Jentzsch G. and Micali A. (1968). The centriole cycle in synchronized Hela cells. *J. Cell Biol.*, **36**, 329–339.

Robbins E. and Shelanski M. (1969). Synthesis of a colchicine binding protein during the Hela cell life cycle. *J. Cell Biol.*, **43**, 371–373.

Rodman T.C. (1967). DNA replication in salivary gland nuclei of D. melanogaster at successive larval and prepupal stages. *Genetics*, **55**, 375–386.

Rodman T.C. and Tahilani S. (1973). The feulgen banded karyotype of the mouse: analysis of the mechanisms of banding. *Chromosoma*, **42**, 37–56.

Roeder R.G. and Rutter W.J. (1969). Multiple forms of DNA dependent RNA polymerase in eucaryotic organisms. *Nature*, **224**, 234–237.

Roeder R.G. and Rutter W.J. (1970a). Specific nucleolar and nucleoplasmic RNA polymerases. *Proc. Nat. Acad. Sci.*, **65**, 675–682.

Roeder R.G. and Rutter W.J. (1970b). Multiple RNA polymerases and RNA synthesis during sea urchin development. *Biochem.*, **9**, 2543–2553.

Romeo G. and Migeon B.R. (1970). Genetic inactivation of the α-galactosidase locus in carriers of Fabry's disease. *Science*, **170**, 180–181.

Roos V.P. (1973). Light and electron microscopy of rat kangaroo cells in mitosis. I. Formation and breakdown of the mitotic apparatus. *Chromosoma*, **40**, 43–82.

Ross A. (1968). The substructure of centriole subfibres. *J. Ult. Res.*, **23**, 537–539.

Ross J., Aviv H., Scolnick E. and Leder P. (1972). In vitro synthesis of DNA complementary to purified rabbit globin mRNA. *Proc. Nat. Acad. Sci.*, **69**, 264–268.

Rowley J.D. and Bodmer W.F. (1971). Relationship of centromeric heterochromatin to the fluorescent banding patterns of metaphase chromosomes in the mouse. *Nature*, **231**, 503–505.

Roy A.K. and Zubay G. (1966). RNA synthesis stimulated by sonicated nucleohistone. *Biochim. Biophys. Acta*, **129**, 403–405.

Rubinstein L. and Clever U. (1972). Chromosome activity and cell function in polytenic cells. V. Developmental changes in RNA synthesis and turnover. *Develop. Biol.*, **27**, 519–537.

Ruddle F.H. (1973). Linkage analysis on man by somatic cell genetics. *Nature*, **242**, 165–169.

Ruddle F.H., Chapman V.M., Ricciuti F., Murnane M., Klebe R. and Khan P.M. (1971). Linkage relationships of seventeen human gene loci as determined by man–mouse somatic cell hybrids. *Nature New Biol.*, **232**, 69–73.

Rudkin G.T. (1965). The relative mutabilities of DNA in regions of the X chromosome of D. melanogaster. *Genetics*, **52**, 665–681.

Rudkin G.T.(1969). Non replicating DNA in Drosophila.*Genetics*, **61**, *suppl.*, 227–238.

Russell L.B. (1961). Genetics of mammalian sex chromosomes. *Science*, **133**, 1795–1803.

Russell L.B. (1963). Mammalian X chromosome action: inactivation limited in spread and in region of origin. *Science*, **140**, 976–978.

Russell L.B. (1964). Another look at the single active X hypothesis. *Trans. N. Y. Acad. Sci.*, **26**, 726–736.

Russell L.B. and Montgomery C.S. (1970). Comparative studies on X autosome translocation in the mouse. II. Inactivation of autosomal loci, segregation, and mapping of autosomal breakpoints in five T (X;1)'s. *Genetics*, **64**, 281–312.

Sachs R.I. and Clever U. (1972). Unique and repetitive DNA sequences in the genome of C. tentans. *Exp. Cell Res.*, **74**, 587–591.

Sadgopal A. and Bonner J. (1970a). Chromosomal proteins of interphase Hela cells. *Biochim. Biophys. Acta*, **207**, 206–226.

Sadgopal A. and Bonner J. (1970b). Proteins of interphase and metaphase chromosomes compared. *Biochim. Biophys. Acta*, **207**, 227–239.

Sager R. (1972). *Cytoplasmic genes and organelles*. Academic Press, New York.

Sakai H. (1966). Studies on sulfhydryl groups during cell division of sea urchin eggs. VIII. Some properties of mitotic apparatus proteins. *Biochim. Biophys. Acta*, **112**, 132–145.

Sakai H. (1968). Contractile properties of protein threads from sea urchin eggs in relation to cell division. *Int. Rev. Cytol.*, **23**, 89–112.

Salas J. and Green H. (1971). Proteins binding to DNA and their relation to growth in cultured mammalian cells. *Nature New Biol.*, **229**, 165–169.

Salim M. and Maden B.E.H. (1973). Early and late methylations in Hela cell ribosome maturation. *Nature*, **244**, 334–336.

Sampson J., Matthews M.B., Osborn M. and Borghetti A.F. (1972). Haemoglobin mRNA translation in cell free systems from rat and mouse liver and Landschutz ascites cells. *Biochem.*, **11**, 3636–3640.

Sanders L.A. and McCarty K.S. (1972). Isolation and purification of histones from avian erythrocytes. *Biochem.*, **11**, 4216–4221.

Sasaki S.S. and Norman A. (1966). DNA fibres from human lymphocyte nucli. *Exp. Cell Res.*, **44**, 642–645.

Sautiere P., Breynaert M.D., Moschetto Y. and Biserte G. (1970). Sequence complete des acides amines de l'histone riche en glycine et en arginine de thymus du porc. *Comptes Rendus Acad. Sci.* (Paris), **271**, 364–365.

Sautiere P., Tyrou D., Moschetto Y. and Biserte G. (1971). Structure primaire de l'histone riche en glycine et en arginine isolee de la tumeur de chloroleucemie du rat. *Biochimie*, **53**, 479–483.

Schachat F.H. and Hogness D.S. (1973). Repetitive sequences in isolated Thomas circles from D. melanogaster. *Cold Spring Harbor Symp. Quant. Biol.*, **38**, 371–381.

Scharff M.D. and Robbins E. (1965). Synthesis of ribosomal RNA in synchronized Hela cells. *Nature*, **208**, 464–466.

Scharff M.D. and Robbins E. (1966). Polyribosome disaggregation during metaphase. *Science*, **151**, 992–995.

Schendl W. (1971). Analysis of the human karyotye using a reassociation technique. *Chromosoma*, **34**, 448–454.

Scherrer K., Spohr G., Granboulan N., Morel C., Grosclaude J. and Chezzi C. (1970). Nuclear and cytoplasmic messenger like RNA and their relation to the active messenger RNA in polyribosomes of Hela cells. *Cold Spring Harbor Symp. Quant. Biol.*, **35**, 539–554.

Schildkraut C.L. and Maio J.J. (1968). Studies on the intranuclear distribution and properties of mouse satellite DNA. *Biochim. Biophys. Acta*, **161**, 76–93.

Schindler R., Grieder A. and Maurer U. (1972). Studies on the division cycle of mammalian cells. VI. DNA polymerase activities in partially synchronous suspension cultures. *Exp. Cell Res.*, **71**, 218–224.

Schochetman G. and Perry R.P. (1972). Characterization of the mRNA released from L cell polyribosomes as a result of temperature shock. *J. Mol. Biol.*, **63**, 577–590.

Schrader W.T., Toft D.O. and O'Malley B.W. (1972). Progesterone binding components of chick oviduct. VI. Interaction of purified progesterone receptor components with nuclear constituents. *J. Biol. Chem.*, **247**, 2401–2407.

Schreck R.R., Warburton D., Miller O.J., Beiser S.M. and Erlanger B.F. (1973). Chromosome structure as revealed by a combined chemical and immunochemical procedure. *Proc. Nat. Acad. Sci.*, **70**, 804–807.

Schreier M.H. and Staehelin T. (1973). Initiation of mammalian protein synthesis: the importance of ribosomes and initiation factor quality for the efficiency of in vitro systems. *J. Mol. Biol.*, **73**, 329–350.

Schutz G., Beato M. and Feigelson P. (1972). Isolation of eucaryotic mRNA on cellulose and its translation in vitro. *Biochem. Biophys. Res. Commun.*, **49**, 680–689.

Schwartz A.G., Cook P.R. and Harris H. (1971). Correction of a genetic defect in a mammalian cell. *Nature New Biol.*, **230**, 5–8.

Seabright M. (1971). The use of proteolytic enzymes for the mapping of structural rearrangements in the chromosomes of man. *Chromosoma*, **36**, 204–210.

Seeber S. and Busch H. (1971). Polypurine sequences as chemical markers of processing of nucleolar 45S RNA. *J. Biol. Chem.*, **246**, 7151–7158.

Sharman G.B. (1971). Late DNA replication in the paternally derived X chromosome of female kangaroos. *Nature*, **230**, 231–232.

Shaw L.M.J. and Huang R.C.C. (1970). A description of two procedures which avoid the use of extreme pH conditions for the resolution of components isolated from chromatins prepared from pig cerebellar and pituitary nuclei. *Biochem.*, **9**, 4530–4541.

Shearer R.W. and McCarthy B.J. (1967). Evidence for RNA molecules restricted to the cell nucleus. *Biochem.*, **6**, 283–289.

Shearer R.W. and McCarthy B.J. (1970). Characterization of RNA molecules restricted to the nucleus in mouse L cells. *J. Cell Physiol.*, **75**, 97–106.

Sheiness D. and Darnell J.E. (1973). Poly-A segment in mRNA becomes shorter with age. *Nature New Biol.*, **241**, 265–268.

Shelanski M.L. and Taylor E.W. (1967). Isolation of a protein subunit from microtubules. *J. Cell Biol.*, **34**, 549–554.

Sheldon R., Jurale C. and Kates J. (1972). Detection of poly-A sequences in viral and eucaryotic RNA. *Proc. Nat. Acad. Sci.*, **69**, 417–421.

Shepherd G.R., Hardin J.M. and Noland B.J. (1971). Methylation of lysine residues of histone fractions in synchronized mammalian cells. *Arch. Biochem. Biophys.*, **143**, 1–5.

Shepherd G.R., Hardin J.M. and Noland B.J. (1972). Dephosphorylation of histone fractions of cultured mammalian cells. *Arch. Biochem. Biophys.*, **153**, 599–602.

Shepherd G.R., Noland B.J. and Hardin J.M. (1972). Turnover of histone acetyl groups in cultured mammalian cells. *Exp. Cell Res.*, **75**, 397–400.

Shepherd J. and Maden B.E.H. (1972). Ribosome assembly in Hela cells. *Nature*, **236**, 211–214.

Sherman M.R., Corvol P.L. and O'Malley B.W. (1970). Progesterone binding components of chick oviduct. I. Preliminary characterization of cytoplasmic components. *J. Biol. Chem.*, **245**, 6085–6096.

Shih T.Y. and Bonner J. (1970a). Thermal denaturation and template properties of DNA complexes with purified histone fractions. *J. Mol. Biol.*, **48**, 469–487.

Shih T.Y. and Bonner J. (1970b). Template properties of DNA polypeptide complexes. *J. Mol. Biol.*, **50**, 333–342.

Shih T.Y. and Lake R.S. (1972). Studies on the structure of metaphase and interphase chromatin of Chinese hamster cells by circular dichroism and thermal denaturation. *Biochem.*, **11**, 4811–4817.

Shin S., Caneva R., Schildkraut C.L., Klinger H.P. and Siniscalco M. (1973). Cells with phosphoribosyl transferase activity recovered from mouse cells resistant to 8-azaguanine. *Nature New Biol.*, **241**, 194–196.

Shyamala G. and Gorski J. (1969). Estrogen receptors in the rat uterus. Studies on the interaction of cytosol and nuclear binding sites. *J. Biol. Chem.*, **244**, 1097–1103.

Sidebottom E. and Harris H. (1969). The role of the nucleolus in the transfer of RNA from nucleus to cytoplasm. *J. Cell Sci.*, **5**, 351–364.

Sieger M., Pera F. and Schwarzacher H.G. (1970). Genetic inactivity of heterochromatin and heteropycnosis in M. agrestis. *Chromosoma*, **29**, 349–364.

Silverman B. and Mirsky A.E. (1973). Accessibility of DNA in chromatin to DNA polymerase and RNA polymerase. *Proc. Nat. Acad. Sci.*, **70**, 1326–1330.

Singer R.H. and Penman S. (1972). Stability of Hela cell mRNA in actinomycin. *Nature*, **242**, 100–102.

Singer R.H. and Penman S. (1973). Messenger RNA in Hela cells: kinetics of formation and decay. *J. Mol. Biol.*, **78**, 321–335.

Skinner L.G. and Ockey C.H. (1971). Isolation, fractionation and biochemical analyses of the metaphase chromosomes of M. agrestis. *Chromosoma*, **35**, 125–142.

Smart J.E. and Bonner J. (1971a). Selective dissociation of histones from chromatin by sodium deoxycholate. *J. Mol. Biol.*, **58**, 651–660.

Smart J.E. and Bonner J. (1971b). Studies on the role of histones in the structure of chromatin. *J. Mol. Biol.*, **58**, 661–674.

Smart J.E. and Bonner J. (1971c). Studies on the role of histones in relation to the template activity and precipitability of chromatin at physiological ionic strengths. *J. Mol. Biol.*, **58**, 675–684.

Smith K.D., Church R.B. and McCarthy B.J. (1969). Template specificity of isolated chromatin. *Biochem.*, **8**, 4271–4277.

Smuckler E.A. and Tata J.R. (1972). Nearest neighbour base frequency of the RNA formed by rat liver DNA dependent RNA polymerase A and B with homologous RNA. *Biochem. Biophys. Res. Commun.*, **49**, 16–22.

Soeiro R., Birnboim H.C. and Darnell J.E. (1966). Rapidly labelled Hela cell nuclear RNA. II. Base composition and cellular localization of a heterogeneous RNA fraction. *J. Mol. Biol.*, **19**, 362–372.

Soeiro R. and Darnell J.E. (1969). Competitive hybridization by pre-saturation of Hela cell DNA. *J. Mol. Biol.*, **44**, 551–562.

Soeiro R. and Darnell J.E. (1970). A comparison between heterogeneous nuclear RNA and polysomal mRNA in Hela cells by RNA–DNA hybridization. *J. Cell Biol.*, **44**, 467–475.

Soeiro R., Vaughan M.H., Warner J.R. and Darnell J.R. (1968). The turnover of nuclear DNA-like RNA in Hela cells. *J. Cell Biol.*, **39**, 112–118.

Solari A.J. (1970a). The spatial relationship of the X and Y chromosomes during meiotic prophase in mouse spermatocytes. *Chromosoma*, **29**, 217–236.

Solari A.J. (1970b). The behaviour of chromosomal axes during diplotene in mouse spermatocytes. *Chromosoma*, **31**, 217–230.

Solari A.J. (1971). Experimental changes in the width of the chromatin fibres from chicken erythrocytes. *Exp. Cell Res.*, **67**, 161–170.

Solari A.J. (1972). Ultrastructure and composition of the synaptonemal complex in spreads and negatively stained spermatocytes of the golden hamster and the albino rat. *Chromosoma*, **39**, 237–263.

Solari A.J. and Moses M.J. (1973). The structure of the central region in the synaptonemal complexes of hamster and cricket spermatocytes. *J. Cell Biol.*, **56**, 145–152.

Sonnenberg B.P. and Zubay G. (1965). Nucleohistone as a primer for RNA synthesis. *Proc. Nat. Acad. Sci.*, **54**, 415–420.

Sotirov N. and Johns E.W. (1972). Quantitative differences in the content of the histone f2c between chicken erythrocytes and erythroblasts. *Exp. Cell Res.*, **73**, 13–16.

Southern E.M. (1970). Base sequence and evolution of guinea pig α-satellite DNA. *Nature*, **227**, 794–798.

Southern E.M. (1971). Effects of sequence divergence on the reassociation properties of repetitive DNAs. *Nature New Biol.*, **232**, 82–83.

Spadari S. and Ritossa F. (1970). Clustered genes for ribosomal RNAs in E.coli. *J. Mol. Biol.*, **53**, 357–368.

Spear B.B. and Gall J.G. (1973). Independent control of ribosomal gene replication in polytene chromosomes of D. melanogaster. *Proc. Nat. Acad. Sci.*, **70**, 1359–1363.

Spelsberg T.C. and Hnilica L.S. (1969a). Studies on the RNA polymerase-histone complexes. *Biochim. Biophys. Acta*, **195**, 55–62.

Spelsberg T.C. and Hnilica L.S. (1969b). The effects of acidic proteins and RNA on the histone inhibition of the DNA-dependent RNA synthesis in vitro. *Biochim. Biophys. Acta*, **195**, 63–75.

Spelsberg T.C. and Hnilica L.S. (1971a). Proteins of chromatin in template restriction. I. RNA synthesis in vitro. *Biochim. Biophys. Acta*, **228**, 202–211.

Spelsberg T.C. and Hnilica L.S. (1971b). Proteins of chromatin in template restriction. II. Specificity of RNA synthesis. *Biochim. Biophys. Acta*, **228**, 212–222.

Spelsberg T.C., Hnilica L.S. and Ansevin A.T. (1971). Proteins of chromatin in template restriction. II. The macromolecules in specific restriction of the chromatin DNA. *Biochim. Biophys. Acta*, **228**, 550–562.

Spelsberg T.C., Steggles A.W., Chytil F. and O'Malley B.W. (1972). Progesterone binding components of chick oviduct. V. Exchange of progesterone binding capacity from target to non target tissue chromatins. *J. Biol. Chem.*, **247**, 1368–1374.

Spelsberg T.C., Steggles A.W. and O'Malley B.W. (1971). Progesterone binding components of chick oviduct. III. Chromatin acceptor sites. *J. Biol. Chem.*, **246**, 4188–4197.

Spirin A.S. (1969). Informosomes. *Europ. J. Biochem.*, **10**, 20–35.

Spirin A.S. Non ribosomal RNP particles (informosomes) of animal cells. In Bosch L. (Ed.), *The mechanism of protein synthesis and its regulation*. North Holland, Amsterdam, pp 515–538.

Spohr G., Granboulan N., Morel C. and Scherrer K. (1970). Messenger RNA in Hela cells: an investigation of free and polyribosome bound cytoplasmic messenger ribonucleoprotein particles by kinetic labelling and electron microscopy. *Europ. J. Biochem.*, **17**, 296–318.

Stanners C.P. and Till J.E. (1960). DNA synthesis in individual L-strain mouse cells. *Biochim. Biophys. Acta*, **37**, 406–419.

Starbuch W.C., Mauritzen C.M., Taylor C.W., Saroja I.S. and Busch H. (1968). A large scale procedure for isolation of the glycine rich arginine rich histone and the arginine rich lysine rich histone in a highly purified form. *J. Biol. Chem.*, **243**, 2038–2047.

Stavnezer J. and Huang R.C.C. (1971). Synthesis of a mouse Ig light chain in a rabbit reticulocyte cell free system. *Nature New Biol.*, **230**, 172–176.

Stebbins C.L. (1966). Chromosomal variation and evolution. *Science*, **152**, 1463–1465.

Stedman E. and Stedman E. (1950). Cell specificity of histones. *Nature*, **166**, 780–781.

Steggles A.W., Spelsberg T.C., Glasser S.R. and O'Malley B.W. (1971). Soluble complexes between steroid hormones and target tissue receptors bind specifically to target tissue chromatin. *Proc. Nat. Acad. Sci.*, **68**, 1470–1482.

Steggles A.W., Spelsberg T.C. and O'Malley B.W. (1971). Tissue specific binding in vitro of progesterone receptors to the chromatins of chick tissues. *Biochem. Biophys. Res. Commun.*, **43**, 20–27.

Stein G. and Baserga R. (1970). Continued synthesis of non histone chromosomal proteins during mitosis. *Biochem. Biophys. Res. Commun.*, **41**, 715–722.

Stein G. and Farber J. (1972). Role of non histone chromosomal proteins in the restriction of mitotic chromatin template activity. *Proc. Nat. Acad. Sci.*, **69**, 2918–2921.

Stein H. and Hausen P. (1970). Factors influencing the activity of mammalian RNA polymerase. *Cold Spring Harbor Symp. Quant. Biol.*, **35**, 709–718.

Stern H. and Hotta Y. (1969). Biochemistry of meiosis. In Lima-de-Faria A. (Ed.), *Handbook of Molecular Cytology*. North Holland, Amsterdam, pp 520–539.

Stevens R.H. and Williamson A.R. (1972). Specific IgG mRNA molecules from myeloma cells in heterogeneous nuclear and cytoplasm RNA containing poly-A. *Nature*, **239**, 143–146.

Stevens R.H. and Williamson A.R. (1973). Isolation of mRNA coding for mouse heavy chain immunoglobulin. *Proc. Nat. Acad. Sci.*, **70**, 1127–1131.

Steward D.L., Shaeffer J.R. and Humphrey R.M. (1968). Breakdown and assembly of polyribosomes in synchronized Chinese hamster cells. *Science*, **161**, 791–793.

Stockert J.C. and Lisanti J.A. (1972). Acridine orange differential fluorescence of fast and slow reassociating chromosomal DNA after in situ DNA denaturation and reassociation. *Chromosoma*, **37**, 117–130.

Stollar B.D. and Ward M. (1970). Rabbit antibodies to histone fractions as specific reagents for preparation and comparative studies. *J. Biol. Chem.*, **245**, 1261–1266.

Straus N.A. (1971). Comparative DNA renaturation kinetics in amphibians. *Proc. Nat. Acad. Sci.*, **68**, 799–802.

Stubblefield E. and Brinkley B.R. (1967). Architecture and function of the mammalian centriole. In Warren K.B. (Ed.), *Formation and Fate of Cell Organelles*. Symp. Int. Soc. Cell. Biol, pp 175–218.

Stubblefield E. and Murphree S. (1967). Thymidine kinase activity in colcemid synchronized fibroblasts. *Exp. Cell Res.*, **48**, 652–657.

Subirana J.A. (1973). Studies on the thermal denaturation of nucleohistones. *J. Mol. Biol.*, **74**, 363–386.

Sugden B. and Keller W. (1973). Mammalian DNA dependent RNA polymerases. I. Purification and properties of an α-amanitin sensitive RNA polymerase and stimulatory factors from Hela and KB cells. *J. Biol. Chem.*, **248**, 3777–3788.

Sumner A.T., Evans H.J. and Buckland R.A. (1971). New technique for distinguishing between human chromosomes. *Nature New Biol.*, **232**, 31–32.

Sung M.T. and Dixon G.H. (1970). Modification of histones during spermiogenesis in trout: a molecular mechanism for altering histone binding to DNA. *Proc. Nat. Acad. Sci.*, **67**, 1616–1623.

Sung M.T., Dixon G.H. and Smithies O. (1971). Phosphorylation and synthesis of histones in regenerating rat liver. *J. Biol. Chem.*, **246**, 1358–1364.

Sutton W.D. and McCallum M. (1971). Mismatching and the reassociation rate of mouse satellite DNA. *Nature New Biol.*, **232**, 83–85.

Sutton W.D. and McCallum M. (1972). Related satellite DNAs in the genus Mus. *J. Mol. Biol.*, **71**, 633–656.

Suzuki Y. and Brown D.D. (1972). Isolation and identification of the mRNA for silk fibroin from Bombyx mori. *J. Mol. Biol.*, **63**, 409–430.

Suzuki Y., Gage L.P. and Brown D.D. (1972). The genes for silk fibroin in Bombyx mori. *J. Mol. Biol.*, **70**, 637–656.

Swan D., Aviv H. and Leder P. (1972). Purification and properties of biologically active mRNA for a myeloma light chain. *Proc. Nat. Acad. Sci.*, **69**, 1967–1971.

Swift H. (1962). Nucleic acids and cell morphology in Dipteran salivary glands. In Allen J.M. (Ed.), *The molecular control of cellular activity*. McGraw Hill, New York, pp 73–125.

Swift H. (1964). The histones of polytene chromosomes. In T'so P. and Bonner J. (Eds.), *The Nucleohistones*. Holden-Day, San Francisco, pp 169–183.

Takagi N. and Sandberg A.A. (1968a). Chronology and pattern of human chromosome replication. VII. Cellular and chromosomal DNA behaviour. *Cytogenetics*, **7**, 118–134.

Takagi N. and Sandberg A.A. (1968b). Chronology and pattern of human chromosome replication. VIII. Behaviour of the X and Y in early S phase. *Cytogenetics*, **7**, 135–143.

Tan C.H. and Miyagi M. (1970). Specificity of transcription of chromatin in vitro. *J. Mol. Biol.*, **50**, 641–653.

Tartof K.D. (1973). Regulation of rRNA gene multiplicity in D. melanogaster. *Genetics*, **73**, 57–70.

Tartof K.D. and Perry R.P. (1970). The 5S RNA genes of D. melanogaster. *J. Mol. Biol.*, **51**, 171–183.

Tata J.R., Hamilton M.J. and Shields D. (1972). Effects of α-amanitin in vivo on RNA polymerase and nuclear RNA synthesis. *Nature New Biol.*, **238**, 161–164.

Taylor E.W. (1965). Control of DNA synthesis in mammalian cells in cultures. *Exp. Cell Res.*, **40**, 316–332.

Taylor J.H. (1957). The time and mode of duplication of chromosomes. *Amer. Nat.*, **91**, 209–222.

Taylor J.H. (1958). Sister chromatid exchanges in tritium labelled chromosomes. *Genetics*, **43**, 515–529.

Taylor J.H. (1963). The replication and organization of DNA in chromosomes. In Taylor J.H. (Ed.), *Molecular Genetics*, volume 1. Academic Press, New York, pp 65–113.

Taylor J.H. (1968). Rates of chain growth and units of replication in DNA of mammalian chromosomes. *J. Mol. Biol.*, **31**, 579–594.

Taylor J.H. (1973). Replication of DNA in mammalian chromosomes: isolation of replicating segments. *Proc. Nat. Acad. Sci.*, **70**, 1083–1087.

Taylor J.H., Adams A.G. and Kwek M.P. (1973). Replication of DNA in mammalian chromosomes. II. Kinetics of H³-thymidine incorporation and the isolation and partial characterization of labelled subunits at the growing points. *Chromosoma*, **41**, 361–384.

Taylor J.H., Woods P. and Hughes W. (1957). The organization and duplication of chromosomes as revealed by autoradiographic studies using tritium labelled thymidine. *Proc. Nat. Acad. Sci.*, **43**, 122–128.

Temple G.F. and Housman D.E. (1972). Separation and translation of the mRNAs coding for α and β chains of rabbit globin. *Proc. Nat. Acad. Sci.*, **69**, 1574–1577.

Teng C.S., Teng C.T. and Allfrey V.G. (1971). Studies of nuclear acidic proteins. Evidence for their phosphorylation, tissue specificity, selective binding to DNA and stimulatory effects on transcription. *J. Biol. Chem.*, **246**, 3597–3609.

Terasima T. and Yasukawa M. (1966). Synthesis of G1 protein preceding DNA synthesis in cultured mammalian cells. *Exp. Cell Res.*, **44**, 669–671.

Terasima T. and Tolmach L.J. (1963). Growth and nucleic acid synthesis in synchronously dividing populations of Hela cells. *Exp. Cell Res.*, **30**, 344–362.

Thomas C.A.jr., Hamkalo B.A., Misra D.N. and Lee C.S. (1970). Cyclization of eucaryotic DNA fragments. *J. Mol. Biol.*, **51**, 621–631.

Thomas C.A.jr. and MacHattie L.A. (1967). The anatomy of viral DNA molecules. *Ann. Rev. Biochem.*, **36**, 485–502.

Thomas C.A.jr., Zimm B.H. and Dancis B.M. (1973). Ring Theory. *J. Mol. Biol.*, **77**, 85–100.

Tobey R.A., Anderson E.C. and Petersen D.F. (1967). The effect of thymidine on the duration of G1 in Chinese hamster cells. *J. Cell Biol.*, **35**, 53–59.

Todaro G.J., Lazar G.K. and Green H. (1965). The initiation of cell division in a contact inhibited mammalian cell line. *J. Cell Physiol.*, **66**, 325–333.

Tsai R.L. and Green H. (1972). Study of intracellular precursors using DNA-cellulose chromatography. *Nature New Biol.*, **237**, 171–173.

Tuan D.Y.H. and Bonner J. (1969). Optical absorbance and optical rotatory dispersion studies on calf thymus nucleohistone. *J. Mol. Biol.*, **45**, 59–76.

Turner F.R. (1968). An ultrastructural study of plant spermatogenesis. *J. Cell Biol.*, **37**, 370–392.

Van den Broek H.W.J., Nooden L.D., Sevall J.S. and Bonner J. (1973). Isolation, purification and fractionation of non histone chromosomal proteins. *Biochem.*, **12**, 229–236.

Vaughan M.H., Warner J.R. and Darnell J.E. (1967). Ribosomal precursor particles in the Hela cell nucleus. *J. Mol. Biol.*, **25**, 235–251.

Verma I.M., Temple G.F., Fan H. and Baltimore D. (1972). In vitro synthesis of DNA complementary to rabbit reticulocyte 10S RNA. *Nature New Biol.*, **235**, 163–167.

Vidali G., Boffa L.C., Littau V.C., Allfrey K.M. and Allfrey V.G. (1973). Changes in nuclear acidic protein complement of red blood cells during embryonic development. *J. Biol. Chem.*, **248**, 4065–4068.

Vidali G., Gershey E.L. and Allfrey V.G. (1968). The distribution of ε-N-acetyllysine in calf thymus histone. *J. Biol. Chem.*, **243**, 6361–6366.

Vosa C.G. (1970). The discriminating fluorescence patterns of the chromosomes of D. melanogaster. *Chromosoma*, **31**, 446–451.

Walker P.M.B. (1971a). Origin of satellite DNA. *Nature*, **229**, 306–308.

Walker P.M.B. (1971b). Repetitive DNA in higher organisms. *Prog. Biophys. Mol. Biol.*, **23**, 145–190.

Walker P.M.B., Flamm W.G. and McClaren A. (1969). Highly repetitive DNA in rodents. In Lima-de-Faria A. (Ed.), *Handbook of Molecular Cytology*, North Holland, Amsterdam, pp 52–67.

Wall R., Philipson L. and Darnell J.E. (1972). Processing of adenovirus specific nuclear RNA during virus replication. *Virology*, **50**, 27–34.

Wallace H. and Birnstiel M.L. (1966). Ribosomal cistrons and the nucleolar organiser. *Biochim. Biophys. Acta*, **114**, 296–310.

Waring M.J. and Britten R.J. (1966). Nucleotide sequence repetition in a rapidly reassociating fraction of mouse DNA. *Science*, **154**, 791–794.

Warner J.R. (1966). The assembly of ribosomes in Hela cells. *J. Mol. Biol.*, **19**, 383–398.

Warner J.R., Girard M., Latham H. and Darnell J.E. (1966). Ribosome formation in Hela cells in the absence of protein synthesis. *J. Mol. Biol.*, **19**, 373–382.

Watson B., Gormley I.P., Gardiner S.E., Evans H.J. and Harris H. (1972). Reappearance of murine hypoxanthine guanine phosphoribosyl transferase activity in mouse A9 cells after attempted hybridization with human cell lines. *Exp. Cell Res.*, **75**, 401–409.

Weinberg R.A. and Penman S. (1970). Processing of 45S nucleolar RNA. *J. Mol. Biol.*, **47**, 169–178.

Weisblum B. and De Haseth P.L. (1972). Quinacrine, a chromosome stain specific for dAT rich regions in DNA. *Proc. Nat. Acad. Sci.*, **69**, 629–632.

Weiss B.G. (1969). The dependence of DNA synthesis on protein synthesis in Hela S3 cells. *J. Cell Physiol.*, **73**, 85–90.

Weiss M.C. and Ephrussi B. (1966). Studies of interspecific (rat x mouse) somatic hybrids. I. Isolation, growth and evolution of the karyotype. *Genetics*, **54**, 1095–1109.

Weiss M.C. and Green H. (1967). Human–mouse hybrid cell lines containing partial complements of human chromosomes and functioning human genes. *Proc. Nat. Acad. Sci.*, **58**, 1104–1111.

Wensink P.C. and Brown D.D. (1971). Denaturation map of the ribosomal DNA of X. laevis. *J. Mol. Biol.*, **60**, 235–248.

Westergaard M. and Von Wettstein D. (1970). Studies on the mechanism of crossing over. IV. The molecular organization of the synaptonemal complex in Neottiella (Cooke) saccardo (Ascomycetes). *Comptes Rendus Lab. Carslberg*, **37**, 239–268.

Westergaard M. and Van Wettstein D. (1972). The synaptonemal complex. *Ann. Rev. Gen.*, **6**, 71–110.

Wetmur J.G. and Davidson N. (1968). Kinetics of renaturation of DNA. *J. Mol. Biol.*, **31**, 349–370.

Whitehouse H.L.K. (1967). A cycloid model for the chromosome. *J. Cell Sci.*, **2**, 9–22.

Whitehouse H.L.K. (1969). *The mechanism of heredity.* Edward Arnold, London.

Widnell C.C. and Tata J.J. (1966). Studies on the stimulation by ammonium sulfate of the DNA dependent RNA polymerase of isolated rat liver nuclei. *Biochim. Biophys. Acta*, **123**, 478–492.

Wilhelm J.A. and McCarty K.S. (1970). The uptake and turnover of acetate in Hela cell histone fractions. *Cancer Research*, **30**, 418–425.

Willems M., Wagner E., Laing R. and Penman S. (1968). Base composition of ribosomal RNA precursors in the Hela cell nucleolus: further evidence of non-conservative processing. *J. Mol. Biol.*, **32**, 211–220.

Williams A.F. (1972a). Deoxythymidine metabolism in avian erythroid cells. *J. Cell Sci.*, **11**, 777–784.

Williams A.F. (1972b). DNA polymerase in avian erythroid cells. *J. Cell Sci.*, **11**, 785–798.

Williams C.A. and Ockey C.H. (1970). Distribution of DNA replicator sites in mammalian nuclei after different methods of cell synchronization. *Exp. Cell Res.*, **63**, 365–372.

Williamson R., Drewienkeiwicz C.E. and Paul J. (1973). Globin messenger sequences in high molecular weight RNA from embryonic mouse liver. *Nature New Biol.*, **241**, 66–68.

Wilson R.K., Starbuck W.C., Taylor C.W., Jordan J. and Busch H. (1970). Structure of the glycine rich arginine rich histone of the Novikoff hepatoma. *Cancer Research*, **30**, 2942–2951.

Winters M.A. and Edmonds M. (1973a). A poly-A polymerase from calf thymus. Purification and properties of the enzyme. *J. Biol. Chem.*, **248**, 4756–4762.

Winters M.A. and Edmonds M. (1973b). A poly-A polymerase from calf thymus. Characterization of the reaction product and the primer requirement. *J. Biol. Chem.*, **248**, 4763–4768.

Wise G.E. and Prescott D.M. (1973). Initiation and continuation of DNA synthesis are not associated with the nuclear envelope in mammalian cells. *Proc. Nat. Acad. Sci.*, **70**, 714–717.

Wobus U., Popp S., Serfling E. and Panitz R. (1972). Protein synthesis in Chironomus thummi salivary glands. *Molec. Gen. Genet.*, **116**, 309–321.

Wu J., Hurn J. and Bonner J. (1972). Size and distribution of the repetitive segments of the Drosophila genome. *J. Mol. Biol.*, **64**, 211–220.

Yamamoto K.R. and Alberts B.M. (1972). In vitro conversion of estradiol receptor protein to its nuclear form: dependence on hormone and DNA. *Proc. Nat. Acad. Sci.*, **69**, 2105–2109.

Yasmineh W.G. and Yunis J.J. (1969). Satellite DNA in mouse autosomal heterochromatin. *Biochem. Biophys. Res. Commun.*, **35**, 779–782.

Yasmineh W.G. and Yunis J.J. (1971). Satellite DNA in calf heterochromatin. *Exp. Cell Res.*, **64**, 41–48.

Yeoman L.C., Olson M.O.J., Sugano N., Jordan J.J., Taylor C.W., Starbuck W.C. and Busch H. (1972). Amino acid sequence of the centre of the arginine-lysine rich histone from calf thymus. The total sequence. *J. Biol. Chem.*, **247**, 6018–6024.

Yogo Y. and Wimmer E. (1972). Polyadenylic acid at the 3' terminus of poliovirus RNA. *Proc. Nat. Acad. Sci.*, **69**, 1877–1882.

Yoshikawa-Fukada M. (1967). The intermediate state of ribosome formation in animal cells in culture. *Biochim. Biophys. Acta*, **145**, 651–663.

Yu F.L. and Feigelson P. (1972). The rapid turnover of RNA polymerase of rat liver nucleolus and of its messenger RNA. *Proc. Nat. Acad. Sci.*, **69**, 2833–2837.

Yunis J., Roldan L., Yasmineh W.G. and Lee J.C. (1971). Staining of satellite DNA in metaphase chromosomes. *Nature*, **231**, 532–533.

Yunis J.J. and Yasmineh W.G. (1971). Heterochromatin, satellite DNA and cell function. *Science*, **174**, 1200–1209.

Yunis J.J. and Yasmineh W.G. (1972). Model for mammalian constitutive heterochromatin. *Adv. Cell Mol. Biol.*, **2**, 1–46.

Zetterberg A. (1966a). Synthesis and accumulation of nuclear and cytoplasmic proteins during interphase in mouse fibroblasts in vitro. *Esp. Cell Res.*, **42**, 500–513.

Zetterberg A. (1966b). Nuclear and cytoplasmic nucleic acid content and cytoplasmic protein synthesis during interphase in mouse fibroblasts in vitro. *Exp. Cell Res.*, **43**, 517–525.

Zetterberg A. (1966c). Protein migration between cytoplasm and cell nucleus during interphase in mouse fibroblasts in vitro. *Exp. Cell Res.*, **43**, 526–536.

Zetterberg A. (1970). Nuclear and cytoplasmic growth during interphase in mammalian cells. *Adv. Cell Biol.*, **1**, 211–232.

Zetterberg A. and Killander D. (1965). Quantitative cytophotometric and autoradiographic studies on the rate of protein synthesis during interphase in mouse fibroblasts in vitro. *Exp. Cell Res.*, **40**, 1–11.

Zimmerman E.F. (1968). Secondary methylation of ribosomal RNA in Hela cells. *Biochem.*, **7**, 3156–3163.

Zimmerman E.F. and Holler B.W. (1967). Methylation of 45S rRNA precursor in Hela cells. *J. Mol. Biol.*, **23**, 149–161.

Zubay G. and Doty P. (1959). The isolation and properties of DNP particles containing single nucleic acid molecules. *J. Mol. Biol.*, **1**, 1–20.

Zylber E.A. and Penman S. (1971a). Product of RNA polymerases in Hela cell nuclei. *Proc. Nat. Acad. Sci.*, **68**, 2861–2865.

Zylber E.A. and Penman S. (1971b). Synthesis of 5S and 4S RNA in metaphase arrested Hela cells. *Science*, **172**, 947–949.

Index